Contents

Harvesting Wild Species

Implications for Biodiversity Conservation

Edited by
Curtis H. Freese

The Johns Hopkins University Press
Baltimore and London

Printed in the United States of America on acid-free recycled paper
06 05 04 03 02 01 00 99 98 97 5 4 3 2 1

The Johns Hopkins University Press
2715 North Charles Street
Baltimore, Maryland 21218-4319
The Johns Hopkins Press Ltd., London

Library of Congress Cataloging-in-Publication Data will be found at the end of this book.
A catalog record for this book is available from the British Library.

ISBN 0-8018-5573-X
ISBN 0-8018-5574-8 (pbk.)

Preface

This book is the product of a three-year study undertaken by the World Wide Fund for Nature (World Wildlife Fund in the United States and Canada, or, more generally, WWF) to explore the link between biodiversity conservation and the commercial consumptive use (CCU) of wild species. The management of this linkage has been a growing and sometimes contentious issue in the field of biodiversity conservation. Consumptive use, particularly when a profit motive is involved, is a double-edged sword for conservationists. While the overexploitation of wild living resources is causing serious population declines of species around the globe, the socioeconomic benefits derived from the use of a wild species give tangible value to that species and thus may provide an incentive for maintaining the species and the ecosystem it inhabits. We also know that there may be no "free lunches" in that conservation approaches based on economic incentives may have their own, sometimes hidden, costs for biodiversity.

In undertaking a study to understand better the conditions that influence which way this sword cuts, and how it cuts, it soon became clear that much more empirical work from on-the-ground management experiences was needed. While considerable work has been done on how economic, social, and ecological conditions influence the sustainability of harvests from wild species populations, much less has been done to understand how use regimes and conditions interact to affect biodiversity. Thus the decision by WWF to commission the 15 case studies presented here.

Brief definitions of a few terms used will provide a framework for the issues covered. A use is *consumptive* when the entire organism or any of its parts is deliberately killed or removed, either as a goal in and of itself (e.g., recreational hunting and fishing), or for a product (e.g., pets, food, leather). Our interest in this study is the consumptive use of species living in the wild, as opposed to captive or domesticated conditions. We define *commercial* in the broad sense to include any use that is driven or greatly influenced by a profit motive for users, managers, or owners of the resource. Finally, we are concerned with the effects of CCU on diversity at the ecosystem, species, and genetic levels.

The case study authors have examined the economic, social, and ecological forces that shape the relationship between CCU and biodiversity conservation. In selecting these studies we cast a wide biogeographic and socioeconomic net and included an array of types of wild species use, so that the CCU-biodiversity linkage is examined under diverse conditions. Eleven of the case studies provide an interdisciplinary examination of the full range of social, economic, and ecological issues that affect this linkage under management programs that range from highly site specific, such as wildlife use in Peru's Tamshiyacu-Tahuayo Communal Reserve (Bodmer et al., Chap. 9), to country-wide, such as the management of Swiss forests (McShane and McShane-Caluzi, Chap. 4). While the organisms considered in nearly all of these 11 cases had experienced a period of overexploitation or habitat degradation, most have now attained, or appear to be on the road to attaining, some level of sound management. This circumstance provides an opportunity to compare within the same setting those conditions conducive to sound and unsound management, and to examine what factors led to improved management.

Four other chapters—on forestry (Hansen, Chap. 6), ungulate management (Teer, Chap. 12), waterfowl management (Callaghan et al., Chap. 14), and salmon management (Francis, Chap. 16)—focus on ecological interactions associated with the harvest and management of wild populations. The principal question addressed by these authors was: Under conditions of sustainable offtake, what, if any, changes in biodiversity and ecosystem function occur because of the offtake itself, the harvest techniques, and associated population/habitat management? In the process of examining this question, some of the chapters begin to explore the inverse question regarding the importance of native ecosystem integrity and biodiversity in maintaining the long-term productivity of the populations being harvested. The subjects of these four chapters were selected according to the criteria of a long, intensive, and relatively well documented history of use and management, and broad taxonomic and ecosystem representation. Conditions of long and intensive management are most likely to reveal the mechanisms by which harvest and management regimes may affect biodiversity and ecosystem function. Because of these criteria, all four studies have a strong temperate zone focus, since the temperate zone is in general where species have been most intensively managed with reasonable documentation.

My first acknowledgment is to the solid institutional and financial support that WWF has provided throughout this study. The vast network of WWF staff, working both at the policy level and in the field, provided a rich and stimulating source of information and ideas. Particular thanks are owed to the following members of the WWF task force assembled for this study: Cleber Alho, Jason Clay, Barry Coates, Steve Cornelius, Anton Fernhout, Jan Habrovsky, Ginette Hemley, Barbara Hoskinson, Kevin Lyonette, Tom McShane, Fulai Sheng, Gordon Shepherd, Michael Sutton, and Magnus Sylven. I am especially grateful to Ginette Hemley and Kevin Lyonette for their vision and perseverance in getting this study launched, and for their confidence in me as the study's principal investigator. Special thanks also go to Maria Boulos and Kimberly Doyle for their attentive and meticulous administrative support throughout the study. Invaluable support for bibliographic research was provided by the Department of Biology, Montana State University. I also gratefully acknowledge the numerous anonymous reviewers who gave generously of their time and wisdom to improve the contributions to this book.

I owe a hearty thanks to Jennifer Swearingen, who worked side by side with me in applying her considerable editorial skills to every contribution and whose efforts were crucial in getting the manuscripts into publishable form. I am grateful to Robert Harington, former science editor at the Johns Hopkins University Press, for taking an early and active interest in the publication of this study, and to Barbara Lamb of the Johns Hopkins University Press and Peter Strupp of Princeton Editorial Associates for their diligence in editing and shepherding the manuscripts through production.

Finally, I wish to thank the authors for their cooperation and for enduring what must have seemed at times like an endless line of inquiry and requests from me. I trust they will find the collective results of their labor worthy of the effort.

Curtis H. Freese

Harvesting Wild Species

ONE

The "Use It or Lose It" Debate

Issues of a Conservation Paradox

Curtis H. Freese

The commercial, consumptive use of wild species is a focal point for much of the current debate regarding the link between sustainable development and biodiversity conservation. More specifically, it is at the center of two often conflicting points of view regarding the best strategy for future conservation efforts. One embraces the "use it or lose it" dictum and the other sees the for-profit motive leading inevitably to overexploitation and biotic impoverishment.

At stake in this debate is how society will manage what is still a major portion of the Earth's land and water surface which is not yet fully converted to urbanization and domestic forms of production or which is securely in protected areas. One view is that more progress will be made toward maintaining biodiversity on this remaining portion by pushing to include a major part of it in fully protected area status (Noss 1991). This view is largely based on the premise that commercial use of wild species in wildlands (the term as used here includes natural aquatic ecosystems) has not yet been broadly demonstrated as a sustainable land-use option that maintains biodiversity and other wildland values such as "wilderness," but rather that it has generally led to biotic impoverishment (Hoyt 1994; Nias 1995). Thus we must depend on fully protected areas and other forms of non-consumptive use as the principal means for maintaining biodiversity and wildland values.

The "use it or lose it" view advocates that, compared to developing more protected areas, more wildlands and biodiversity will be conserved by making use of the living resources in those wildlands that are not strictly protected. A major part of such use has traditionally been consumptive use, wherein the organism or any of its parts are harvested (killed or removed from the population). The logic behind this proposition is that revenues generated by the commercial, consumptive use of wild species will provide economic incentives for sound management of the harvested population(s). This in turn implies that the target population's habitat will be protected, thereby benefiting the broader goals of biodiversity conservation (McNeely 1988; Benson 1992; WWF 1993). This position is staked out by Janzen (1994) with respect to tropical habitats when he claims that the "use it or lose it" strategy "envisions 80%–90% of tropical terrestrial biodiversity conserved on 5%–15% of the tropics," as compared to continuing a traditional approach to protected areas, in which "10%–30% . . . will be conserved on 1%–2% of the tropics." Advocates of this approach do not necessarily believe that the national parks approach has failed, but rather that significant additional conservation gains using that approach are not possible in much of the world.

The above may unfairly suggest a highly polarized set of views in the conservation community. Most conservationists would probably contend that the best strategy is a mixture of these two approaches, with one or the other to be favored depending on the circumstances. Regardless of what conservationists think, much of the world depends on wild species for an array of products, whether for food, fiber, or medicine. Thus, in many cases, the question is not *whether* to use wild species, but rather how to move from a system of use that is clearly not sustainable toward one that is better.

It was within this context that most of the contributors to this volume were asked to examine the array of social, economic, and ecological questions that confront commercial, consumptive use, its sustainability, and its effects on biodiversity. In addition, in four of the chapters—waterfowl management (Callaghan et al., Chap. 14), forest management (Hansen, Chap. 6), ungulate management (Teer, Chap. 12), and salmon management (Francis, Chap. 16)—the authors were asked to focus strictly on the ecological questions regarding how management programs that have achieved some level of sustainability in terms of offtake may affect biodiversity and ecosystem function. The types of use examined in the chapters represent a wide array of socioeconomic and ecological conditions. This broad-spectrum approach was taken with the goal of elucidating both those questions and issues that are universal regardless of the conditions and those that are more variable.

I review here the major social, economic, and ecological issues and trends that emerge from these studies and others. I conclude by examining how the interplay of social, economic, and ecological factors influences the divergent tendencies of economic specialization and ecosystem management, and what this means for biodiversity. Though by no means exhaustive of the issues surrounding wild species use, this introduction provides a framework for analyzing the array of management experiences and issues presented in this volume.

The Sustainability of Commercial, Consumptive Use: Overview

Skepticism about the "use it or lose it" strategy is understandable. During the last two hundred years, as the human population has continued to grow exponentially and technology for harvesting wild species and their products has advanced at an even faster rate, the majority of wild species resources for which any significant market exists have been overexploited and their ecosystems highly altered and generally simplified. While the perverse effects of economic forces on exploitation of species and habitats are widely evident, well-documented examples of a positive linkage between commercial use and biodiversity conservation have been limited (Kremen et al. 1994; Rasker and Freese 1995). Moreover, concerns have been raised about the tradeoffs that consumptive use implies for biodiversity, even if the offtake is sustainable (Mangel et al. 1993; Robinson 1993).

Nothing illustrates this trend better than timber harvesting, beginning two hundred years ago with the overexploitation of temperate forests (e.g., Williams 1989; Koch and Kennedy 1991; McShane and McShane-Caluzi, Chap. 4), and more recently tropical forests, particularly in Southeast Asia (e.g., Gillis 1992a,b; Poffenberger 1992). Marine fisheries have fared no better, as stocks of one species after another have been depleted throughout the world (FAO 1988; Ludwig et al. 1993; Norse 1993; Rosenberg et al. 1993; Weber 1994), with attendant large losses and alterations of biodiversity and ecosystem structure in many coastal marine waters (Dayton et al. 1995). Similar problems caused by commercial use can be cited for other groups of organisms, from parrots (Thomsen and Brautigam 1991) and river turtles (*Podocnemis* spp.) (Alho 1985) to palm heart (*Euterpe* spp.) (Clay, Chap. 8) and rattan (*Calamus* spp.) (Peluso 1992).

Perhaps no form of consumptive use, however, has had more widespread impoverishing effects on native biota than overgrazing by domestic livestock of native grasslands. The United Nations Environment Program estimates that 73 percent of the world's 3.3 billion hectares (ha) of dry rangeland are at least moderately desertified, largely due to overgrazing (Durning and Brough 1992).

Despite this dismal record, there are examples of consumptive use where at least the offtake has been sustainable. One of the most sustainable forms during this century, under varying degrees of commercialization, has been sport hunting of large game, particularly cervids, in North America and Europe. Populations of all cervids except the caribou (*Rangifer tarandus*) (subject to mostly traditional harvesting) have increased since 1900, and caribou numbers are now increasing also (Gill 1990). Similarly, offtake from sport hunting and commercial culling of large game in some regions of southern Africa also appears to be sustainable (Luxmoore 1985; Cumming 1989; Crowe et al., Chap. 10). While hunting or culling of large herbivores may occasionally be a useful biodiversity management tool (Teer, Chap. 12), offtake and management resulting from the big game hunting market may also have negative consequences for biodiversity (McNab 1991; Geist 1995; Teer, Chap. 12; and see the section in this chapter on "Ecological Issues").

Many wild species use programs demonstrate the tendency for good management to emerge only after a period of obvious overuse and population decline. Works in this volume on the markhor goat (*Capra falconeri*) in Pakistan (Johnson, Chap. 11), the American alligator (*Alligator mississippiensis*) (Joanen et al., Chap. 13) and striped bass (*Morone saxatilis*) in the United States (Upton, Chap. 15), and forests in Switzerland (McShane and McShane-Caluzi, Chap. 4), Mexico (Kiernan and Freese, Chap. 3), and India (Singh et al., Chap. 2) demonstrate this pattern.

Other types of commercial, consumptive use, especially of nontimber plant products from forests, are much less researched and the sustainability of their harvest unknown. For some, offtake may be sustainable without strong management because demand may not yet exceed productivity, such as Brazil nuts (*Bertholletia excelsa*) in the Amazon basin (Clay, Chap. 7) and berries in the Nordic countries (Kardell 1986; Salo 1995). However, there is mounting evidence that many nontimber plant products are being overexploited, even in highly traditional systems (Edwards and Bowen 1993; Hall and Bawa 1993). Similarly, increasing levels of hunting of forest wildlife for commerce in nearby urban centers, particularly in tropical forest regions, appear to be causing declines in many populations (Steel 1994; Bodmer 1995; Fa et al. 1995; Bodmer et al., Chap. 9).

Economic Issues

Failure of Markets to Internalize Costs and Benefits

The failure of markets to internalize both the benefits of biodiversity and wildlands and the costs of their use is a widely recognized issue (see reviews by McNeely 1988, and Swanson and Barbier 1992). People who benefit most from wild species use are often not those who pay the costs of maintaining the species and its ecosystem. Many wild species and ecosystems are undervalued because there are no markets to compensate those who help generate some of the benefits, called positive externalities, that others reap. Thus grain farmers who maintain wetlands on their property for migratory waterfowl (and thereby forgo profits from expanding their grain fields) may benefit duck club owners who live elsewhere, possibly in another country. It is difficult to quantify this benefit, however, and even if one could, there are few if any efficient mechanisms for duck club owners to pay farmers for it. Duck hunters' contributions to a nonprofit organization, which then invests those funds in wetland easements by the farmers, is one mechanism. Another is through taxes or royalties paid by the duck club owners to the government, which then allocates funds, sometimes through international agreements, to compensate the farmer for wetland conservation. If such payments are not made, the duck hunters are, in economic language, "free riders."

The flip side of the above scenario is negative externalities. The full costs to other users of the resource or to society at large are not factored in when grain farmers drain their wetlands. Their action may result in lost revenues for the duck club owner, poorer fishing in streams formerly fed by the wetlands, lower water tables and higher costs for farmers who irrigate, and the loss of a favorite bird-watching site for local residents.

The values cited above fall into the categories of either "use values" or "functional values" (Vatn and Bromley 1994). In addition, wildlands and biodiversity also carry various "nonuse values." These include (1) the "option value," based on an individual's desire to maintain the option to use the resource in the future; (2) the "existence value," based on the knowledge that the resource simply exists; and (3) the "bequest value," based on the individual's desire to pass the resource on to future generations (Loomis et al. 1991). The challenge lies in how to estimate these values better and, once estimated, how to incorporate them into decision making about land use so that the value of wildland and biodiversity does not rest solely on the shoulders of marketable wild species commodities.

Economic valuation of the goods and services provided by wildlands and biodiversity is difficult because most functional and nonuse benefits are impossible to quantify reasonably. Consequently there is widespread aversion to attempting such calculations and to incorporating them into decision making (Bingham et al. 1995). Further, there is considerable inertia in existing political and economic systems that caters to well-established economic stakeholders at the expense of new ones, particularly those involving environmental goods and services. As Rivlin (1993) states, "forces opposed to policy change tend to have the resources and organization to capture the political system, while the beneficiaries of change do not." Decision making is thus bound by "the tyranny of the status quo" (Vatn and Bromley 1994). The result is that decision makers, whether the local landowner or government policymaker, continue to make land-use decisions that are largely based on commodity values, and that are inimical to the broader societal benefits of biodiversity and ecosystem services.

Although economic valuations are only indicative, they can improve decision making by expanding the bounds of economic values to include some indirect, nonmarketed functions of wild species and ecosystems. Such economic criteria, however, are only a subset of the information and considerations used in decision making. As Vatn and Bromley (1994) caution, "there is nothing in economics in general—

or in hypothetical valuation in particular—that can address the optimal level of air or water quality, or of land devoted to parks and wilderness." Further, the tenets of orthodox economic logic require that each of us be perfectly selfish and rational, neither of which is necessarily present in decision making (Parson and Clark 1995). Thus decisions should include the best economic estimates available, but monetary measures are only a subset of an array of currencies—votes, appeals to spiritual values, traditions, common sense—that enter into decision making.

Forest management is a prime example of this problem: because of the ease of calculating the economic value of timber and the difficulty of quantifying nontimber values, timber harvesting emerges by default as the best and most productive economic use of the forest. Appasamy (1993) illustrates this situation in India where nontimber forest products used by poor people "do not enter the system of national accounts, which results in the undervaluation of the forest wealth of the country." Gillis (1992a) notes that because the "assignment of monetary values to the protective services provided by forests . . . is much more difficult than for productive services . . . in dozens of nations from Southeast Asia to Latin America to Africa, the owners of property rights to the natural forest have placed far heavier value on the productive rather than the protective resources provided by the forest." McShane and McShane-Caluzi (Chap. 4) illustrate how the estimation and incorporation of other forest values in Switzerland, where wood production is shown to represent only 5 percent of the total value, are leading to a restructuring of forest policies that should benefit biodiversity.

The above example highlights the need to develop mechanisms for the stakeholders of the nonuse values of biodiversity to help pay the costs of biodiversity conservation. An example of this is the German Development Assistance Agency's efforts to negotiate, as a stakeholder interested in biodiversity, a buyout payment to the government of the Central African Republic to stop logging in the Dzanga-Sangha Reserve (Telesis 1991). Thus much of what is called financial "assistance" or "subsidies," both of which convey the sense of a handout, are, in fact, payments from willing-to-pay consumers who hold nonuse values for biodiversity and wildlands.

"Green marketing" provides a mechanism for internalizing the costs of sustainably managing a consumptive use, and enables consumers to pay for broader biodiversity values (Johnson and Cabarle 1993). Green marketing is based on the premise that the costs of good stewardship of a harvested species can be passed on to the consumer, who is willing to pay a premium for the product if it is certified as being sustainably produced without undue effects on the natural environment. Clay (Chap. 8) notes that "the values generated by the harvesting, processing, and sale of wild species are traditionally captured far from the ecosystem that produced it. Green marketing is one way to return at least a portion of the value added to wild-harvested products to the producers themselves." The Plan Piloto Forestal in Mexico (Kiernan and Freese, Chap. 3) has attempted to use green markets to expand foreign markets for its hardwood tropical timber, though with limited success to date.

Failure to Fully Value Future Benefits

Under traditional market mechanisms, the economic value of a wild species or its products is greatest for current benefits and decreases for future benefits. The rate of decrease is called the "discount rate." The higher the discount rate, the less likely one will forgo revenues from harvesting a resource now for revenues from later harvests (Randall 1981; Barbier 1992). As Lee (1993) contends, "efficient markets tend to allocate resources to the current generation at the expense of later ones" and "if resources are traded in markets, the value of conserving them for ecologically significant lengths of time is set by markets, not by biology; usually, biological conservation turns out to be worth very little."

Clark (1973) demonstrated the perversity of the discount rate for wild species management by showing that for species with a low annual growth rate (e.g., whales), the rational economic decision is to harvest the entire population and put the revenues in an investment with a higher annual return. Thus Norgaard (1995) notes that "high interest rates encourage transformation of natural ecosystems toward faster-growing species or other uses of land." In addition, uncertainty about tenure, markets, and population levels of commercial species creates high discount rates because a premium is placed on exploiting resources and earning revenues now rather than risk losing them later.

The study of wildlife use in the Peruvian Amazon by Bodmer et al. (Chap. 9) shows how poverty and the "have-to-eat-today" principle can create a high discount rate. Poverty compels people to overharvest today to meet current needs and thereby forgo future returns from sound resource stewardship. Similarly, overcapitalization and the "ratchet effect" commonly seen in marine fisheries (Ludwig et al. 1993), whereby new capital investments are made during periods of good fishing, places a premium on maximizing harvests and revenues now at the expense of future returns. Further, without alternative sources of economic support to maintain socioeconomic viability during the recovery period for an overharvested population, the combination of poverty/debt and an overexploited resource creates a dilemma in which one must choose between either short-term socioeconomic sustainability or recovery of the target population. In the state of Quintana Roo, Mexico, overexploitation of mahogany (*Swietenia macrophylla*) during an era of government concessions to a lumber company has left standing stocks depleted. Without outside support to sustain the communities economically during the long period required for replenishment of mahogany stocks, the only alternative is to continue to harvest the remaining large mahoganies at what may be an unsustainable rate. Though this may lead to even greater depletion of mahogany stocks in 50 to 75 years, the opportunity costs of alternative land uses such as citrus groves, sugarcane, and cattle pasture continue to be offset and the forest cover maintained (Kiernan and Freese, Chap. 3).

The perverse effect of the discount rate is another case in which markets, left unfettered, do not lead to sustainability. Clearly, "discounting the future" is at odds with option values, bequest values, and intergenerational equity as decisions are made regarding the distribution of benefits from biodiversity. For society to overcome these odds, there is a "need for economic reasoning to work in conjunction with ethical criteria—a larger system of values" (Norgaard 1995).

Product Demand and Sustainability

Changing demand is one way of changing rates and types of consumption. Influencing demand therefore represents a potential management tool. Demand can be influenced by the availability and price of substitutes and the consumer's buying power and desires, and the quantity demanded can be influenced by changing the price of the product. Government policies and regulations may control the extent to which consumers are able to exercise their demand.

The growth in demand for wild living resources, fueled by rapid population growth in the South and high per capita consumption in the North, is outpacing the productivity of wild living resources on a global scale (Postel 1994; Vitousek et al. 1986). This presents an overarching issue for any discussion of demand and how to manage it. Globally, sustainable management of wild species and maintenance of healthy ecosytems require that growth in the human population be reduced to replacement level or below and that overall consumption of many species be reduced. If we fail to do so, victories of apparent sustainability are destined to be temporary and local in scale.

Some types of end uses and the demands they create are more manageable than others. For example, demand for luxury goods should be easier to control than for wild species products that meet basic human needs. Luxury markets are more amenable to consumer awareness programs, and their control via policies and regulations is potentially easier because public pressure should rise more rapidly to challenge such uses when they endanger populations. The precipitous drop in demand for elephant ivory in the United States following publicity about the effects of the illegal ivory trade on elephant populations is an example (O'Connell and Sutton 1990). Conversely, as noted below, luxury markets for many wild species uses are being promoted in attempts to add value to wildlands that are competing with alternative land uses.

Are there price thresholds above and below which sustainable offtake becomes much more problematic? We might expect that below a certain price/profit the use becomes economically unsustainable because insufficient revenues are generated for reinvestment in management, or alternative uses of the species or wildland are economically more attractive. This is arguably the case for many forests in Latin America, particularly where species that fetch a high price, such as mahogany, are rare (either naturally or from logging) and demand and prices are low for lumber of so-called "lesser-known species," and nonexistent for most species (Perl et al. 1991; Verissimo et al. 1992). Pricing by governments of tropical wood resources below their commercial—not to mention their social—values has had pernicious effects on tropical forest management around the world (Gillis 1992b).

These factors have stimulated efforts to create markets and increase the demand for lesser-known timber species as a conservation strategy (Buschbacher 1990; Johnson and Cabarle 1993; Kiernan and Freese, Chap. 3). Similar efforts to create markets and thus increase total forest value are under way for many nontimber forest products (Clay 1992; Stiles 1994). The same logic is behind efforts to expand markets for sport hunting and other wildlife uses on private land in North America (Rasker and Freese 1995), in Pakistan (Johnson, Chap. 11), and in southern African countries such as Zimbabwe (Metcalfe 1994), South Africa (Crowe et al., Chap. 10), and Namibia (Lungu 1990). The logic, as noted before, is that alternative land uses will prevail if greater economic revenues are not generated by native species. This, together with social equity objectives, also underpins efforts to increase the revenue share that reaches the local stakeholders. Interestingly, many of these strategies are based on luxury uses for international markets.

High prices/profits may often be crucial for offsetting the opportunity costs of competing land uses, but if too high they may undermine sustainability. Four social and economic problems may arise: (1) under an open-access regime higher prices may make it profitable to go after an already depleted population, which can escalate the problem as demand for a decreasing supply pushes prices still higher; (2) even if property rights are well defined, their enforcement becomes more difficult because the payoff for a successful poacher is so high; (3) high prices/profits facilitate bribery and may create corruption at various levels; and (4) unusually high prices/profits may be correctly viewed as temporary, creating high discount rates, and thereby creating a strong incentive to cash in on high earnings before the price falls or the population is depleted. In short, unusually high prices/profits may create a destabilizing environment for good management. Rhino horn and bluefin tuna provide obvious examples of this problem (Gaski 1993; Milliken et al. 1993), as does the collapse of numerous coastal fisheries following invasion by national and international markets (Ruddle and Johannes 1983; Weber 1994).

Social Issues

Key Stakeholders and Their Motives

An array of individuals and institutions have a stake in the commercial use of wild species and its effects on wildlands and biodiversity.

Human stakeholders (as opposed to nonhuman organisms for which stakeholder interests could also be defined) can be divided into four major groups:

1. those with a direct interest in the species/product being harvested and resulting revenues;
2. those with values, economic and noneconomic, related to the target species and its ecosystem that are affected by the externalities of use and management;
3. those concerned with humanitarian issues (sensu Nash 1989), that is, human rights and social equity, particularly for local people, and the humane treatment of animals; and
4. agencies and authorities that are outside the flow of benefits and externalities, but with management rights or substantial influence (legitimate or otherwise) over commercial, consumptive use programs, and how the interests of the first three categories of stakeholders are addressed.

In large part, the first category represents stakeholders whose property rights are clearly defined for the land being used and the product being harvested and sold (Randall 1981; van Kooten 1993). This includes the various players in the marketing chain of a commercial product, from the resource owner and harvester, to intermediaries and the end-use consumer, to government agencies that derive revenues from taxes and fees imposed at various links in the chain. These stakeholders are influenced by and able to express their interest through traditional market mechanisms. As Clay (Chaps. 7, 8) demonstrates, in examining markets for Brazil nuts and palm hearts, the relationship between these stakeholders and how revenues are distributed among them significantly influences incentives for sustainable resource use.

The second category of stakeholders largely represents those whose benefits are based on public goods and services from the ecosystem being managed, and who are therefore incidentally affected by consumptive use activities (Randall 1981; van Kooten 1993). These may range from individuals whose economic well-being is directly affected, such as fisheries communities downstream from a logged watershed, to those that attach noneconomic values to the ecosystem under management. These latter stakeholders may include local tribal peoples with strong spiritual ties to a wildland, international conservation organizations and their donors, or scientists with academic interests. Although this second category of stakeholders may receive substantial material and nonmaterial benefits from the ecosystem's biodiversity and functions, the absence or inefficiency of markets through which they can pay for those benefits, and the resultant invisibility of such values in decision making, inhibit their influence on how an ecosystem is managed.

The third category brings in a set of values and motivations that are largely distinct from those operating in the first two. It therefore may involve influence groups, such as human and animal rights organizations, whose primary motivation is neither revenues nor biodiversity conservation. Human rights issues, whether the rights of indigenous peoples, economic development for the poor, or the role of women, yield an array of principles and policies to guide how decisions about wild species use are made and how the resulting socioeconomic benefits are distributed (e.g., IUCN/UNEP/WWF 1991). Meanwhile, animal rights interests are having an increasing influence on the consumptive use of wildlife (e.g., Hoyt 1994; Nash 1989). Animal and human rights interests often clash, however, such as when wildlife use restrictions based on humane reasons cause socioeconomic hardships for people that have traditionally depended on such use (IUCN/UNEP/WWF 1991).

The fourth category of stakeholders cuts across the other three. The most common examples are elected officials and other politicians who are not directly affected by the economic benefits or externalities of commercial, consumptive use programs, and governmental resource management agencies whose budgets do not depend on taxes or fees from commercial, consumptive use. Nevertheless, the actions and well-being of politicians and management agencies are obviously influenced by the other stakeholders' satisfaction with how well their interests are attended to.

Clearly, the pivotal stakeholder in this array of interests is the individual or institution that ultimately decides and controls the fate of a wildland and its biodiversity (generally the landowner or the institution that controls access to and use of the resource). A major challenge in linking the consumptive use of wild species with biodiversity conservation is to understand how the decisions of this pivotal stakeholder are affected by the interests and influence of all the other stakeholders.

Resource Rights versus Government Control

Increasing attention has recently been given to improving resource tenure rights for local people (Larson and Bromley 1990; Poffenberger 1992; IIED 1994; Western and Wright 1994; and various contributions in this volume). This may occur through efforts to define existing ownership in a better legal manner, through outright transfer of ownership from the state to individuals or groups, or through the issuance of usufruct rights for public land resources. Assigning resource rights to individuals or communities, it is argued, should provide a strong incentive for good resource stewardship. In much of the developing world, the emphasis has been on communal management systems under some form of common property ownership. This trend is well illustrated by participatory forest management in India (Singh et al., Chap. 2), the return of forest use rights to communal landowners in the Yucatan Peninsula of Mexico (Kiernan and Freese, Chap. 3), and the development of the Tamshiyacu-Tahuayo Communal Reserve in Amazonian Peru (Bodmer et al., Chap. 9). The conditions under which communal or common property management is effective in creating incentives for good resource stewardship, compared to individual ownership, remain an area of considerable debate (Berkes 1989; Bromley and Cernea 1989; Hodson et al. 1995; Mendelsohn and Balick 1995).

Various factors may undermine the effectiveness of privatization for sound resource management and biodiversity conservation, and thus regulation, influence, or assistance by government or other outside agencies that represent broader public interests is often necessary. The question is what balance to strike between local control, government oversight, and the role and influence of other stakeholders (e.g., nongovernmental conservation organizations).

The mobility of some wild species makes it difficult or impossible to assign use rights or ownership to individuals or local communities, with the result that no one individual or community has a strong economic incentive to harvest sustainably. Without oversight by government agencies, often under terms of international agreements, such species are highly prone to the "tragedy of the commons" (Hardin 1968). Highly migratory fish, such as salmon (Francis, Chap. 16) and striped bass (Upton, Chap. 15), and migratory waterfowl (Callaghan et al., Chap. 14) illustrate this problem.

Marine fisheries provide the most vexing problem regarding assignment of resource use rights since neither the migratory nature of many species harvested nor their aquatic habitat lends itself to defining such rights easily. Some community-based coastal fisheries have evolved relatively effective management systems without government intervention (Berkes 1987; Acheson 1989; Ruddle 1989). However, high prices due to strong national and international markets, as well as new fisheries technologies, can cause a breakdown in these systems and erode the sustainability of the fisheries (Weber 1994). Where community-based traditional systems are absent or do not work, there is considerable uncertainty about what systems for allocating fishing rights are most effective under a given set of conditions. Many systems are based on an annual quota of landings, with open access to any and all who wish to fish. Upton (Chap. 15) cites the example of the state of Maryland, U.S.A., in which the number of fishers in the gill net fishery for striped bass jumped from 453 in 1991 to more than 900 in 1994. The state had to cut the number of fishing days by nearly half over this period because the quota was caught more quickly. Increasing numbers of fishers, improved gear, and ever-faster fulfillment of quotas eventually lead to "derby fishing," such as in the Alaska halibut fishery, where regulators were forced to restrict the season to two or three 24-hour periods per year (Weber 1994). An array of management and market problems emerge under these conditions (Weber 1994; Upton, Chap. 15).

Individual transferable quotas (ITQs) have emerged as an alternative management technique (Sissenwine and Rosenberg 1993). Under ITQs, each of a predetermined number of fishers receives a share of the annual quota, which he or she may then harvest as fish populations and markets dictate. ITQs may be purchased, sold, or leased just like property. Difficulties include decisions about who should receive the original allocation of ITQs, and concerns that, without some restriction on the number of ITQs per individual or company, highly capitalized fishing operations may buy out small holders and eventually control the fisheries (Weber 1994). New Zealand has applied ITQ management on a more comprehensive scale than any other nation, with mixed results to date (Sissenwine and Mace 1992).

The open-access nature of marine fisheries outside of the Exclusive Economic Zone is particularly problematic since control depends solely on international accords and the difficult policing of the vast marine realm. International agreements such as the U.N. Conference on the Law of the Sea (UNCLOS), the Convention on the Conservation of Antarctic Marine Living Resources (CCAMLR), the International Convention for the Regulation of Whaling (ICRW), and the International Commission for the Conservation of Atlantic Tuna (ICCAT) have had varying degrees of effectiveness in regulating marine fisheries, and many problems remain with implementation and compliance (Norse 1993; Safina 1993).

Problems also emerge when management jurisdictions cut across ecosystem functions. Gopinath and Gabriel (Chap. 5) describe a classic example of this in Malaysia, where the Matang Mangrove Forest Reserve is managed by the Forest Department principally for wood production. The management regime, however, may affect the filtering, detrital output, and fish nursery functions of the mangrove, with potentially negative effects on nearshore fisheries which fall under the authority of the Fisheries Department.

As discussed in the concluding section, privatization of wild living resources may facilitate specialization in species with high commodity values at the expense of other, nonmarketable biodiversity values. This can lead to biotic impoverishment even though the offtake of the target species is sustainable. Though government ownership may be subject to the same profit-based incentives, some level of government oversight (whether local or national) will often be required to protect the broader public benefits of biodiversity. This can create conflicts within or between government agencies since such agencies themselves are also often dependent on revenues derived from commodity production on wildlands.

Revenue Distribution

Two major questions arise regarding revenue distribution and its effects on resource management. The first concerns the proportion of the total retail price that is captured by the resource owner, the harvester, intermediaries, and the retailer. A commonly raised concern, particularly in the developing world, is that the resource owners/harvesters receive a small fraction compared to the intermediaries and retailers. Gillis (1992a), for example, notes that in many tropical countries rents captured from logging accrue primarily to the relatively wealthy rather than the poor. Clay (Chaps. 7, 8) shows that harvesters in Amazonian Brazil receive, at best, 1 percent of the retail price of Brazil nuts, and 1–4 percent of the retail price of palm hearts, paid by buyers in the United States. In such circumstances, harvesters may have neither the economic incentive nor the financial means to implement sound management practices, and the value of the forest may be insufficient to deter alternative uses of the land. As Repetto (1988) concludes, inappropriate forest revenue systems, including the loss of enormous economic rents to timber concessionaires and other timber exploiters, "have created economic incentives that powerfully accelerate the rate of deforestation."

Another result of the great price differential between the ends of the market chain, so common in internationally traded products, is the leverage it provides for green marketing. For Brazil nuts, for example, a $0.05 premium charged to consumers for ensuring that the Brazil nuts they buy are sustainably harvested would add only one-half of 1 percent to their specialty store price. If, however, this premium is fully passed on to the harvester in the forest, it would more than double his income.

The second question concerns the distribution of revenues between local resource owners, or those with usufruct rights, and the government agency charged with oversight and management responsibilities. There are wide disparities in how revenues are divided. For example, local communities in West Bengal, India, receive less than 25 percent and the government more than 75 percent of gross revenues from the sale of sal (*Shorea robusta*) poles (Poffenberger 1994), whereas in Quintana Roo, Mexico, the communities receive 100 percent of the price of the lumber sold (Kiernan and Freese, Chap. 3). Perhaps not coincidentally, communities in West Bengal have increasingly directed the management of their forest toward nontimber forest products for which there is no government fee (Singh et al., Chap. 2). Kiss (1990) notes that government bureaucracies in Africa often intercept wildlife-related revenues that are meant to be distributed to local communities. A less transparent revenue-distribution system, in the form of corruption (e.g., payoffs to government officials for permitting excessive or illegal harvest levels), may also evolve under systems of co-management with generally pernicious effects on sustainability and biodiversity.

Ecological Issues

Uncertainty, Variability, and Scale

Natural systems are much less stable than is generally believed over both the short term (measured in years or decades) and long term (measured in centuries or millennia), and we are not yet very good at predicting when, why, how, and at what rate changes in populations and ecosystems occur. The result is that one year's abundant and highly harvestable population may, without warning, drop to low and unharvestable levels a year or two hence (Hilborn and Ludwig 1993). A wide array of economically important wild species resources show such boom-and-bust cycles, from seed production of tropical trees (Janzen and Vásquez-Yanes 1991; Clay, Chap. 7) to marine fish stocks (Wise 1991; Francis, Chap. 16; Upton, Chap. 15). Further, what we perceive to be a static "natural" diversity and ecosystem structure is often, in fact, one stage in a long transition or cycling from one ecosystem configuration to another due to both anthropogenic and nonanthropogenic forces (Botkin 1990; Sprugel 1991; Francis, Chap. 16).

Uncertainty emerges both from inherent stochasticity in ecosystems and from our incomplete knowledge of how they work and how they respond to human intervention. Our recognition that ecosystem and population fluctuations are greater and less deterministic than we once thought raises several management issues. Botkin et al. (1993) suggest that a virtually universal problem with management of most wild species is the assumption by biologists and managers that "given enough research, exact numbers can be determined for population size, components of population dynamics, and the responses of populations to given harvest levels." However, they note, "because of the effects of environmental uncertainty, exact or correct numbers are rarely if ever possible to obtain, and the result is (1) wrong numbers and harvest levels are recommended, or (2) biologists and managers delay recommendations until they have 'more information.' The latter case leads decision makers to set their own limits, based on economic or political considerations instead of scientific ones. This almost always leads to . . . degradation of the resource."

Uncertainty and large population fluctuations are a particularly vexing problem for the management of marine fisheries (Wise 1991; Ludwig et al. 1993; Rosenberg et al. 1993). Large and unpredictable annual variations in striped bass recruitment (Upton, Chap. 15) and longer-term fluctuations in Pacific salmon stocks (Francis, Chap. 16) exemplify the difficulties that arise when the marketplace's objective for a steady and predictable level of production meets ecological reality. Large investments in harvesting capabilities during good years make it difficult to reduce harvesting pressure during low population years. This, coupled with the open-access nature of many marine fisheries, uncertain scientific information, and risk-prone management decisions, has led to the depletion of marine fish stocks around the world (Sissenwine and Rosenberg 1993).

Forest ecosystems that have been managed for wood production are also beginning to reveal how uncertainty, natural disturbances (e.g., wildfires, blowdowns, pest outbreaks), and long-term cycling in ecosystem structure and processes may be at odds with traditional silvicultural goals and techniques. For example, traditional forest management suppresses natural disturbance, which reduces species diversity and may reduce productivity (Roberts and Gilliam 1995; Hansen, Chap. 6). Kiernan and Freese (Chap. 3) show how management of a disturbance-dependent species such as mahogany is compromised by traditional silvicultural approaches. Mladenoff and Pastor (1993) describe how a reciprocal linkage between carbon and nitrogen in the northern hardwood and conifer region of North America may impose cyclic patterns of productivity as hardwoods and conifers succeed one another. In the same forest region, silvicultural attempts to develop a spruce monoculture free of spruce budworm (*Choristoneura fumiferana*) infestations, by interfering with the natural cycling between fir (*Abies balsamea*) and aspen (*Populus* spp.) stand dominance, led to reduced productivity and reduced biodiversity (Baskerville 1988). Such long-term, landscape-level processes led Mladenoff and Pastor (1993) to propose that "rather than managing for a sustained level of a particular target population . . . managers may consider sustaining the cyclic nature of populations at the ecosystem level while maintaining a sustained yield of a target population at the regional level."

The need to manage ecosystems at much larger scales than has traditionally been practiced in both fisheries and forestry is well illustrated in this volume. Hansen (Chap. 6), for example, points out that "the flows of materials, energy, and organisms among ecosystems are key determinants of ecosystem function," and thus forest management must place much greater emphasis on landscape- and regional-level factors than does traditional forestry. Similarly, Francis (Chap. 16) emphasizes that sound salmon management must deal with the vastness and variability (in time and space) of the salmon ecosystem, and that "salmon may themselves be an important 'glue' that holds their ecosystem together in that they serve as a nutrient pump from the marine to the freshwater parts of the system."

Species Life Histories and Risks of Overexploitation

Management must be sensitive to those life history characteristics of individual species that make them susceptible to overharvest and population decline, as well as to the broader ecosystem effects of management programs. This requires a more precautionary approach to the management of many species. Long-lived, slow-reproducing species, such as whales, sharks, elephants, and primates, may be particularly vulnerable to overharvesting (Mangel et al. 1996). In some cases, whole communities of organisms, such as those found in deep-sea habitats, may be characterized by low reproductive capacities and thus may be particularly vulnerable to overharvesting (Dayton et al. 1995). The large population declines caused by hunting of such species, compared to species with higher reproductive rates, are well demonstrated in both African and neotropical forest wildlife (Bodmer 1995; Fa et al. 1995; Fitzgibbon et al. 1995; Bodmer et al., Chap. 9). However, to the extent that slow-reproducing species exhibit density-dependent reproductive strategies, managers should be better able to predict population responses to offtake and thus harvest protocols should be easier to design than for species that are more density independent. The difficulty lies in the implementation—the maintenance of very low harvest levels. Kirkwood et al. (1994), in an examination of fisheries, suggest that longer-lived species "retain a resilience to catastrophes that is not available to the short-lived species." They caution, however, that though short-lived species are capable of producing high sustainable yields, they "can be particularly vulnerable to a combination of high exploitation and occasional environmental events that devastate the spawning stock."

The determination of sustainable harvest levels is perhaps least understood where some part of the organism, rather than the organism itself, is harvested. Because the effects on survivorship and reproduction in the population are less direct, changes are difficult to monitor and detect. Nontimber forest products, such as fruits and latex, are most prominent in this category (e.g., see Boot and Gullison 1995; Bodmer et al., Chap. 9; Clay, Chaps. 7, 8; Singh et al., Chap. 2). Peters (1994), for example, cautions that the collection of commercial quantities of fruits and seeds can cause changes in the structure and dynamics of a tree population and, if uncontrolled, can result in its gradual extinction.

One largely untested management method being widely proposed to mitigate the risk of overharvest, and to maintain biodiversity in consumptive use programs, is the establishment within harvest areas of reserves that are off-limits to harvesting. In marine fisheries, such refugia may serve various functions, such as providing sources of recruits and protecting key ecological processes (Bohnsack 1993; Roberts and Polunin 1993; Agardy 1994). Bohnsack (1993) suggests that 10–20 percent of the continental shelf should be in nonfishing reserves, while Clark (1996) cites recent work that suggests that marine reserves may need to cover 50 percent or more of the area occupied by the stock as a hedge against overexploitation. In forestry, reserves within logged areas may help maintain key ecosystem processes as well as serve as refugia for species that require forest interior habitats (Hansen, Chap. 6). In waterfowl hunting, nonshooting zones located within hunting areas may be crucial as rest and feeding areas for both game and nongame species (Callaghan et al., Chap. 14).

One logical response to resource fluctuations, both natural and human induced, is opportunistic or pulsed harvesting. A repeating pattern of overexploitation–population recovery–overexploitation may result. This has arguably been an effective strategy, at least under low human population densities, for slash-and-burn migratory agriculture and hunting systems in the lowland tropics, that is, deplete the local resources, move on to more fertile and productive lands while the exhausted ones recover, and so on (Hart 1978; Hart and Hart 1986). Such a strategy may be viable where human population density and consumption are low relative to the geographic scale of the resource. Unfortunately, few, if any, such frontier conditions exist anywhere in the world (Postel 1994), and thus this strategy would seem to be rarely feasible.

Effects of Biodiversity on Ecosystem Function

Closely linked to our understanding of uncertainty and variability are questions regarding the link between biodiversity and ecosystem functions. How much ecological redundancy is there among species in a given ecosystem? More precisely, what components of biodiversity must be maintained in the process of optimizing production of economically important species? At what point does directing management toward production and offtake of these species begin to compromise key ecological functions?

Considerable uncertainty remains regarding the link between biodiversity and commodity production for even our best-studied and intensively harvested ecosystems. Hansen (Chap. 6), in reviewing management of forests of the Pacific Northwest in the United States, concludes that it is unclear how ecologically based management strategies, such as maintenance of structural complexity and biodiversity, will successfully maintain long-term ecological productivity. A simulation study that compared a silvicultural regime of clear cutting and short-rotation cycle with a regime that maintained greater structural complexity via green tree retention and longer rotations indicated that, though the latter regime's biodiversity would be greater, its wood production and revenues would be lower (Hansen et al. 1995). Marine fisheries face similar concerns. Hammer et al. (1993), in assessing the problem of fish stock depletion and control, suggest that "much of the problem lies in maintaining biodiversity and balance among various organisms and the associated resilience of the marine ecosystem and its ability to continue providing valued fish resources and ecological services."

Limited theoretical and empirical evidence suggests considerable ecological redundancy among species within ecosystems (Lawton and Brown 1993; Schulze and Mooney 1993; Vitousek and Hooper 1993; Walker 1995). Lawton and Brown (1993) suggest that "the absolute minimum level of species richness necessary to maintain particular ecosystem functions . . . may be far below pristine levels." Indeed, McNaughton (1993) found situations in grassland and successional communities where primary productivity was inversely related to biodiversity.

Not all species are created equal when it comes to ecological processes (Lawton and Brown 1993). The keystone species concept suggests that some species are particularly important for maintaining certain ecological processes (Paine 1969; Schulze and Mooney 1993) or the species composition of a community (Paine 1969; Bond 1993). Keystone species, from elephants to nitrogen-fixing fungi, should receive special attention when managing an ecosystem for the consumptive use of select species.

Pimm (1993) cautions, however, that the conclusion that "numerous species" are redundant is based on limited evidence. Tilman and Downing (1994) found greater resistance to drought in grasslands with greater biodiversity, and concluded that most species are not ecologically redundant. Experimental work by Naeem et al. (1994) indicated that reduced biodiversity may alter several ecosystem processes. Biodiversity may be particularly important in conferring stability and resilience to perturbations in some ecosystems (McNaughton 1993; Pimm 1993; Hansen, Chap. 6), and ecosystems that lack their full complement of species may be subject to invasion and disruption by other species (Woodward 1993). Indeed, the most important argument for maintaining species diversity when managing for wild species commodities may be to maintain ecosystem resilience and adaptability to infrequent but large perturbations and long-term environmental change (Shulze and Mooney 1993; Vitousek and Hooper 1993; Walker 1995).

Specialization, Biodiversity, and Ecosystem Management

From Diversity to Simplicity: A Slippery Slope

Janzen (1994) notes that for as long as there have been humans, we have sought to simplify and homogenize the natural world. Salwasser (1994) captures this concern with his admonition that the new paradigm of ecosystem management faces a dismal future if we allow it to be perverted toward one or more specialized uses. This section focuses on how specialized uses are grease for the slippery slope between biodiverse wildlands and the nondiverse agroscape.

A basic economic principle is that commercialization favors economic efficiency which leads to specialization in production (Randall 1981; Swanson 1993). Within any given biodiverse landscape, some components (genes, species, functional services) of that diversity will have greater economic value than others, leading humans to "simplify and homogenize" to increase production of those more highly valued resources. Economic development and an increasingly international marketplace greatly facilitate such specialization since no one region must produce all the goods and services it requires or wants. Thus each region can specialize in selected components of biodiversity or wildland values to be marketed anywhere in the world (Norgaard 1987; Ekins et al. 1994).

Motives for specialization are not limited to stakeholders who directly profit from the use. Biodiversity conservation stakeholders may support specialization in order to generate sufficient revenues to offset the opportunity cost of an alternative land use. This may be particularly so in biodiverse ecosystems where economically important species occur at low densities. In such cases, the strategy is to specialize production and thereby conserve the populations/species and their habitats, but with some potential costs to "naturalness" and biodiversity. The dictum now slips to "*specialize* and use it or lose it."

Management interventions to increase productivity of commercialized species will often come at the expense of other species. Those that have a competitive, parasitic, or predatory relationship with the target species are often directly eliminated. Other species may be indirectly affected by such practices, and management of abiotic factors that favor the commercial species may incidentally create inimical conditions for other components of biodiversity.

Specialization takes four forms, each with potential costs for biodiversity: (1) increase the number or biomass harvestable—more trees, bigger fish; (2) improve and standardize the quality of the product—straighter tree boles, better fighting fish; (3) simplify the system so that there are fewer competing, predatory, or parasitic species to interfere with management objectives; and (4) improve the harvesting technique, including both the efficiency of the equipment and the skill of the harvester. As we shall see, however, selective harvesting and short-sighted management objectives may yield results that are the opposite of those desired.

The degree to which ecosystems are manipulated or simplified to favor the production of target wild species ranges from the control of just one or a few "undesirable" species in the habitat to extensive physical modification of the habitat. Wolf (*Canis lupus*) numbers have been controlled to augment moose (*Alces alces*) populations that are hunted in Alaska, U.S.A. The overall effects on biodiversity of such practices are unknown, though there is evidence in the case of moose and wolves that attempts to replace wolf predation with human offtake may lead to larger population swings in moose (Gasaway et al. 1983; van Ballenberghe and Ballard 1994). Along Canada's Atlantic coast the government has not only sanctioned, but subsidized, the harvest of harp seals (*Phoca groenlandica*) to decrease seal predation of and, it is argued, thereby facilitate recovery of overfished populations of the North Sea cod (*Gadus morhua*). However, because harp seals eat illex squid (*Illex illecebrosus*) which in turn eat young cod, such control measures may, in fact, depress the recovery of cod populations (Mackenzie 1996).

The management of wetlands, such as water level manipulation, for harvestable wildlife such as waterfowl and alligators is purported to be largely synonymous with their management for biodiversity (Wentz and Reid 1992; Joanen et al., Chap. 13). This proposition, however, has not been rigorously researched, a surprising state of affairs given the long history and extent of wetland management in North America and Europe and the volumes of associated research (e.g., see Callaghan et al., Chap. 14). Management to increase the density and productivity of economically desirable trees, including enrichment plantings, is common practice in silviculture, with the effects on biodiversity ranging from mild, where the forest is largely left intact, to severe, where monocultures are created (Ledig 1992). The studies by Gopinath and Gabriel (Chap. 5) on mangrove management and by McShane and McShane-Caluzi (Chap. 4) on Swiss forest management demonstrate how forest integrity and biodiversity are compromised when management is directed toward commercial wood production.

Managing for nontimber forest products may also have consequences for biodiversity, though these effects are poorly understood. Hall and Bawa (1993) warn that "harvesting of seeds and fruits decreases the availability of food for frugivore (fruit-eating animals) populations, limiting the number of organisms and perhaps decreasing the diversity of frugivores in a given community. These changes in turn may alter other trophic (food web) relationships and thus negatively affect other species in the community. Even something as simple as removal of dead wood and leaves may devastate detritivore (decomposers, e.g., microbes, fungi, etc.) communities critical to the cycling of the nutrients which support the vegetation in the forest." Bodmer et al. (Chap. 9) indicate that the harvesting of palm fruits in the Peruvian Amazon has probably depressed populations of forest mammals that eat them. Anderson et al. (1995) evaluated how forest biodiversity on three islands in the Amazon River of Brazil was affected by management by local communities for primarily the production of nontimber forest products, such as the palm *Euterpe oleracea*, whose fruits and hearts are marketed. On the two islands of greatest forest diversity, the number of native tree species was reduced by roughly half in managed forests.

Perhaps the most extreme case of habitat modification and specialization to serve an international commodity market is represented by the extensive conversion of mangrove forests to ponds for shrimp mariculture in much of Latin America and Southeast Asia (Southgate 1992; Cheng 1994; Gopinath and Gabriel, Chap. 5).

Genetic manipulation of native species is common in commercial enterprises. Such "genetic improvement" through silvicultural selection for desirable characteristics or hybridization between stocks or species has been widespread in forestry (Kitzmiller 1990; Millar et al. 1990; Riggs 1990; Burton et al. 1992; Ledig 1992). As Ledig (1992) notes, "domestication involves conscious, directional selection and is very effective in causing divergence from the wild-type." The effects on global biodiversity depend, in part, on the extent to which stands of wild genotypes are displaced in the process. Though much less extensive as a management tool, hybridization for stock improvement is also a management technique in some fisheries (Marnell 1986). In both forestry and fisheries, however, dysgenic effects caused by misguided harvesting practices have probably been more extensive (see below).

If attempts at genetic improvement do not yield the desired results, the logical next step is to introduce entirely exotic stocks or species. This has been widespread in freshwater ecosystems for both recreational and food fisheries. Stocking of hatchery fish or exotic species of fish frequently results in hybridization with consequent genetic alteration or total genetic swamping of native species. Predation and competition by introduced species can also lead to extinction of native species (Marnell 1986; Moyle et al. 1986; Nelson and Soule 1987). Introduction of the Nile perch into Lake Victoria for sport fishing may eventually eliminate more than half of the three hundred-plus species of endemic cichlids (Miller 1989; Barel et al. 1991). Native salmon stocks have also been replaced or suppressed by the introduction of hatchery stocks in many areas of the Northeast Pacific along the coasts of the United States and Canada (Francis, Chap. 16) and in Scandinavia (Hindar 1992). In North America (Canada, the United States, and Mexico), at least 140 species of freshwater fish have had their ranges expanded through introductions (Moyle et al. 1986), with at least 11 species successfully introduced from outside the continent (Courtenay and Kohler 1986). Moyle et al. (1986) conservatively estimate that 25–50 percent of freshwater fish caught in the continental United States are from introduced populations. The introduction of exotic fish for recreational and food fisheries is a widespread and ongoing international enterprise with destructive consequences for native ecosystems and species (Welcomme 1984).

Though of lesser magnitude than in fisheries management, significant introductions of exotics have also been stimulated by recreational hunting, as evidenced by waterfowl management (Callaghan et al., Chap. 14) and ungulate management (Teer, 1995; Chap. 12).

Management and harvesting methods may also cause unintentional changes in biodiversity that are contrary to management goals. Mangel et al. (1993) caution that even "apparently sustainable exploitation can have profound effects on genetic, species, and ecosystem diversity." Though the genetic effects are hard to quantify, Ledig (1992) characterizes many timber harvest regimes, such as "creaming" and "high-grading," as "ranging from selection against the most valuable forms to selection for the poorest." Further, harvest regimes may alter mating systems and thus the genetic structure of populations, and directly eliminate locally adapted populations (Ledig 1992). In fisheries, both theoretical (Brown and Parman 1993; Policansky 1993) and empirical work (Rowell 1993) indicate that selective harvest of large fish can lead to reduced age and size at maturation. Though cause-and-effect relationships are not yet established in most instances, changes in age or size at maturity have been documented for several exploited stocks of marine fish (Rowell 1993; Brown and Parman 1993).

Among large mammals, selective harvesting of tusked female African elephants (*Loxodonta africana*) led to a rapid increase in tuskless females born into the population (Jachmann et al. 1995), and the selective harvest of large-horned rams in bighorn sheep (*Ovis canadensis*) has been hypothesized to lead to decreased genetic variability and fitness (Fitzsimmons et al. 1995). (See Teer, Chap. 12, for further discussion regarding genetic effects in large mammals.)

The effects that sustainable offtake of a given species and associated management practices may have on species diversity are complicated by questions of temporal and geographic scale, and depend on the measure of species diversity used (e.g., richness or heterogeneity, and the taxa being measured). Recent research in both tropical and temperate forest management reveals that species diversity may respond negatively, positively, or with no discernable effects to more sustainable harvest and management practices (Frumhoff 1995; Halpern and Spies 1995; Hansen et al. 1995; Roberts and Gilliam 1995; Salick et al. 1995; Kiernan and Freese, Chap. 3; Hansen, Chap. 6).

Teer (Chap. 12) reviews conditions where, because of the loss of habitat or of natural controls such as predators, increasing populations of large mammal species such as white-tailed deer (*Odocoileus virginianus*), elk (*Cervus elaphus*), and African elephants have caused declines in species diversity. Reduction of populations through hunting may thus lead to increases in biodiversity, though the effectiveness of recreational hunting as a tool to keep ungulate numbers in check may be limited. Macnab (1991), however, asserts that while commercial use of wildlife on private lands in southern Africa may promote the conservation of the commercially important species, large predators are not tolerated and habitats are managed to the detriment of many other species. As Teer (Chap. 12) suggests, the management of keystone species, such as large ungulates, may have a major impact on ecosystem diversity and function.

The impacts on biodiversity caused by overharvesting in marine fisheries have been well documented (Dayton et al. 1995), but how the sustainable offtake of fish stocks affects biodiversity and ecosystem functions is poorly understood (e.g., see Upton, Chap. 15). Nicol and de la Mare (1993) provide a hypothetical model for how this question could be addressed in managing harvest of the Antarctic krill (*Euphausia superba*), another keystone species. The harvest levels would be set at some level below that necessary for stable recruitment in the krill in order to protect other species in the food chain that depend on krill.

Francis (Chap. 16) notes that while salmon may be a keystone food resource for vertebrate predators and scavengers, perhaps more significantly, spawning salmon are an important vehicle for transporting marine nutrients into their natal watersheds. Thus the human removal of "surplus" stock may have significant effects on the nutrient pool in these watersheds. The replacement of wild spawners with hatchery-produced fish, he warns, only serves to exacerbate this problem.

The harvest technology and skill and care of the harvester are also key variables affecting biodiversity. Substantial incidental structural damage to nearby vegetation is caused by some selective logging practices in tropical forests, though the overall effects on biodiversity are not well understood (Frumhoff 1995). More serious impacts on wildlife are often due to the ready access to previously remote areas that logging roads provide for hunters and other forms of human disturbance (Putz 1993; Paquet and Hackman 1995).

Callaghan et al. (Chap. 14) document how waterfowl hunting, under conditions of apparently sustainable offtake, can undermine biodiversity. Incidental effects include the disturbance and dispersal of target and nontarget bird species from hunting areas, the killing of nontarget species by hunters, and the toxicity of lead shot in the food chain. Further, because waterbirds often have a significant influence on nutrient dynamics and energy flow in wetlands, their management can have broad repercussions for both ecological functions and biodiversity of wetlands.

Specialization in harvest technology can be a double-edged sword for biodiversity. New technologies can reduce incidental catches (e.g., turtle excluder devices in fisheries) and incidental damage (e.g., highlining in logging). However, the economy of scale in harvesting can also lead to large and highly unselective harvesting technologies, as exemplified by shrimp trawls where, on a global level, 5.2 metric tons of bycatch are discarded for every metric ton of shrimp landed (Alverson et al. 1994). The damage to benthic marine ecosystems caused by bottom-fishing devices is perhaps one of the most insidious and potentially damaging effects of inappropriate technology developed to increase "efficiency" (Dayton et al. 1995).

The Relation between Socioeconomic Dependency and Biodiversity

Unless we can find and sustain a diversity of uses, both consumptive and nonconsumptive, many of much lower economic value than others, specialization will provide a slippery slope to ecosystem simplification. This is particularly so in highly diverse systems such as tropical forests where high-value species (e.g., macaws, mahogany) occur at low densities. Oligarchic wildlands, where one or a few highly valued species dominate (e.g., pine forests, palm forests, rivers with massive salmon runs), are inherently more compatible with economic specialization (Peters 1992), and thus may require fewer biodiversity trade-offs. The risk, however, is that overexploitation of an oligarchic species can readily lead to degradation of the entire ecosystem.

In many systems, it will continue to be domesticated species (i.e., once-wild populations/species that have undergone the specialization process a few millennia earlier) that best meet the need for production efficiency, and they will continue to compete with native biodiversity for use of the remaining wildlands, particularly terrestrial ones. A modern Mexican cornfield represents the ultimate degree of specialization—extreme genetic alteration of a crop grown in a highly simplified ecosystem which has largely displaced wild stands of corn (Mangelsdorf et al. 1964). Where wildlands and wild species become a major platform for socioeconomic development, there will be a strong tendency for those species and their habitats to look increasingly less natural and more domesticated. The wildland manager's job, then, is to find "a compromise between maximizing monetary profits and maximizing biological diversity" (Kuusipalo and Kangas 1994).

Certain uses and values—nature tourism, option values, existence values, scientific research, pharmaceutical prospecting—favor natural landscapes and biodiversity. Further, the rural poor are often highly dependent on diverse ecosystem products and services from nearby wildlands (Gillis 1992a; Singh et al., Chap. 2). However, there are various stakeholders and interests for whom specialization and simplification, based on traditional commodity-based values, provide the path to "best" use. Though such dependency on one or a few resources should provide incentives for their sustainable management, specialized systems may often be inherently less stable, both ecologically and socioeconomically, than more diversified systems. Product diversity helps buffer against both ecological and market fluctuations that affect the supply of and demand for individual products. As Janzen (1994) states, "if the area is conserved for its value on just one or a few axes, then it is in the same risk zone as the country that depends on a monoculture agroscape."

In the United States, for example, about 75 percent of the costs of wildlife management at the state level are paid from hunting-related revenues (Sparrowe 1993)—a system thus highly dependent on and specialized in one form of use. Davis (1985) notes that "public land budgets are limited, forcing biologists-managers into cost analysis in order to get the most production from limited resources and leading public agencies to user fees and charges to enhance operating budgets." The mismanagement of the black duck in the eastern United States has been cited as an example of this perverse incentive at work in public agencies. As Gilbert and Dodds (1987, p. 134) comment in explaining the failure to reduce bag limits despite rapidly falling black duck populations, "the black duck is a major bag species on the flyway, and management agencies derive their revenue primarily from the sale of licenses."

The economic ratchet is cranked up several more notches when heavy financial investments in equipment, leased land, accumulated national debt, and so on, are made that must be paid off. As noted earlier, overcapitalization in new equipment and technology in marine fisheries and the economic trap it creates for fishers are major causes of overfishing (Ludwig et al. 1993; Upton, Chap. 15).

Burton et al. (1992) demonstrate the changing values of tree species over time in British Columbia, Canada, and conclude that "stand management activities designed to promote more valuable species at the expense of less valuable species may backfire because of shifts in log values during a rotation." They estimate that the past silvicultural suppression of one former "weed" species, the red alder (*Alnus rubra*), in Oregon and Washington, has resulted in an annual revenue loss of US$50 million, plus the loss of indirect benefits such as nitrogen fixation. More broadly, McShane and McShane-Caluzi (Chap. 4) show how the economic values of Swiss forests have shifted over time from consumptive to nonconsumptive uses. Burton et al. (1992) conclude that "managing forests for diverse taxa, structures, and functions may be the best investment strategy under a scenario of unpredictable changing values." This applies equally to other ecosystems and to both consumptive and nonconsumptive use values. According to this viewpoint, we might revise the dictum to read "*diversify* use or lose it."

Biodiversity Standards and Monitoring

An overarching question regarding specialization and biodiversity is what trade-offs, if any, does sustainable offtake (as opposed to obvious overexploitation) imply for biodiversity? More broadly, if the goal of ecosystem management is to meet a diversity of "environmental, economic, and social benefits" (Salwasser 1994), how can we ensure that consumptive use does not degrade ecosystem components that provide these diverse benefits? These questions have received little attention, even in some of the most common and best-studied wild species uses, such as salmon management (Francis, Chap. 16), waterfowl management (Callaghan et al., Chap. 14), big game management (Teer, Chap. 12), and temperate forest management (Hansen, Chap. 6).

Perhaps the most problematic part of this question from the perspective of biodiversity conservation is the search for standards of "naturalness," including natural fluctuations and change, against which to measure human-induced change, whether at the ecosystem or genetic level. In ecosystem management this elusive goal has been referred to as "native ecosystem integrity" (Grumbine 1994). Conservationists have done little to define such standards or benchmarks, and few efforts have been undertaken to monitor such change (Kremen et al. 1994; Hansen, Chap. 6). The task is not easy. The more we learn about both intrinsic change in ecosystems and historic human impacts on previously labeled "pristine" ecosystems (Botkin 1990; Sprugel 1991; Denevan 1992), the more difficult it is to define such standards in terms of biodiversity conservation objectives. Most terrestrial ecosystems have long been altered by human influence (e.g., Denevan 1992). In reviewing marine systems, Dayton et al. (1995) lament that "*all* (italics in original) efforts to evaluate bycatch and environmental effects of heavy fishing on natural systems are too late because most sensitive species have long been impacted, leaving no concept of natural relationships or patterns."

Thus our conservation objectives are affected by how prepared we are to accept humans and their influence on biodiversity and ecosystems as part of the "natural" process. Can genetic change in a prey species caused by selective harvest by humans be considered in the same light by the biodiversity conservationist as change caused by another predator? Given this context, the task of defining objectives for biodiversity conservation must consider ecosystem and evolutionary processes and change as much as static measures of biodiversity, and it will require a heavy dose of human values and subjectivity.

The Paradox of Consumptive Use in Biodiversity Conservation

Commercial, consumptive use presents the proverbial double-edged sword for conservationists. If well managed, it can be a tool for nature conservation; if poorly managed, it can readily lead to overexploitation and biotic impoverishment. Even if relatively well managed, however, it is unwise to expect consumptive use to shoulder the full burden for biodiversity conservation. First, overreliance on consumptive use values alone, while ignoring other biodiversity-based values, may fail to offset the opportunity costs of alternative uses of the land. Second, even if consumptive use revenues do offset such opportunity costs, specialization in consumptive use products may entail substantial trade-offs in biodiversity in the wildland under management.

Linking commercial, consumptive use with biodiversity conservation involves two basic management strategies. One is to mitigate the biotic degradation being caused by current overexploitation and unsound management practices. Some forms of use will clearly never be sustainable in today's world and need to be stopped, such as tiger bones in the medicinal trade (Mills and Jackson 1994). Others require that we work to move from a system that is clearly unsustainable toward one that is more ecologically benign, such as bushmeat hunting for local markets in the tropical forest regions of the world (Fa et al. 1995; Bodmer et al., Chap. 9) or the management of the world's forests and fisheries. This requires that we give attention both to improving on-the-ground management and to influencing demand. Any success in the latter depends, ultimately, on curtailing both per capita consumption patterns and human population growth.

The second strategy is to define how commercial, consumptive use can be a conservation tool and to actively promote such approaches. Three potential conservation benefits of this strategy can be defined. (1) Revenues from commercial, consumptive use may help offset the opportunity costs of alternative land uses that would degrade or eliminate the native ecosystem (e.g., monocrop agriculture or estuaries used as sinks for industrial pollutants). (2) Production of the commercially used wild species product may be more ecologically benign than substitutes that would be used if the wild species product did not exist. What synthetic substitutes derived from nonrenewable resources might be used if lumber were not available for construction? Where and how would substitutes for fish protein be produced if some marine fisheries were eliminated? (3) The use of wild species and their products may serve to maintain awareness of the link between human welfare and natural ecological systems.

The last two are beyond the scope of this volume. The first, however, is most prominent regarding the conservation benefits of consumptive use, and it is apparent in several wild species use programs examined in this volume. Crowe et al. (Chap. 10), in a detailed comparison of cattle ranching with game ranching on the Rooipoort Estate in South Africa, demonstrate that game ranching is economically more viable. This advantage could be enhanced if other revenue-generating activities compatible with game ranching, such as gamebird hunting and tourism, are considered. Joanen et al. (Chap. 13) credit the commercial use of alligators, together with several other wetland species, for helping to protect wetlands in Louisiana. Revenues from trophy hunting of caprinids in Pakistan show promise as an incentive for local people to restore and maintain caprinid populations and their habitats (Johnson, Chap. 11). Three chapters describe programs where forest areas under communal management have been set aside and restored for their consumptive use values (Bodmer et al., Chap. 9; Kiernan and Freese, Chap. 3; Singh et al., Chap. 2). Gopinath and Gabriel (Chap. 5) credit the commercial use of the mangrove forest in Malaysia's Matang Mangrove Reserve for helping to deflect the spread of shrimp aquaculture. Upton (Chap. 15) notes that both sport and commercial fishers along the Atlantic coast of the United States are beginning to express concerns about habitat conservation in the estuaries and rivers used by striped bass and other commercial species. The commercial importance of salmon has clearly played a key role in efforts to restore both salmon populations and their freshwater habitats in North America (Lee 1993; Francis, Chap. 16). Finally, despite its incidental effects on biodiversity (Callaghan et al., Chap. 14), revenues generated by waterfowl hunting have protected an estimated 40 million ha of wetlands in North America (Heitmeyer et al. 1993).

Many of these studies and others in this volume, however, reveal the two primary pitfalls—specialization and the inability to compete with alternative land uses—that come with overreliance on commodity production from wildlands as a biodiversity conservation tool. As alluded to in previous sections, two fundamental changes in the way we account and manage for biodiversity, one based on human values and the other on ecological functions, are required if we are to implement successfully any approximation of ecosystem management (Slocombe 1993; Grumbine 1994; Salwasser 1994). (1) We must better account for and give greater weight in decision making to the full range of use and nonuse values of biodiversity and native ecosystems. A diversified set of values will underpin management for a biodiverse and healthy ecosystem. (2) We must improve our understanding of, and better incorporate into management, the functional importance of biodiversity and ecosystem integrity in maintaining not only consumptive use values, but other values as well. Native species and natural processes of ecosystems may often be crucial for maintaining the productive capacity of ecosystems for commodity production, as well as for fulfilling the broader array of nonuse values.

An emphasis on longer time horizons in planning and management will serve these objectives. Maintaining biodiverse wildlands is the best way to hedge our bets against the uncertainties presented by the future. Today's values are seldom tomorrow's; today's weed is tomorrow's miracle plant. Negative ecological feedback in overly specialized commodity production systems in wildlands may require decades to be felt. A cornerstone to ecosystem management must be to maintain ecosystem resilience and adaptability in the face of unpredictable catastrophic events and long-term environmental change.

The cultural and socioeconomic foundation for biodiversity conservation can be strengthened by not only better recognizing and expanding existing biodiversity values, but also by fostering new ones through research and education. Neither emotional and spiritual attachment nor more quantifiable economic values will develop for those components of biodiversity that are unknown and unexplained. The strategy requires more biodiversity research, whether prospecting for pharmaceuticals or inventorying butterflies, that is conducted in ways to make it useful and accessible to the wildland manager, the land-use planner, and the public.

We will fail to stem the tide of biotic degradation if the consumptive use values of biodiversity are not complemented by nonconsumptive values in decision making. To do so, we must identify the array of stakeholders who now enjoy a free ride, often on the back of consumptive uses that currently provide the justification for maintaining wild species and wildlands. We are quick to point the finger at those who are most visible—the illegal poacher or fisher—among the free riders. However, the individual who opts to buy the less-expensive tropical hardwood chair that came from a mined forest rather than pay the premium for one from a well-managed forest is as much a free rider as the clandestine logger. So is the individual who enjoys nature programs on television, who benefits from medicines derived from wild species, or who through ecological research seeks advancement in academia, but who contributes little or nothing (whether in money, votes, or other means of influence) toward conserving biodiversity and wildlands. Greater attention should be given to identifying and educating these more subtle free riders about their stakeholder interests and responsibilities.

In many cases, however, stakeholders may legitimately view the so-called free ride as more a "right" for which they should not have to pay. Clean air and water may be the most universal examples, but others may place biodiversity in this same category. Where the benefits from biodiversity and wildlands are broadly perceived as a public right, and where transfer payments between those who value biodiversity and those who bear the cost of maintaining it are not possible, then governments or other higher authorities must play a strong role in representing and protecting societal interests. Yet governments are apt to exercise this role only when society demands it, and thus the responsibility falls back on those who value biodiversity and recognize its societal benefits to pay their dues via votes, influence, education, and other means. For some members of society, however, particularly for the rural poor that depend on wildland resources but are marginalized from the political process, neither monetary nor non-monetary avenues for giving voice to their interests are readily available. Given that areas of great biodiversity value are often used and inhabited by the rural poor, incorporating their interests within a broader framework of biodiversity conservation is crucial.

Thus we must rely on greater responsibility by a broad spectrum of stakeholders to help bear the costs of biodiversity conservation. Ecologically sustainable resource use is but one important mechanism among many for bearing those costs. But where market mechanisms fall short, as they often will, we must look to public authorities to play a strong role in protecting societal interests if biodiversity is not to be lost to economic specialization, resource homogenization, and alternative land uses.

Acknowledgments

M. Tundi Agardy, Cleber Alho, Jason Clay, Barry Coates, Steven Cornelius, Holly Dublin, Anton Fernhout, Ginette Hemley, Barbara Hoskinson, Thomas McShane, and Michael Sutton participated in long discussions and shared many ideas and references that were invaluable in preparing this paper. Jennifer Swearingen's editorial skills were crucial for getting the manuscript into publishable form. I am grateful to WWF–International and WWF–US for both financial and various kinds of institutional support. Maria Boules and Kim Doyle were particularly helpful with administrative affairs. The Department of Biology, Montana State University, provided generous support for bibliographic research.

References

Acheson, J. M. 1989. Where have all the exploiters gone? Co-management of the Maine lobster industry. In *Common property resources: Ecology and community-based sustainable development,* ed. F. Berkes, 199–217. London: Belhaven Press.

Agardy, M. T. 1994. Advances in marine conservation: The role of marine protected areas. *TREE* 9:267–270.

Alho, C.J.R. 1985. Conservation and management strategies for commonly exploited Amazonian turtles. *Biological Conservation* 32:291–298.

Alverson, D. L., M. H. Freeberg, S. A. Marawski, and J. G. Pope. 1994. *A global assessment of fisheries bycatch and discards.* FAO Fisheries Technical Paper 339. Rome: U.N. Food and Agricultural Organization.

Anderson, A. B., P. Magee, A. Gély, and M. A. Gonçalves J. 1995. Forest management patterns in the floodplain of the Amazon estuary. *Conservation Biology* 9:47–61.

Appasamy, P. P. 1993. Role of non-timber forest products in a subsistence economy: The case of a joint forestry project in India. *Economic Botany* 47:258–267.

Barbier, E. B. 1992. Economics for the wilds. In *Economics for the wilds: Wildlife, wildlands, diversity and development,* ed. M. Swanson and E. B. Barbier, 15–33. London: Earthscan.

Barel, C.D.N., W. Ligtvoet, T. Goldschmidt, F. Witte, and P. C. Goudswaard. 1991. The haplochromine cichlids in Lake Victoria: An assessment of biological and fisheries interests. In *Cichlid fishes: Behaviour, ecology and evolution,* ed. M.H.A. Keenleyside, 258–279. London: Chapman and Hall.

Baskerville, G. L. 1988. Redevelopment of a degrading forest system. *Ambio* 17:314–322.

Benson, D. E. 1992. Commercialization of wildlife: A value-added incentive for conservation. In *The biology of deer,* ed. R. D. Brown, 539–553. New York: Springer-Verlag.

Berkes, F. 1987. Common-property resource management and Cree Indian fisheries in Subarctic Canada. In *The question of the commons: The culture and ecology of communal resources,* ed. B. J. McCay and J. M. Acheson, 66–91. Tucson: University of Arizona Press.

Berkes, F., ed. 1989. *Common property resources: Ecology and community-based sustainable development.* London: Belhaven Press.

Bingham, G., R. Bishop, M. Brody, D. Bromley, E. Clark, W. Cooper, R. Costanza, R. Hale, G. Hayden, S. Kellert, R. Norgaard, B. Norton, J. Payne, C. Russell, and G. Suter. 1995. Issues in ecosystem valuation: Improving information for decision making. *Ecological Economics* 14:73–90.

Bodmer, R. E. 1995. Managing Amazonian wildlife: Biological correlates of game choice by detribalized hunters. *Ecological Applications* 5:872–877.

Bohnsack, J. A. 1993. Marine reserves: They enhance fisheries, reduce conflicts, and protect resources. *Oceanus* Fall:62–71.

Bond, W. J. 1993. Keystone species. In *Biodiversity and ecosystem function,* ed. E.-D. Schulze and H. A. Mooney, 237–253. Berlin: Springer-Verlag.

Boot, R.G.A., and R. E. Gullison. 1995. Approaches to developing sustainable extraction systems of tropical forest products. *Ecological Applications* 5:896–903.

Botkin, D. B. 1990. *Discordant harmonies: A new ecology for the twenty-first century.* New York: Oxford University Press.

Botkin, D. B., K. Cummins, T. Dunne, H. Regier, M. Sobel, and L. M. Talbot. 1993. *Status and future of anadromous fish of Western Oregon and Northern California: Rationale for a new approach.* Center for the Study of the Environment, Research Report 931001. Portland, Oregon, and Santa Barbara, California: Center for the Study of the Environment.

Bromley, D. W., and M. M. Cernea. 1989. *The management of common property resources: Some conceptual and operational fallacies.* World Bank Discussion Paper 57. Washington, D.C.: World Bank.

Brown, J. S., and A. O. Parman. 1993. Consequences of size-selective harvesting as an evolutionary game. In *The exploitation of evolving resources,* ed. T. K. Stokes, J. M. McGlade, and R. Law, 248–261. Berlin: Springer-Verlag.

Burton, P. J., A. C. Balisky, L. P. Coward, S. G. Cumming, and D. D. Kneeshaw. 1992. The value of managing for biodiversity. *Forestry Chronicle* 68:225–237.

Buschbacher, R. J. 1990. Natural forest management in the humid tropics: Ecological, social, and economic considerations. *Ambio* 19:253–258.

Cheng, C. H. 1994. *Mangroves in jeopardy: The economics of prawn aquaculture and its implications on mangrove management in Perak.* WWF–Malaysia Project Report, Selangor, Malaysia.

Clark, C. W. 1973. The economics of overexploitation. *Science* 181:630–634.

Clark, C. W. 1996. Marine reserves and the precautionary management of fisheries. *Ecological Applications* 6:369–370.

Clay, J. 1992. Some general principles and strategies for developing markets in North America and Europe for non-timber forest products: Lessons from Cultural Survival Enterprises, 1989–1990. In *Non-timber products from tropical forests: Evaluation of a conservation and development strategy,* ed. D. C. Nepstad and S. Schwartzman, 101–106. New York: New York Botanical Garden.

Courtenay, W. R., Jr., and C. C. Kohler. 1986. Exotic fishes in North American fisheries management. In *Fish culture in fisheries management,* ed. R. H. Stroud, 401–413. Bethesda, Maryland: American Fisheries Society.

Cumming, D.H.M. 1989. Commercial and safari hunting in Zimbabwe. In *Wildlife production systems: Economic utilisation of wild ungulates,* ed. R. J. Hudson, K. R. Drew, and L. M. Baskin, 148–169. Cambridge: Cambridge University Press.

Davis, R. K. 1985. Research accomplishments in wildlife economics. *Transactions of the North American Wildlife and Natural Resources Conference* 50:392–404.

Dayton, P. K., S. F. Thrush, M. T. Agardy, and R. J. Hofman. 1995. Environmental effects of marine fishing. *Aquatic Conservation: Marine and Freshwater Ecosystems* 5:205–232.

Denevan, W. M. 1992. The pristine myth: The landscape of the Americas in 1492. *Annals of the Association of American Geographers* 82:369–385.

Durning, A. T., and H. B. Brough. 1992. Reforming the livestock economy. In *State of the world 1992,* ed. L. R. Brown, 66–82. New York: W. W. Norton.

Edwards, D. M., and M. R. Bowen, eds. 1993. *Focus on Jaributi.* Proceedings of the non-timber forest products seminar held in Kathmandu on May 12, 1993. Forest Research and Survey Centre Occasional Paper 2/93, Kathmandu, Nepal.

Ekins, P., C. Folke, and R. Costanza. 1994. Trade, environment and development: The issues in perspective. *Ecological Economics* 9:1–12.

Fa, J. E., J. Juste, J. Perez del Val, and J. Castroviejo. 1995. Impact of market hunting on mammal species in Equatorial Guinea. *Conservation Biology* 9:1107–1115.

Fitzgibbon, C. D., J. Mogaka, and J. H. Fanshawe. 1995. Subsistence hunting in Arabuko-Sokoke Forest, Kenya, and its effects on mammal populations. *Conservation Biology* 9:1116–1126.

Fitzsimmons, N. N., S. W. Buskirk, and M. H. Smith. 1995. Population history, genetic variability, and horn growth in bighorn sheep. *Conservation Biology* 9:314–323.

Food and Agricultural Organization. 1988. *Review of the state of world fishery resources.* FAO Fisheries Circular 710, revision 7. Rome: FAO.

Frumhoff, P. C. 1995. Conserving wildlife in tropical forests managed for timber. *BioScience* 45:456–464.

Gasaway, W. C., R. O. Stephenson, J. L. Davis, P.E.K. Shepherd, and O. E. Burris. 1983. Interrelationships of wolves, prey, and man in interior Alaska. *Wildlife Monographs* 84:1–50.

Gaski, A. L. 1993. *Bluefin tuna: An examination of the international trade with an emphasis on the Japanese market.* Cambridge: TRAFFIC International.

Geist, V. 1995. North American policies of wildlife conservation. In *Wildlife conservation policy,* ed. V. Geist and I. McTaggart-Cowan, 77–129. Calgary: Detselig Enterprises.

Gilbert, F. F., and D. G. Dodds. 1987. *The philosophy and practice of wildlife management.* Malabar, Florida: Robert E. Krieger Publishing.

Gill, R. 1990. *Monitoring the status of European and North American cervids.* Global Environment Monitoring System, Information Series no. 8. Nairobi: United Nations Environment Program.

Gillis, M. 1992a. Economic policies and tropical deforestation. In *Non-timber products from tropical forests: Evaluation of a conservation and development strategy,* ed. D. C. Nepstad and S. Schwartzman, 129–142. New York: New York Botanical Garden.

Gillis, M. 1992b. Forest concession management and revenue policies. In *Managing the world's forests: Looking for balance between conservation and development,* ed. N. P. Sharma, 139–175. Dubuque, Iowa: Kendall/Hunt.

Grumbine, M. E. 1994. What is ecosystem management? *Conservation Biology* 8:27–38.

Hall, P., and K. Bawa. 1993. Methods to assess the impact of extraction of non-timber tropical forest products on plant populations. *Economic Botany* 47:234–247.

Halpern, C. B., and T. A. Spies. 1995. Plant species diversity in natural and managed forests of the Pacific Northwest. *Ecological Applications* 5:913–934.

Hammer, M., A. M. Jansson, and B.-O. Jansson. 1993. Diversity change and sustainability: Implications for fisheries. *Ambio* 22:97–105.

Hansen, A. J., S. L. Garman, J. F. Weigand, D. L. Urban, W. C. McComb, and M. G. Raphael. 1995. Alternative silvicultural regimes in the Pacific Northwest: Simulations of ecological and economic effects. *Ecological Applications* 5:535–554.

Hardin, G. 1968. The tragedy of the commons. *Science* 168:1243–1248.

Hart, J. A. 1978. From subsistence to market: A case study of the Mbuti net hunters. *Human Ecology* 6:325–353.

Hart, T. B., and J. A. Hart. 1986. The ecological basis of hunter-gatherer subsistence in African rain forests: The Mbuti of eastern Zaire. *Human Ecology* 14:29–55.

Heitmeyer, M. E., J. W. Nelson, B.D.J. Batt, and P. J. Caldwell. 1993. Waterfowl conservation and biodiversity. *Midwest Fish & Wildlife Conference 1993,* Columbia, Missouri.

Hilborn, R., and D. Ludwig. 1993. The limits of applied ecological research. *Ecological Applications* 3:550–552.

Hindar, K. 1992. Conservation and sustainable use of Atlantic salmon. In *Conservation of biodiversity for sustainable development,* ed. O. T. Sundland, K. Hindar, and A.H.D. Brown, 168–185. Oslo: Scandinavian University Press.

Hodson, T. J., F. Englander, and H. O'Keefe. 1995. Rain forest preservation, markets, and medicinal plants: Issues of property rights and present value. *Conservation Biology* 9:1319–1321.

Hoyt, J. A. 1994. *Animals in peril: How "sustainable use" is wiping out the world's wildlife.* Garden City Park, New York: Avery Publishing Group.

IIED. 1994. *Whose Eden? An overview of community approaches to wildlife management.* London: International Institute for Environment and Development.

IUCN/UNEP/WWF. 1991. Caring for the earth: a strategy for sustainable living. Gland, Switzerland: The World Conservation Union.

Jachmann, H., P.S.M. Berry, and H. Imae. 1995. Tusklessness in African elephants: A future trend. *African Journal of Ecology* 33:230–235.

Janzen, D. H. 1994. Wildland biodiversity management in the tropics: Where are we now and where are we going? *Vida Silvestre Neotropical* 3:3–15.

Janzen, D. H., and C. Vásquez-Yanes. 1991. Aspects of tropical seed ecology of relevance to management of tropical forested wildlands. In *Rain forest regeneration and management,* ed. A. Gomez-Pompa, T. C. Whitmore, and M. Hadley, 137–157. Man and the biosphere series, no. 6. Paris: UNESCO; Canforth, England: Parthenon Publishing Group.

Johnson, N., and B. Cabarle. 1993. *Surviving the cut: Natural forest management in the humid tropics.* Washington, D.C.: World Resources Institute.

Kardell, L. 1986. Occurrence and berry production of *Rubus chamaemorus* L., *Vaccinium oxycoccus* L. & *Vaccinium microcarpum* Turca, and *Vaccinium vitis-idaea* L. on Swedish peatlands. *Scandinavian Journal of Forest Research* 1:125–140.

Kirkwood, G. P., J. R. Beddington, and J. A. Rossouw. 1994. Harvesting species of different lifespans. In *Large-scale ecology and conservation biology,* ed. P. J. Edwards, R. M. May, and N. R. Webb, 199–227. London: Blackwell Scientific Publications.

Kiss, A. 1990. Principles and issues. In *Living with wildlife: Wildlife resources management with local participation in Africa,* ed. A. Kiss, 5–15. Washington, D.C.: World Bank.

Kitzmiller, J. H. 1990. Managing genetic diversity in a tree improvement program. *Forest Ecology and Management* 35:131–149.

Koch, N. E., and J. J. Kennedy. 1991. Multiple-use forestry for social values. *Ambio* 20:330–333.

Kremen, C., A. M. Merenlender, and D. D. Murphy. 1994. Ecological monitoring: A vital need for integrated conservation and development programs in the tropics. *Conservation Biology* 8:388–397.

Kuusipalo, J., and J. Kangas. 1994. Managing biodiversity in a forestry environment. *Conservation Biology* 8:450–460.

Larson, B. A., and D. W. Bromley. 1990. Property rights, externalities, and resource degradation: Locating the tragedy. *Journal of Developmental Economics* 33:235–262.

Lawton, J. H., and V. K. Brown. 1993. Redundancy in ecosystems. In *Biodiversity and ecosystem function,* ed. E.-D. Schulze and H. A. Mooney, 255–270. Berlin: Springer-Verlag.

Ledig, F. T. 1992. Human impacts on genetic diversity in forest ecosystems. *Oikos* 63:87–108.

Lee, K. N. 1993. *Compass and gyroscope: Integrating science and politics in the environment.* Washington, D.C.: Island Press.

Loomis, J. B., M. Hanemann, and B. Kanninen. 1991. Willingness to pay to protect wetlands and reduce wildlife contamination from agricultural drainage. In *The economics and management of water and drainage in agriculture,* ed. A. Dinar and D. Zilberman, 411–429. Boston: Kluwer Academic.

Ludwig, D., R. Hilborn, and C. Walters. 1993. Uncertainty, resource exploitation, and conservation: Lessons from history. *Science* 260:17, 36.

Lungu, F. B. 1990. Zambia: Administrative Design for Game Management Areas (ADMADE) and Luangwa Integrated Development Project (LIRDP). In *Living with wildlife: Wildlife resource management with local participation in Africa,* ed. A. Kiss, 115–122. Washington, D.C.: World Bank.

Luxmoore, R. 1985. Game farming in South Africa as a force in conservation. *Oryx* 19:225–231.

Mackenzie, D. 1996. Seals to the slaughter. *New Scientist* March 16:34–39.

Macnab, J. 1991. Does game cropping serve conservation? A reexamination of the African data. *Canadian Journal of Zoology* 69:2283–2290.

McNaughton, S. J. 1993. Biodiversity and function of grazing ecosystems. In *Biodiversity and ecosystem function,* ed. E.-D. Schulze and H. A. Mooney, 361–383. Berlin: Springer-Verlag.

McNeely, J. A. 1988. *Economics and biological diversity: Developing and using economic incentives to conserve biological resources.* Gland, Switzerland: IUCN.

Mangel, M., R. J. Hofman, E. A. Norse, and J. R. Twiss, Jr. 1993. Sustainability and ecological research. *Ecological Applications* 3:573–575.

Mangel, M., L. M. Talbot, G. K. Meffe, M. T. Agardy, D. L. Alverson, J. Barlow, D. B. Botkin, G. Budowski, T. Clark, J. Cooke, R. H. Crozier, P. K. Dayton, D. L. Elder, C. W. Fowler, S. Funtowicz, J. Giske, R. J. Hofman, S. J. Holt, S. R. Kellert, L. A. Kimball, D. Ludwig, K. Magnusson, B. S. Malayang III, C. Mann, E. A. Norse, S. P. Northridge, W. F. Perrin, C. Perrings, R. M. Peterman, G. B. Rabb, H. A. Regier, J. E. Reynolds III, K. Sherman, M. P. Sissenwine, T. D. Smith, A. Starfield, R. J. Taylor, M. F. Tillman, C. Toft, J. R. Twiss, Jr., J. Wilen, and T. P. Young. 1996. Principles for the conservation of wild living resources. *Ecological Applications* 6:338–362.

Mangelsdorf, P. C., R. S. MacNeish, and W. C. Galinat. 1964. Domestication of corn. *Science* 143:538–545.

Marnell, L. F. 1986. Impacts of hatchery stocks on wild fish populations. In *Fish culture in fisheries management,* ed. R. H. Stroud, 339–347. Bethesda, Maryland: American Fisheries Society.

Mendelsohn, R., and M. Balick. 1995. Private property and rainforest conservation. *Conservation Biology* 9:1322–1323.

Metcalfe, S. 1994. The Zimbabwe Communal Areas Management Programme for Indigenous Resources (CAMPFIRE). In *Natural connections: Perspectives in community-based conservation,* ed. D. Western and R. M. Wright, 161–192. Washington, D.C.: Island Press.

Millar, C. I., F. T. Ledig, and L. A. Riggs. 1990. Conservation of diversity in forest ecosystems. *Forest Ecology and Management* 35:1–4.

Miller, D. J. 1989. Introductions and extinctions of fish in the African Great Lakes. *Trends in Ecology and Evolution* 4(2):56–59.

Milliken, T., K. Nowell, and J. B. Thomsen. 1993. *The decline of the black rhino in Zimbabwe.* Cambridge: TRAFFIC International.

Mills, J. A., and P. Jackson. 1994. Killed for a cure: A review of the worldwide trade in tiger bone. Cambridge: TRAFFIC International.

Mladenoff, D. J., and J. Pastor. 1993. Sustainable forest ecosystems in the northern hardwood and conifer forest region: Concepts and management. In *Defining sustainable forestry,* ed. G. H. Aplet, N. Johnson, J. T. Olson, and V. A. Sample, 145–180. Washington, D.C.: Island Press.

Moyle, P. B., H. W. Li, and B. A. Barton. 1986. The Frankenstein effect: Impact of introduced fishes on native fishes in North America. In *Fish culture in fisheries management,* ed. R. H. Stroud, 415–426. Bethesda, Maryland: American Fisheries Society.

Naeem, S., L. J. Thompson, S. P. Lawler, J. H. Lawton, and R. M. Woodfin. 1994. Declining biodiversity can alter the performance of ecosystems. *Nature* 368:734–737.

Nash, R. F. 1989. *The rights of nature: A history of environmental ethics.* Madison: University of Wisconsin Press.

Nelson, K., and M. Soule. 1987. Genetical conservation of exploited fishes. In *Population genetics and fishery management,* ed. N. Ryman and F. Utter, 345–368. Seattle: University of Washington Press.

Nias, R. C. 1995. *Using it and losing it: The commercial exploitation of wildlife in Australia.* World Wide Fund for Nature Australia Discussion Paper and Position Statement. Sydney: WWF–Australia.

Nicol, S., and W. de la Mare. 1993. Ecosystem management and the Antarctic krill. *American Scientist* 81:36–47.

Norgaard, R. B. 1987. Economics as mechanics and the demise of biological diversity. *Ecological Modelling* 38:107–121.

Norgaard, R. B. 1995. Ecology, politics, and economics: Finding the common ground for decision making in conservation. In *Principles of conservation biology,* ed. G. K. Meffe and C. R. Carroll, 439–465. Sunderland, Massachusetts: Sinauer Associates.

Norse, E. 1993. *Global marine biological diversity: A strategy for building conservation into decision making.* Washington, D.C.: Island Press.

Noss, R. F. 1991. Sustainability and wilderness. *Conservation Biology* 5:120–122.

O'Connell, M. A., and M. Sutton. 1990 (June). *The effects of trade on international commerce in African elephant ivory: A preliminary report.* Washington, D.C.: World Wildlife Fund and The Conservation Foundation.

Paine, R. T. 1966. Food web complexity and species diversity. *American Naturalist* 100:65–75.

Paine, R. T. 1969. A note on trophic complexity and community stability. *American Naturalist* 103:91–93.

Paquet, P., and A. Hackman. 1995. *Large carnivore conservation in the Rocky Mountains.* Toronto: WWF-Canada; Washington, D.C.: WWF-US.

Parson, E. A., and W. C. Clark. 1995. Sustainable development as social learning: Theoretical perspectives and practical challenges for the design of a research program. In *Barriers and bridges to the renewal of eco-*

systems and institutions, ed. L. H. Gunderson, C. S. Holling, and S. S. Light, 428–460. New York: Columbia University Press.

Peluso, N. L. 1992. The rattan trade in East Kalimantan, Indonesia. In *Non-timber products from tropical forests: Evaluation of a conservation and development strategy,* ed. D. C. Nepstad and S. Schwartzman, 115–127. New York: New York Botanical Garden.

Perl, M. A., M. J. Kiernan, D. McCaffrey, R. J. Buschbacher, and G. J. Batmanian. 1991. *Views from the forest: Natural forest management initiatives in Latin America.* Washington, D.C.: Tropical Forestry Program, World Wildlife Fund.

Peters, C. M. 1992. The ecology and economics of oligarchic forests. In *Non-timber products from tropical forests: Evaluation of a conservation and development strategy,* ed. D. C. Nepstad and S. Schwartzman, 15–22. New York: New York Botanical Garden.

Peters, C. M. 1994. *Sustainable harvests of non-timber plant resources in tropical moist forests: An ecological primer.* Washington, D.C.: Biodiversity Support Program, WWF-US.

Pimm, S. L. 1993. Biodiversity and the balance of nature. In *Biodiversity and ecosystem function,* ed. E.-D. Schulze and H. A. Mooney, 347–359. Berlin: Springer-Verlag.

Poffenberger, M. 1992. *Sustaining Southeast Asia's forests.* Research Network Report no. 1. Berkeley: Center for Southeast Asian Studies, University of California.

Poffenberger, M. 1994. The resurgence of community forest management in Eastern India. In *Natural connections: Perspectives in community-based conservation,* ed. D. Western and R. M. Wright, 53–79. Washington, D.C.: Island Press.

Policansky, D. 1993. Fishing as a cause of evolution in fishes. In *The exploitation of evolving resources,* ed. T. K. Stokes, J. M. McGlade, and R. Law, 2–14. Berlin: Springer-Verlag.

Postel, S. 1994. Carrying capacity: Earth's bottom line. In *State of the World 1994,* 3–21. World Watch Institute. New York: W. W. Norton.

Putz, F. E. 1993. *Considerations of the ecological foundation of natural forest management in the American tropics.* Durham, North Carolina: Center for Tropical Conservation, Duke University.

Randall, A. 1981. *Resource economics: An economic approach to natural resource and environmental policy.* New York: John Wiley & Sons.

Rasker, R., and C. Freese. 1995. Wildlife in the marketplace: Opportunities and problems. In *On fundamental policies in wildlife conservation,* ed. V. Geist and I. McTaggart-Cowan, 177–204. Calgary: Detselig Publishers.

Repetto, R. 1988. *The forest for the trees: Government policy and the misuse of forest resources.* Washington, D.C.: World Resources Institute.

Riggs, L. A. 1990. Conserving forest resources on-site in forest ecosystems. *Forest Ecology and Management* 35:45–68.

Rivlin, A. M. 1993. Values, institutions, and sustainable forestry. In *Defining sustainable forestry,* ed. G. H. Aplet, N. Johnson, J. T. Olson, and V. A. Sample, 255–259. Washington, D.C.: Island Press.

Roberts, C., and N. Polunin. 1993. Marine reserves: Simple solutions to managing complex fisheries? *Ambio* 22:363–368.

Roberts, M. R., and F. S. Gilliam. 1995. Patterns and mechanisms of plant diversity in forested ecosystems: Implications for forest management. *Ecological Applications* 5:969–977.

Robinson, J. G. 1993. The limits to caring: Sustainable living and the loss of biodiversity. *Conservation Biology* 7:20–28.

Rosenberg, A. A., M. J. Fogarty, M. P. Sissenwine, J. R. Beddington, and J. G. Shepherd. 1993. Achieving sustainable use of renewable resources. *Science* 262:828–829.

Rowell, C. A. 1993. The effects of fishing on the timing of maturity in North Sea Cod (*Gadus morhua* L.). In *The exploitation of evolving resources,* ed. T. K. Stokes, J. M. McGlade, and R. Law, 44–61. Berlin: Springer-Verlag.

Ruddle, K. 1989. Solving the common-property dilemma: Village fisheries rights in Japanese coastal waters. In *Common property resources: Ecology and community-based sustainable development,* ed. F. Berkes, 168–184. London: Belhaven Press.

Ruddle, K., and R. E. Johannes, eds. 1983. *The traditional knowledge and management of coastal systems in Asia and the Pacific.* Papers presented at a UNESCO-ROSTEA Regional Seminar, December 5–9, 1983. Jakarta: UNESCO Regional Office for Science and Technology for Southeast Asia.

Safina, C. 1993. Bluefin tuna in the West Atlantic: Negligent management and the making of an endangered species. *Conservation Biology* 7:229–233.

Salick, J., A. Mejia, and T. Anderson. 1995. Non-timber forest products integrated with natural forest management, Rio San Juan, Nicaragua. *Ecological Applications* 5:878–895.

Salo, K. 1995. Non-timber forest products and their utilization in the Nordic countries. In *Multiple use forestry in Nordic countries,* ed. M. Hytönen, 117–155. Vantaa, Finland: Finnish Forest Research Institute.

Salwasser, H. 1994. Ecosystem management: Can it sustain diversity and productivity? *Journal of Forestry* August:6–10.

Schulze, E.-D., and H. A. Mooney. 1993. Ecosystem function of biodiversity: A summary. In *Biodiversity and ecosystem function,* ed. E.-D. Schulze and H. A. Mooney, 497–510. Berlin: Springer-Verlag.

Sissenwine, M. P., and P. M. Mace. 1992. ITQs in New Zealand: The era of fixed quota in perpetuity. *Fishery Bulletin* 90:147–160.

Sissenwine, M. P., and A. A. Rosenberg. 1993. Marine fisheries at a critical juncture. *Fisheries* 18(10):6–14.

Slocombe, D. S. 1993. Implementing ecosystem-based management. *BioScience* 43:612–622.

Southgate, D. 1992. *Shrimp mariculture development in Ecuador: Some resource policy issues.* Department of Agricultural Economics, report no. 5, Ohio State University, Columbus, Ohio.

Sparrowe, R. 1993. What is wise use of waterfowl populations? In *Waterfowl and wetland conservation in the 1990s: A global perspective,* ed. M. Moser, R. C. Prentice, and J. U. van Vessem, 85–86. IWRB Special Publication no. 26. Slimbridge, England: IWRB.

Sprugel, D. G. 1991. Disturbance, equilibrium and environmental variability: What is "natural" vegetation in a changing environment? *Biological Conservation* 58:1–18.

Steel, E. A. 1994. Study of the value and volume of bushmeat commerce in Gabon. Unpublished report. World Wide Fund for Nature, Gland, Switzerland.

Stiles, D. 1994. Tribals and trade: A strategy for cultural and ecological survival. *Ambio* 23:106–111.

Swanson, T. M. 1993. Economics of a biodiversity convention. *Ambio* 21:250–257.

Swanson, T. M., and E. B. Barbier. 1992. *Economics for the wilds.* London: Earthscan.

Teer, J. G. 1995. Exotic animals: Conservation implications. In *Wildlife conservation policy,* ed. V. Geist and I. McTaggart-Cowan, 235–246. Calgary: Detselig Enterprises.

Telesis. 1991. *Sustainable economic development options for the Dzanga-Sangha Reserve: Executive summary.* Report prepared for the World Wildlife Fund and the PVO-NGO/NRMS Project. Washington, D.C.

Thomsen, J. B., and A. Brautigam. 1991. Sustainable use of neotropical parrots. In *Neotropical wildlife use and conservation,* ed. J. G. Robinson and K. H. Redford, 359–379. Chicago: University of Chicago Press.

Tilman, D., and J. A. Downing. 1994. Biodiversity and stability in grasslands. *Nature* 367:363–365.

van Ballenberghe, V., and W. B. Ballard. 1994. Limitation and regulation of moose populations: The role of predation. *Canadian Journal of Zoology* 72:2071–2077.

van Kooten, G. C. 1993. *Land resource economics and sustainable development: Economic policies and the common good.* Vancouver, British Columbia: UBC Press.

Vatn, A., and D. W. Bromley. 1994. Choices without prices without apologies. *Journal of Environmental Economics and Management* 26:129–148.

Verissimo, A., P. Barreto, M. Mattos, R. Tarifa, and C. Uhl. 1992. Logging impacts and prospects for sustainable forest management in an old Amazonian frontier: The case of Paragominas. *Forest Ecology and Management* 55:169–200.

Vitousek, P. M., P. R. Ehrlich, A. H. Ehrlich, and P. A. Matson. 1986. Human appropriate of the products of photosynthesis. *BioScience* 36:368–373.

Vitousek, P. M., and D. U. Hooper. 1993. Biological diversity and terrestrial ecosystem biogeochemistry. In *Biodiversity and ecosystem function,* ed. E.-D. Schulze and H. A. Mooney, 3–13. Berlin: Springer-Verlag.

Walker, B. 1995. Conserving biological diversity through ecosystem resilience. *Conservation Biology* 9:747–752.

Weber, P. 1994. *Net loss: Fish, jobs, and the marine environment.* Worldwatch Paper 120. Washington, D.C.

Welcomme, R. L. 1984. International transfers of inland fish species. In *Distribution, biology, and management of exotic fishes,* ed. W. R. Courtenay, Jr., and J. R. Stauffer, Jr., 22–40. Baltimore: Johns Hopkins University Press.

Wentz, W. A., and F. A. Reid. 1992. Managing refuges for waterfowl purposes and biological diversity: Can both be achieved? *Transactions of the 57th North American Wildlife and Natural Resources Conference*:581–585.

Western, D., and R. W. Wright, eds. 1994. *Natural connections: Perspectives in community-based conservation.* Washington, D.C.: Island Press.

Williams, M. 1989. *Americans and their forests.* Cambridge: Cambridge University Press.

Wise, J. P. 1991. *Federal conservation and management of marine fisheries in the United States.* Washington, D.C.: Center for Marine Conservation.

Woodward, F. I. 1993. How many species are required for a functional ecosystem? In *Biodiversity and ecosystem function,* ed. E.-D. Schulze and H. A. Mooney, 271–291. Berlin: Springer-Verlag.

WWF. 1993 (August). *Sustainable use of natural resources: Concepts, issues, and criteria.* A World Wide Fund for Nature International Position Paper. Gland, Switzerland: WWF International.

TWO

Participatory Forest Management in West Bengal, India

Samar Singh, Avenash Datta, Anil Bakshi, Arvind Khare, Sushil Saigal, and Navin Kapoor

An innovative strategy is evolving in India to counter forest degradation. Participatory forest management (PFM) is a concept by which government forest departments and local communities jointly manage state forests. The core idea embodied in PFM is recognition of the crucial role of forest-dependent communities in the forest management process. PFM is based on the premise that India's degraded forests contain abundant rootstock that could regenerate rapidly if protected and given respite from continuous biotic interference. If motivated and compensated for their opportunity costs, local communities could protect and regenerate these forests. The PFM approach goes beyond nominal "participation." Local communities and forest departments work together as "partners." PFM involves joint management plans and agreements that consider local people's needs and priorities. Wherever possible, additional support activities and wage employment are made available to compensate for the initial opportunity costs of forest protection. India has a history of some one hundred fifty years of "planned and scientific" forest management centered around timber production to fulfill the "national" needs. However, this management approach led to alienation of local community groups from the resources on which they had depended for centuries. It also proved to be ineffective in curbing degradation. The situation was compounded by the development of forest-based industries and increasing pressure from livestock. Scarcity of agricultural land, human encroachment, and the heavily subsidized flow of raw materials from state forests to industry accelerated the degradation process.

Realizing that earlier efforts to curb the exhaustion of forests were not successful, in 1976, forest lands, which were formerly only a state subject, were made a subject under the jurisdiction of both the national and state governments. The national government acquired the power to pass laws concerning forests and wildlife. A first major step was the passing of the Forest Conservation Act in 1980.

However, forest degradation continued, at an estimated rate of 47,300 hectares (ha) per year between 1980 and 1985. A program of social forestry was launched on nonforest lands to meet people's subsistence needs. The purpose was to ease the pressure on forests, but it failed to alleviate the relentless battle between foresters and the local users of forest products. The antagonism of people became endemic, and traditional forestry as a management regime was ineffective. Many foresters themselves began to question the methods and practices of the forest departments. They began to realize that successful forestry management would require the active participation of local communities.

Meanwhile, many communities initiated local efforts to protect forests adjoining their villages. Some enthusiastic forest officers in West Bengal also started involving the fringe communities in forest management. The results in all cases were dramatic in terms of forest regeneration.

It was in this context that the National Forest Policy of India was formulated in 1988. The policy, recognizing the intimate relationship between people and forests, envisages people's involvement in the development and protection of forests. The needs of local forest communities (fuelwood, fodder, and building material) are to be given first priority. Revenue objectives have been given secondary importance, and industrial requirements are to be met primarily from the farm forestry sector.

The objectives of PFM are explicitly different from traditional, "custodial" forestry objectives (Table 2.1). The emphasis is on matching the needs of the people with those of the forest. As A. K. Banerjee,

Table 2.1. The Shift from Custodial to Participatory Forest Management

Custodial forest management	Participatory forest management
Centralized management	Decentralized management
Revenue orientation	Resource orientation
Production motives	Sustainability
Single products	Multiple products
Large working plans	Microplanning
Target orientation	Process orientation
Unilateral decision making	Participatory decision making
Punitive rules	Self-abnegation rules
Controlling people	Facilitating people
Department bureaucracy	People's institutions
Assumed homogeneity (cultural and biological)	Recognizing diversity
Achieving single, preset objectives	Fulfilling multiple, need-based objectives
Area management, timber production, single technical package, fixed procedures, plantation as first option	Site-specific management, multiple products combined with biodiversity, menus of options, experimentation and flexibility, low-input management, natural regeneration
Single species	Multiple species and multitier plantations

Source: Arora and Khare 1994.

who initiated the PFM experiment in Arabari, West Bengal, argued, in poor countries forest production is less an ecological issue than one of survival. Since people participate in joint management programs for economic and not emotional reasons, they will continue to participate only if they realize gains in return. Problems are complex, composed of a web of ecological, economic, institutional, and political strands that cannot be separated. This requires new approaches, skills, and methods of working in the forest departments. Decentralization and empowerment of local communities are seen as essential strategies for the successful management of natural resources.

This case study looks at the key changes during the two decades of participatory forest management in South West Bengal and investigates the linkages between local participation in forest management and its various ecological, social, and institutional impacts.

The Setting: West Bengal

West Bengal, in the eastern part of India with a total area of 8.87 million ha, encompasses a complex physiography with the Bay of Bengal on the south, the Eastern Himalayas on the north, Bangladesh on the east, and Nepal, Bihar, and Orissa on the west. The state comprises 2.7 percent of the total geographical area of India but is home to 8.3 percent of the country's population. With a population density of 791 persons per square kilometer (km^2) (Government of India 1991), it is one of the most densely populated states in the country. The decade 1981–91 showed an increase of 24.7 percent in the total population of West Bengal.

There are three main geographical divisions of the state. About two-thirds of the total geographical area consists of the Lower Ganga Plain, a flat or gently undulating alluvial plain with elevation up to 30 meters (m). The western portion of the state, the Chhotanagpur Plateau, consists of uplands where crystalline rocks extend from Bihar. The extreme north of the state, the Eastern Himalayan Region, consists of steep, hilly country forming a part of the Himalayan zone. The state experiences a hot and humid monsoonal climate.

Agriculture

West Bengal is a predominantly agriculture-based state with about 95 percent of its rural population engaged in agriculture (Gangopadhyay 1991) and approximately 63 percent (5.57 million ha) of its geographical area under cultivation. Approximately 9 percent of the geographical area of the state is designated as forest area (Forest Survey of India 1993), while the remaining area is either fallow or is not available for cultivation. The latest available statistics classify 23 percent (2.11 million ha) of the total area of the state as wastelands (Government of India 1989).

Although agriculture is the mainstay of the majority of rural people, the nature and distribution of rainfall render agricultural production in unirrigated areas unreliable and suboptimal. Unimodal rainfall (75–80% of it received during the four months of the monsoon season, i.e., June to September) implies that only one crop per year is possible on unirrigated lands, leaving the land fallow and cultivators unemployed for the greater part of the year. During the period 1952–53 to 1974–75, the harvest was either affected by droughts or floods in 14 of the 23 years (Singh and Bhattacharjee 1991).

The agriculture sector has to be viewed in light of population growth in the state. Population pressure is high, with a land/person ratio of 0.15 ha/person as compared to the national average of 0.32 ha/person. Area per agricultural worker is only 0.69 ha against the national average of 1.12 ha. Between 1961 and 1987, the production of food grains in the state increased 62 percent, but the per capita availability decreased 9 percent (Bhowik et al. 1988).

General Economy

The situation would not be so critical if the other sectors of the state economy were performing well. Per capita income in West Bengal declined from Rs. 2,485 (Government of West Bengal 1988) to Rs. 1,882 in 1990–91 (Choudhary 1994). The growth of the state's economy has been slower than national growth because of a decline in industrial investment and an increase in population. The economy is still mainly dependent on the primary sectors. An estimate of the state domestic product in 1986–87 revealed that agriculture, forestry, and fisheries provided 42.2 percent of the state's income.

In a situation where the industrial sector is stagnating and there is high population pressure on agricultural lands, it is not surprising that 68 percent of the population is designated as nonworkers and 45 percent of the people live below the poverty line. The social and ecological repercussions of such a situation are bound to be negative. In the absence of alternate sources of livelihood, especially in the nonagricultural season, the population living in and around forests tends to overutilize the only resource available to them. For the tribal population (8.9% of the total state population), the forests are virtually the only income earning/subsistence resource during the nonagricultural season. As the resource becomes degraded, people start migrating, initially as seasonal migrants in search of wage labor in the nearby agricultural areas, and later as slum dwellers in the cities.

The forest lands, already under pressure from the local communities, also suffer from revenue demands put on them by the state. As revenue from the industrial sector declines, the state attempts to bolster its revenue from other sectors of the economy, including the forestry sector. Thus the revenue needs of the state come into conflict with the people's subsistence needs. This in turn results in a conflict between the people and the forest managers which contributes to the degradation of the forest resource.

South West Bengal

South West Bengal (SWB) encompasses the districts of Midnapore, Bankura, Purulia, Burdwan, and parts of Birbhum. This region of undulating topography occupies 43.7 percent of the area of the state of West Bengal (38,791 km^2). The soils are red loam with the presence of subsoil Kankar pan. The climate is generally hot and dry, with the rainy season extending from June to September. Average rainfall is around 1,300 millimeters (mm). Apart from having 37.9 percent of the recorded forest lands of the state, the region also accounts for 69.6 percent of the total wastelands of the state (Bagchi and Phillip 1993).

The population of SWB numbers 21,967,823 (32.3% of the state), of which 80 percent are engaged in agriculture, half of them as laborers. The density is 566.3 persons per km^2. Of the total population, 23.3 percent belong to the group of scheduled castes and 8.9 percent form scheduled tribes (categories recognized as socially and economically weaker sections of the society in the Constitution of India). The main tribes in the region are Santhal, Oraon, Munda, Bhumij, Maheli, Ho, Kora, Mru, Lodha, Lohra, and Chakma.

The sal (*Shorea robusta*) forests of SWB are of coppice origin and belong to Tropical Dry Deciduous Sal Forests as per the classification of forest types (B.K.B. Roy 1991). The major associates of sal are *Pterocarpus marsupium* (peasal), *Madhuca latifolia* (mahua), *Diospyros melanoxylon* (kendu), *Schleichera oleosa* (kusum), *Terminalia tomentosa* (asan), *Holoptelia integrifolia* (challa), *Aegle marmelos* (bel), *Bombax ceiba* (semul), and *Cleistanthus collinus* (parashi). There are also some patches of mixed forests, especially in Purulia and Bankura.

These forests are of considerable importance as they are located near the industrial belt of West Bengal and Bihar, and they play an important protective role in the watersheds of the Damodar and Subarnarekha rivers (S. Roy n.d.; Sen n.d.).

Forest Management in South West Bengal: The Swift Approach

The ownership pattern and the management of forests of SWB can be divided into several distinct phases. These will be addressed as the precolonial, colonial, postindependent, and subsequent modern periods.

Precolonial Period

Little documentary evidence is available about the forests or their management in the precolonial period. However, some authors suggest that certain social institutions and cultural mechanisms were developed by the local people for the sustainable use of resources (Gadgil 1985; Gadgil and Malhotra 1983). Various cultural constraints (such as a ban on hunting for some days, sacred species and groves, and so on) were enforced to maintain a symbiotic relationship between people and forests. These cultural practices helped to promote prudent use of the natural resources (Gadgil 1989; Malhotra and Deb 1991). Shifting cultivation was widely practiced, with a fallow period of about 15 to 20 years (Malhotra et al. 1991). Although far from perfect and operating with less population pressure, indigenous systems generally recognized the close interrelationship between agriculture and forestry, and some effort was made to create a balance between what was taken from the forests and forest regeneration (Lurie 1991).

Colonial Period

The forests of SWB (known as *jungle mahals*) were brought under British rule through the East India Company during the last decade of the eighteenth century.* The takeover of these forests by the British government put a check on their traditional use by the local people. It also resulted in clearance of vast tracts of forest, and the land was brought under cultivation in order to pay taxes (Palit 1990).

*The British, in need of timber for the Royal Navy (Stebbing 1982) and for the expanding railway network, designated 23 percent of Indian territory as government forests (Lurie 1991).

Forest destruction gained momentum after the opening of the Ajay-Sainthia and the Sainthia-Tinapahar railway lines (1860) and the main line of the Bengal-Nagpur through Kharagpur, Jhargram (1898), and Midnapore (1903), which made the interior areas accessible (Malhotra 1991a,b). This immediately increased the value of forest products. The *zamindars* (feudal lords) started managing these forests under the coppice system with a rotation of five years or even less (against a normal rotation of 80 to 120 years) in order to meet the increasing demand for wood products. The resulting forest degradation led to passage of the Bengal Private Forest Act, 1945 (amended in 1948), which required landowners to prepare a forest working plan and have it approved by the forest department. The act also provided for voluntary or compulsory vesting of the forest lands with the government.

Postindependence Period

After gaining independence, India followed a policy of nationalization of forest resources. A new National Forest Policy was drawn up in 1952. It declared that the government-owned forests should be used primarily to produce timber for industry and commerce rather than for meeting subsistence needs of the people (Lurie 1991). In pursuance of the policy of nationalization, the Estate Acquisition Act, 1953, was enacted, and all forest lands were acquired by the forest department in 1955. This policy of nationalization intensified the process of alienation of local people from the forests. The authority of the populace was undermined and their needs neglected. Furthermore, many owners quickly harvested the forests in anticipation of acquisition by the state.

After acquisition of these forests, the forest department initiated "scientific management," and working plans were drawn up to manage the forests. Relations deteriorated further between the staff "guarding" the forest and people who were dependent on it for sustenance. Meanwhile, the human population of SWB swelled, mainly because of the migration of refugees resulting from the partition of Bengal in 1947 and the wars of 1969 and 1971. The administrative measures adopted for protection of forests did not have the desired impact, and forest degradation continued.

The disadvantaged groups of people living on the forest peripheries inadvertently became both agents and victims of this destructive pattern. The forests became virtually an open-access resource. This led to widespread corruption and unholy alliances between contractor-staff and contractor-people. People dependent for their livelihood on the disposal of illegally collected timber and fuelwood conducted their business in *hats* (local markets). A contractor-collector channel emerged, draining out the forest resources to urban areas. The forest department raided these local markets with the help of police, but the threat of beatings, fines, and arrests failed to deter people desperate for fuelwood or those earning an income from stripping the forest.

The government agencies lacked the resources to enforce restrictions on a massive scale (Palit 1993). Consequently vast forest tracts of the region (especially Midnapore and Purulia districts) suffered serious depletion, and the villagers lost all interest in management or protection. Due to repeated felling, much of the sal forests were reduced to scrub condition with a few trees of *Diospyros melanoxylon* and *Madhuca latifolia,* which were seldom cut. By the mid-sixties, the acute shortage of various forest products resulted in en masse migration of rural poor to other areas in search of livelihood. Also, the forest department incurred a loss of revenue. Conventional forestry practices had clearly been unable to protect and regenerate the forests of South West Bengal.

Turning Point

Around this time a forest officer, Ajit Kumar Banerjee, was appointed as silviculturist at a small research station at Arabari in Midnapore district. Here experiments were being conducted on sal, teak, eucalyptus, and other timber species. Efforts to study the growth and regeneration of trees were failing because local people kept grazing their cattle on the research plots and cutting the saplings for self-consumption or sale. Frustrated by the constant disruptions, Banerjee began meeting with the local people of the surrounding villages to explore the possibility of obtaining their cooperation. After extensive discussions, he concluded that any effort to protect and regenerate forests had to address the connection between poverty and deforestation. He surmised that unless the village people got a better livelihood by protecting the forest than by pilfering it, it would be difficult to motivate them to cooperate with the forest department in any forest protection program, as forests were the only source of livelihood for the poor people, particularly during the lean agriculture season. The communities were offered a deal: in exchange for leaving the trees alone and forming protection committees, the local people would be given employment. Further consultations with village people led to a new arrangement guaranteeing their continued access to nontimber forest products (NTFPs), for example, fruit, leaves, mushrooms, twigs, and fodder grass, for consumption and for sale to generate household income. Also, local people would receive a portion of the revenue from the harvest of the sal forests after they had regenerated.

In 1972, a conference on "Problems of Protection" was held by the forest department. The conference further reinforced Banerjee's ideas of PFM (P. Guhathakurta, pers. comm. 1994). It was recommended that the forest department (1) "encourage setting up of the local forest protection committees with suitable recognition, conferring powers of honorary forester on members and such other concessions as may be made available, and (2) identify the needs of forest products of the local population, particularly where such committees are set up and meet their needs before any sale to market through open auction is done."

The Arabari Experiment

After his initial discussions with the local people, Banerjee initiated the Arabari experiment (Socioeconomic Forestry Project) in the East Midnapore forest division in 1972. It had the following broad objectives:

1. Provide employment to forest fringe dwellers.
2. Allow them to collect subsistence products from the forests.
3. Give them the right to a portion of the sale proceeds from the harvest of the forest rehabilitated with their cooperation.

Approximately 1,272 ha of degraded government forests were selected for revival in 11 *mouzas* (revenue villages). Of these, seven had degraded sal forests totaling more than 700 ha, about 185 ha were refractory areas unfit for tree farming, and the rest was either fully degraded sal forests or blanks (areas with little or no root stock) available for reforestation.

During 1972, a contact program of village meetings was launched and about five hundred families with a total population of 2,500 agreed to participate in the program. Villagers were told about the benefits that would accrue to them in the form of employment, forest products, and cash income. The people agreed to protect the forests near their villages. The forest department employed them in activities undertaken to regenerate the sal forests with good root stock and in planting totally degraded and blank areas with quick-growing species such as eucalyptus and *Acacia auriculiformis,* and cash crops such as *Agave* spp. and cashew. Poles were also provided to the people for purposes such as house repair at one-fourth the market price. Participants were given exclusive rights over NTFP collection for self-consumption as well as for sale.

A great degree of openness was maintained by the forest department officials during the course of program implementation. A clear message was given to the people that they could approach higher authorities in the department if they faced any problem in the program (Banerjee, pers. comm. 1994). The first formal meeting of the villagers took place in 1977 when a committee was constituted with an elected president and secretary to coordinate the PFM program.

By 1985, a survey by the forest department revealed that the number of participating families had risen to 618 with a total population of 3,607. A review of the project in 1986 revealed that good sal coppice forest had regenerated more than 700 ha, while 300 ha had plantation crops (eucalyptus, *Acacia auriculiformis,* agave, and cashew). A government order issued on 7 March 1987 (G.O. no. 1118-For/D/GM-76/85) allowed harvest of one-tenth of the restocked forest and plantation area, and distributed 25 percent of the net earned revenue among the participating families.

Benefit Sharing

In 1987–88, 97 ha of sal coppice forests and eucalyptus plantations were harvested in the Arabari experiment for the first time through the West Bengal Forest Development Corporation (WBFDC) Ltd., involving the local people in the harvesting operation. A total revenue of Rs. 1,799,500 (against a projected value of Rs. 2,300,000) was realized through the sale of 2,540 cubic meters (m^3) of fuelwood and 92,947 poles. The forest department incurred the sum of Rs. 741,500 toward operational expenses and depot maintenance (against projected expenses of Rs. 198,000). With this, Rs. 121,638 were also spent as proportionate costs of the raising and maintenance of forests and plantations over 97 ha during the project period (1971 onwards). The net sale proceeds thus obtained by the forest department were Rs. 936,550. A total sum of Rs. 234,137 was received by 618 families (i.e., Rs. 379 per family) as their 25 percent of the net sale proceeds, and Rs. 370,750 (i.e., Rs. 600 per family) was received as wages during the harvesting process (based on the assumption that 50 percent of the harvest cost is paid as wages). Thus the total income to the 618 families from 97 ha of harvested forest was Rs. 604,887 (i.e., Rs. 979 per family). The net revenue realized by the forest department was Rs. 702,413 (B.K.B. Roy 1991).

These results laid the foundation for the future partnership between the people and the forest department. After the first harvest of 97 ha in 1987–88, an additional 181 ha of sal coppice forests and 63 ha of plantations were harvested during three harvest seasons from 1988–89 to 1990–91. Over the first four years (1987–88 to 1990–91), an average of 690 poles and 25 m^3 of fuelwood were obtained per hectare with a total sale value of Rs. 14,950. The forest department incurred a total sum of Rs. 6,000 per hectare as harvesting costs and Rs. 1,254 per hectare as the cost of raising enrichment plantations. The people received Rs. 4,684 per hectare in the form of employment and Rs. 1,932 as their share of the net sale proceeds per hectare. A revenue of Rs. 5,764 per hectare accrued to the forest department as their share (Sen n.d.).

In terms of per family benefits, the accruals from the sale proceeds and wage labor worked out to be Rs. 1,938 for the four-year period. Rs. 1,618 also accrued to each family from the sale of other forest products, such as sal leaves, kendu leaves, cashews, fruits and flowers, and mushrooms.

Benefit/Cost Ratio

The benefit/cost ratio (BCR) of the operations can be calculated using the above-mentioned figures and assuming that (1) the planting and maintenance costs were incurred in equal sums over the first three years of the project, (2) the benefits obtained from 1988–89 to 1990–91 were spread equally over the period, and (3) the interest rate rested at 10 or 12 percent. The BCR is found to be 1.68 (at 10% interest) and 1.52 (at 12% interest). In addition to the above, people harvested fuelwood and fodder on payment of a nominal royalty (at half the scheduled rate), which yielded a total revenue of Rs. 284,000 for the forest department. Also, 290 ha of plantations of quick-growing species were raised along with 65 ha of cashew plantations and 12 ha of agave plantations. The forest department spent Rs. 1,641,000 (between 1971–72 and 1984–85) in carrying out periodic silvicultural operations, from which 220 thousand man days (i.e., 27 man days per family per year) of employment was generated (B.K.B. Roy 1991).

While a successful example of participatory forest management was emerging in Arabari, it remained an isolated case for about a decade, with little effect on the conventional forest management system in the state. Nonetheless, it demonstrated that opening communications with the local forest communities could reduce the decades-long people–forest department conflict and lead to regeneration of degraded forests. This novel approach helped the forest department to develop effective control mechanisms of forest exploitation and to identify terms for effective management partnerships with the people.

The Expansion Phase

The Arabari project led to improved forest department-people relationships, resuscitation of sal forests, and a simultaneous increase in income for rural people and the forest department. The villagers gained confidence and were able to evolve mechanisms to restrict access in order to protect and regenerate more than 1,270 ha of forest land which was hitherto in a degraded condition. The encouraging results of the Arabari experiment laid the foundation for the PFM program in SWB and the growth of experimental Forest Protection Committees (FPCs). By 1989, when the formal state government order was issued, more than 1,200 FPCs were managing 152,000 ha of forest land, that is, more than 37 percent of forest land under the Western Circle of the West Bengal Forest Department. (A *circle* comprises a number of forest divisions; Western Circle includes the forest divisions of Bankura, Purulia, Midnapore, and Burdwan.) The district coverage in Midnapore, Bankura, and Purulia was 39.5 percent, 35.3 percent, and 38.5 percent, respectively (Palit 1989). Government recognition also came in the form of national awards such as the Indira Priyadarshini Awards given to Arabari villages in 1987, to some Panchayat members in 1988, and to one FPC and a Social Forestry range in 1989. (Each forest division comprises a number of smaller administrative units or *ranges*.)

Chandra and Poffenberger (1989) suggest that rapid growth of FPCs may have reflected the desire of forest communities to have their rights recognized during this period of authority reallocation.

A survey (Chandra and Poffenberger 1989) of five FPCs in the three districts (Bankura, Purulia, and Midnapore) of SWB to explore the patterns of FPC formation revealed the following:

1. In all cases, information regarding the program came from forest department field staff or neighboring communities.
2. The forest department representatives, be it the range officers or the beat officers, played a crucial role in the formation of the committees. They repeatedly visited the villages and held discussions with villagers on the problems associated with deforestation. In these meetings, the local *Pradhan* (Panchayat head) was also involved in all the FPCs.
3. The "spread effect" contributed to the success of the program. There were cases when people were quite skeptical even after the forest department officials approached them. They agreed only when people from another FPC convinced them. It took communities some time (up to many months) before they reached consensus.
4. Another possible factor was that the villages that already had FPCs thought that with the formation of FPCs in the neighboring villages, conflicts over forest use could be reduced, as the villages nearby would then have their own sources of fuel, fodder, and NTFPs.

In almost all cases the pattern of FPC formation appeared to be more or less the same. After the consensus was reached, a list of all participating households was drawn up and officers were selected. In multicaste villages, representatives from all castes were usually included in the executive committee of the FPC, which led to greater community participation. It appears that most FPCs seriously took up the task of protection, through day patrols and fine systems, with a large degree of success. It did not function well in villages where only a few families were involved in the FPC. Also, the system of keeping paid watchers did not yield desired results.

Peer group pressure, small fines, and social sanctions were commonly used for control by the FPCs. In difficult situations and where outsiders were involved, the backup authoritative support of the forest department was sought. People felt that their authority was undermined in situations when the forest department did not support them.

In such cases there was loss of faith in the forest department, and the relationship between the FPCs and the department became strained. This desire for formal recognition and authority was also reflected in the demand for issuance of identity cards by many FPCs. This issue of recognition was partially resolved in 1989 when the Government of West Bengal issued a government resolution giving formal recognition to FPCs in SWB and clearly laid down their rights, duties, and responsibilities. Through a series of orders in 1990, the PFM program was extended throughout the state.

Current Status

By 1994, the program, which was initiated in 1972 as an experiment over an area of 1,272 ha involving 618 households, had spread to 390,919 ha in 2,423 FPCs, out of which 301,781 ha (77%) were within SWB (Government of West Bengal 1994). In 1991, a total of roughly 2,690 forest department staff were engaged in PFM.

Since its inception in 1972, the PFM program has gradually developed into a movement. The major achievement has been in qualitative terms, that is, the positive shift in forest department-people relationships, better quality of life, and rejuvenation of forest ecosystems. The experience in West Bengal demonstrates the efficacy of the *sui generis* PFM approach and has become a model for similar efforts elsewhere in the country.

The Legal and Political Environment of PFM

The legal framework of the PFM program can be broadly divided into two categories, the national level and the state level. This framework emerged concurrently with the development of the PFM program in SWB and in turn facilitated its implementation and spread. The national level legal framework comprises mainly the National Forest Policy (1988) and the Central Government Circular (1990) regarding the involvement of local communities and voluntary agencies in the regeneration of degraded forests. The various state government schemes, resolutions, and orders, issued from time to time, comprise the state level legal framework. Many of these orders were pioneering in nature (e.g., sharing of revenue with the people) and were possible only because of the supportive and conducive political environment prevailing in the state. The leftist coalition government of West Bengal has been quite progressive and committed to land reforms and other social programs to empower poor rural communities. It is mainly because of this factor that the various ministers took keen interest in the program, and the resolutions and orders that facilitated PFM development were quickly passed.

Land Reforms and the Social Forestry Program

West Bengal is known for its progressive land reforms. Starting in 1977, sharecroppers and landless persons were given 99-year leases (*pattas*) on land declared surplus under the Land Ceiling Act. In 1979, the government of West Bengal vigorously implemented "Operation Barga." Under this program, the *bargadars* (sharecroppers) had their names recorded in the revenue records in the course of meetings organized at the village level by officials of the revenue department. This program resulted in the provision of land to the landless and security of tenure to sharecroppers and small landholders. In 1980, the state government's "New Directives on Forest Management," which was followed by operational guidelines, gave privileges and concessions to the tribal population living in the vicinity of the forests. These initiatives provided economic and social benefits to the rural population, especially the tribal people and poorer populace. However, more to the point, the implementation of the programs demonstrated the will and the commitment of the state government to ameliorate the lot of the poorer sectors of rural society and thus prepared the people to participate with greater confidence in the other programs of rural development.

It was in this context that the Social Forestry Project was taken up for implementation in the state in 1980. The project encouraged the establishment of forest plantations on revenue wastelands, farmlands, and degraded forest areas.

Overexploitation of patta lands had severely reduced their productivity for raising agricultural crops. A study in Midnapore district revealed that average yields of various crops on these lands ranged from 38.8 percent to 56.1 percent of the average yields on the farmer-owned lands (average of five years) (Singh and Bhattacharjee 1991). The state government strongly promoted group-farm forestry. Groups of farmers having 20 ha or more of vested land or having their lands in a compact block were motivated to plant trees by way of free government distribution of seedlings, fertilizers, and pesticides. In addition, incentives were offered in the form of Re. 0.10 and Re. 0.14 per surviving plant at the end of the first and second year, respectively (Shah 1988). In a sense, the land reforms and the Social Forestry Project complemented and supported each other. Given the facilitating conditions, the Social Forestry Project did well in the state. It achieved 108 percent of the target of 93,000 ha.

Several facilitative orders were issued under the social forestry scheme which helped in building a conducive environment for the PFM program as well. In order to involve local people and Panchayats in the social forestry program, a government order was issued in 1985 (FD memo no. 4544 dated 9 Oct. 1985) under which the divisional forest officer, range officer, and beat officer were made members of the Bhumi Sanskar Sthayee Samity, a statutory body of Zilla Parishad/ Panchayat Samity (district and block level local self-government bodies). One of the duties of this body was "to identify the beneficiaries from amongst scheduled castes and scheduled tribes to reap the benefit from the plantations created under the Social Forestry Project and Integrated Tribal Development Program." The Department of Panchayat and Community Development renamed this samity as Bon-O-Bhumi Sanskar Sthayee Samity (BOBS) in 1986 and increased its purview to include forest matters as well.

A number of bold steps were taken in 1986 which increased the confidence of the people in the forest department. The forest department agreed to share 50 percent of usufruct from the plantations raised by it on vested wastelands and other public lands under the RLEGP (Rural Landless Employment Guarantee Program), with the rural poor identified by BOBS. The Panchayat Samity was made responsible for protection of these plantations (FD memo no. 1925 dated 24 Apr. 1986). Another order, regarding plantations raised under the Rehabilitation of Degraded Forests program of the Social Forestry Project, specified that 25 percent of the usufruct would be distributed to the local people as selected by BOBS (FD memo no. 2379 dated 11 June 1986). Similarly, strip plantations created by the forest department under the Social Forestry Project were handed over to Panchayats for maintenance and protection. They were given authority to identify poor people who would receive usufructuary benefits from these plantations

(FD memo no. 2914 dated 22 July 1986). All these moves created a conducive atmosphere for further development of forest department-people relationships.

National Forest Policy, 1988

A new National Forest Policy was issued by the government of India in 1988. It completely reversed the policies pursued to that time and redefined the objectives of forest management. Conservation and basic needs of the people were given priority over industrial and commercial objectives. The policy aimed at meeting basic needs of rural and tribal people and maintaining the intrinsic relationship between forests and people by protecting their customary rights. It stressed involving people in programs of protection, conservation, and management of forests. This policy gave legitimacy to the PFM program.

The 1990 Order

In June 1990, the government of India issued guidelines to all state governments to involve local communities as well as voluntary agencies in the regeneration of degraded forests. This order formed the basis for the formulation of the PFM programs by many state governments. Many novel concepts such as sharing of usufruct with the communities, participatory microplanning, and involvement of nongovernmental organizations (NGOs) were introduced in this order. Although the 1990 order formed the basis for PFM programs in many states, West Bengal had already issued a number of facilitating resolutions, which had, in fact, influenced the government of India's order.

The Role of Nontimber Forest Products in PFM

Nontimber forest products (NTFPs) refer to all forest products other than timber, including plants, animals, and their parts or products. These were formerly referred to as minor forest products (MFPs) as the major object of forest management was timber production. However, after it was realized that these products play a very important role in village and tribal economies, the term NTFP gained preference over MFP.

These products provide a substantial part of household income for forest fringe dwellers. According to one estimate, NTFP income accounts for 55 percent of the total employment in the forestry sector with a potential to generate 1.85 million person years of employment (Pachauri n.d.).

The local communities in West Bengal, especially tribal people, use a number of NTFPs. Sal trees have religious significance. Most villages have a small patch of sal trees which is preserved as a sacred grove; leaves and branches are not collected from sacred groves. Sal twigs and flowers are used in Salui worship in the spring. The Santhal tribe puts up sal poles in front of their houses on religious festival days. Karam (*Adina cardifolia*) twigs are used in the worship of Karam, a tribal deity. Siuli (*Nyctanthes arbortristis*) flowers are used for worship of Saraswati and also for ornamental purposes.

Fruits such as bel, kendu, sal, amlaki (*Emblica officinalis*) and tamarind (*Tamarindus indica*) are generally collected. Oil is extracted from sal seeds and bhela (*Semicarpus anacardium*) fruits are used for treating certain ailments of cattle. Leafy vegetables such as kurol sag and ban-pui (*Basella alba*) are also collected; these fruits and vegetables provide important nutrients during midsummer which is a food scarcity period. Parashi leaves are collected for use as agricultural pesticide. Mushrooms and dioscoria tubers are also collected. Bakhar roots and bark of kurchi (*Holarrhena antidysentrica*) are collected for making bakhar balls (used in fermenting country liquors). Apart from their use as leaf plates, some sal leaves are also used as chutah (country cheroots). Leaves of bhurru (*Gardenia qummifera*), khod (*Syzigium cumini*), and kari or ame are used as fodder. Kharang grass (*Aristida* sp.) is collected for making brooms and palui (*Phoenix acaulis*) leaves for weaving mats. Kharang grass is also used for stitching sal leaves for making leaf plates. Kendu leaves are collected in large numbers for sale to government cooperatives (Malhotra et al. 1991; Pal n.d.). Other important NTFPs include mahua flowers for distillation, medicinal herbs, tassar silkworms, and honey. The total value of these products in the country is estimated to exceed US$1 billion annually (Poffenberger 1990).

NTFPs have played an important role in the success of the PFM program in West Bengal. The participating communities have been allowed under the PFM program to collect NTFPs such as grass, fallen twigs, fruits, flowers, and seeds free of cost. However, a few important

NTFPs such as sal seeds (for oil) and kendu leaves (for cheroots) cannot be sold in the open market, but must be deposited with the West Bengal Tribal Development Cooperative Corporation through government-promoted cooperatives called LAMPS (large area multipurpose societies). The collectors receive wages for these products. In the case of other NTFPs, people collect them for self-consumption or for sale. These NTFPs may be sold as such, or with some value added, either directly to the consumers or to traders and their agents.

NTFPs as Subsistence Goods

There is mounting evidence (from the microstudies of various FPCs) to suggest that the multiple products coming from protected, biodiverse forests under PFM are a major source of sustenance for the poorer segments. These people's survival would be threatened if these forests were converted into monoculture plantations.

A survey conducted in 1991 (Malhotra et al. 1991) involving 216 households randomly selected from 12 FPCs in the Jamboni range in Midnapore district found the following:

1. A total of 214 wild plant species were observed in the regenerating sal forests. Of these, 155 species (72%) are used by the local communities for the purposes of food, fuel, fiber, fodder, medicine, construction, commerce, household articles, religious use, ornamental use, and recreation. Seventy plants are used frequently and regularly. In addition, two animal species (karkut ants and silk cocoons) and three plant species found only in plantation areas are also used regularly. Thus a total of 75 species are used regularly by the people.
2. Eleven parts of plants are used for various purposes. More than one part is collected and used, thus the total number of parts used from 75 species totals 109. The most frequently used parts are twigs of 35 plants, fruits of 17 species, and leaves of 11 species. Fifteen species (11 mushrooms, 2 insects, and 2 fibrous plants) are used as a whole.
3. Thirty-eight plants/parts are consumed for food especially during the food scarcity period. Fifteen plants are used frequently for medicinal purposes.
4. Some ethnic variation was observed in NTFP use, for example, mahua flowers, bakhar roots, and karkut ants are used exclusively by tribal people.

5. Sixty-six species are available during the whole year except the rainy months (June and July), when the number is reduced to 38.

A study of the Raigarh FPC in the Ranibandh range of South Bankura forest division shows that the FPC members greatly depend on forestry and allied production systems for subsistence. Agriculture (mainly rainfed paddy) is secondary as agricultural productivity is poor and is dependent on the vagaries of rainfall. It was found that a total of 43 NTFPs were collected. Of these, 27 NTFPs were common, whereas 16 were uncommon or rare.

In the case of medicinal NTFPs, *Andrographis paniculata* (kalmegh) ranks highest in terms of collection at the primary collectors' level. Other important NTFPs are bark of *Symplocos racemosa* (lodh), flowers of *Woodfordia floribundata* (dhadki), fruits of *Aegle marmelos* (bel), and bark of *Holarrhena antidystentrica* (kurchi), in descending order of collection at the primary collector's level (Pal n.d.).

Fuelwood is the most important subsistence item collected by the people from the forests. Many families were dependent on fuelwood sale for income, but the formation of FPCs has reduced its importance. However, people still collect twigs, branches, and leaves as these are the only source of fuel for cooking.

Tubers are considered to be one of the more important products collected. Though they have low market value (Rs. 3 per kg), they are important sources of starch and nutrients during times of food scarcity. Many kinds of tubers are used, though their preparation is time-consuming. Many wild mushrooms are also used as food items by the people and are an important source of nutrition.

People take grass and leaf fodder from the forest area. Forest leaf fodder (sal and others) is especially important during the dry season (April and May) (Poffenberger 1993). People also grow sabai grass (*Eulaliopsis binata*) on forest blanks and other wastelands. This grass is mainly used for rope making. It is estimated that 40 percent of the total sabai production of the study area comes from crops raised on forest lands (Pal n.d.).

NTFPs as Income Source

NTFPs not only provide direct sustenance to the forest fringe dwellers but also play a crucial role in the generation of income throughout the year for local people. Malhotra et al. (1991) report that the most important commercial NTFPs are sal leaves and seeds, kendu leaves and fruits, fuelwood, fodder, mahua flowers, dioscoria tubers, medicinal plants, tassar, and mushrooms. The average annual NTFP incomes for tribal and caste Hindu households were Rs. 2,523 and Rs. 2,738, respectively (at 1991 prices). This did not include income from medicinal plants, several fruits, and animals gathered/hunted occasionally. NTFP incomes in tribal and caste households contributed 22 percent and 16 percent, respectively, to the total family income. The contribution of fuel and fodder to NTFP income was significant, being 79 percent of the NTFP income for tribals and 74 percent for the caste Hindus. They estimated that income flow from NTFPs in a household was seven times greater than the amount the household would get as their 25 percent share of revenues generated from the harvest of 10-year-old regenerated sal forests.

A study by Chandra and Poffenberger (1989) in Pukuria village in the South Bankura district found that after six years of protection a 130-ha sal forest had regenerated rapidly and started yielding substantial NTFP income to 93 families of the village. The study also found that while men worked on paddy lands and migrated to nearby districts, women spent four hours or more daily collecting NTFPs for self-consumption or sale. Women earned an estimated Rs. 7 to Rs. 10 per day from 1 ha of mixed sal forest after five to six years of protection. On the other hand, income from the 25 percent share in revenues generated from the sal forest harvest provides an average daily income of Rs. 1.4 per ha.

Chandra and Poffenberger (1989) reported that Pukuria village exported approximately 700,000 leaf plates per month during the harvest season and 10 metric tons of sal seed annually. Each family collected 150 to 200 kg of seeds for oil extraction, generating an income of Rs. 240 per season. The study found that women collect up to 12 tassar cocoons per day during August to September and December to January, which they sell for up to Re. 1 each. These are from the larvae of wild or semiwild types of moths belonging to the family Saturniidae; *Antheraea mylitta* is the main species found in sal forests of SWB. Kendu leaves are collected during April and May. The village women collect up to 10,000 leaves in a day for which they receive Rs. 20 from LAMPS. Another study shows that an average of 200 leaves are collected per person daily during the months of March and April (Malhotra et al. 1991).

A study of Raigarh FPC in South Bankura (discussed earlier) revealed that an average of 430 man-days of employment (per family per year) is generated through NTFP collection. The contribution of medicinal NTFPs to the family income in Raigarh FPC was found to be Rs. 750 per year (Pal n.d.).

Mushrooms are locally known as *chatu* in SWB. These appear with the onset of the monsoon and are generally available until the last week of September. In Midnapore, 15 kinds of mushrooms grow in the forest, of which five or six are edible. In West Midnapore, five species were collected for sale by the local people. These were sold in the local village market for prices ranging from Rs. 5/kg to Rs. 20/kg, depending on the variety (Malhotra et al. 1991).

Mushrooms are highly seasonal, some available for only a week, and are collected by almost all village people for household consumption purposes. However, it is mainly tribal people and those belonging to the lower castes who are engaged in selling them to earn income. Women and children are generally involved in mushroom collection; a survey in Midnapore indicated that women, men, and children are involved in the collection activity in the ratio of 60:25:15. On average, three to four hours are spent by the collectors to collect about 3 kg per day. Collection on individual days may go up to 10 or 12 kg. These are marketed locally; in towns these have to compete with cultivated varieties (Suneja et al. 1992).

A study carried out by Mishra and Sinha (1992) found that kash (*Saccharum* spp.), nal (*Phragmites* spp.), and hogla grasses are important income sources for local people in the Baikunthapur forest division (North Bengal). Kash is mainly used for thatching purposes and to provide shade in the nurseries; its major demand is from tea estates in the region. Nal is used in making house walls and in house fencing. Hogla is used for making sleeping mats. Most of the collectors are male members of the family. Nal collection during the study yielded Rs. 17/person/day, kash collection yielded Rs. 21/person/day, and hogla collection yielded Rs. 30/person/day. In comparison, agricultural labor yielded Rs. 12.50/person/day (although the prescribed minimum wage rate as laid down by the government was Rs. 28.96/man/day) (Mishra and Sinha 1992).

Increased biodiversity in the sal forests regenerated by the PFM program not only clearly provides subsistence goods to the people, but also enhances their income generation. Moreover, these products provide income to the people in a situation in which agriculture offers them lower and less secure returns. The annual flow of NTFPs is more important for these people than the one-time terminal income from the harvest of sal forests and the sale of timber in a 10-year cycle (Moench 1991).

Marketing of NTFPs

The success of PFM in large part depends upon the accrual of regular benefits to the local people by collection and sale of various NTFPs from forests. However, some village communities adjoining forests are not making sal plates as they do not have marketing facilities (Dutta and Adhikari 1991). This may well be true of other NTFPs. There is a need, therefore, to address the problems concerning the marketing of NTFPs.

Presently NTFPs are primarily collected by the poor (marginal farmers and landless), who do not have any other source of income especially during the lean season (Das 1993; Dutta and Adhikari 1991). However, this activity is not very remunerative, and people are forced to search for other avenues of income such as farm labor and migration to agriculturally advanced areas or urban centers. It is a different matter that there, too, local people generally get paid much less than the official minimum wage rate. Increasing the returns from the collection and sale of NTFPs would provide relief to the poorer tribal people.

Any corrective measure should take into consideration the seasonal nature of NTFP markets. Seasonality plays an important role in changes in the demand and supply of various NTFPs. The supply is influenced by the agricultural calendar. Being a secondary occupation, the collection level drops during the sowing and harvesting seasons (June to August and November to December, respectively), causing the prices to rise. A study of the Raigarh FPC in South Bankura shows that the lowest collection of NTFPs is during the rainy season and the highest between September and January (Pal n.d.). At least for five months in a year, the male members cannot afford to continue this activity, as they either become engaged in their own farm activities or are employed as daily wage laborers. During this period, collection as well as marketing activities are carried on only by the female members of the family, which leads to lower collection. On the other hand, the prices of NTFPs fall during the agriculturally lean season, due to an increase in collection. The purchase price also varies with the seasonal variation in demand, which increases during the marriage season (second half of January, the first part of March, April, early August to 15 November, and the first half of December) and during religious occasions (Pachauri n.d.).

The NTFPs collected from the sal forests can be divided into two major groups: nonperishables and perishables.

Nonperishables

Among the nonperishable NTFPs that can be processed, sal leaf is the most important (Malhotra 1991a,b). Village people, mainly women, spend an average of six to seven hours in leaf collection (Das 1993; Dutta and Adhikari 1991; Pachauri n.d.). Typically a household earns Rs. 13 a day by producing 700–800 plates (a household has been taken to consist of five members), whereas the official wage rate in the state is Rs. 28 per man-day. Even the prevailing market wage rates vary from Rs. 8 to Rs. 10 per man-day plus a meal (Dutta and Adhikari 1991).

The earnings of the household vary with the mode of selling of sal leaf plates. If the villagers sell directly to rural consumers, they earn Rs. 3.00 to Rs. 3.50 per 100 plates, whereas the retail markets of district towns offer Rs. 4 to Rs. 5 per 100 plates (Das 1993). But the latter involves six to seven hours of travel, and all their plates may not be sold. The villagers, therefore, sell their products to the local retailers who pay Rs. 2.50 to Rs. 3.00 per 100. However, if the production is substantial, then the preferred market is wholesalers in nearby areas, who offer Rs. 16 to Rs. 22 per 1,000 plates depending upon quality and demand.

Male members sell the plates outside the villages during the nonagricultural season, whereas during the agricultural season, women are left with the only alternative of selling the plates to commission agents appointed by wholesalers. These agents visit the villages and purchase the plates at a price of Rs. 13–20 per 1,000. In turn, they sell them on a commission of Re. 0.50 to Re. 1.00 per bundle of 1,000 to district wholesalers. The district wholesalers might sell their products directly to the wholesalers of Calcutta at the rate of Rs. 31 to Rs. 32 a bundle (1,000 plates) or to a contractor (Pachauri n.d.). Villagers obtain a low return because they have to sell most of the plates to commission agents at low prices because of the difficulty of visiting the local wholesaler. The households' efforts, from plate making to selling, are individualistic, and thus there is no economy of scale.

The sal plates of West Bengal face competition from the plates arriving from the neighboring states of Orissa and Bihar. Orissa plates are preferred because of generally superior quality and are also cheaper due to lower wage rates and lower electricity charges in Orissa (Das 1993). In addition, machine-made plates, which are in high demand, have entered the market recently.

The marketing trend of medicinal NTFPs follows a much simpler route. In Bankura, NTFPs are sold either directly or through village middlemen to the wholesalers as per the demands placed on village people by the traders. The wholesalers, on average, sell medicinal NTFPs at five times the purchase price. This activity is not very remunerative as it is highly labor intensive. Thus only 20 of 111 families in the Raigarh FPC (Ranibandh range, South Bankura) were engaged in medicinal NTFP collection. (Eight families having specific knowledge of medicinal species depend exclusively on this activity, while others engaged in collection in addition to other employment activities [Pal n.d.].)

Perishables

Products such as mushrooms offer a contrasting picture, mainly because of their limited availability and perishable nature. The mushrooms are sold either directly to the consumers (if there is a district market nearby) or to a village merchant. The collectors receive Rs. 5 to Rs. 6 per kg on average. The village merchant in turn sells them to brokers and middlemen at a margin of Rs. 2 to Rs. 3 per kg. The brokers sell the same to the traders at a profit of Re. 0.50 to Re. 1 per kg. From the traders, the mushrooms find their way to Calcutta and Bihar. In the process of transportation, either by train or by truck, the mushrooms lose weight by 10–15 percent. The price paid by the consumers in Calcutta varies from Rs. 16 to Rs. 25 per kg, whereas in Bihar the consumers pay an average of Rs. 30 per kg. The share received by the collector is one-fifth to one-fourth the price paid by the consumer.

The perishable nature of mushrooms puts a constraint on the collectors. Distress selling tends to take place because of the low bargaining power of the collectors and poor access to transport facilities. The price variation in different markets is governed by the socioeconomic composition of consumers. Thus the mushroom that fetches a price of Rs. 30 per kg in Kharagpur sells at Rs. 16 to Rs. 20 in rural markets and Rs. 25 per kg in Midnapore. In Calcutta, the same species is sold in canned form at a rate of Rs. 82.5 per kg. Prices also tend to vary in relation to the demand at the time of arrival in the market.

Analysis

In earlier times agriculture was more remunerative as the land-man ratio was higher and people were able to sustain their livelihood. With the increase in population and simultaneous decrease in land productivity, the dependence of people on forest resources gradually increased. NTFPs became more rewarding than rainfed agriculture in terms of input/output ratio. However, poor marketing facilities for NTFPs result in insufficient income generation, and people must migrate to find employment.

The labor investment in the NTFP business shows an incongruent relationship with the income from it. Lack of processing at the collectors' level and operation in isolation contribute to low returns. The limited access to markets and dependence on intermediaries result in low returns to the collectors (Khare 1993). The market is controlled and dictated by the traders because of their traditional credit links with the consumers, both off and during the season. The only reasons for the villagers to take up this activity are zero investment and low opportunity cost of labor.

In an effort to protect the collectors from unscrupulous traders, in 1980 the government granted monopoly rights of collection and disposal of kendu leaves and sal seeds to the West Bengal Tribal Development Cooperative Corporation Ltd. (WBTDCC) for subsequent allotment and operation through the LAMPS on payment of a modest royalty. However, this new marketing system has reduced the number of legal buyers (due to the granting of monopoly collection rights to WBTDCC and LAMPS) to whom the primary collectors can sell. This reduction in the free flow of NTFPs has also opened the door to clandestine traders in the market.

There is an urgent need to improve the efficiency of the present marketing system to sustain the interest of people in the PFM program. Improved marketing of NTFPs has a critical role to play because the final harvest of sal is done over a 10-year cycle; in the interim it is NTFPs from the regenerated sal areas that provide subsistence and income to the people. Further, it is feared that if even 10 percent of eligible FPCs were to harvest their production of pole timber in 1994, the market would collapse. The discussion with FPC members by the authors of this case study revealed that they were aware of the poor market for poles and were more interested in sale of NTFPs.

The focus of the government should be on the regularization of private trade rather than on its elimination. It should provide required market information to the collectors and protect their interests. Sustained political effort and an efficient bureaucratic machinery are required to carry on a monopoly purchase by the government, a difficult proposition over a long period (Saxena and Gulati 1993). Attempts should be made to change the "employer-employee" relationship between the LAMPS and people, and FPC members should receive a share of the revenue from kendu leaves and sal fruits collected from their area, instead of just the collection wages that they now receive.

It is clear that NTFPs have played a key role in sustaining the interest of local people in the PFM program. At the same time, some NTFPs may be currently overexploited as no attention is paid to their regeneration or propagation (Pal n.d.). More research on NTFPs is clearly needed.

Looking to the future, it appears that increased productivity of NTFPs and improved marketing facilities will have a crucial role to play in involving people in sustaining and expanding the PFM program. The forest department's new focus on NTFPs, rather than on timber, is also crucial for the program's success.

Impact of the Program

The impact of the PFM program can be seen in the vegetation dynamics of regenerating sal forests, the livelihood patterns of fringe communities, and the attitude and functioning of the forest department. Of the 6,418 km^2 of degraded forest in West Bengal, 3,909 km^2 are managed jointly with communities through 2,423 FPCs under the new management system. Forest Survey of India's latest report (FSI 1993) cites SWB as the area in West Bengal where, compared to the previous

FSI report in 1991, 41 km^2 of degraded scrub forest (less than 10% canopy cover) has been upgraded to the category of open forest (10–40% canopy cover). Recent analysis of land satellite images shows that closed forest cover in Midnapore district alone has increased from 11 percent to nearly 20 percent of total land area in the past six years. These regenerating forests now produce a wide variety of medicinal, fiber, fodder, fuel, and food products for participating rural communities. In the following pages, we bring together information scattered in numerous microstudies to develop an overall picture.

Ecological Impact

In revegetating the degraded forests, two approaches were followed by the forest department. Natural regeneration occurred in areas where rootstock was available, and quick-growing species were planted in totally degraded sites and forest blanks. In the latter approach, mostly monocultures of eucalyptus and akashmoni (*Acacia auriculiformis*) were raised, although recently some mixed plantations have also been undertaken.

A study of vegetation dynamics carried out in an FPC in Bankura, which had protected 170 ha of degraded forest for six years, revealed that predominant species such as sal, bijasal, kendu, asan, and mahua were vigorously coppicing in nature. The occurrence of herbs was found to be low because, in part, the microclimate required for the regeneration of some NTFPs, especially medicinal ones, had not yet developed, and commercial overexploitation had retarded their regeneration. Commercialization of the forests has also led to a sharp decline in the numbers of mammals, reptiles, and migratory birds (Malhotra et al. 1991).

India is one of the 10 most biologically diverse countries in the world. The various roles and functions of biodiversity are being increasingly recognized and incorporated into new programs. The foresters of West Bengal are also well aware of the importance of biodiversity. The current chief of social forestry, S. Palit, writes, "in a natural forest ecosystem the in-built defense mechanism of nature comes into force to control the perturbations, whereas in monocultures, tackling one problem leads to the creation of another" (Palit 1993).

There has been an increase in the biodiversity of the forests protected by the communities as is revealed by many microstudies. In a survey carried out in 12 FPCs under the Jamboni range in Midnapore (Malhotra et al. 1991), a total of 255 species of plants, vertebrates, and invertebrates were observed in the area (regenerating forests, plantations, and settlement areas); 84 percent of these were found in regenerating sal forests. Species richness was relatively lower in the young forests regenerating from sal coppices than in the older forests. The species diversity was found to be relatively poor in the eucalyptus plantations. The village people were of the view that mushrooms and medicinal herbs do not grow on the plantation soil. A hardy species capable of thriving on the soil of plantations, kendu was found in high abundance in both plantations and sal forests. In places, sal was found to invade the plantation areas along with many companion species, an indication of the superior vigor and sustainability of naturally regenerating forests over plantations.

Some of the most important advantages emerging from forest regeneration have been improved groundwater infiltration and slowed runoff. The reestablishment of forest near the village has also allowed a large number of birds to nest in the area. The birds play an important role in controlling insect pests that attack the rice crop. FPC members also felt that there had been a decrease in the incidence of disease since forest regeneration (Poffenberger 1993).

Another interesting impact of forest regeneration in SWB is the return of elephants to these tracts. Up to about 1987, only a few solitary elephants remained in the Ajodhya hills, the Bundwan range of Purulia district, and the Banaspahari area of Midnapore district. A herd of wild elephants from the Dalma wildlife sanctuary in Bihar used to visit these areas between October and December; their movement was restricted to west of the Kangsabati River. A large herd of about fifty elephants entered East Midnapore division in 1987 and stayed primarily in the Arabari range until March 1988. During 1988–90, this tract was frequently visited by elephants. Attempts to drive them away were not very successful. This pattern of visits continued through 1991 and 1992 (Dey 1992), and in 1993 resulted in national coverage as the elephants strayed close to human habitations and caused a scare among the native population. Indeed, the extended presence of the herd has led to the loss of about 15 human lives every year. The compensation paid by the government has increased from a figure of Rs. 1.5 lakhs (1 lakh = Rs. 100,000) in 1986 to Rs. 13 lakhs in 1991 and an estimated figure of Rs. 30 lakhs during 1992. Serious thought needs to be given to this issue to avoid a potential setback to the PFM program.

Sal Leaf Harvesting

It has already been mentioned that one of the major NTFPs collected from the forests under PFM is sal leaves, which are used for making disposable leaf plates and cups and as packing material. Sal leaf harvesting has come under criticism from some senior foresters because it reduces photosynthesis and the generation of wood biomass, and thus has an adverse impact on the health of the trees. It may also deprive the forests of leaf litter and reduce humus in the soil. This in turn reduces the activity of microflora and their ability to recycle phosphates to the trees (Chaturvedi 1992).

Deb (1990), however, reports that the total number of leaves on a sal tree is 12–20 times the suitable leaves within reach of an average harvester, and that the actual harvest is even less as many suitable leaves within reach are overlooked (approximately 90% are harvested). Insects were found to reduce tree productivity more than leaf harvesting. In the absence of insect damage, leaf harvesting, especially before the rainy season, may have beneficial effects similar to pruning. Thus the impact of leaf extraction on sal trees has yet to be established firmly and a threshold limit needs to be determined.

Short Coppice Rotation

There is a growing debate over the impact of silvicultural practices adopted under PFM on the sal forests of SWB. This debate centers on the sustainability of the current short rotation management system. At present, sal forests are harvested every 10 to 15 years, with regeneration from the coppiced stumps. Foresters against this short rotation system argue that it leads to drier soils and erosion because of excessive exposure of the forest floor. They also claim that only low-quality sap wood, not heartwood, is produced in the first 15 years. Other foresters argue that sal forests should not be coppiced at all, because fungi-infected stumps often lead to poor quality regrowth.

Proponents of the current management system, however, point out that it was adopted only after the failure of earlier management systems, which had led to complete forest destruction. They argue that only when there is a steady flow of wood (i.e., from the short rotation coppice system) will there be an incentive for forest protection. They also maintain that this system is completely sustainable if correct management practices are followed. If this is true, the short rotation system represents an inexpensive and widely replicable technology that could be used to reforest much of India's 18 million ha of degraded forest land (Guhathakurta 1992).

Socioeconomic Impact

The PFM program has affected the participating communities in significant ways. A few are discussed below.

Employment Pattern. The PFM program has had a positive impact on the employment availability for the local communities, as has already been described.

Involvement of Women. Women play a key role in management of natural resources as well as in the domestic economy (Guhathakurta et al. 1992). In tribal communities, where women are especially active in forest product collection and processing, their contribution to the subsistence and cash generating needs of the family is substantial (S. B. Roy 1992). In one case, 71 species were exclusively collected by women, as compared to 23 species exclusively collected by men and 10 collected by both (Deb as quoted by Chatterjee n.d.).

Women maintain closer physical ties with the natural resource base, thereby acquiring an in-depth knowledge and understanding of the properties and uses of various species. The impact of degradation is felt more by women who reside permanently in the area, whereas men are recognized as intermittent dwellers (Chatterjee n.d.). Deforestation means longer excursions into the forest in search for fuel and fodder, and closure of forests for protection also affects women more adversely than men. In addition, enforcement must be handled differently in the cases of men and women. In the Indian context it is very difficult for a male (even forest department staff) to take strict action against a woman offender. Any use of physical force can bring accusations of molestation and a violent reaction from the kin of the woman.

This problem can be resolved to a great extent if women are involved in greater number in FPCs and if they also start taking an active part in patrolling (Chatterjee n.d.). Recognizing these problems, the government of West Bengal modified its earlier resolution to emphasize women's involvement in the program. The concept of joint membership was introduced, which provides that if the husband becomes a member of the FPC, his wife automatically becomes a member. However, the situation has not improved significantly despite these legislative measures. A survey in Midnapore district shows that only 3–6 percent of the FPC members are women. Women reported that they participate less in the program because it is considered to be the men's domain and men represent the views of the entire household. No specific attempts were made by the authorities to involve women. It was felt by the women members that more women should become members of the executive committee of the FPC (Guhathakurta and Bhatia 1992).

Women's noninvolvement in the FPC can be attributed to two causes: social constraint and physical constraint. The former is mainly due to traditional conventions and beliefs which prevent women from attending meetings and sharing knowledge in the presence of the elder male members of their family. The physical constraint is imposed by the multifarious duties of women in the household, on account of which they rarely find time to attend meetings. Also, only male members are invited by the beat officer to the meetings (Chatterjee n.d.).

Efforts to improve women's participation have been initiated, and even some all-women FPCs have been constituted (e.g., Brindabanpur Women FPC in Bankura, which has been formed with women who were earlier forest offenders). However, more efforts are required at this stage to translate legislative measures into field reality.

Institutional Changes

A remarkable degree of attitudinal change is occurring in the people as well as in forest department personnel in West Bengal. In this process, various institutional mechanisms have evolved.

Forest Protection Committees

The Indian Institute of Bio-Social Research and Development (IBRAD), an association of anthropologists and sociologists who had previously worked with the state health department to strengthen participation in community health programs, is documenting the ways in which the FPCs emerge, their impact on communities, and the problems faced by participating villagers and local foresters. A survey of 42 FPCs in West Midnapore division conducted by IBRAD found considerable heterogeneity among FPCs in terms of ethnic composition, number of villages per FPC, proportion of village households participating in the program, and effectiveness of the FPC. The ethnic composition of the villages appears to influence the effectiveness of the FPC. In general, tribal communities are better managers of forests than other caste groups. This may be due to their greater knowledge of the forests as well as greater economic and spiritual ties with the forests. Another factor is that the social organization among tribal people is generally based on community institutions. Also, single village FPCs are generally more effective than FPCs comprised of two or more villages (B.K.B. Roy 1991).

FPCs have developed a number of institutional arrangements for protecting and managing their forests. Several protection and patrolling systems have evolved in different areas. Generally, patrolling is done on a voluntary basis by small teams in which almost all active male members of the FPC must participate by turn. Women generally do not take part in patrolling, but they do participate whenever women from surrounding areas are involved in illicit felling. Boys who herd cattle also keep a vigil on the forest for many FPCs. In a few cases, paid guards were also kept, but the system was not found to be very successful (Malhotra and Poffenberger 1989).

A number of systems to impose fines have been established by FPCs to check illicit felling. Many FPCs follow a system in which an offender is simply warned in the case of minor offenses, but for major offenses the FPC hands over the offenders to the forest department or even to the police. In some FPCs, a system of monetary fines and social sanctions is also followed, often with a different set of norms for members and outsiders. However, heavy monetary fines are generally not imposed because FPC members believe that imposition of heavy fines could result in retaliatory cutting by the offender (Malhotra and Poffenberger 1989). Most confrontations with outsiders occur during the first and second years of protection, after which the restrictions and rights of the protecting village are generally recognized (Poffenberger n.d.).

Forest Department

A shift from purely departmental functioning to participatory management has brought about a series of new responsibilities for the forest department, such as organizing people, microplanning, and implementation of support activities. Repercussions are seen in terms of the change in work burden on the one hand and acquisition of new skills on the other. Earlier, the index of success was determined by physical and financial parameters; institutional or social parameters were not given much importance. This approach has changed with the implementation of the PFM program. Forest staff have responded positively, and they feel that PFM is the only chance to restore the health of the forest (S. B. Roy 1992).

A unique feature of West Bengal's PFM program is the emphasis on microplanning. Detailed guidelines on microplanning direct the forest staff to adopt a flexible process of planning, intimately involving local people and their needs in the program. Information regarding drinking water, roads, and other public facilities is collected, and an attempt is made to meet the most pressing needs either through the forest department or through programs of other departments. There is, however, still a tendency among many forest department personnel to look at microplanning as a form-filling exercise rather than a true participatory understanding of problems and solutions.

Although the major emphasis of the PFM program is on meeting the needs of the local communities, the forest department has also gained financially from the program. The forest department now receives revenues from areas where previously little or nothing was obtained. In the Arabari area, for example, the forest department received a net revenue of Rs. 702,413 from 97 ha of sal coppice forests and eucalyptus plantations in 1987–88. The figures for the 1988–89 to 1990–91 period from the same area show the forest department earned Rs. 5,764 per ha as its share of revenue on an additional 84 ha of forest and plantation. The revenue data of the Bankura district shows that it increased from Rs. 2,448,347.90 in 1987–88 to Rs. 4,215,764.45 in 1991–92 (Jha n.d.).

Nongovernmental Organizations

The PFM program provided an opportunity for NGOs to get involved far more closely in forestry through microplanning, information dissemination, training, and research. There is room, however, for NGOs to play an expanded role in the PFM program.

IBRAD has developed a "sensitization training" methodology which has been found effective in orienting foresters as well as village communities toward PFM. Through its training programs, IBRAD is preparing field officers of the forest department for their new role.

Rama Krishna Mission (RKM), a leading rural development organization in West Bengal involved in vocational training programs, has provided support teams to work with foresters and villagers in the design of resource development plans. RKM is also helping the forest department explore alternative self-employment activities such as silkworm rearing and cultivation of cashew trees. The Center for Women's Development Studies is engaged mainly in studies of gender issues and employment generation activities in West Bengal.

The NGOs are screened by the forest department before they can participate, as the program is conducted on state-owned forest lands. There is a provision for screening of NGOs at the state, regional, and district levels (V. K. Singh 1993). In the Bankura district, for example, of a total of 38 NGOs, three have been identified as district NGOs and one as a training center (Jha n.d.). Three state level NGOs are also part of the Working Group which has been constituted in an effort to institutionalize partnerships between the forest department, NGOs, and the community. It allows the forest department, including senior officers, to obtain inputs directly from NGOs and academicians while making policy decisions (S. B. Roy, 1991).

Another positive impact of the program has been the involvement of various academic and research organizations. Previously, forestry had been confined mainly within the realms of government departments. This has now changed, and institutions such as the Indian Institute of Technology, Indian Institute of Forest Management, North Bengal University, Indian Statistical Institute, and Indian Institute of Sciences are assisting the forest department by conducting research and developing management options for the new program.

The foregoing account shows that, starting as a small experiment in Arabari in the early 1970s, the PFM program has expanded in the last twenty years, especially since 1989, and has been eminently successful in regenerating vast areas of sal forests in SWB. In its evolution, the program has also become more multifaceted and multidisciplinary. The program's consultative and participatory approach has helped to reorient the forest department and the village communities to work in harmony and to their mutual advantage. The lessons learned in the implementation of the program are expected to be of great use in extending PFM to North Bengal, the Darjeeling Hills, and the Sundarbans region.

Emerging Issues

The PFM program has raised a number of issues which have relevance and implications beyond SWB. These issues can be broadly classified into three categories: (1) forest management issues, (2) marketing related issues, and (3) institutional issues.

Forest Management Issues

The experience in West Bengal demonstrates that sal forests have a remarkable ability to regenerate if protected. The crucial aspect for PFM success is the maintenance of interest by local communities in forest conservation over the long run. This requires a continuous flow of benefits from the regenerating forest, which comes from multiple products of the forests (rather than timber alone). The need for outside support (e.g., employment) is greatest in the early period when access to the forest is restricted and the people are adjusting to the new system of management. Thus there is a need to shift from timber to multiple product management. In this light, questions of rotation, optimal harvest of sal leaves, and management of NTFPs need to be addressed. Systems integrating timber with NTFPs need to be developed, and the approach must differ according to ecological and socioeconomic conditions. In areas where sal stumps have also been extracted for fuelwood, flows of forest products will regenerate more slowly. In such areas, enrichment planting and other employment generation activities will be required. Similarly, the approach must be different where the forest area per household is low. In a village where forest area per household is 0.5 ha or less, people may have little incentive to protect the forest unless some additional benefits are provided to them. The forest department should focus its research on development and management aspects of these new production systems.

Marketing Issues

It has become obvious that economic incentives play an important role in sustaining the interest of people in forest conservation activities. Inefficient marketing systems reduce the returns to the people. The present system seems quite deficient as the earnings from the forest products do not adequately compensate people for their efforts. Nationalization of certain NTFPs has, in fact, reduced the number of legal buyers, choked the free flow of goods, and increased the exploitation of primary collectors by unscrupulous private traders. The government's decision to grant monopoly rights of collection and disposal of kendu leaves and sal seeds to WBTDCC in 1980 has been of little or no benefit to communities.

Because NTFPs, unlike timber, are not shared with the government, their importance to local people is relatively enhanced. There is a need to adopt an integrated approach at the village level, reduce the number of intermediaries, and shift from elimination to regularization of private trade. FPC members should be encouraged to operate as a group to improve their bargaining power.

Another important issue is the marketing of products from the sal forests. There is no assured market for sal poles, and these will have to compete with eucalyptus and *Acacia auriculiformis* poles which are coming into the market because of the maturation of social forestry plantations. If the market collapses, people will lose interest and the program might be set back. Market research is needed to understand market forces and the commercial potential for various farm forestry products and NTFPs. There is an urgent need to develop market-related skills in the forest department.

Institutional Issues

It is becoming increasingly clear that the PFM program can be sustained only if strong institutions are developed. This can be achieved if sufficient attention is paid to crucial issues such as equity, empowerment of local communities, reorientation of forest department staff, gender roles and needs, and linkages among various organizations.

The village is not a homogeneous entity. Mechanisms have to be developed to ensure that the benefits and costs are distributed equitably within the FPC. The benefits relinquished by the communities from the hitherto open-access forests affect various segments differently. In many places, women are having to travel further to collect the same amount of fuelwood or shift to inferior fuels such as leaves. The disadvantaged sectors need to be compensated adequately by providing employment and intermediate benefits until gains from the regeneration start flowing. The questions of intervillage equity, of late entrants and their usufructuary rights, of differences in per household forest area from FPC to FPC, and of the shift of pressure from FPC to neighboring nonFPC areas also need attention. A transparent system of equity in terms of sharing responsibilities and benefits is the *sine qua non* for effective community participation in the long run. Transparency is also essential in financial matters. The system of sharing between the department and the FPC and within FPCs should be simple and clear to everyone. At present the forest department shares only *net* revenue with the people and deducts all its operational costs. The people may not understand the complex calculation of administrative costs and may develop mistrust toward the forest department. It would be better if the sharing were done on a gross returns basis. The issue of the relationship between Panchayats and FPCs is also an important one, especially since the introduction of the new Panchayati Raj Act by the government.

The question of preexisting user rights in the forests is also very important. In many places, communities and Panchayats already possess a range of usufructuary rights over forests granted under earlier forest acts. There is a danger of conflict if new rights are granted to a particular FPC (it may represent only a few of the original right holders) without considering the existing rights. There may not be a problem at the time of FPC formation because the forest is degraded, but once benefits start flowing, the problems are likely to emerge.

Up to now, the forest department has adopted the PFM approach only in degraded areas. Involvement of people in management of better stocked areas also needs attention. It should not appear that people are being merely used for regenerating state forest lands; there should be a concept of equal partnership. The community should be empowered to make decisions. Under the present arrangement, the forest department has an upper hand and can even dissolve the FPCs. There is a need to enable the FPCs to develop their own identity and to remove the feeling that their existence depends largely on the goodwill of the forest department.

There is also a need for greater sensitivity to gender issues. Although a beginning has been made by modifying the government order, a conscious effort is lacking at the field level. Women should be involved in the FPC meetings, where their participation should be actively encouraged. As it is difficult for rural women to interact freely with male staff members, more female officers and staff should be recruited. At present the West Bengal Forest Department has only three women employed as officers and none engaged as field staff. This has proved to be a major constraint in soliciting participation by women as well as in dealing with women offenders.

The changed system of management has brought new responsibilities for the forest department. There is a need for greater efficiency within the department and for training and reorientation to acquire new skills. The number of FPCs in a forest ranger's beat in West Bengal varies from 1 to 30 or more. For effective linkages, a beat officer needs to be in close contact with each FPC once a week. This implies that the number of FPCs he can effectively handle is six. It is imperative that the beats and ranges be reorganized into smaller units (Palit 1993). There is a need to introduce more flexibility in the functioning of the department, and toward this end the staff performance appraisal parameters need to be suitably altered.

Another emerging issue is related to microplanning. Microplanning is an interactive process which should focus on finding solutions to local problems instead of unduly emphasizing procedural aspects. Community needs identified through the microplanning process often require involvement by other government departments. In PFM areas the forest department will inevitably have to play a lead role in influencing other departments to respond to these needs. Involvement of NGOs and Panchayats is yet another issue of importance. On the whole, there is a need to develop institutional arrangements through which the forest department can play a greater role in affecting coordination and consultation among the relevant agencies at the local level than is presently the case.

Conclusions

The West Bengal experience has shown that the process of natural forest degradation can be reversed. It has been demonstrated that technical issues, though important, are not paramount. The regenerating sal forests are able to produce NTFPs on a continuous or seasonal basis which can be sustainably harvested by people for consumption and sale. If appropriate economic returns are available, local communities can effectively take on forest management responsibilities. These in turn depend on an efficient marketing system for forest products. If people are empowered and there is a shift in orientation of government resource management departments, regenerating forests, increasing biodiversity, and sustainable harvests can provide the right economic incentive to people to conserve their forests.

References

Arora, H., and A. K. Khare. 1994. Experience with the recent joint forest management approach (draft). New Delhi: Society for Promotion of Wastelands Development.

Bagchi, S. K., and M. Phillip. 1993. *Wastelands in India: An untapped potential.* New Delhi: UPALABDHI.

Bhowik, K. L., M. G. Som, and D. K. Dasgupta. 1988. Status of production of foodgrains in West Bengal. In *Agricultural Situation in India* 13(8): 507–508. New Delhi: Directorate of Economics and Statistics, Department of Agriculture and Co-operation, Ministry of Agriculture.

Chandra, N. S., and M. Poffenberger. 1989. Community forest management in West Bengal: FPC case studies. In *Forest regeneration through community protection: The West Bengal experience,* ed. K. C. Malhotra and M. Poffenberger, 5–8. Calcutta: West Bengal Forest Department.

Chatterjee, M. n.d. *Women in joint forest management: A case study from West Bengal.* Technical paper no. 4. Calcutta: Indian Institute of Bio-Social Research and Development (IBRAD).

Chaturvedi, A. N. 1993. Sustainability of community participation in forest management. *Wastelands News* 8(2):53–56.

Choudhary, K. 1994. Going for broke: An economic crisis in Bihar. *Frontline,* February 25, 1994.

Das, J. K. 1993. *Sal leaf plate industry in West Bengal.* JFM Study Series. Bhopal: Indian Institute of Forest Management.

Deb, D. 1990. *An estimation of the possible effect of community leaf harvesting on the productivity of sal.* Calcutta: IBRAD.

Dey, S. C. 1992. Asian elephant and its status in South West Bengal. *West Bengal* 34(19):399–404. Calcutta: Government of West Bengal.

Dutta, M., and M. Adhikari. 1991. *Sal leaf plate making in West Bengal: A case study of cottage industry in Sabalmara, West Midnapore District.* Working Paper no.2. Calcutta: IBRAD.

Forest Survey of India. 1993. *The state of the forest report 1993.* New Delhi: Ministry of Environment and Forests.

Gadgil, M. 1985. Social restraints on resource utilization: The Indian experience. In *Culture and conservation,* ed. J. A. McNeely and D. Pitt, 135–154. London: Croom Helm.

Gadgil, M. 1989. The Indian heritage of a conservation ethic. In *Conservation of the Indian heritage,* ed. B. Allehin, E. R. Allehin, and B. K. Thapar, 13–22. New Dehli: Cosmo.

Gadgil, M., and K. C. Malhotra. 1983. Adaptive significance of the Indian caste system: An ecological perspective. *Annals of Human Biology* 10:465–478.

Gangopadhyay, S. 1991. West Bengal. In *Agroclimatic zone specific research: Indian perspective under NARP,* ed. S. P. Ghosh, 438–462. New Delhi: Indian Council of Agricultural Research.

Government of India. 1989. *India statistical abstract 1989.* New Delhi: Central Statistical Organization, Department of Statistics, Ministry of Planning.

Government of India. 1991. *Census of India.* New Dehli: Registrar General and Census Commissioner.

Government of West Bengal. 1988. *Technology manual for forest management with people's participation.* Calcutta: Social Forestry Wing, Forest Directorate, West Bengal.

Government of West Bengal. 1994. *Role of forest protection committees in West Bengal.* Calcutta: Department of Forests, Government of West Bengal.

Guhathakurta, P. 1992. Is management of coppice sal forests on short rotations sustainable? *Wastelands News* 8(1):31–33. New Delhi: Society for Promotion of Wastelands Development.

Guhathakurta, P., and K. S. Bhatia. 1992. A case study on gender and forest resources in West Bengal. Unpublished.

IBRAD. n.d. *Study of vegetation dynamics in regenerating sal forest of Moupal Beat.* Working paper no. 4. Calcutta: IBRAD.

Jha, A. K. n.d. Experience in joint forest management and microplanning in Bankura South Division. Unpublished.

Khare, A. K. 1993. *Forest product marketing.* Working paper no. 4. Madhya Pradesh Integrated Forestry Sector Project. Washington, D.C.: World Bank.

Lurie, T. 1991. Saving the forests: India's experiment in co-operation. *The Ford Foundation Letter* 22(1). New York: Ford Foundation.

Malhotra, K. C. 1991a. *People, biodiversity and regenerating tropical sal forests in West Bengal, India.* Working paper no. 7. Calcutta: IBRAD.

Malhotra, K. C. 1991b. People, biodiversity and regenerating tropical sal (*Shorea robusta*) forests in West Bengal, India. Paper prepared for the International Symposium on Food and Nutrition in the Tropical Forest. Biocultural Interactions and Applications to Development, UNESCO, Paris, September, 1991.

Malhotra, K. C., and D. Deb. 1991. *History of deforestation and regeneration/plantation in Midnapore District of West Bengal, India.* Working paper no. 13. Calcutta: IBRAD.

Malhotra, K. C., D. Deb, M. Dutta, T. S. Vasulu, G. Yadav, and M. Adhikari. 1991. *Role of non-timber forest produce in village economy: A household survey in Jamboni Range, Midnapore District, West Bengal.* Calcutta: IBRAD.

Malhotra, K. C., and M. Poffenberger, eds. 1989. *Forest regeneration through community participation: The West Bengal experience.* Calcutta: West Bengal Forest Department.

Mishra, A., and N. K. Sinha. 1992. *Study of marketing of nal, kash, and hogla in the forest-based ecosystem of North Bengal Dooars in respect to joint forest management.* Bhopal: Indian Institute of Forest Management.

Moench, M. 1991. *Training and planning for JFM: Sustainable forest management.* Working paper series. New Delhi: Ford Foundation.

Pachauri, R. n.d. *Sal plate processing and marketing in West Bengal.* Working paper no. 12. New Delhi: Ford Foundation.

Pal, S. K. n.d. *Status of NTFPs collection and marketing at the primary collectors and wholesalers level: A case study of Raigarh FPC in Ranibandh range of South Bankura Forest Division of West Bengal.* Narendrapur: Ramakrishna Mission Lokaşiksha Parishad.

Palit, S. 1989. Present status of forest protection committees. In *Forest regeneration through community participation: The West Bengal experience,* ed. K. C. Malhotra and M. Poffenberger. Calcutta: West Bengal Forest Department.

Palit, S. 1990. Sal forest from feudal lords to forest protection communities. Paper presented at the workshop on Sustainable Forestry, September 10–12, 1990, New Delhi.

Palit, S. 1993. *The future of Indian forest management: Into the twenty-first century.* Working paper no. 15. New Delhi: Society for Promotion of Wastelands Development.

Poffenberger, M. 1990. *Joint management of forest lands: Experiences from South Asia.* New Delhi: Ford Foundation.

Poffenberger, M. 1993. The resurgence of community forest management in Eastern India. In *Natural connections: Perspectives in community-based conservation,* ed. D. Western and R. M. Wright, 53–79. Washington D.C.: Island Press.

Poffenberger, M. n.d. *Joint forest management in West Bengal: The process of agency change.* Sustainable Forest Management Working Paper Series no. 9. New Delhi: Ford Foundation.

Roy, B.K.B. 1991. *Arabari experience: A model for forest management with people's participation.* Calcutta: West Bengal Forest Department.

Roy, S. n.d. Participatory forest management in south western tract of West Bengal. Unpublished.

Roy, S. B. 1991. *Forest protection committees in West Bengal, India: Emerging policy issues.* Working paper no. 6. Calcutta: IBRAD.

Roy, S. B. 1992. Forest protection committees in West Bengal. *Economic and Political Weekly* 27(29):1528–1530. Bombay.

Saxena, N. C., and M. Gulati. 1993. *Forest management and recent policy changes in India. Proceedings of workshop on policy and legislation in community forestry, January 27–29, 1993,* ed. K. Warner and H. Wood, 121–128. RECOFTC report no. 11. Bangkok: Regional Community Forestry Training Center.

Sen, S. n.d. *A note on the socio-economic forest managed at Arabari, Midnapore.* Midnapore: Divisional Forest Office, Silvicultural Division.

Shah, T. 1988. Gains from social forestry and lessons from West Bengal. *Wastelands News* 3(3):17–24. New Delhi: Society for Promotion of Wastelands Development.

Singh, K., and S. Bhattacharjee. 1991. *Privatisation of common pool resources of land: A case study in West Bengal.* Anand: Institute of Rural Manangement.

Singh, V. K. 1993. *Operational guidelines for participatory management of degraded forests in West Bengal.* Calcutta: Directorate of Forest, Government of West Bengal.

Stebbing, E. P. 1982. *The forests of India,* vol. 2. New Delhi: A. J. Printers.

Suneja, V., S. B. Roy, and J. K. Das. 1992. *Marketing of mushrooms.* Working paper no. 14. Calcutta: IBRAD.

THREE

Mexico's Plan Piloto Forestal

The Search for Balance between Socioeconomic and Ecological Sustainability

Michael J. Kiernan and Curtis H. Freese

Natural forest management for timber and other goods and services is increasingly proposed as an option for conserving the biodiversity of tropical forest ecosystems (Frumhoff 1995; Palmer and Synnott 1992; Perl et al. 1991). The proposal is based on a simple premise: forest owners and managers have a vested interest in maintaining the forest if they receive tangible and secure economic returns from it. The relative species abundances and overall species composition may shift compared to the "native" forest, but this is deemed preferable to a complete change in land use that might entirely eliminate a diverse forest ecosystem.

Though this logic seems both simple and compelling, there are few examples of successful natural forest management in either the tropics or temperate regions (Poore et al. 1989; Botkin and Talbot 1992). Further, we poorly understand to what degree biodiversity is important for long-term forest productivity, or how biodiversity may be affected by natural forest management methods (Frumhoff 1995; Hansen, Chap. 6). Thus two broad questions confront attempts at natural forest management: (1) What ecological and socioeconomic conditions are a prerequisite to successful natural forest management? (2) Can natural forest management provide an effective means to offset the opportunity costs of alternative land uses, while at the same time remain compatible with the goals of biodiversity conservation?

We address these and other questions through an analysis of one of the most promising, community-based natural forest management initiatives under way in the tropics today—the Plan Piloto Forestal of Quintana Roo, Mexico.

The Quintana Roo Forest

The tropical, semievergreen forest of the Yucatan Peninsula is divided among the countries of Mexico, Guatemala, and Belize. Its northeastern limits lie in the state of Quintana Roo, whose 5 million hectares (ha) occupy the northeastern portion of the Yucatan Peninsula. Like most of the peninsula, the soils of Quintana Roo are derived from a massive limestone platform. Its topography is flat and karstlike, with sinkholes distributed across the landscape. Only one river, the Rio Hondo, flows in the state, forming the southern border with Belize (Barrera de Jorgenson 1993).

The region has a generally humid climate with an annual mean temperature of approximately 25°C and average annual precipitation of 1,200 millimeters (mm). Rains are concentrated in the months of May to October, with a pronounced dry season from November through April (Negreros-Castillo 1991; Barrera de Jorgenson 1993).

Although 90 percent of Quintana Roo was forested at the beginning of this century, only 50 percent has been identified as being of potential forest use today (Flachsenberg 1993a). More than 100 tree species have been identified in the state, with an average of 40 species per ha. Densities range from 420 to 1,060 trees per ha with diameters greater than 10 centimeters (cm) at breast height (dbh) (inventories from Noh Bec ejido, H. Flachsenberg, pers. comm. 1994). The most common of the larger tree species include sapodilla (*Manilkara zapota*) and breadnut (*Brosimum alicastrum*). Mahogany (*Swietenia macrophylla* King) is among the largest of species, though not among the most common.

A Brief History of Forest Use in Quintana Roo

Pre-Columbian Use, the Ejido System, and the MIQRO Concession

The Yucatan Peninsula and nearby regions of Mesoamerica have been occupied by first the Olmec and then the Mayan cultures for at least three thousand years. The Mayan civilization reached its peak in this region around one thousand one hundred to one thousand seven hundred years ago, when its population density may have been as high as four hundred to five hundred individuals per square kilometer (km^2) in rural areas (Gómez-Pompa and Kaus 1990). That compares to a density of 9.6 per km^2 today (INEGI 1992). Evidence suggests that large areas of forest in the Yucatan were cleared by the Mayans for shifting agriculture and that the forests themselves were actively managed for a variety of plant and animal products (Gómez-Pompa and Kaus 1990; Gómez-Pompa 1991).

Following the Spanish conquest, Mayan groups continued as the principal inhabitants of what is now Quintana Roo. Mahogany extraction began as early as 1770 for the European market, which eventually led to concessions from the Mexican government to English and American companies in the southern portion of the territory which lasted until about 1914. These events helped forge the basic patterns of urban settlement, and land and resource use, that characterize Quintana Roo to the present (Galletti 1994).

In the 1930s, colonists from other regions of Mexico began moving into what is now Quintana Roo, drawn by the opportunities offered by the extraction of *chicle* from the sapodilla tree. Demand for Mexico's chicle grew steadily from the 1880s, with Quintana Roo as the primary source by the early 1900s. From then until the 1930s, the territorial population remained constant at about ten thousand inhabitants. With the onset of World War II and the promotion of chewing gum among troops, demand for chicle sharply increased, and the population of the state nearly tripled by 1950 (Barrera de Jorgenson 1993).

Meanwhile, throughout the 1930s and 1940s, the Mexican government distributed usufruct rights over "public" lands to colonists under the *ejido* system. Ejido lands were in turn allocated for use by and among the individual members, known as *ejidatarios*. Mexico's ejido system, established in the Mexican Constitution of 1917, is a form of communal land tenure that grew out of Mexico's revolution (1910–17). The main tenets of the system were that (1) land and natural resources belong to the nation, (2) use of the land and resources is granted to Mexican citizens either as private property or social property (ejidos and Indian communities), (3) land in ejidos and communities cannot be sold or transferred, and (4) corporations cannot own land. With potentially significant implications for forest management, in 1992 the constitution was amended to allow ejidos and communities to sell, lease, or mortgage their land, and to develop joint ventures with corporations (Vargas 1992).

As a general rule, ejido lands put to agriculture were allocated in lots of about 20 ha per ejidatario. In the case of Quintana Roo, however, where chicle extraction was the primary economic activity and little agriculture was practiced beyond the subsistence level, allocations of 400 forested ha per ejidatario were the rule. Hence settlement in Quintana Roo came to be characterized by large forested ejidos with low population densities, a critical precondition for the eventual development of the Plan Piloto Forestal. As the principal economic activity in these forested ejidos, chicle extraction was organized into cooperatives by the Mexican government in the 1930s, and by the end of the decade, 48 chicle cooperatives were operating in Quintana Roo (Galletti and Argüelles 1987). The Federación de Cooperativas de Quintana Roo (Federación) was organized in 1940 to group these under one "umbrella." Since then, the Federación has been a key player in the chicle industry of the state.

Three other events contributed significantly to the shaping of the Quintana Roo landscape in the late 1950s and early 1960s. The first of these occurred in 1954 when a 460,000-ha timber concession in the mahogany-rich central portion of the territory was granted to the state-owned enterprise, Maderas Industriales de Quintana Roo (MIQRO). The second event struck in the form of Hurricane Janet in 1955, leveling 330,000 ha of the richest forest in the concession. MIQRO then obtained new lands to the north and south of the affected area, resulting in the superimposition of the concession over most of the chicle ejidos (Galletti 1994).

The third event was a wave of new colonization in the 1960s, spurred by a Mexican government policy to promote large-scale farming and cattle ranching in the region. When this was over, the number of ejidos in the concession area had grown from about six to more than thirty, with many organized around agricultural production. Though the existing 400-ha ejido allotments based on forest-based chicle tapping were left intact, new ejidatarios were each granted the mere 20 ha deemed necessary for agriculture (Galletti 1989, 1992, 1994; Galletti and Argüelles 1987).

Meanwhile, MIQRO's harvesting of the concession continued. Though extensive and targeted to only two species (mahogany and cedar [*Cedrela odorata*]), MIQRO's extraction was based on a rough forest inventory, minimum diameter cuts, a cutting cycle of 25 years (the period of the concession), and the principle of annual harvesting blocks. An early attempt to incorporate lesser-known species in the harvest mix was abandoned. These management practices were applied in the ejidos with high-hectare endowments established before the concession. Today these are the most forestry oriented and economically profitable ejidos. In those ejidos established after the concession, and in which farming activities predominated, cutting by MIQRO often exceeded allowable annual amounts. Mahogany and cedar were usually harvested from the entire ejido in one or two years, and large tracts of the forest were cleared by the ejidatarios for agriculture (Galletti 1992, 1994; Galletti and Argüelles 1987).

Regardless of when they were established, logged ejidos received little or no benefit from the concession. Though MIQRO paid a stumpage fee into a special fund for the ejidatarios, access to the fund required the formulation of a socially oriented project for review and approval, which rarely happened. Any employment of ejidatarios by MIQRO was sporadic and low paying. With no incentives among the local players (local population, private enterprise, government officials) to conserve the forest, by the time the concession expired in 1983, half of the forested land under the concession in Quintana Roo had been cleared (Galletti and Argüelles 1987; Vargas 1992). Thus, in 1983, MIQRO and the government found little support from the forested ejidos for renewal of the concession. Rather, several important factors converged to set the stage for a new approach to forestry in the state, reflecting trends also under way at the national level.

A New Forestry Policy

Upon expiration of the concession, MIQRO was a financially and politically weak enterprise. Concurrently, the state government, led by a new governor, became increasingly receptive to the needs of its ejidatario constituency, perhaps as a result of important demographic changes occurring in the state. In the north, the development of tourism around Cancún was resulting in massive inflows of population and investment, and alternative (nontimber) sources of income to the state government.

Within this context, and aided by political and technical support from the Deutsche Gesellschaft für Technische Zusammenarbeit (GTZ) under an agreement with the Mexican government referred to as the Acuerdo México-Alemania (AMA), an alternative to the traditional, concession-based approach to forest extraction in the state was formulated. The opening of a political "space" by the AMA was a crucial precondition for helping ejidatarios assume more direct management of their forests under assistance from a technical team supported by the AMA (Janka and Lobato 1994). Hence, in 1983, the MIQRO concession was not renewed and the Plan Piloto Forestal was born.

The primary goal of the Plan Piloto Forestal is to empower ejido residents and increase the economic returns they receive from the forest. Forest conservation does not figure explicitly as the most important goal (Snook 1991; González Cortés 1993). As a foundation for the Plan Piloto Forestal, Mexican and AMA advisors worked closely with the ejidos to prepare comprehensive socioeconomic analyses of the ejidos, and helped gather information on timber markets. These served as the starting point for a negotiation process between ejidatarios and buyers of forest products, particularly timber products, that has been ongoing since. This process, in effect, allows the ejidatarios to decide how mahogany and cedar roundwood, sawnwood, and furniture products are produced and sold. Creating the socioeconomic and political conditions ("space") for such direct negotiation and independent decision making, largely free of government intervention, preceded any intensive efforts to inventory the forest resources (Janka and Lobato 1994). With time, this process has become more comprehensive as forest use intensifies and new tree species are added to the product mix.

The Structure of the Plan Piloto Forestal

The first 10 ejidos to participate in the Plan Piloto Forestal coalesced under a political aegis forged by the AMA and the governor of Quintana Roo to negotiate an effective twentyfold increase in the net value received from the sale of mahogany to MIQRO. The association of these 10 ejidos was first informal, but in 1986 they organized under a legally constituted association, the Sociedad de Productores Forestales Ejidales de Quintana Roo (SPFEQR). The total land area and Permanent Forest Areas of the SPFEQR are presented in Table 3.1. Permanent Forest Areas are areas of forest that must be designated by each ejido in the Plan Piloto Forestal for communal management exclusively for forestry production. The structure of the SPFEQR, and of

Table 3.1. The Ejidos of the Sociedad de Productores Forestales Ejidales de Quintana Roo (SPFEQR)

Ejido	Total area (ha)	Permanent forest area (ha)	Number of ejidatarios
Francisco Botes	18,900	5,000	290
Caoba	68,553	30,000	286
Chacchoben	18,530	6,000	320
Los Divorciados	12,000	5,000	86
Noh Bec	23,300	17,055	200
Nuevo Guadalajara	28,100	6,000	293
M. Avila Camacho	12,969	3,500	204
Petcacab	54,000	24,000	191
Plan de la Noria	9,500	5,000	52
Tres Garantias	44,520	20,000	103
Total	290,372	121,555	2,025

Source: H. Flachsenberg, pers. comm. 1994.

other forestry associations that other ejidos subsequently formed, consists of a General Assembly made up of all ejidatarios of the member ejidos (Flachsenberg 1993b).

Building on the experience of SPFEQR, as well as on sociopolitical events beyond the scope of this chapter, four additional associations of ejidos involved in forest management have been formed. After only 10 years of operation, there were five associations bringing together a total of 51 ejidos in the state, whose combined Permanent Forest Areas totaled roughly 500,000 ha (Argüelles 1993). The forestry associations and ejidos are so varied that generalizations are difficult. Some ejidos have generous endowments of forest land, which may or may not be rich in mahogany. Some produce railroad ties from lesser-known species, some produce chicle, and some are more dependent on agriculture than others. Finally, population densities vary widely.

Each association receives technical assistance in forestry from a Technical Unit staff that is hired and paid for, in large part, by the association itself out of revenues from timber production. The Technical Unit is, in essence, the equivalent of the state forestry service, though it is located administratively (and politically) outside of the Secretaria de Medio Ambiente, Recursos Naturales y Pesca (SEMARNP), the Mexican federal agency with jurisdiction over forests and forestry. Apart from the forestry association, each ejido has its own distinct and separate governing structure as the maximum decision-making authority for that ejido. The Technical Unit works with each of the member ejidos to plan and carry out forestry operations within that ejido (Flachsenberg 1993b).

A variety of sources fund this large and complex initiative. The official coordinator of the Plan Piloto Forestal is an employee of the state government and SEMARNP, with some support received from the AMA. Until recently, advisors of the AMA received funding from GTZ and the Mexican government. Operations of the forestry associations are funded primarily through the sale of timber, though staff of the Technical Units also receive support from a variety of sources. In 1990, about 40 percent of the total cost of the Technical Units was covered by the ejidos themselves. Other sources included the Mexican federal government (25%), the AMA (15%), the state government of Quintana Roo (10%), the Federación de Chicleros (5%), and various international foundations (5%).

Though the Technical Unit specialists officially carry out their responsibilities on behalf of SEMARNP, their administrative independence was attained by political maneuvering on the part of the AMA. Currently the five Technical Units have 32 technicians with an average coverage of roughly 12,000 ha of Permanent Forest Area per technician (Argüelles 1993).

We focus here on the experience of the first-established forestry association, SPFEQR, and its 10 constituent ejidos. We examine (1) the market forces that drive forest use in the Plan Piloto Forestal; (2) the management and harvesting of timber and chicle and their impacts on biodiversity and the forest ecosystem; (3) the sociopolitical context in which the Plan Piloto Forestal operates; and (4) the socioeconomic factors affecting its success. We conclude with a discussion of the Plan Piloto Forestal in the context of the goals of sustainable wild species use and biodiversity conservation.

Markets

As noted earlier, the Plan Piloto Forestal is essentially an inititative to gain control over several production chains organized around the delivery to market of value-added timber products. Access to associated revenues is sought incrementally, link by link. Concurrent with this, the Plan Piloto Forestal seeks to intensify use of the Permanent Forest

Areas by developing or accessing markets for new forest products. Here we review the markets for products from the tree species of the Plan Piloto Forestal ejidos. These are classified into three categories: (1) high-value hardwood timbers (mahogany and cedar); (2) all other softwood and hardwood timbers; and (3) chicle.

Mahogany

From the market's perspective, mahogany and cedar are essentially alike, and both meet largely with the same end uses. Of the two, mahogany is more abundant in the forests of Quintana Roo, but cedar is also present in limited quantities. Henceforth we refer only to mahogany with an implied reference to both species, a practice followed by the Plan Piloto Forestal.

Few other timbers are as valuable and sought after as mahogany, evidenced by a market that has existed for more than half a millennium. This popularity is based on the fact that, in many ways, mahogany is the most workable of all woods. It has extremely high dimensional stability and takes well to glue, producing the strong joints required in the manufacture of furniture. Mahogany is resistant to decay, has low moisture absorption, and is subject to minimum shrinkage, swelling, and warpage (Lamb 1966).

On the ejidos of the Plan Piloto Forestal, mahogany trees are harvested from the wild according to a minimum diameter restriction of 55 cm. Trees are felled, and then limbed and bucked on site to yield well-shaped boles that are clear of branches and deformities, suitable for sawnwood and veneer. Small limbs, branches, and misshapen boles are usually left in the forest. Boles are loaded onto trucks and removed to the sawmill where they are sawn to dimension or sliced to produce veneer. For the most part, both sawnwood and veneer are used in the production of quality furniture or in interior woodwork.

Given the strong market, ejidatarios have developed uses for sawnwood scrap and by-products in the production of furniture for the local market. Occasionally, scraps are also used for rough-cut fencing or for the construction of walls in ejido homes, and it is not unusual to see piles of mahogany scraps in front of homes, awaiting use as domestic cooking fuel.

While some mahogany-producing countries export the majority of their product, all sales of Quintana Roo mahogany occur within

Mexico to buyers located largely in Mexico City. Traditionally, the state of Quintana Roo has been the largest producer of mahogany in Mexico. The state's mahogany production in 1943 was 40,000 cubic meters (m^3), or fully 66 percent of the national production (Rendón Trujillo 1945). Today the state accounts for virtually 100 percent of all the mahogany legally harvested and produced in Mexico.

Since the export of roundwood from the state of Quintana Roo is prohibited by law, 24 sawmills in the state, some of which are portable, process the nearly 40,000 m^3 of roundwood harvested for all species. Of these, 13 belong to Plan Piloto Forestal ejidos (Plan Piloto Forestal 1993). Non–ejido-owned mills tend to be located away from the production forest, and therefore the source of roundwood, rendering them potentially less cost efficient than ejido-owned mills, which tend to be located close to the Permanent Forest Areas. Several of the non-ejido mills were equipped to saw the vast supplies of large-diameter mahogany that are no longer available. Because of the reduced supply of mahogany under the Plan Piloto Forestal, the Quintana Roo mills depend heavily on imports of mahogany from the forests of Belize and Guatemala, some legal, some not, with 45,000 m^3 of mahogany imported from these sources in 1992 (Plan Piloto Forestal 1993). In 1988, the price for mahogany from Quintana Roo was 15–20 percent higher than for imported mahogany from Belize and Guatemala—a reflection of the fact that the latter comes from largely unmanaged forest. Thus, in part, the conversion of much of Quintana Roo's forest to a more sustainable and community-based management system resulted in a transfer of more damaging mahogany extraction practices to neighboring countries (Galletti 1992).

A cubic meter of mahogany is currently priced at about US$230, delivered to the mill. After deduction of harvesting and transport costs, net income to the producer is about US$130 per m^3. With a sawmill efficiency rate of 50 percent, 1 m^3 can be sawn to yield about 230 board feet that in turn sell for between US$1.50 and US$1.75 per board foot. The net income per board foot is about US$0.88 following deduction of processing expenses. These levels of revenue are considerably higher than those of other timber species.

The price elasticity of supply for mahogany is low. That is, increases in the availability or supply have little dampening effect on the prices paid for the timber because no substitutes of equal quality are effectively available. In recent years, the large influx of imported mahogany to Quintana Roo from Guatemala and Belize has not lowered the prices obtained by the ejidos for sale of their product in Mexico, though it has precluded any increase. Even with the elimination of trade barriers under the North American Free Trade Agreement, and the potential of increased timber imports from the United States and Canada, it is unlikely that the price of mahogany would be affected, given the strong preference for this timber at the higher-end furniture and wood products markets.

The success of the Plan Piloto Forestal to date lies squarely with mahogany. A strong market for mahogany has meant that management of the Permanent Forest Areas is economically viable and attractive. Without mahogany, or another species just as valuable, it is doubtful that the Plan Piloto Forestal would have achieved much success in promoting natural forest management.

Other Hardwood and Softwood Timbers

In addition to mahogany and cedar, markets currently exist for about 20 tree species found in the forests of Quintana Roo. Based on physical properties of the wood, these are classified as either hardwoods or softwoods (not to be confused with temperate conifers). Of these 20, 15 are relatively abundant in the Permanent Forests, with two softwood species in particular, sac chaca (*Dendropanax arboreus*) and amapola (*Pseudobombax ellipticum*), having the strongest existing markets. Following that, lesser markets exist for about five of the hardwood species. When markets do exist, the parts of the species used and the process of physically harvesting and preparing them for processing is essentially the same as that of mahogany, though in the case of some hardwoods, only the heartwood is used. Overall, from 1986 to 1992, the demand for hardwoods and softwoods declined, with the result that annual production from Quintana Roo ejidos decreased more than threefold (Flachsenberg 1992).

In general, softwood species tend to have domestic markets and enjoy a greater market acceptance. Uses include the manufacture of veneer, plywood, and other laminants, cellulose, packing crates, pencils, toothpicks, and popsicle sticks. Softwood species also find use in interior construction. Most of the market for Plan Piloto Forestal softwoods is focused on *D. arboreus* and *P. ellipticum,* and virtually all of the amounts harvested can be sold. Sales in this case are limited by lack of sufficient abundance in the Permanent Forests.

Markets for hardwood species tend to be weaker, with lesser volumes in demand. Hardwood markets in Mexico have traditionally existed for end uses including interior and exterior construction, hardwood and parquet flooring, carpentry, boats, railroad ties, and tool handles. Potential does exist for increased domestic sales of lesser-known species for use in parquet flooring and wall paneling, but to obtain access to these markets the ejidos must increase the quality of their product while maintaining a level price. An AMA study found that the trend in use of hardwoods is toward substitution with pine, oak, and other materials, some of which is imported. This is due in part to hardwood prices that are 60 percent higher than pine on average, and to other problems such as inconsistent quality and supply (Foerster et al. 1992). Softwood imports under the North American Free Trade Agreement can be exported to reinforce this disparity. In one of the strongest markets for hardwoods, nearly 10,000 m^3 are harvested each year by ejidatarios to produce railroad ties for sale within Mexico. A total of 10 species are used, though chechem (*Metopium browni*) accounts for more than 80 percent of all consumption. Aside from the 20 species of softwood and hardwood for which some type of market exists, an additional 50 species are having commercial potential.

On average, lesser-known species sell for about US$0.77 per board foot in Mexico, and for about US$1.15 to US$4.00 per board foot on export (based on June 1994 sales contracts to buyers in the United States; H. Flachsenberg, pers. comm. 1994). Much of the effort to develop markets for lesser-known species is currently focused on the export of hardwood timbers to international markets, used primarily in the production of instruments, fine furniture, and paneling. Though overseas buyers will pay higher prices for lesser-known hardwoods, several obstacles face the ejidos in their development of these markets. Overseas buyers tend to have standards of quality that are difficult to meet. In many cases, the properties of the species are relatively unknown, or the skills and technology required to saw and dry the woods are not yet available locally. Since much of the milling capacity in Quintana Roo was set up for large-dimension mahogany, many mills are not equipped to process the smaller dimensions that are often required for lesser-known species (Plan Piloto Forestal 1993). As such, developing markets will require retooling and acquisition of new processing equipment.

Currently the Plan Piloto Forestal ejidos export seven to eight containers of hardwoods per year, or about 10 percent of the ejidos total hardwood production. Further development of hardwood production and marketing is caught in a vicious cycle. As long the ejidos lack the equipment required to offer a high-quality product, there will be few buyers and few sales; but as long as the sale of hardwoods represent insignificant income for the ejidos, they will not invest in better technology (Flachsenberg 1994).

Chicle

Chicle is the latex extracted from the sapodilla tree, *M. zapota,* a tall, shade-tolerant, broadleaf evergreen most easily recognized in the forests of Quintana Roo by the crisscross pattern of scars etched in the bark by chicle extractors. Chicle is a significant source of income for many ejidatarios (Ramírez Aguilar 1992); from 1988 to 1992 it represented 20 percent of revenues from forest production activities (A. Argüelles, pers. comm. 1994). Collected by ejidatarios known as *chicleros,* it is processed to remove excess moisture and sold to the Federación de Cooperativas Chicleras (Federación). Though chewed by the Maya and Aztecs, modern markets for chicle date back to the end of the last century when exports to the United States were initiated from the state of Veracruz, Mexico. The market structure has always been relatively simple, with few producers and buyers. Before World War II, the United States was the principal buyer, with Mexico and Belize the principal producers. Traditionally Quintana Roo produced about 50 percent of Mexico's total exports. With the development of a synthetic, petroleum-based substitute after the war, exports of chicle from Mexico to the United States came to a halt (Konrad 1987; Barrera de Jorgenson 1993). The annual average production in the state is about 400 tons, with a value of about US$1.5 million. Year-to-year production varies widely, largely due to problems of financing and internal organization among cooperative members and the Federación (Flachsenberg 1992).

Currently only four buyers in Japan and Italy account for all purchases of Mexico's natural chicle. Processed chicle is exported from Quintana Roo through a marketing system that is corrupt and inefficient. Chicle extraction occurs in about half of the ejidos associated with the Plan Piloto Forestal, with chicleros organized into cooperatives at the ejido level. With 3,500 members, the 27 cooperatives are grouped together into the Federación, which advances cash to chicleros for the purchase of supplies used in chicle tapping. The Federación sells most of the production to the state-owned entity, Impulsora y Exportadora Nacional (IMPEXNAL), for around US$6.00 per kilogram (kg). Net pay to the tapper, after deductions, is about 60 percent of this (Aldrete and Eccardi 1993).

Chicle is a newcomer to the Plan Piloto Forestal, which until recently had targeted only timber. Studies are now under way to understand the chicle market better and to improve its potential for providing economic returns to some ejidatarios. In this sense, it is interesting to note the difference in the organization of revenue sharing from the production of chicle and timber in Quintana Roo. Timber production is organized at the ejido level, and as such, all ejidatarios participate in the distribution of benefits. Chicle, on the other hand, is organized into cooperatives composed of individual chicleros within each ejido. Thus, whereas all ejidatarios benefit from the sale of timber regardless of whether or not they work in forest management activities, only those ejidatarios that are chicleros make an income from the sale of chicle.

The Management Regime: An Ecological Perspective

Forest Inventories

The Plan Piloto Forestal began its first three years, 1983–85, with only limited annual inventories of the Permanent Forest block to be harvested. The first objective was to win the interest and participation of the ejidos from an economic perspective. After three years, when the forest's economic value was more tangible to ejidatarios, detailed forest inventories became more feasible, both socially and financially.

The first detailed inventories were carried out by the SPFEQR ejidos from 1986 to 1989 at a sampling intensity of 2 percent. To promote ejido commitment and participation, the ejidatarios themselves collected the data in revolving brigades, and, under technical assistance from AMA staff, they helped process and analyze the data. A cutting cycle of 25 years was established using mahogany as the guide species, based on research carried out in Puerto Rico. Populations of all species were grouped into four diameter classes, and annual harvest volumes for mahogany, hardwoods, and softwoods were established. These volumes were used by the forerunner of SEMARNP to authorize extraction quotas.

To systematize harvest plans and inventories, each Permanent Forest Area is subdivided into Five-Year Blocks, each of which in turn is subdivided into five Annual Harvesting Blocks. Each Annual Harvesting Block is also subdivided into Minimum Extraction Units of 25 ha on most ejidos. Planning of all extraction and silviculture occurs at the Minimum Extraction Unit level.

The complex design of the initial inventory, coupled with lack of technical experience on the part of the data collectors, led to errors in estimating volumes. The resulting problems were first recognized on the Noh Bec ejido when, to obtain the quantity of mahogany authorized for a given year, the Annual Harvesting Block had to be expanded beyond the area originally defined. In 1992, a better-designed inventory was launched in Noh Bec in the second Five-Year Block. The new inventory was carried out by trained and permanent brigades, as opposed to rotating shifts, of ejidatarios (Argüelles et al., n.d.). Following this design, the Five-Year Block slated for harvest is inventoried at the beginning of the five-year harvest period, and an annual extraction quota for mahogany and other species is set for each Annual Harvest Block. This system helps spread out the harvest of commercial-size mahoganies over the five-year period within each Five-Year Block.

Harvest Volumes

SEMARNP and its predecessor government agency have authorized annual harvest volumes since 1983 for each ejido based on the inventory data for three principal categories: mahogany, hardwoods, and softwoods. The designation of a quota does not imply that the entire volume authorized will be harvested, since that depends primarily on the markets, species, and volumes for which purchase orders exist. In the case of mahogany, virtually all of the authorized volumes are harvested and sold. For hardwoods and softwoods, however, the total volumes authorized are seldom harvested. Volumes authorized and harvested from 1991 to 1993 by the 10 ejidos of SPFEQR are given in Table 3.2. The remainder of this section looks at each category and describes the potential impacts of harvesting on each.

Table 3.2. Volumes of Timber Authorized and Harvested (m^3) by the Ejidos of SPFEQR, 1991–1993

	1991		1992		1993	
	Authorized	Harvested	Authorized	Harvested	Authorized	Harvested
Mahogany	6,400	6,520	6,502	6,390	6,635	5,934
Softwoods	12,330	2,930	11,537	2,229	14,196	6,016
Hardwoods	8,402	4,529	12,420	3,282	25,859	4,115

Source: H. Flachsenberg, pers. comm. 1994.

Mahogany

Mahogany ranges from southern Mexico to northern Bolivia, with an eastward extension along the southern limit of the Amazon forest in Brazil. Of the three species of *Swietenia,* only *S. macrophylla,* the bigleaf mahogany, occurs in the forests of Quintana Roo. According to IUCN (1992), the conservation status of mahogany in Mexico is considered "rare." Most of the remaining commercial stands in Mexico are found in the Permanent Forests of the Quintana Roo ejidos.

Mahogany occurs in a number of different associations and tolerates a wide range of soil conditions, from well-drained alkaline to limestone uplands. As a shade-intolerant species, regeneration is best on sites exposed to full light and where leaf litter is absent (Lamb 1966; Snook 1993). Seedlings compete poorly with sprouts and advance regeneration seedlings, and appear to be favored by disturbed mineral soil (Snook 1993). This preference for disturbed sites and the frequency of even-aged stands indicate that mahogany is a colonizing or pioneer species. The best existing stands of mahogany in the forests of Quintana Roo appear to be the result of fire-induced disturbances. This suggests a management regime for mahogany that mimics large-scale disturbances, similar to those created by forest fires and hurricanes (Snook 1991, 1993). The wind-dispersed seeds of mahogany are not viable beyond one rainy season, and thus the location and timing of clearings or other disturbances relative to dispersal from seed trees may also be important in management. Growth rings appear not to be annual, and there are few data on growth rates in natural forest (Snook 1994). Little is known of its pollinating agents (Snook 1991, 1993).

As a general rule, mahogany is selectively logged throughout its range. The Tropical Forest Foundation notes that "*S. macrophylla* is sensitive to selective logging for several reasons: (1) selective logging removes mahoganies allowing competing vegetation to invade the liberated growing space; (2) a single harvest may remove all the seed trees from an area; (3) harvesting schedules frequently do not coincide with fruiting—trees are frequently cut before they have produced seed; (4) the stock of trees in the 10–45-cm class (stock for the next cut) are low due to high natural mortality rates; and (5) selective logging, because it focuses on the best specimens, tends to reduce genetic stock and decrease the ability to resist pest attacks and other disturbances" (Tropical Forest Foundation 1992, p. 2).

Harvesting Rates

Current extraction on the ejidos is scheduled according to a 25-year cutting cycle and implicit 75-year rotation (Snook 1991). All mahogany individuals found in the commercial age class (i.e., 55 cm and larger) are selectively harvested from each Minimum Extraction Unit. Over the first 25 years, all the mahogany individuals in the commercial class will be harvested on each ejido. It is assumed that by the beginning of the next 25-year cutting cycle, all the individuals in the next smaller class will have grown into the commercial class, and these will be harvested in the second cutting cycle. The process is repeated in the third cutting cycle to complete the 75-year rotation (Snook 1991).

The viability of the 25-year cutting cycle is dependent on the accuracy of the estimated growth rate of 0.7 cm per year. H. Flachsenberg (pers. comm. 1994) cites an average rate of around 0.6 cm per year, with site conditions exercising a strong influence on the actual rate (e.g., some stands in Noh Bec show a 1.1-cm annual rate of growth). Data gathered by Snook (1993) indicate that growth rates vary by age, with average rates quickly dropping to 0.3 to 0.4 cm per year, suggesting that cutting volumes may need to be further adjusted downward (Snook 1991, 1993; H. Flachsenberg, pers. comm. 1994). Snook (1993) contends that, based on her research, "over 90% of the mahoganies in this forest (Noh Bec) grow too slowly to attain the 55-cm commercial diameter limit over the course of the current 75-year rotation." She concludes that "if the objective of management is to produce 55-cm-diameter veneer logs, the rotation length must be nearly doubled from 75 to about 120 years. To provide for sustained yields, the volumes and areas harvested each year must be reduced in the same proportion."

Gap Size and Disturbance

Logging is mechanized on the ejidos. Chainsaws, skidders, and large trucks are used, and access roads are constructed. The average size of the gap opened by the removal of one tree is 80–90 square meters (m^2). Skid trails and log yards account for the opening of another 800–4,000 m^2 per ha. With an average of about 0.7 gaps per ha, harvesting of commercial species results in the opening of about 6 percent of the forest cover (Flachsenberg, n.d.). To ensure sufficient light for mahogany seedlings, the minimum size required of each disturbed area or gap has been estimated to be anywhere from 400 m^2 (H. Flachsenberg, pers. comm. 1994) to 2,000 m^2 (L. Snook, pers. comm. 1995). Thus the gaps created by tree removal and skidding are probably inadequate, in both size and effect on competitive vegetation, for mahogany regeneration, but the log yards probably are suitable (Snook 1993). In general, it appears that mahogany regeneration will be poor under current harvesting practices (Snook 1991, 1994; H. Flachsenberg, pers. comm. 1994).

The removal of all commercial-size mahoganies during the cutting cycle probably has a major impact on seed supply for regeneration. Though mahogany begins to produce seed at a relatively young age (roughly 30 cm dbh), these younger trees are generally not canopy emergents. This may limit wind dispersal of their seeds compared to older trees (H. Flachsenberg, pers. comm. 1994). Seed dispersal opportunities are further limited by the fact that mahoganies are generally removed before other commercial species, and thus the gaps created by removal of the latter are not available for seed dispersal from the former (Snook 1993).

To address the problem of mahogany regeneration, the ejidos are planting roughly 10 mahogany seedlings in the logging yards for every mahogany tree harvested. With a 70 percent survival rate, plus natural regeneration, and control of competing vegetation, these areas should lead to full replacement of the harvested population (H. Flachsenberg, pers. comm. 1994). An evaluation of enrichment planting by Negreros-Castillo (1994) in the Maya zone of Quintana Roo, however, revealed an average survivorship of 13 percent after three years for seedlings planted between 1986 and 1993. Regardless of survivorship, given that significant mahogany regeneration is at best limited to that 5–10 percent of the forest subject to logging, and that there is virtually no reproduction of mahogany in the 90–95 percent that is unlogged, vast areas of forest are excluded from commercial production (H. Flachsenberg, pers. comm. 1994). As Snook (1991, 1993) and Flachsenberg (1993b) note, migratory agriculture would be an excellent way to promote regeneration in this polycyclic forestry system, though its promotion within Permanent Forest Areas would likely be rejected by ejido members.

When grown in plantation, young mahogany trees are highly susceptible to attacks from the shoot borer, *Hysipila grandela*. Because of this, attempts to cultivate mahogany in plantations have often failed in the neotropics, though there are now some apparent successes in Puerto Rico and elsewhere. In Southeast Asia, where the shoot borer is absent, large mahogany plantations now exist (Figueroa Colón 1994).

Sustainability of the Offtake

It is unlikely that the current harvest of mahogany on the ejidos of the Plan Piloto Forestal is sustainable. Current levels and methods of harvest will probably not lead to a level of regeneration and repopulation that will fully replace the volume of mahogany being harvested during the current cutting cycles. The irony is that any lower level of offtake, without other means to create disturbance sites for mahogany regeneration, will also likely lead to lower recruitment and eventually a lower number of trees for harvest. Under lower or no harvest levels, and in the absence of other anthropogenic disturbance, one has to rely on the vagaries of hurricanes and forest fires to create a level of recruitment that would permit harvest levels as high or higher than current ones. This would depend on the presence of seed sources which, under current practices, are being largely removed.

The Plan Piloto ejidos, however, are moving in the direction of a more sustainable harvest through an incremental process of adjustment. When harvesting of mahogany first began in 1984, a total of 12,936 m^3 were harvested on the ten ejidos of the SPFEQR. Based on data from the initial forest inventories, volumes were adjusted downward in 1993 to 5,935 m^3. In comparison, 20,500 m^3 of mahogany were harvested by MIQRO in 1982. (Though the area of the MIQRO cut is not identical to the SPFEQR area, it is comparable.) Bringing about these changes, and future downward adjustments that will probably be needed in harvest levels, require the acceptance and commitment of the ejidos, also achieved incrementally. To have further reduced the level of offtake of mahogany at the beginning of the program may very well have destroyed any economic incentives for the ejidatarios to establish and maintain large Permanent Forest Areas. The crucial political support by the state government may have also quickly evaporated. To the extent that economic returns from other forest resources or values can be increased from each hectare of forest, consensus is more easily obtained for reducing the level of mahogany harvest, and the potential for undertaking management to improve mahogany regeneration is increased. A more intensive use of the Permanent Forest Areas is thus critical from both economic and ecological perspectives, leading again to the subject of lesser-known species.

Other Hardwoods and Softwoods

Both the economic and silvicultural viability of the Plan Piloto Forestal depend in large part on several lesser-known species of hardwoods and softwoods. The harvest of a wider range of species can create the larger gaps required to ensure the regeneration of mahogany and secure its role as the keystone species of ejido forest management. The creation of clearings is not economical, however, if the harvested individuals cannot be sold. Yet the economic feasibility of other management measures is even more remote.

Since mahogany is the "guide species" around which forest management in the Plan Piloto Forestal is organized, harvesting of other hardwoods and softwoods in each year should theoretically occur within the Annual Harvesting Block set for that year. Like mahogany, cutting volumes for other species are determined from the inventory data. Yet, when an order exists for a certain species, though harvesting may begin in the current Annual Harvesting Block for mahogany, it may continue into contiguous or noncontiguous blocks not slated for the current harvest, until the order for that species is filled. In this sense, the management regime in the Permanent Forest applies mainly to mahogany, and less so to other species. The degree to which harvesting is permitted in noncurrent blocks depends very much upon established policy of the individual ejido.

One overarching problem that faces the harvesting and management of lesser-known species is that harvesting permits are issued by category only, that is, for "softwoods" or "hardwoods," with species not specified. Ejidos can theoretically harvest any species for which purchase orders exist, up to the volume authorized for that *category* of wood, regardless of the actual abundance of the harvested species in the forest. As reflected in Table 3.2, though, the ejidos rarely harvest the full amounts authorized at present, a situation that may change if and when new markets for lesser-known species are developed.

The development of markets for lesser-known species is not without risk, and "species snatching" is a potential problem. Buyers are normally interested in securing large and regular quantities, orders which are not easily filled from a species-rich forest in which any one species generally occurs at a low density. Large orders of a nonabundant species can lead to the harvesting of virtually all individuals of that species. In the case of the Plan Piloto Forestal, siricote (*Cordia dodecanda*), machich (*Lonchocarpus rugosus*), and pukte (*Bucida burseras*), three of the most promising species from the perspective of new markets, are also among the rarest species in the forest.

Railroad tie production is also an activity with potential negative impacts on noncurrent harvesting blocks. Only two of the SPFEQR ejidos produce railroad ties, though the activity is very significant on many other Plan Piloto Forestal ejidos. The Permanent Forests of the ejidos are often physically located far from the ejido village or settlement, since these were the areas where good forest remained at the time of designation. Ties are harvested and extracted by ejidatarios by hand. While human-traction is potentially the lowest-impact type of harvesting, the ties must be transported to the nearest road on the back of the ejidatario, leading to a preference for extraction from areas close to roads, regardless of where the current Annual Harvesting Block of the ejido is located.

Another impact related to railroad tie production results from the Mexican government purchasing a set quota of ties, regardless of the volume that this entails in terms of extraction. For example, it may be that only 250 m^3 of hardwoods are authorized for extraction under the SEMARNP permits. If the government announces that it will buy 1,000 ties, the production of which might require extraction of 350 m^3, the ejidatarios harvest and produce the 1,000 ties, exceeding the annual allowable cut for that category. Also, a minimum diameter limit of 35 cm exists for all hardwood and softwood species, though given the normal dimensions of ties, trees with diameters as small as 24 cm are sometimes cut. Finally, though up to 10 species are normally sought for ties, 80–90 percent of all production is from chechem (H. Flachsenberg, pers. comm. 1994).

Sustainability of Offtake

As suggested above, the offtake of some lesser-known species may not be sustainable. Though inventories are carried out at the species level, management decisions are made by category. Managing by category of species (i.e., softwood and hardwood), as opposed to a species-by-species basis, clearly carries risks. Yet, at this stage of development of the Plan Piloto Forestal, both the limited biological information available for each lesser-known species and socioeconomic constraints require either the current approach to management or that there be little or no harvest of the few lesser-known species for which significant markets exist. Furthermore, as noted above, for both economic and management reasons there is a strong need to develop new uses and markets for lesser-known species, though sound information about the management needs of any species for which a market develops may lag years behind. Again, the short-term need for socioeconomic sustainability within the ejidos, and for political support and the consequent maintenance of the Permanent Forest Areas, is given precedence over long-term concerns about impacts on individual species.

Chicle

M. zapota is the most common commercial tree species in the Permanent Forests of the Plan Piloto Forestal, with up to 147 trees per ha (Barrera de Jorgenson 1993), and representing, on average, more than 30 percent of the basal area on some ejidos (Aldrete and Eccardi 1993). According to the Plan Piloto Forestal inventories, most *M. zapota* trees are small- to medium-sized, and trees with diameters larger than 45 cm are uncommon. Despite its abundance, *M. zapota*'s range in Mexico is primarily limited to the Yucatan forests because of deforestation elsewhere.

M. zapota trees are tapped during the rainy season from mid-July to mid-February, an activity that provides income during the period when timber harvesting has ceased. The extraction process begins with the setting of harvesting quotas by SEMARNP. Chicleros can tap up

Table 3.3. Sapodilla Abundance and Chicle Production in the Ejidos of SPFEQR

Ejido	Sapodilla trees per ha	Annual production potential (kg)	Average annual extraction 1980–1992 (kg)
Francisco Botes	9.8	1,658	0
Caoba	25.5	12,106	3,603
Chacchoben	20.9	3,530	3,936
Los Divorciados	24.7	3,623	0
Noh Bec	34.4	33,830	20,564
Nuevo Guadalajara	21.4	3,133	0
M. Avila Camacho	14.7	2,506	0
Petcacab	15.7	13,375	19,280
Plan de la Noria	25.2	2,198	143
Tres Garantias	15.1	6,927	4,231

Source: Ramírez Aguilar 1992.

to the amount authorized. Chicle prices determine how much of the quota the chicleros will extract, with a high price normally leading to higher production.

Trees are tapped every five to seven years on a revolving basis (Aldrete and Eccardi 1993; Barrera de Jorgenson 1993). Though larger trees are preferred, trees with diameters as small as 12 cm are sometimes tapped (Barrera de Jorgenson 1993). However, nine ejidos of the Plan Piloto chicle cooperatives have agreed to a minimum of 30 cm dbh for tapping (H. Flachsenberg, pers. comm 1994). Chicle is an open-access resource to ejido members, and chicleros are free to tap anywhere in the ejido, with no limit on the number of trees they can tap. Rights over individual trees are not held by chicle tappers. To be a cooperative member and participate in a special type of security fund, a chiclero must extract a minimum quota set internally by the ejido.

Chicle production ranges from 0.7 to 2.0 kg per tree (Aldrete and Eccardi 1993; Barrera de Jorgenson 1993). According to Barrera de Jorgenson (1993), there are too few big *M. zapota* trees in the forests of Quintana Roo; many are old, and yields are declining. Table 3.3 presents the number of trees per ha on each of the SPFEQR ejidos, as well as annual chicle production potential and volume extracted from 1980 to 1992.

Despite the historic economic importance of *M. zapota,* little research has been done on the effects of tapping methods and frequency on chicle production in subsequent years, or on tree survivorship. Barrera de Jorgenson noted that, "if sapodilla trees are tapped properly, latex can be harvested every three to five years without apparently causing permanent damage to the trees. Chicle extraction, however, has not been managed on the basis of even the limited scientific information available. Rather, consumer demand has been the basis for establishing harvesting quotas" (Barrera de Jorgenson 1993, p. 69). Barrera de Jorgenson further noted that "properly tapped trees suffer no damage to the vascular cambium. . . . After one day's tapping, the tree is generally allowed to rest three or more years. However, if incisions are made too deeply, the tree becomes more susceptible to attacks by insects and fungi" (p. 79). A 1993 meeting of chicleros from Mexico, Belize, and Guatemala revealed no current management plans or reforestation programs for *M. zapota* in these regions (Aldrete and Eccardi 1993).

The Impact of Harvesting and Management on Biodiversity

Two facets of the current forest management program may affect forest biodiversity: (1) offtake and management that affect the abundance of different size classes of the commercially important species, particularly mahogany, and (2) forest openings and related disturbances that accompany tree cutting and extraction.

A major objective of the silviculture practiced in the Plan Piloto Forestal is to increase the relative abundance of mahogany. At what point such intensive management for mahogany would begin to reduce forest biodiversity (measured, say, by species richness) is not known. At its extreme, new silvicultural investments and technologies may lead to the development of mahogany and cedar plantations that replace natural forest, with impoverishing effects on biodiversity. On the other hand, plantations of mahogany and cedar could appear on cleared lands that have proved less suitable for agriculture, with potentially beneficial effects on biodiversity for the region. Both of these potential scenarios seem currently remote. Of more immediate concern is the reduced number of large mahoganies in the forest due to harvesting. Pollinators, seed predators, cavity hole nesters, and other organisms that depend on food, shelter, or other services provided by mature mahoganies may be affected. Snook (1993) reports that hole-nesting birds may already be negatively affected by the removal of large trees. Such effects, however, remain to be investigated.

Mahogany exhibits different ecotypes throughout its range (Bascope 1992; Figueroa Colón 1994), and it has probably adapted to the biophysical conditions of the Yucatan (Snook 1994). The selective removal of mahoganies or enrichment plantings of selected mahogany stock could affect the genetic diversity and adaptiveness of native stands in the Yucatan. However, the extent of any ecotypic adaptations by mahogany in the Yucatan, and the potential for genetic alteration due to current management practices, remain unstudied.

Roads and skidder trails, built to access cutting blocks, appear to have minor effects on overall forest structure. Natural tree falls in the forest of the Plan Piloto Forestal, particularly in the ejido of Noh Bec where a study of natural and logging gaps is being conducted (Dickinson 1993), are much less frequent and smaller in size than in wetter forests studied elsewhere. The relatively small scale of the additional openings created by logging operations (5–10 percent of the forest cover) might therefore be expected to increase structural diversity of the forest.

Dickinson (1993), based on studies in the Noh Bec ejido, reports that species composition of vegetation in natural gaps is similar to that found under a closed canopy, while species composition in logging gaps has a greater abundance of pioneer species. Natural gap regenerators tend to be from advance regeneration or trunk sprouts of topped or downed stems, whereas logging gaps appear to favor seedlings and root sprouts. If one discounts the one species (*Guettarda combsii*) that is most common in both natural and logging gaps, the next most abundant species in natural gaps all grow well under low light conditions. In contrast, in logging gaps the next three most abundant species do not do well in low light conditions.

A recent study of the effects of logging on migratory birds in the Noh Bec ejido has found no difference in the numbers or species composition of migrants between three unlogged control sites and two sites that were logged in 1992 and 1990. Nor does there appear to be any difference in the composition of resident species. Neither migrants nor residents appear to have obligate gap specialists in the forest of the northern Yucatan (including Quintana Roo). Tree fall gaps, however, may allow some species to sort out by different age/sex classes. For example, in continuous semievergreen forest, tree fall gaps are the main microhabitat of female hooded warblers (*Wilsonia citrina*), whereas wintering males usually occupy closed canopy forest (Lynch and Whigham 1995). In a similar study in Belize, selective logging of mahogany and Spanish cedar appeared to have little effect on forest bird diversity and the frequency of individual species (Whitman et al. 1994).

A major impact on forest wildlife sometimes results from the construction of roads that provide easy access to hunters. Ejidatarios of SPFEQR that we inteviewed and other individuals working in the Plan Piloto Forestal reported, however, that hunting pressures do not appear to have increased in the areas of road construction.

Regardless of future developments, currently, when viewed from the broader landscape perspective, the Permanent Forests of the Plan Piloto Forestal form a regional matrix of forest cover of some 500,000 ha that apparently harbors the full array of floral and faunal diversity indigenous to that region. In addition, they serve as buffers to the 680,000-ha Sian Ka'an Biosphere Reserve on the east and the 700,000-ha Calakmul Biosphere Reserve to the southwest. Though not entirely contiguous, these forests may also serve as a biological corridor between the two protected areas.

The forests of Quintana Roo provide a challenge to those who wish to define some benchmark of "native" biodiversity against which to measure anthropogenic change. The forest community one sees today is the product of more than three thousand years of often substantial human use and intervention, and of infrequent but severe natural catastrophic events in the form of hurricanes and fires. For the goals of biodiversity conservation, one can probably define a basic biodiversity measure, the native composition of species, that should be maintained in the region as a whole. Beyond that, however, conservation goals must accommodate an array of both anthropogenic and nonanthropogenic forces that have shaped and will continue to transform a highly dynamic system.

Socioeconomic Incentives and Controls

The most important motivating factors affecting forest management in the Plan Piloto Forestal are probably economic. Economic factors, however, depend in large part on the prevailing social structure and systems of revenue distribution. We examine here the social factors that characterize the Plan Piloto Forestal and follow this with an analysis of the economic benefits that accrue to ejidatarios.

Social Factors

Tenure is widely recognized as a key factor in forest management (Perl et al. 1991; Johnson and Cabarle 1993). The Plan Piloto Forestal serves as evidence of the importance of tenure in securing long-term commitments to resource management. Effective tenure implies control over land and resources, and with strong ejidatario control over new members and access to resources, settlement pressures from colonists are virtually nonexistent. Moreover, the phenomenal growth of Cancún to the north has served as a "job machine" that helps diffuse both internal and external pressures for growth in the ejidos. The effect that these factors have had on maintaining a low population density relative to the resource base has been key to the success to date of the Plan Piloto Forestal.

Another key social factor is the manner in which resource users are collectively organized to make decisions about production and, perhaps more important, the distribution of revenues. In contrast to regions where new colonists may have more individualistic approaches to securing their livelihood, the Plan Piloto Forestal was built on the ejido foundation with its decades-old approach to communal decision making. Rather than attempt to replace this with a structure organized expressly for forest production, the Plan Piloto Forestal worked within the ejido structure, with the participation of virtually all ejidatarios guaranteed and present from the outset. However, the participatory approach does not always provide a highly efficient form of decision making and organization for forest production (Flachsenberg 1993b). As the forest-based enterprises of the ejidos evolve, it is more evident that the ejidos may need to find ways to make decisions more quickly and effectively if they are to compete successfully.

Achievements of the Plan Piloto Forestal have brought the ejidos a degree of recognition both in and outside of Mexico, serving as a source of pride and motivation. In this regard, the provisional designation of "Well Managed" conveyed on selected ejidos by an overseas certifier has helped to reinforce progress to date toward achieving sustainability. On the economic side, a favorable designation could help to open markets for lesser-known species.

Economic Factors

There are several levels on which an ejidatario can participate in the economic returns from timber. The most direct are the net revenues

Table 3.4. Estimated 1993 Net Revenue (US$) from Sale of Roundwood by the Ejidos of SPFEQR

Ejido	Revenue from mahogany	Revenue from other species	Total revenue	Cost of production	Net revenue	Per capita revenue
Francisco Botes	59,000	71,847	130,847	N/A	N/A	N/A
Caoba	169,800	43,993	213,793	N/A	N/A	N/A
Chacchoben	18,400	105,260	123,660	64,821	58,839	200
Los Divorciados	51,800	54,753	106,553	42,875	63,678	341
Noh Bec	327,600	386,820	714,420	427,379	287,041	1,435
Nuevo Guadalajara	22,000	N/A	N/A	N/A	N/A	N/A
M. Avila Camacho	33,600	41,787	75,387	35,478	39,909	211
Petcacab	299,800	112,380	412,180	189,300	222,880	1,077
Plan de la Noria	N/A	N/A	N/A	N/A	N/A	N/A
Tres Garantias	205,000	109,833	314,833	187,371	127,462	1,214

Source: H. Flachsenberg, pers. comm. 1994.
N/A, not applicable.

from the extraction and sale of roundwood. In 1993, these ranged from a per capita income of US$200 for Chacchoben ejido to US$1,435 for Noh Bec. Net and per capita revenues from the sale of roundwood in 1993 are summarized for the 10 ejidos in Table 3.4.

Per capita revenues in Table 3.4 are not necessarily all in the form of cash distributions to ejidatarios. In ejidos where forestry income is high, revenues are often used to purchase or improve public infrastructure such as roads, schools, and clinics. The remainder may be distributed evenly among ejidatarios in cash. In ejidos with lower income from forestry, revenues are more likely to be distributed among ejidatarios without prior deductions for community investments. Whichever the case, it is the ejidatarios themselves who decide how any and all revenues from forestry are invested or distributed. Yet even on ejidos with the highest incomes from forestry, actual cash distributed per ejidatario is less than US$800, often representing the sole or most important source of cash (Flachsenberg 1994).

Additional economic returns can come though direct employment. In all ejidos, some ejidatarios are employed in timber harvesting. In ejidos with larger production forests, ejidatarios are also often employed in value-added processing of timber. Both harvesting and processing activities pay wages that are currently twice those of the Mex-

Table 3.5. Ejidatarios Occupied in Forest Activities in the Ejidos of SPFEQR

	Ejidatarios		
Ejido	Total	Occupied[a]	Percentage
Francisco Botes	299	20	6.7
Caoba	300	150	50.0
Chacchoben	330	40	12.1
Los Divorciados	87	30	34.5
Noh Bec	116	90	77.6
Nuevo Guadalajara	293	60	20.5
M. Avila Camacho	206	30	10.5
Petcacab	202	110	54.5
Plan de la Noria	92	92	100.0
Tres Garantias	105	90	85.7

Source: SPFEQR 1994.
[a]Oscillating figure according to working volume

ican minimum wage (Sociedad 1994). As noted earlier, not all male residents of an ejido are necessarily ejidatarios, and one need not be an ejidatario to work in forest management or processing, though many are. Table 3.5 reflects the importance of the forest industry in generating employment among ejidatarios on each of the 10 ejidos of the SPFEQR. Finally, on ejidos that also process roundwood, ejidatarios can receive income from the distribution of net revenues from value-added processing.

Though revenues currently cover most of the ejidos' costs of production, they are still insufficient to cover the costs of silviculture, technical assistance, monitoring, and policy-related activity. As Flachsenberg (1993b) has argued, a new strategy of mixed financing must be developed to compensate for these, to avoid a switch by ejidatarios to other forms of land use. However, he goes on to caution that it would be illusory to believe that drastic changes will occur in the markets for lesser-known timber species; most of these new revenues will have to come from elsewhere.

For some ejidos, one possibility is chicle. Income from chicle is distributed quite differently from timber income. Until the 1960s, chicle was the principal source of income for the rural population (Aldrete and Eccardi 1993). Currently, chicleros earn a net of about US$12 per day after deductions of 40 percent from gross revenues paid to the

Table 3.6. Chicle Production and Revenues (US$) for the 1991–1992 and 1993–1994 Seasons in the Ejidos of SPFEQR That Harvested Chicle

Ejido	1991–92 volume (kg)	1991–92 gross revenue (US$)	1993–94 volume (kg)	1993–94 gross revenue (US$)
Caoba	1,569	3,661	7,857	26,190
Chacchoben	1,143	2,667	3,021	10,068
Noh Bec	12,107	28,249	19,760	65,865
Petcacab	18,992	44,315	16,344	54,480
Plan de la Noria	1,716	4,004	0	0
Tres Garantias	8,499	19,830	8,283	27,610

Source: H. Flachsenberg, pers. comm. 1994.

Federación de Cooperativos Chicleros (Barrera de Jorgenson 1993). All net revenues accrue only to the individual chiclero engaged in tapping, with the total amount depending upon the volume extracted. Not all SPFEQR ejidos are engaged in tapping. Of those that are, not all ejidatarios participate. Table 3.6 presents the volumes extracted on those ejidos during the 1991/92 and 1993/94 seasons and the gross revenues generated.

Two other forest products—meat from wild animals and firewood—deserve mention here. Both are harvested primarily for direct consumption or for local markets within ejidos, yet their economic importance is substantial. A recent study of the three ejidos of Tres Garantias, Noh Bec, and M. Avila Camacho showed an annual harvest of between 3,000 and 4,000 kg of meat from wild birds and mammals in each ejido. At a price of around US$3 per kg (the local price for wildlife meat), that amounts to US$9,000–$12,000 worth of wildlife harvested per year (Ehnis Duhne 1993). For ejidos such as Tres Garantias and Noh Bec, this is less than 10 percent of net revenues obtained from the sale of roundwood, but for M. Avila Camacho it is roughly 25 percent of roundwood sales. Research by Jorgenson (1994) on the impact of subsistence hunting on wildlife in the Maya zone found that extraction of most species was occurring at unsustainable levels. Tres Garantias and Noh Bec are beginning to develop wildlife management programs that include restoration of depleted populations (e.g., the ocellated turkey, *Agriocharis ocellata*), the possible development of fee hunting, and wildlife viewing for nature tourists (Ehnis Duhne 1993).

The other major forest products are firewood for energy, and wood and palm thatch for the construction of houses, mainly for direct use within the ejidos. Firewood cutting is limited to abundant understory species whose rate of reproduction exceeds the rate of extraction (Flachsenberg et al. 1992). Palm leaves are increasingly harvested to thatch the roofs of tourist facilities in the region, a usage that could impact the abundance of certain palm species within the forest. We located no sources that document the extent to which ejidatarios use medicinals from the forest, though our interviews with ejidatarios indicated that this was not extensive.

Significant socioeconomic changes have occurred since the beginning of the Plan Piloto Forestal in 1983. In contrast to pre-1983 when only chicle provided occasionally significant revenues to those ejidatarios engaged in the activity, timber is now the most important source of revenues on most ejidos. The future economic success of forest management in Quintana Roo will depend on various competing economic interests and market factors, including many outside the influence of the Plan Piloto Forestal. Other sources of wood for the international and domestic markets will have a strong influence. Galletti (1994) argues that timber production in Guatemala and Belize is less controlled than in the Plan Piloto Forestal, making wood from these countries cheaper. Competition from the less expensive operations of uncontrolled timber exploitation undercuts attempts at sustainable management, a condition that could be further exacerbated by free trade agreements and the globalization of Mexican markets. González Pacheco (1991, cited by Vargas 1992) estimates, for example, that the internal price of precious woods in Mexico is more than double the price for equivalent products from Malaysia.

Current markets for forest resources and benefits are not sufficiently developed in breadth or depth to outcompete categorically the use of land for crop agriculture and cattle pasture, particularly when the latter are subsidized by government, as has often been the case in Quintana Roo (Edwards 1986). Though various economic and noneconomic incentives may favor one land use over another, one advantage of long-term forest production is that, particularly given the soils and climate of Quintana Roo, timber productivity should be more stable than agricultural productivity. Also, in contrast to agricultural products, mahogany provides one of the most stable markets in the world.

Conclusions

The ejidos of SPFEQR, and of the Plan Piloto Forestal in general, have established an entirely new social and political approach to forest management in Quintana Roo, one that is spreading to other states in southern Mexico. It is an approach that emphasizes a local, bottom-up, rather than central-government, top-down, approach as the best bet for sound forest management. The latter has clearly failed in Mexico (Galleti and Argüelles 1987; Janka and Lobato 1994). This bottom-up approach can be characterized as resting on four main tenets:

1. Secure land and resource tenure for local people are prerequisites to the operation of socioeconomic incentives for sound forest management.
2. Those who hold such tenure must have the power to negotiate directly the economic rent and other benefits they receive from forest resources and services.
3. The capture and distribution of those rents and benefits, as well as acceptance of responsiblities for managing the resource, must be equitable under a system of communal ownership and management.
4. Local communities, through the development of their own expertise and decision-making capabilities, with modest but essential outside financial and technical input, can practice a form of adaptive resource management under which adjustments in forest management can be made as conditions and new information require.

Under this approach and the umbrella of the Plan Piloto Forestal, the ejidos of the SPFEQR have, over the last 11 years, developed one of the most promising efforts under way anywhere in the tropics to produce a system of forest production that is both socioeconomically and ecologically sustainable. Their short-term success has been catalytic, as now some 6,607 rural families, representing five forestry associations, are participating in a new form of rural development based on the wise use of forest resources (Argüelles 1991). Theirs has been, and will continue to be well into the future, an uphill effort because of a synergistic mix of ecological, historical, and economic hurdles that they have inherited with the forest that impede efforts at sustainable forest management.

The economic hurdle is that only two species, mahogany and cedar, currently have markets strong enough to generate significant revenues—

a weak foundation on which to base management in species-rich tropical forests like that of Quintana Roo. The ecological hurdle is that those species live in a highly diverse forest, with the result that they occur at low densities, an unfavorable condition for deriving significant returns per hectare of land. This is compounded by a reproductive strategy of the most important species, mahogany, that is incompatible with the low-impact logging practices required to extract it, and that, because of the parasitic shoot borer, does not lend itself to high-density plantings. The historical hurdle is that SPFEQR inherited a forest that had already been largely mined of mahogany, and thus they were left to restore the remnants of past practices while at the same time having to make a living from them.

The management of most forests, and particularly those of Quintana Roo, is a balancing act between socioeconomic, political, and ecological sustainability. At times one must be favored over the others, but none can ever be fully sacrificed for the sake of the others, or the effort will ultimately fail. Strategists and practitioners of the Plan Piloto Forestal had to secure the socioeconomic incentives and political support (which created the space for them to operate) before many of the ecological concerns could be addressed. With more detailed inventories and greater ecological knowledge, better silvicultural practices can be implemented. However, the more detailed inventories of mahogany and greater knowledge of its reproductive biology have only served to tighten the noose around the neck of socioeconomic sustainability: harvest levels may have to be further curtailed, and more intensive and expensive silvicultural practices must be implemented, if the offtake of mahogany is to continue for perpetuity at equal or greater levels.

The risk, however, is that the noose may be tightened too much, and ejidatario and political support for the Plan Piloto Forestal, and more specifically, for maintenance of the Permanent Forest Areas, may begin to fade. If current management practices continue, there exists a strong possibility that at the end of the 75-year rotation no mahogany will remain to be cut. In the process, however, one has bought time (up to 75 years, if necessary) for the forest, for innovations in the management of mahogany, and for other species uses and forest values to help pick up the tab of socioeconomic sustainability and forest conservation.

Where is forest management in the SPFEQR heading, and will forest uses and management be able to avert land uses that are inimical to biodiversity? While current harvesting and silvicultural practices appear to have minor effects on forest biodiversity, how will biodiversity be affected if use and management intensify? The stakeholders who possess the greatest decision-making power over the future disposition of forest land are the ejidatarios, and for them, first and foremost, the forests are managed to produce a better livelihood. Benefits to biodiversity conservation have been in large part a by-product of that goal. To date, the ejidatarios' livelihood is almost entirely derived from wood and chicle, but near-term opportunities for expanding revenues from these sources appear limited. For now, any significant expansion of mahogany production faces severe technical and economic constraints. Indeed, the returns from mahogany may even diminish in the future. The possibility of expanding uses and markets for the lesser-known species of softwoods and hardwoods, and the development of value-added wood industries in the ejidos, are also fraught with problems and considerable unpredictability. And although ejidos may eventually obtain a greater share of chicle revenues, it is unlikely that chicle markets will significantly expand.

Many other stakeholders have a less direct but nevertheless potentially significant influence on how the forests and lands of Quintana Roo are used. These range from marketers and consumers of wood products to state politicians whose popularity may depend on how those lands are managed; from people who may have an interest in converting those forested lands to other uses to nature conservationists who wish to see the forests and their biodiversity preserved.

As it is, the fate of the forest and its biodiversity rests almost solely on the production value of the forest. The opportunity costs of alternative land uses have not been determined for the Permanent Forest Areas, but current revenues from forest products may be insufficient for forest management to compete successfully with alternative land uses such as farming or cattle grazing. Although there is some outside financial input based in part on biodiversity interests, by and large the biodiversity stakeholders in the forest of Quitana Roo are getting a free ride on the shoulders of timber production, and most of that from one species. If the biodiversity of Quintana Roo's forest is to have a more certain future, then a more diversified market for the forest's products and services is required. The efforts of the Plan Piloto Forestal currently focus on diversifying wood uses and timber markets, and these efforts are important. But biodiversity stakeholders must be willing to pay for what they value in Quintana Roo's forests, and new markets and benefit-sharing mechanisms must be established so that the ejidatarios are fairly compensated. Such biodiversity markets may be based on diverse uses and values: nature tourism; bird watchers in the United States and Canada who value the winter habitat of migratory birds; nonprofit conservation groups; the research community that values biodiverse research areas; the value of carbon sequestration by Quintana Roo's forest; wood users and biodiversity prospectors who wish to maintain the option values provided by Quintana Roo's forest, and so on.

To the extent that these biodiversity values are made tangible in socioeconomic terms to the people of Quitana Roo, they are more likely to maintain and potentially expand the Permanent Forest Areas, and to incorporate biodiversity objectives into their management plans. Without diversifying, the future of Quitana Roo's forests rests on a shaky foundation of single-commodity specialization within an environment of competing land uses.

Acknowledgments

We thank Manuel Aldrete, Alfonso Argüelles, Amanda Barrera de Jorgenson, Ronnie de Camino, Matt Dickenson, Daniel González, Henning Flachsenberg, Rene Foerster, Hugo Galletti, Helmutt Janka, Laura Snook, Meg Symington, Alberto Vargas, and the leaders of SPFEQR and the Noh Bec and Manuel Avila Camacho ejidos for providing information, enlightening discussions, and constructive comments on the manuscript. Financial support for this work was provided by World Wildlife Fund–U.S. and World Wide Fund for Nature–International. Curtis Freese also thanks the Library and Department of Biology of Montana State University for their invaluable services in bibliographic research.

References

Aldrete, M., and F. Eccardi. 1993. Background documentation to "Primer Encuentro Regional de Productores de Chicle Guatemala-Belice-México," June 23–26, 1993. Chetumal, Quintana Roo, Mexico.

Argüelles, A. 1991. Experiencias en el desarrollo rural: el caso del Plan Piloto Forestal, Quintana Roo, México. In *Proceedings of the Humid Tropical*

Lowlands Conference: Development strategies and natural resource management, Vol. III: *Promising timber management strategies,* ed. J. C. Dickenson III, 1–14. Bethesda: Development Alternatives, Inc.; Gainesville: Tropical Research and Development.

Argüelles, A. 1993. Conservación y manejo de selvas en el estado de Quintana Roo, México. Unpublished report of Acuerdo México-Alemania, Chetumal, Quintana Roo, Mexico.

Argüelles, A., S. Gutierrez, E. Ramírez, and F. Sánchez Román. n.d. *Un modelo de inventario forestal para apoyar las operaciones de extracción, de silvicultura e industriales de las empresas forestales ejidales de Quintana Roo.* Sociedad de Productores Ejidales Forestales de Quintana Roo, S.C. and Acuerdo México-Alemania, Chetumal, Quintana Roo, Mexico.

Barrera de Jorgenson, A. 1993. Chicle extraction and forest conservation in Quintana Roo, México. MA thesis, University of Florida, Gainesville, Florida.

Bascope, F. 1992. Mahogany (*Swietenia macrophylla*) in Bolivia: Its sites, industrial development, and actual situation. Paper presented at the Tropical Forest Foundation Mahogany Workshop: Review and Implications for CITES. Washington, D.C., February 3–4, 1992.

Botkin, D. B., and L. M. Talbot. 1992. Biological diversity and forests. In *Managing the world's forests: Looking for balance between conservation and development,* ed. N. P. Sharma, 47–74. Dubuque: Kendall/Hunt.

Dickinson, M. 1993. *End of year progress report.* Smithsonian Institution Scholarly Studies Program. Noh Bec Ejido, Quintana Roo, Mexico.

Edwards, C. R. 1986. The human impact on the forests of Quintana Roo, Mexico. *Journal of Forest History* 30:120–127.

Ehnis Duhne, E. A. 1993. *Informe de actividades del programa "Manejo de Fauna Silvestre en Ejidos Forestales de Quintana Roo e Inventario de Poblaciones Silvestres," de la Sociedad de Productores Forestales Ejidales de Quintana Roo, S.C. durante el año 1993.* Chetumal, Quintana Roo, Mexico.

Figueroa Colón, J. C. 1994. An assessment of the distribution and status of big-leaf mahogany (*Swietenia macrophylla* King). Unpublished manuscript. Puerto Rico Conservation Foundation and International Institute of Tropical Forestry.

Flachsenberg, H. n.d. *Regeneración en patios de concentración de trocería y el incremento medio anual de la caoba.* Acuerdo México-Alemania, Chetumal, Quintana Roo, Mexico.

Flachsenberg, H. 1992. Descripción general del recurso forestal de Quintana Roo. Unpublished report of Acuerdo México-Alemania, Chetumal, Quintana Roo, Mexico.

Flachsenberg, H. 1993a. *Diagnostico de la actividad forestal en Quintana Roo.* Acuerdo México-Alemania, Chetumal, Quintana Roo, Mexico.

Flachsenberg, H. 1993b. *Aspectos socio-culturales, técnicos, económicos y financieros en el manejo de selva.* Acuerdo México-Alemania, Chetumal, Quintana Roo, Mexico.

Flachsenberg, H. 1994. Problemas de transferencia de tecnología. Paper presented at the Workshop on Techniques for Sawing and Drying Tropical Hardwoods, Chetumal, Quintana Roo, Mexico, April 19–23, 1994.

Foerster, R., M. del Angel, F. Montalvo, and R. Terrón. 1992. *Estudio del mercado nacional de maderas corrientes tropicales.* Acuerdo México-Alemania, Chetumal, Mexico.

Frumhoff, P. C. 1995. Conserving wildlife in tropical forests managed for timber. *BioScience* 45:456–464.

Galletti, H. 1989. Economía política de la planificación comunal del uso del suelo en areas forestales tropicales: Una experiencia de caso en Quintana Roo, México. Paper presented at the Symposium on Multiple Use Systems and Methods for Soils, Universidad Autónoma de Nuevo León, Linares, Nuevo León, Mexico, December 14–16, 1989.

Galletti, H. 1992. Aprovechamiento e industrialización forestal: Desarrollo y perspectivas. In *Quintana Roo: Los retos del fin de siglo,* 101–153. Cambio XXI Fundación Quintanaroense. Chetumal, Mexico: Centro de Investigaciones de Quintana Roo.

Galletti, H. 1994. Las actividades forestales y su desarrollo histórico. In *Estudio integral de la frontera México-Belize,* vol. 1, ed. A. César Dachary, D. Navarro, and S. M. Arnaiz-Burne, 109–170. Chetumal, Mexico: Centro de Investigaciones de Quintana Roo.

Galletti, H., and L. A. Argüelles. 1987. La experiencia en al aprovechamiento de las selvas en el estado de Quintana Roo, México: Del modelo forestal clásico a un modelo forestal alternativo. Paper presented at the International Workshop on Silviculture and Management of Tropical Forests, SARH-COFAN-FAO, Chetumal, Mexico, May 11–20, 1987.

Gómez-Pompa, A. 1991. Learning from traditional ecological knowledge: Insights from Mayan silviculture. In *Rainforest regeneration and management,* ed. A. Gómez-Pompa, T. C. Whitmore, and M. Hadley, 335–341. Paris: UNESCO; Canforth, England: Parthenon Publishing Group.

Gómez-Pompa, A., and A. Kaus. 1990. Traditional management of tropical forests in Mexico. In *Alternatives to deforestation: Steps toward sustainable use of the Amazon rainforest,* ed. A. B. Anderson, 45–63. New York: Columbia University Press.

González Cortés, D. 1993. *Red de sociedades civiles ejidales de Quintana Roo: Recuperación de experiencia.* Chetumal, Quintana Roo, Mexico: Plan Piloto Forestal.

Instituto Nacional de Estadística Geografía e Informatica (INEGI). 1992. *Quintana Roo, perfil sociodemográfico, XI censo general de población y vivienda, 1990.* Aguascalientes, Mexico: INEGI.

IUCN. 1992. *Analyses of proposals to amend the CITES appendices.* Prepared by IUCN/SSC Trade Specialist Group, World Conservation Monitoring Centre, and the TRAFFIC Network for the eighth meeting of the Conference of the Parties. Gland, Switzerland: IUCN–the World Conservation Union.

Janka, H., and R. Lobato. 1994. Alternativas para enfrentar la destrucción de las selvas tropicales: Algunos aspectos de la experiencia del Plan Piloto Forestal de Quintana Roo. Paper presented at the Workshop on Government Policies and Forest Resources, Washington, D.C., June 1–3, 1994.

Johnson, N., and B. Cabarle. 1993. *Surviving the cut: Natural forest management in the humid tropics.* Washington, D.C.: World Resources Institute.

Jorgenson, J. 1994. La cacería de subsistencia practicada por la gente Maya en Quintana Roo. In *Madera, chicle, caza y milpa,* ed. L. K. Snook and A. Barrera de Jorgenson, 19–46. Proceedings from a 1992 conference sponsored by PROAFT, INIFAP, USAID, and WWF.

Konrad, H. W. 1987. Capitalismo y trabajo en los bosques de las tierras bajas tropicales mexicanas: El caso de la industria del chicle. *Historia Mexicana* 36:465–506.

Lamb, B. 1966. *Mahogany of tropical America: Its ecology and management.* Ann Arbor, Michigan: University of Michigan Press.

Lynch, J. F., and D. F. Whigham. 1995. The role of habitat disturbance in the ecology of overwintering migratory birds in the Yucatan Peninsula. In *Conservation of neoptropical migratory birds in Mexico,* ed. A. S. Estrada and M. Wilson, 195–210. Maine Agricultural and Forest Experiment Station Miscellaneous Publication 727.

Negreros-Castillo, P. 1991. Ecology and management of mahogany (*Swietenia macrophylla* King) regeneration in Quintana Roo, Mexico. Doctoral dissertation, Iowa State University, Ames, Iowa.

Negreros-Castillo, P. 1994. Evaluation of direct planting and post-harvesting natural regeneration in Quintana Roo, Mexico. Report to the Tropical Ecosystems Directorate of U.N. Man and the Biosphere Program. Unpublished manuscript.

Palmer, J., and T. Synnott. 1992. The management of natural forests. In *Managing the world's forests: Looking for balance between conservation and development,* ed. N. P. Sharma, 337–373. Dubuque: Kendall/Hunt.

Perl, M. A., M. J. Kiernan, D. McCaffrey, R. J. Buschbacher, and G. Batmanian. 1991. *Views from the forest: Natural forest management initiatives in Latin America.* Washington, D.C.: World Wildlife Fund.

Plan Piloto Forestal. 1993. Cuadro de demanda de la industria forestal del estado de Quintana Roo, Chetumal, México. Unpublished manuscript.

Poore, D., P. Burgess, J. Palmer, S. Rietbergen, and T. Synnott. 1989. *No timber without trees: Sustainability in the tropical forest.* London: Earthscan.

Ramírez Aguilar, G. A. 1992. Aprovechamiento del latex de chicozapote (*Manilkara zapota*) y potencial productivo en Quintana Roo. Thesis, Universidad Juarez del Estado de Durango, Venecia, Durango, Mexico.

Rendón Trujillo, C. 1945. La producción de la caoba en México y el problema de su exportación. Thesis, Escuela Nacional de Agricultura, Chapingo, Mexico.

Sociedad de Productores Forestales Ejidales de Quintana Roo, S.C. 1994. *Manejo del recurso forestal: Indices económicos de las anualidades entre 1991 y 1992.* Chetumal, Quintana Roo, Mexico.

Snook, L.K. 1991. Opportunities and constraints for sustainable tropical forestry: Lessons from the Plan Piloto Forestal, Quintana Roo, Mexico. In *Proceedings of the Humid Tropical Lowlands Conference: Development strategies and natural resource management,* Vol. V: *Secondary forest management,* ed. L.B. Ford, 65–83. Bethesda: Development Alternatives, Inc.; Gainesville: Tropical Reseach and Development.

Snook, L. K. 1993. Stand dynamics of mahogany (*Swietenia macrophylla* King) and associated species after fire and hurricane in the tropical forests of the Yucatan Peninsula, Mexico. Doctoral dissertation, Yale School of Forestry and Environmental Studies, New Haven. University Microfilms International, Ann Arbor, Michigan, 9317535.

Snook, L. K. 1994. *Mahogany: Ecology, exploitation, trade and implications for CITES.* Washington, D.C.: World Wildlife Fund.

Tropical Forest Foundation. 1992. *Executive summary. Mahogany Workshop: Review and Implications of CITES, Washington, D.C., February 3 and 4, 1992.*

Vargas, A. 1992. Economic, ecological and social aspects of forest management in peasant organized communities in Quintana Roo, Mexico. Dissertation proposal, University of Wisconsin, Madison.

Whitman, A. A., J. M. Hagan III, and N.V. L. Brokaw. 1994. Effects of selective logging on birds in a tropical forest in Belize. Unpublished manuscript.

FOUR

Swiss Forest Use and Biodiversity Conservation

Thomas O. McShane and Erica McShane-Caluzi

Switzerland, although a small country with a population of approximately seven million, is one of the wealthiest on the planet. There are few natural resources other than wood and water in Switzerland. Its wealth comes from the service sector—mainly banks, insurance, and tourism—and from industry.

At the end of the last Ice Age, forests completely covered the Swiss Plateau region. Over the next few millennia, the forests were cleared as settlement and agriculture expanded; by the year 1200 the forest cover had been reduced to 40 percent of its original range (von Fellenberg 1981). Historically, concern for the forest arose long ago. The first known forest protection law is dated 1298 (Perrig and Fux 1945). It was in no way comprehensive, because it concerned only a single small community (St. Maurice in the Valais). With the passage of time, other, mostly montane, communities decreed regulations concerning forest use. In spite of these laws, forests were overused and improperly exploited. During the industrial revolution, wood was the primary fuel (as raw logs or in the form of charcoal), and consumption increased dramatically. Switzerland exported wood for fuel and construction to the Netherlands, northern Italy, and the Mediterranean coast of France.

Apart from wood harvesting, forests were exploited in a diversified manner: forage for domestic animals, resin production, charcoal, and so on. As a result, the massive deforestation of forest land led to catastrophic events, such as avalanches, landslides, and flooding, resulting in human casualties and economic loss (Tromp 1980a).

Realizing that these events were linked to forest loss, a major reforestation program was begun 100 to 130 years ago, and it was mandated that the country's forest cover should not diminish further. Today the results of these reforestation efforts are reflected in abnormally structured forests with trees of the same age (cf. Mayer 1976). In many cases, native trees have been planted on ecologically inappropriate sites (M. Spinnler, pers. comm. 1995). This strategy was a success in terms of quantity of wood produced, and since 1876 the Swiss forests have been exploited on a "sustainable" basis in terms of wood production. Annual growth is estimated at 6.8 million cubic meters (m^3), while 4.2 million m^3 are exploited (Office fédéral des questions conjoncturelles 1991). Today forests are being used for timber, protection against natural disasters, hunting, and recreation.

Ecosystem function has been treated, until recently, as less important, and the belief that forests could not fulfill their functions without exploitation went unchallenged. However, the production-driven philosophy has begun to change, and now parameters other than the forests' protective function in mountain areas and maximized wood production are being taken into consideration. Increased awareness of the importance of biological diversity, greater interest in recreation, and heightened concern over the impact of pollution resulted in a new forestry law. The revised law gives equal weight to the forests' multiple functions: protective (biodiversity and landscape protection), social (recreation through equal access), and commercial (timber production) (OFEFP 1995).

We explore here the process of institutional change and the demands on Swiss forests, taking into account economic, ecological, social, political, and historical factors. We review changes in forest management in Switzerland and evaluate their impact on biodiversity conservation.

Forest Composition and Structure

Forests cover 1,204,000 hectares (ha) of Switzerland. This corresponds to approximately 29 percent of the country (OFEFP 1993a), which

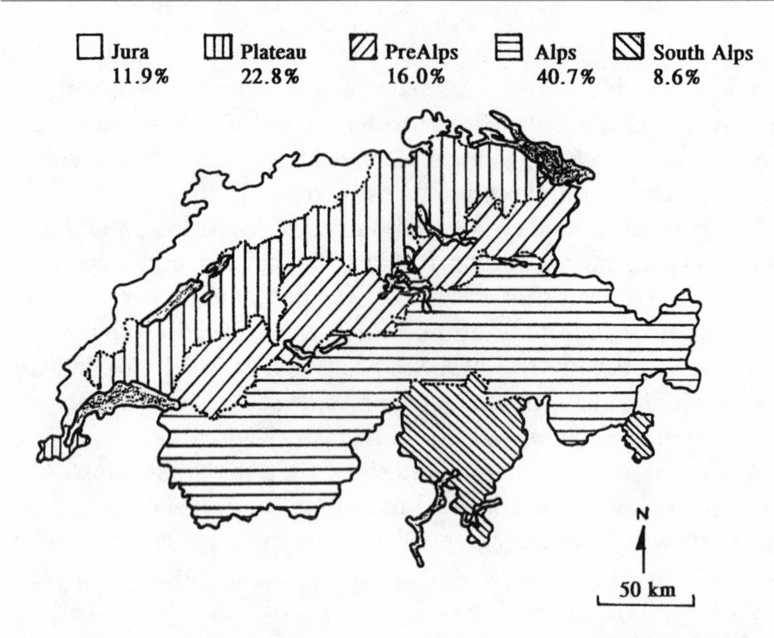

Figure 4.1. Five forest regions of Switzerland. (*Source:* OFEFP 1989, 1993a.)

compares favorably to that found in adjacent countries (Austria 44.8%, Germany 29%, France 27.8%, Italy 26.8%). Public ownership accounts for 68.5 percent of the total forest area, with more than half in traditional and political communes. Individuals own another 28 percent, and companies only 3.5 percent. Forests in Switzerland range from 200 meters above sea level (asl) in the south Tessin to 2,300 meters asl in the alpine valleys (OFEFP 1993a). More than half of the forested land in the country is over 1,000 meters asl. By region, approximately 39.7 percent of the Jura is classified as forest, 24.2 percent of the Plateau, 32.8 percent of the Prealps, 22.7 percent of the Alps, and 46.4 percent of Switzerland south of the Alps (Tessin and part of Graubünden). Figure 4.1 shows the distribution of land regions within Switzerland. The Alps have the lowest amount of forest cover because of the large amount of land area found above tree-line, followed by the Plateau region where the majority of the country's population is found and where the land has been converted to other uses.

Fifty-three tree species, 12 coniferous and 41 broad-leaved, occur in Swiss forests (cf. Ellenberg and Kloetzli 1982; Leibundgut 1984).

Table 4.1. Forest Surface by Degree of Mix (in 1,000 ha)

Degree of mix	Jura	Plateau	Prealps	Alps	South Alps	Total
Pure conifers (91–100%)	52.9	86.4	117.6	238.6	44.8	540.3
Mixed conifers (51–90%)	52.0	52.4	44.7	29.4	8.7	187.2
Mixed broadleaf (11–50%)	33.3	30.1	18.5	17.8	6.9	106.6
Pure broadleaf (0–10%)	48.2	49.8	16.7	28.1	67.0	209.8
Pasture	7.6	8.6	13.9	19.8	4.0	53.6
Total	194.0	227.0	211.4	333.7	131.4	1097.5

Source: IFN 1991, table 323.

Three tree species dominate in coverage: spruce (*Picea abies* [L.] H.Karsten), with 49 percent of basal coverage, beech (*Fagus sylvatica* L.), with 16 percent of basal coverage, and fir (*Abies alba* Miller), with 15 percent of basal coverage (OFEFP 1993a). These three species form 80 percent of the volume of the forest. Spruce dominates throughout almost all of Switzerland with the exception of the southern Tessin where beech, chestnut (*Castenea sativa* Miller), and other broad-leaved trees make up the forest. Beech is found throughout the Plateau, in the Jura, and on the Prealps, as well as south of the Alps. Table 4.1 classifies the Swiss forest by coniferous-broadleaf mix.

Stand structure is an important aspect of Swiss forests. High forest (*Hochwald*) is characterized by sexually reproduced stands in the same state of development and approximately the same age class (cf. Mayer 1976). This forest type is the most extensive in the country, making up 58 percent of the Swiss forest area (OFEFP 1989). High forest is best represented on the Plateau. These stands are the result of past intensive silvicultural practices with the goal of maximizing wood production, with little thought to alternative forest outputs. Stands rich in broad-leaved species with stems sprouting from stumps left from earlier harvesting are known as coppice forests. These forests make up 4 percent of the country's forest area. Many of these stands were harvested at 200 years for high-quality hardwood and are now exploited on a short rotation of every 10 to 30 years for fuelwood and construction materials (OFEFP 1989). In the Alps, the forest is more multi-layered with stands in different stages of development, including coppice forest. This structure is the result of these forests being managed primarily for their protection role in contrast to wood production. These stands represent 17 percent of the Swiss forest area and are considered under little threat (OFEFP 1989). It is evident that stand structure has been closely related to forest use.

Types of Forest Values

The outputs considered by classical microeconomics are private goods whose value is determined in the competitive marketplace at the equilibrium between supply and demand (Price 1990). Many forest outputs fall within this category, notably timber and other tree products, such as Christmas trees and leaves for forage. Forage from shrubs, forbs, and grasses can be valued in terms of the value added to grazing herds. Similarly, the water used for irrigation can be valued in terms of the value added through increased crop yields. Game animals and fish may also be valued in terms of their contribution to the economy as a source of food. Finally, the use of recreational facilities that depend on forests, such as ski areas or private campgrounds, takes place within the market economy (Price 1990).

However, many of the joint products of the forest cannot be valued in the marketplace: they are not market goods. Some of the outputs mentioned above display nonmarket characteristics, and their value in real markets may be changed by various types of market intervention (e.g., subsidies, taxes). At the other end of the spectrum from market goods are pure public goods. These can be defined as goods whose use by an individual has no impact on its subsequent use by another, such as clean air, sunshine, and scenic amenities (Samuelson 1954). In Switzerland an important example of this concept is the protection afforded by forests from floods or avalanches, which demonstrates the fact that the avoidance of a "public bad" (e.g., avalanche destroying property) is a public good (Price 1990). Another public good is the *existence value* of a particular forest landscape, free-flowing stream, or wilderness area—the mere knowledge that it exists. In this case, as with the value of preserving a landscape or the gene pool of a forest ecosystem, consumers do not have to be present in either space or time to derive benefits. The preservation of a resource

Table 4.2. Classification of Swiss Forest Outputs

Output	Type of good		
	Private (market)	Impure public	Pure public
Ecosystem diversity			Option/existence
Fish	Input (to be sold)	Recreational (permits sold)	
Forage	Marketed grazing permits	Community use	
Game	Input (to be sold)	Recreational (permits sold)	
Genetic diversity			Option/existence
Hazard protection		Individuals' life/-property	Public safety
Landscape		Limited access viewpoints	Public viewpoints
Recreation	Developed (ski areas/ campgrounds)	Undeveloped trails, campsites, other	
Water quality	Industrial, municipal, domestic use	Recreational use	Perception
Water quantity	Industrial, irrigation	Recreational use	Perception
Wilderness		Environment for recreation	Existence value
Wood	Sold on market (stumpage fees/ products)	Community use (local public good)	Long-term security

Source: Price 1990.

for unknown long-term benefits provides *option values* (Krutilla and Fisher 1985).

Between market goods and pure public goods are a wide range of other goods. These may be described as impure public goods (Cornes and Sandler 1986). The characteristics of such outputs are that their benefits are partially rival or excludable. The use of forest for recreation and as wilderness provides examples of impure public goods. Up to a certain level of use, the benefits are equal for all consumers. However, past this level, one or more individuals perceive that congestion is occurring (i.e., the social carrying capacity has been reached) (Heberlein 1977). Thus one person's use affects another's use (rival benefits). To avoid congestion, fees or permits can be used to limit use (excludable benefits). Most forest outputs, in some sense, are impure public goods, including water quality, landscapes that can be viewed from viewpoints with limited access, and hazard protection that benefits individuals' lives, safety, and property rather than public facilities (Price 1990). Table 4.2 classifies the various outputs emanating from Swiss forests.

Demands on Forests

Wood Production

Determination of Exploitation Levels. The Swiss forest industry operates on the principle of sustained-yield management whereby timber is not exploited more quickly than it is replaced. Under this principle a number of parameters are measured to determine the possible, and usually the maximum, amount of wood that can potentially be harvested. These parameters include growing stock, forest area, forest composition, and growth increment. This information is fairly well known in Switzerland, and as a result the harvest potential can be evaluated with relatively good precision. Swiss forests have an average stocking rate of 333 m^3/ha, and even though wood harvest increased by 30 percent between the 1920s and 1980s, it is believed that the forest can still produce more wood than it currently is (OFEFP 1989, 1993a). The forest industry and the National Forest Inventory (IFN) estimate that the Swiss forests could reasonably be expected to sustain an annual offtake of 7 million m^3 of roundwood (including logs, industrial wood, and firewood). This represents a 56 percent increase over current harvest levels. Long term, the volume of wood harvested could surpass 7.7 million m^3 (OFEFP 1993a).

Harvesting Methods. Almost all harvesting in Switzerland's forests is selective. Clear felling of large tracts of land is prohibited under law. Harvesting requires a certain level of surface infrastructure and development to be economically viable. The two extraction methods most commonly used in Switzerland are logging roads and skidder trails on accessible terrain, and cable logging systems on difficult-to-access terrain (OFEFP 1989).

Limitations on Harvesting. In Switzerland, the conditions under which harvest takes place vary greatly from one region to another. The primary reason for this is the vast differences in topography. Additionally, wood harvest is limited by other constraints such as stand structure, species composition, and transportation network. For example, the mountain forests of the Jura, the Prealps, and the Alps make up the most important wood reserve in the country (76% in volume). These forests, however, contribute only half of the wood harvested in Switzerland (OFEFP 1989). Situated on steep slopes, and exposed to extreme climatic conditions, management is expensive and difficult to justify. These are the areas that are most often not cut, and by consequence retain the conditions and species mixes most like that of "natural" forest. These "less profitable" forests have played an important function in protection against erosion, flood, and avalanche.

Access and Transportation Networks. Switzerland possesses approximately 26,000 kilometers (km) of forest roads accessible by truck (OFEFP 1989). The average density of the network is 24 meters of road per hectare (m/ha), though this varies greatly from region to region. In the Alps, road density is not more than 10 m/ha, while on the Plateau it attains 65 m/ha (Wilhelm 1988). In general, as the altitude increases, the density of roads decreases. Public forest land is better served by road than private forest land with the exception of south of the Alps, where the best-forested terrain is near villages and under private ownership. Access and transportation networks are important for other uses as well. They allow access for people intent on using the forest for such activities as hiking, picnicking, camping, skiing, and hunting.

Exploitation Levels

Statistics on wood production in Switzerland date back to 1913 (Fig. 4.2). Since 1932, Switzerland has been harvesting 3–4 million m^3 of wood per year. There have been two major increases in wood production during this time as demand for firewood increased during the two world wars. Since 1963, wood production has continually increased, and 1986, with a harvest of 4.6 million m^3, marked the greatest volume since World War II (OFEFP 1989).

Wood harvesting probably has the greatest impact on the forest ecosystem in Switzerland in terms of changes in forest structure, composition, and dynamics, and, in turn, biodiversity (cf. Mayer 1976; Bruderer and Thoenen 1977; Bruderer and Luder 1982; Gonseth

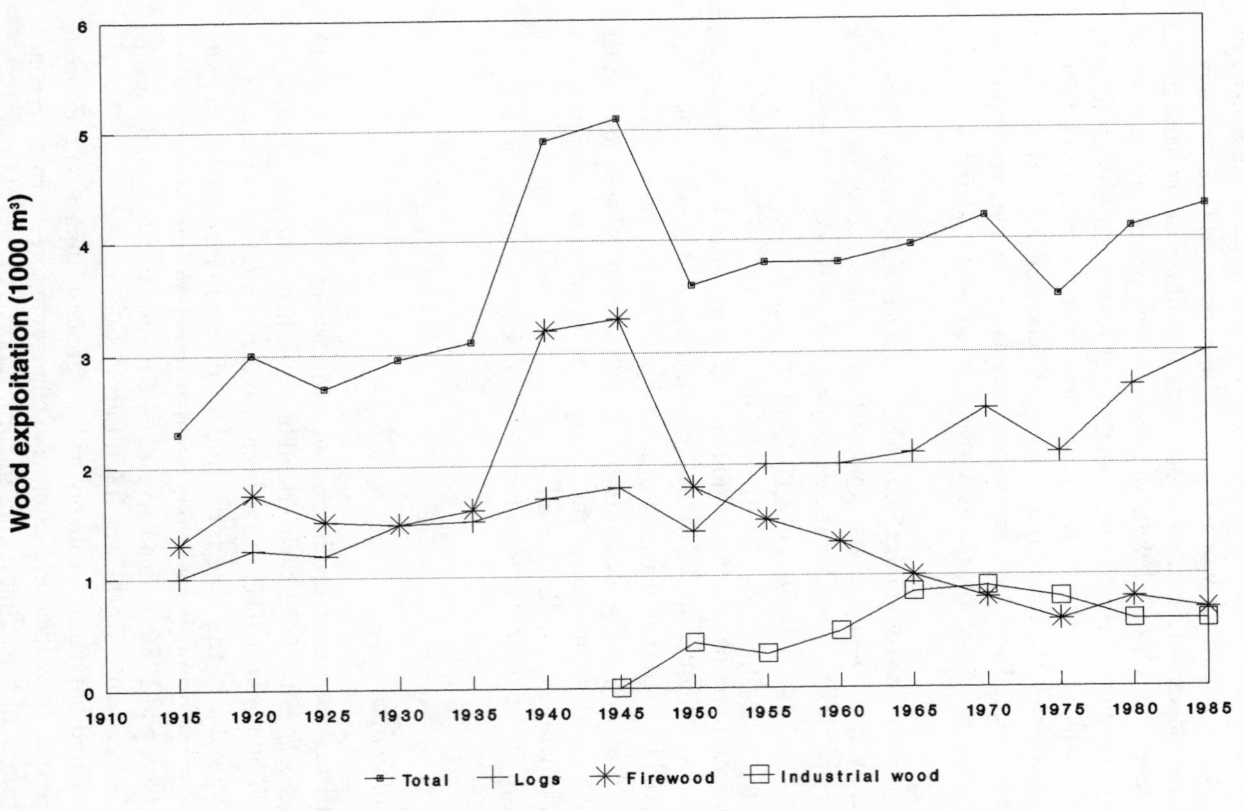

Figure 4.2. Wood exploitation in Switzerland. (*Source:* OFEFP 1989.)

Table 4.3. Wood Production from Swiss Forests

Year	Millions of m^3/year
1963–66	3.5
1973–77	3.8
1983–87	4.3
1988–89	4.5

Source: Office fédéral des questions conjoncturelles 1991.

1987; Grossenbacher 1988; Amann 1994). At the same time, wood production is seen by government, communities, and private individuals as an important component of economic livelihood. The exploitation and processing of wood supported an estimated 85,000 jobs throughout the country in 1993 (OFEFP 1995). This traditional view of the forest continues to be the primary factor in the decision-making process in terms of forest policy and management in Switzerland.

In the course of the last 25 years, the exploitation of Swiss forests has increased in parallel with increasing demand for wood (Table 4.3) and with the wood reserve in the forest. A variety of reasons account for the growth in harvest levels: improved silvicultural techniques; better transportation networks in the mountain regions and in private forests; enhanced extension services to small, private forest landowners; large volumes of timber recently damaged during storms and from snow; improved harvest techniques (i.e., mobile cable harvesting systems, skidders, harvesters) reducing the cost of exploitation; increased wood prices from 1973 to 1980, leading to higher harvest levels on private lands in particular; and public subsidies and compensation available for both the private and public sectors (Office fédéral des questions conjoncturelles 1991).

Switzerland produces 4.5 million m^3 of wood annually, or about 65 percent of the estimated potential sustained yield of the forest (OFEFP 1993a). Based on 1991 figures, public forest produced 77 percent of the wood (it represents 68.5% of the forest area), while private forest produced 23 percent. Conifers accounted for 76 percent of deliveries or about 3.5 million m^3 (OFEFP 1993a). Broad-leaved trees made up about one-fourth of deliveries amounting to 1 million m^3. Roundwood/logs command the highest price and make up about 60 percent of deliveries (almost 3 million m^3); industrial wood and firewood make up, in almost equal parts, the balance of deliveries.

The market demand for wood is estimated to be about 7 million m^3 of roundwood (Office fédéral des questions conjoncturelles 1991; OFEFP 1993a). More wood is not harvested because of the high harvest cost of Swiss wood in comparison to prices on the world market (i.e., it is cheaper to import wood than to cut it locally), and wood from Swiss forests does not satisfy the demand for quality wood met by imports from neighboring countries.

Since the beginning of the 1990s, Switzerland has annually exported more roundwood than it imports by about 0.9 million m^3 (OFEFP 1993a). However, imports dominate finished or worked products. Due to the high labor costs, some of the roundwood exported to neighboring countries is imported back into Switzerland as finished products (F. Jaquet, pers. comm. 1994). Expressed in roundwood equivalents, imports exceeded exports by 3.3 million m^3 in semifinished products and by 0.5 million m^3 of paper and carton (OFEFP 1993a).

Reforestation

Replanting policies have changed species composition throughout the forest zone. Part of this is due to plantation methods of replanting leading to single-species, even-aged stands of timber (cf. Mayer 1976). These forests bear little resemblance to the "natural" forests that they replaced. Early in the century, deciduous hardwoods were preferred due to the relatively high demand for fuelwood. As a result, hardwoods were planted extensively. After World War II the growth in demand for building materials led to an increase in the planting of conifers (F. Jacquet, pers. comm. 1994).

Recent changes in forest legislation place more emphasis on forest diversity than earlier legislation (Hauselmann 1993). As a result, new methods of replanting that mimic natural succession are being tested.

Recreational Activities

The time available for leisure in Switzerland has effectively doubled over the last one hundred years (OFEFP 1993b). Relaxation and leisure activities take a number of forms (e.g., walking, hiking, mountain biking, cross-country skiing, mushroom collecting, nature photography). The increasing density of forest roads together with sports such as mountain biking, snowshoeing, and cross-country skiing, among others, have penetrated areas of forest that previously had little impact or visitation from recreational use. Many of these areas are of vital importance to species sensitive to disturbance. Some recreational activities can have a major impact on the forest ecosystem and biodiversity. For example, the International Centre for Alpine Environments considers downhill skiing the single most damaging human activity in alpine environments (Covington 1993). The sport causes massive erosion, leaving mountainsides scarred and denuded of vegetation. Mushrooms in Switzerland have traditionally been a free resource and could be collected by all, largely as a recreational activity. As a result of increased interest by the public leading to overexploitation, collection limits such as weight per day and collection days per week have been instituted in many cantons (U. Breitenmoser, pers. comm. 1995). As demand for outdoor recreation grows in Switzerland, so will, no doubt, pressure on its forests.

Hazard Protection

The protective function of the forests in Switzerland is most important in the mountain regions. These areas account for approximately three-fourths of the country's forests. Not only are villages and towns below the slopes protected by forest cover from erosion, landslides, and avalanches, but also ski slopes, roads, and railways, which are commercially and strategically important. For example, more than 58,000 trains pass through the Gotthard tunnel in the Alps each year, carrying 5.5 million passengers and 12 million tons of goods (Mehr 1989).

The health of Swiss forests is critical to their role as protection against catastrophic events. In 1985, 42 percent of the trees in the Swiss Alps and 27 percent of the trees in nonmountain regions showed signs of unexplained defoliation. In 1987, the figure rose to 60 percent in the alpine areas and 48 percent on the Plateau (Mehr 1989). The proportion of trees that exhibited defoliation levels greater than 25 percent tripled from 8 percent in 1985 to 23 percent in 1994 (OFEFP 1995). While no direct link can be proven between this phenomenon and environmental pollution, it is now generally accepted that the problem stems from different types of stress, varying from one locality to another (OFEFP 1995).

Studies are looking into the influence of climate, site conditions, ozone, sulphur dioxide, secondary insect attack, disease, and other environmental factors. Nitrogen emission levels in Switzerland in the present decade have occasionally exceeded critical loads specified for forest ecosystems (OFEFP 1995). Tests have shown that excessive amounts of nitrogen can have direct negative effects on trees. Ozone is also a threat to forest health. In the 10 years or more that ozone concentrations in Swiss forests have been monitored, the threshold values required for forest protection have been consistently exceeded (OFEFP 1995). Trees in weakened states are susceptible to secondary insect attack (OFEFP 1993a). For example, in 1992, defoliation by bark beetles resulted in the salvage harvest of 0.5 million m^3 of wood. Alarmed by the threat of ever-increasing defoliation levels, the Swiss government has been working to develop strategies and methods to control insect infestation as well as emissions of sulphur and nitrogen oxides. Switzerland has some of the strictest rules in Europe on automobile pollution, and its consequent restrictions on truck transport through the Alps have negatively affected its relations with the European Union.

As forests continue to be degraded in critical areas, danger from rockslides and avalanches increase, while flooding becomes more severe. The disastrous floods in Brig in autumn 1992 serve as an example, while greatly raising the awareness of the consequences of lost forest cover.

Swiss Forests and Biodiversity

Forest Structure and Composition

Swiss forests have been extensively and intensively exploited, and despite deforestation at the turn of the century, the forest area has remained relatively constant at about 29 percent of the nation's territory for a number of decades. Nevertheless, the loss of a number of species within the forest reveals problems within the ecosystem. While forest cover is not being reduced by habitat conversion, ecological processes are being modified or perturbed by human activities, most notably forest exploitation (Mader 1979, 1981; Brown et al. 1980; Schütz 1982; Amann 1994).

The emphasis on wood production in this century led management to focus on a limited number of economically valuable species and cutting at specific ages (Table 4.4). The result was a simplified

Table 4.4. Average Age of Exploitation of Selected Species in Swiss Forests

Species	Average age at exploitation	Natural longevity
Broad-leaved		
Quercus sp.	180–300 years	700 years
Acer pseudoplatanus	120–140 years	400 years
Ulmus sp.	120–140 years	400 years
Fraxinus sp.	100–140 years	300 years
Populus nigra	30–50 years	300 years
Fagus sylvatica	120–140 years	250 years
Acer platanoides	120–140 years	150 years
Conifers		
Picea abies	80–120 years	600 years
Pinus sp.	100–120 years	600 years
Larix decidua	100–140 years	800 years
Abies alba	90–130 years	600 years

Source: Broggi and Willi 1993.

forest structure of even-aged trees with few species, and reduced light penetration in the forest (Mayer 1976). Broad-leaved forests, formerly the dominant vegetation type on the Swiss Plateau, have been reduced to less than 40 percent of their natural potential because of extensive planting of conifers (Leibundgut 1984).

These forest management practices effectively blocked natural ecological cycles and altered biodiversity in the forest. Current silvicultural techniques have ignored critically important aspects of forest development, including pioneer phases following natural disaster or harvest, natural regrowth, and the long phases of senescence with the accumulation of old trees and dead wood. In short, the techniques effectively stopped the process of natural succession within the forest. Indicative differences between "natural" and "production" forests are presented in Table 4.5.

In Switzerland, forest reserves come closest to "natural" in structure and composition. Reserves make up only .08 percent of the country's forest area. There are currently 39 forest reserves covering a total of just over 1,000 ha (Broggi and Willi 1993). These range in size from 244.8 ha to less than 1 ha, the majority being less than 10 ha. As a result of their small size, reserves play a limited role in biodiversity conservation (cf. MacArthur 1984).

Table 4.5. Forest Structure and Composition of Natural and Production Forests in Switzerland

Natural forest	Production forest
Trees of old age and large size	Cutting rotation of 100–150 years
Few young trees	High proportion of young trees in number and volume
Large differences in age in limited area	Even age stands frequent
Diversity of plant and animal species	Preference given to homogeneous stands of economic value
Elevated proportion of dead wood, decomposed organic material, and downed wood	Elimination of dominant trees predestined to become dead wood
Mixed structure, stump coppicing, knotty or gnarled trees mixed with those having straight stems	Selective thinning favoring trees with good growth

Source: Broggi and Willi 1993.

Impacts on Biological Diversity

Three factors have been suggested to be particularly important in the loss of biodiversity in Swiss forests: the fragmentation of the forest, the reduction in ecotones, and the systematic elimination of old-growth trees (Debrot and Meyer 1989; Broggi and Willi 1993). Other, related factors may include the absence of dead and downed wood (Harmon et al. 1986) and the loss of functional diversity (Franklin et al. 1981, 1986).

Fragmentation of the Forest

Forests in Switzerland are becoming increasingly isolated due to agricultural and urban activities. Additionally, forest exploitation roads have cut large areas of forest into subunits effectively isolating forest blocks.

Mader (1979, 1981) studied the effects of a forest road on the surrounding area and its faunal composition. The presence or absence of predacious ground beetles (Carabidae) was calculated as an indicator. Carabidae are a family of predacious ground beetles containing 581 species in 82 genera in Switzerland. They are mostly active predators found in litter or on vegetation. They are typical forest species rarely found in cleared or grassy areas.

Changes in abiotic factors within 6 m of the road were examined. Wind and evaporation increased in a 72-m band, while dust and auto emissions were measurable up to 300 m into the surrounding forest. Other abiotic factors changed as well, including temperature, soil humidity, and water drainage patterns. These changes were then correlated with impacts on faunal composition.

Mader (1981) concluded from reduced Carabidae numbers and changes in the composition of species that the band of disruption created by forest roads (i.e., impact on the structure of the forest) was approximately 26 m. As a result, 3-m wide roads at a density of 65 m/ha (the average value on the Plateau) impact 36 percent of the forest, though the roads occupy not more than 1.9 percent of the area. Throughout Switzerland the density of forest roads is about 24 m/ha (Wilhelm 1988), and most cantons have plans to extend their road networks. This will lead to further fragmentation of the forest.

Mader (1979, 1981) also measured the effect of the roads as dispersal barriers to ground beetles and small mammals (yellow-necked mice [*Apodemus flavicollis*] and bank voles [*Clethrionomys glareolus*]). He concluded that forest roads were effective barriers for these species and may block the exchange of genetic material in what were once single populations.

Reduction in Ecotones

Gonseth (1987) found that of the 120 species of threatened diurnal butterflies in Switzerland, 40 are negatively impacted by changes in forest openings and 30 by intensive silvicultural practices. The majority of these species are closely associated with small clearings and openings in the forest endowed with a diversity of plant species. For example, *Lopinga achine* is a species of butterfly characteristic of low-altitude mesophilic forest clearings. The species was relatively common some 30 years ago, but has become rare and is no longer found throughout most of the lower altitude forests of Switzerland which have undergone changes due to intensive forestry practices. These highly managed forests do not experience gaps caused by old-growth senility and death and similar types of disturbance. Schütz (1982) demonstrated that modern methods of silviculture, in particular intensively managed forests, short-circuit these stages of forest dynamics and natural succession, effectively altering forest structure.

Old-Growth Trees

Generally, under traditional silvicultural practices, trees are harvested when they are at their economic optimum. At about the time a tree becomes economically exploitable, it begins to play new ecological roles in the forest ecosystem. A natural forest, with a diversity of age classes represented, shelters a richer floral and faunal diversity. Barth (1987) lists 55 bird species in a forest of mixed-age classes ranging from 0 to 400 years; less than 50 species in a forest with age classes ranging from 100 to 150 years; 40 bird species for trees 30 to 60 years old; and between 20 and 25 bird species in a forest of less than 30 years in Germany. His study detailed the impoverishment of the forest by the absence of old trees. Indications from other forest studies, most notably from the western United States (Bowles 1963; Mannan 1982), are that the number of bird species filling specific niches, the number of individuals per species, and the species diversity in general is greatest in older forests. In particular, bird species that rely upon cavitied trees and foraging in and around bark are best represented in these older forests.

The black woodpecker (*Dryocopus martius*) provides an example of a cavity-dependent species. In Switzerland, 90 percent of the bird's nests can be found in beech trees older than 120 years (Debrot and Meyer 1989), the age at which these trees are intensively harvested (Table 4.4). Fewer nesting holes are excavated, which reduces nesting sites for species that rely on holes after use by the black woodpecker (secondary use). Several of these species are considered endangered, including the stock dove (*Columba oenas*), the hoopoe (*Upupa epops*), and the little owl (*Athene noctua*). Furthermore, diverse Hymenoptera such as hornets and half of the indigenous bat species in Switzerland rely on cavities in trees (Gebhard 1985; Barth 1987).

Dead and Downed Wood

Standing dead trees and fallen logs are essential to many organisms and biological processes within forest ecosystems (Harmon et al. 1986), yet such structures have rarely been maintained under Swiss forest management regimes. Ellenberg et al. (1986) demonstrate the importance of downed wood in a German forest as habitat for mushrooms, mosses, lichens, and other plant species. They found that the older the decomposition process, the greater the number of species associated with the tree. The removal of dead wood may thus contribute significantly to forest species impoverishment (Ellenberg et al. 1986).

Dead and dying wood provides one of the two or three greatest resources for animal species in a natural forest, and if fallen timber and slightly decayed trees are removed, the whole system becomes impoverished and may lose as much as a fifth of its total fauna (cf. Schütz 1982; Gebhard 1985; Barth 1987; Marti et al. 1988; Amann 1994). The drastic reduction in species directly growing on or in wood (lignicoles) has been noted in areas where old and dead wood no longer exist (Speight 1989). In addition to its role as habitat for land animals, woody debris also provides habitat, structure, energy, and nutrients for aquatic ecosystems (Harmon et al. 1986). Furthermore, it provides sites for nitrogen fixation, sources of soil organic matter, and sites for the establishment of other higher plants, including tree seedlings (Harmon et al. 1986).

Functional Diversity

Maintaining nitrogen-fixing organisms within the Swiss forest landscape is an example of managing for functional diversity. Many nitrogen-fixing species of plants are associated with the early stages of succession. Others, such as lichens, are associated with late-successional stages, and still others (microbial) are associated with woody debris (Franklin et al. 1981, 1986). Additionally, maintaining large-volume, complex crown structures improves the forest's effectiveness in removing moisture and particulate materials from the atmosphere.

In the last fifty years, ungulate populations, notably red deer (*Cervus elaphus*), roe deer (*Capreolus capreolus*), chamois (*Rupicapra rupicapra*), ibex (*Capra ibex*), and wild boar (*Sus scrofa*), have increased significantly. Some of these species are now causing considerable browsing damage. Trees such as fir and larch (*Larix decidua* Miller) are threatened with local extinction as regeneration is suppressed. The loss of predator populations, such as lynx (*Lynx lynx*), wolf (*Canis lupus*), and brown bear (*Ursus arctos*), means that natural checks on ungulate populations no longer exist (U. Breitenmoser, pers. comm. 1995). Although all of these species are hunted or culled, the offtake is insufficient to compensate for the loss of natural controls. Forest and wildlife management activities in Switzerland have tended to eliminate these coefficients of biodiversity to minimize competition from noncrop species and speed the development of simplified systems (Debrot and Meyer 1989).

Indicator Species: The Case of the Capercaillie

The capercaillie (*Tetro urogallus*), the largest grouse in Europe, serves as an excellent indicator of forest health and disturbance in Switzerland. A forest-dependent species, its habitat requirements are indicative of forest cover that is as close to natural as is possible in Switzerland today (Marti et al. 1988).

The capercaillie prefers older forests with a good proportion of pine, fir, and possibly beech. The forest structure must be relatively open (a maximum of 50–70% canopy cover) and layered with passages through the forest allowing flight. The bird requires a high proportion of old trees as well as snags and fallen dead wood, and a well-developed shrub layer dominated by Myrtillaceae (Marti 1993). The bird feeds largely on pine needles but also eats aspen (*Populus tremula* L.) and berries. During periods of nesting and rearing young, the species also requires small openings in the forest, areas of tall herbaceous plants, and mixed habitats including peat bogs, ponds, and prairies.

The Swiss population of the capercaillie is threatened. In 1970, fewer than 1,100 male birds were recorded during the mating season; 15 years later, the population was estimated at about half the 1970 level (Marti 1992). The bird's range in the country has also diminished. In 1900, it was possible to see capercaillie in forests on the Plateau. Today the bird's range is limited to the Jura, the Prealps, and some alpine valleys.

Though hunting of the species is not permitted in Switzerland, the capercaillie is sensitive to activities such as logging, as well as nonconsumptive uses such as various recreational activities (Marti et al. 1988). Intensive forest management practices that result in single-species stands (most notably of spruce) and the elimination of vegetation at ground level have degraded the species' habitat. Recreational use of the forest can negatively impact the bird's behavior. Males will abandon breeding areas for long periods of time following disruption.

Young may suffer from hypothermia and diseases as a result. A diverse range of activities within the forest has been shown to cause disturbance including mushroom and wild fruit collection, mountain biking and motorbiking, cross-country skiing, and the presence of observers and photographers at mating sites. Access to capercaillie habitat has been made much easier by the opening of forest roads into areas that were previously inaccessible.

Management for the complex habitat requirements of the capercaillie, involving systems and mosaics rather than just one preference, can have serious consequences for the current methods of forestry in Switzerland. If capercaillie habitat is to be maintained, one of the requirements is that following harvest the forest must be reconstituted in a layered and varied structure. Swiss federal law regarding nature protection and landscapes specifically states that silvicultural methods and transport networks must respect the needs of the capercaillie. Specific management guidelines have been developed by the Swiss Federal Office of Environment, Forests and Landscape (OFEFP), and the Swiss Ornithological Station (Marti 1993). The challenge remains how best to link policies with practice.

Socioeconomic Incentives and Controls

Legislation and Policy

The inhabitants of the Swiss Alps have long been aware of the importance of their forests, not only as a source of wood, but as protection for people and property from natural hazards such as avalanches, floods, and landslides. The first documented evidence of this dates to 1298 when the Valaisan commune of St. Maurice passed an order regulating the utilization of forests and pastures (Perrig and Fux 1945).

The growing awareness of the need to regulate forest utilization is demonstrated by a growing number of communal decrees from the early fourteenth century on. Many of these were likely restatements of earlier proclamations. The most common form of decree, dating from 1337, was the *Bannbrief,* which prohibited certain uses in specific areas of forest owned by the commune. Eventually more than three hundred Swiss forests were subject to this type of regulation. For some of these forests, wardens were employed to enforce the regulations, the first in 1530 (Tromp 1980b).

In spite of the recognition of the importance of the forests and growing numbers of regulations, the increasing requirements of the population for food and forest products led to extensive grazing in, and clearing of, the forests, particularly from the fifteenth century on. As a result of these activities, the timberline was lowered 200–300 m (Langenegger 1984), and some high valleys were completely deforested. Attempts at reforestation were rare and generally unsuccessful (Tromp 1980a).

Consumption continued to increase not only for construction, fuel, and agriculture, but also in neighboring countries as populations expanded and industrialization developed in the seventeenth, eighteenth, and nineteenth centuries (Auer 1956). Swiss alpine timber was used for fuel in the manufacture of glass, charcoal, lime, and steel, and also for shipbuilding throughout Europe (Tromp 1980a). Throughout this period, natural disasters, particularly floods, increased in number and severity (Tromp 1980a). Independent communal regulation of the forest effectively ended in 1876 when the federal government took regulatory responsibility from both the communes and private concerns.

The result of this action was the passage of a Federal Forest Law (FFL) in 1876. The early versions of this law focused on wood production and the protection role of the forest in the mountains. The FFL has been in effect for more than one hundred years. The principal tenets of this law, which is still in effect today, consist of the following:

—Swiss forest area must not be diminished.

—Forest clearing must be compensated by replanting in the same region.

—Sustainable wood production is to be maintained as a high priority.

—Reforestation, protection, and transportation infrastructure for forest management is to be subsidized (OFEFP 1993a).

The most common forest management conflict in Switzerland has been between wood production and forest protection (OFEFP 1993b). In general, the two main protagonists have been forest "owners," be they private or public (most often communes) who traditionally have earned revenue from wood production, and groups (usually nongovernmental organizations) that support protecting the forest for its existence and option values—biodiversity, landscapes, and wilderness.

As a result of increased environmental awareness throughout Switzerland, along with a general trend toward urbanization, a greater value is being assigned to the recreational and biological roles of the forest in contrast to wood production. These factors, along with sudden signs of reduced forest health and the increasingly poor state of the finances of forest owners, pressed politicians to update the FFL extensively in 1991 (Hauselmann 1993). The law became effective at the beginning of 1993.

Inherent in the new forest law is the increased importance of biological diversity. The FFL actively sets out to change the traditional forms of silviculture that have been practiced in Switzerland over the last century or so. For example, replanting of sites that have been cut will take place only in areas where natural regrowth will not result in the desired natural diversity, such as where monocultures presently dominate. In these cases, the species to be planted are selected on the basis of known associations of soils, trees, and shrubs to establish regeneration as close as possible to the natural state. The 1993 law requires that linkages be created between isolated forest patches through the development of biological corridors. Where roads bisect large forest areas, artificial biological corridors are required to ensure that forest animals can move between forest blocks (OFEFP 1995).

Subsidies and Compensation

Since 1902, subsidies have represented a tactical means of directing forest management in Switzerland (OFEFP 1993a). For a landowner to receive support for forest management, government directives must be followed. The primary areas of support included plantation development and forest protection, transportation networks, forest salvage, and silvicultural restoration. The basic principle embodied in the FFL that *sustained wood harvest guarantees that the forest fulfills all of its diverse functions* (also known as the Kielwasser principle) has resulted in federal monies primarily being allocated to production activities. There are those who feel that this principle is outmoded and inappropriate to forest management requirements today (LSPN 1989; Hauselmann 1993).

Under provisions of the new law, compensation will be made available by the federal authorities for forest care, and indemnities will be allocated for management costs not covered by forest income (essentially from wood production). These changes mean that activities oriented toward forest quality and the promotion of biodiversity now become available for federal support (OFEFP 1993a).

Enforcement and Institutional Structures

Application of the Law

For the FFL to be enforced, an application ordinance is required. In the case of the new law, the ordinance states that in order to qualify for subsidies, the forest owner (private, commune, canton) must submit a management plan that includes a "natural" inventory of the area and a definition of management goals and their priorities. Management plans are developed at the field level, and are obligatory only to qualify for subsidies.

The new forest law also allows cantonal foresters to apply management prescriptions, such as which species to plant, to both public and private lands. While the policies and prescriptions laid out in the new law are generally accepted at the federal and cantonal levels, and by leaders of owners' associations, there is resistance among private landowners at the local level because they believe it intrudes on their property rights. Hauselmann (1993), in his study for the Forest Stewardship Council, found it impossible to lay his hands on a forest management plan and experienced resistance to the idea at the field level.

Institutional Structures

The Swiss structure for forest management is built on a federal model with weak central powers (OFEFP 1993a). The responsibility and technical capacity to implement laws and policies rest with the cantons. The responsibility of the federal government is to ensure that actions undertaken at these lower or decentralized levels conform with the law. The Swiss Forest Service consists of two levels distinct in their responsibilities—a federal level and a cantonal level.

At the federal level, the Federal Council oversees the development and application of policies, laws, and executive orders (OFEFP 1993a). The Federal Department of the Interior authorizes the clearing of all forest land over 5,000m^2, grants subsidies, and advises the political authorities on all policy questions concerned with forestry, the timber trade, and hunting. The Federal Forest Service provides training services and collects information on the state of Swiss forests.

At the cantonal level, the distribution of tasks and responsibilities is susceptible to variation depending on the canton. In general, the cantonal forest department supervises all forest activities at the cantonal level, and is primarily responsible for planning and coordination. The cantonal forest service operates as an extension service providing information, technical advice, and training. All cantonal forestry staff come under the forest service, including district foresters and forest guards who work with landowners in forest management.

As a result of the new FFL, forest policy in Switzerland becomes the product of consensus-seeking debate between the federal administration, various environmental protection organizations, the forestry services (canton and commune), and those that commercially exploit the forest (OFEFP 1995). The Forests Working Community is a group formed to bring these stakeholders together. Major issues that continue to raise debate include game animals, forest recreation, forest access, the expansion of the forest road network, and exploitation methods and levels.

Opportunity Costs

Benefit Values

A study is currently in progress in Switzerland attempting to determine the value of the basic outputs of the forest (Rauch-Schwegler 1994). The primary outputs identified include wood production (based on commercial value), recreational function (based on replacement value), protection function (replacement value), species diversity (replacement value), and game (commercial value). The initial results of the study are presented in Table 4.6.

By capitalizing the annual interest on this amount (approximately CHF 9 billion) at a rate of 1–2 percent, a basic rate for natural resources, the annual value of the forest in Switzerland is CHF 450–900 million (H. Schelbert-Syfrig, pers. comm. 1994, quoted in Rauch-Schwegler 1994).

The amounts determined for each output were arrived at by different methods and different studies. Recreational values were esti-

Table 4.6. Total Value of Benefits in Swiss Forests (in billions of CHF)

Wood production	0.45
Recreation	1.60
Protection function	3.50–4.10
Species diversity	2.80
Game production	0.01
Total	8.35–8.96

Source: Rauch-Schwegler 1994.

mated in Canton Zürich based on how much people were willing to pay to use the forest. In the Tessin, a study concluded that the loss of the forest around Lugano would represent a direct loss of CHF 200 million in revenues from forest recreation users as well as a loss of CHF 50 million in existence value (Schelbert-Syfrig 1988).

The forest's value for avalanche control is based on the finding that without the forest the national economy would have had to pay out CHF 106 billion in property damage claims over the last 30 years or CHF 3.5 billion per year (Altweg 1988). Using the cost of constructing protective devices to replace the role the forest plays results in a figure of CHF 125 billion over 30 years, or CHF 4.1 billion per year.

For every hectare of forest that disappears, with it go animals, plants, microorganisms, and, more generally, a host of ecosystem services. Though the Swiss forest covers only 29 percent of the country, it holds an estimated 70 percent of the 20,000 species recorded or thought to exist in Switzerland. Studies to determine the cost of reintroducing extirpated species estimate that the theoretical replacement cost of 14,000 forest species would be around CHF 2.8 billion, assuming all species would be returned (Vester 1986).

Some 48,000 roe deer, red deer, and wild boar were hunted in Switzerland in 1992 (Rauch-Schwegler 1994). The value of game meat in the country is estimated to be CHF 10 million per year. Herbs, natural teas, and fruits in the forest are freely accessible.

Some major values of the forest have not been calculated. The forest acts as an important watershed. Large quantities of water are filtered and stored by the forest, and used in electrical power generation. The city of Lausanne gets 25 percent of its water from forested watershed.

Option Values

Subsidies have long been used to direct landscape management in Switzerland, and it is unlikely that this will change in the near future. Because of the country's high living standards, most of its agricultural and natural resource products are not competitive on world markets. To maintain rural lifestyles that were the basis of the development of the Swiss Confederation, government has traditionally supplied a high level of financial inputs. This policy, along with strict zoning laws, has had the effect of maintaining the landscape as a mosaic of forest, agricultural land, and villages; urban sprawl is almost unknown. The use of subsidies has allowed the federal and cantonal governments to move resource management in specific directions.

The current budget deficit in Switzerland has resulted in higher taxes and reduced outlays for a number of sectors. Additionally, the recent completion of the Uruguay Round of the General Agreement on Tariffs and Trade and pressure from the European Union have resulted in a reexamination of current subsidy levels. Coupled with this, the deficit among public forest enterprises has been increasing regularly over the last five to eight years (Fig. 4.3). From a deficit of less

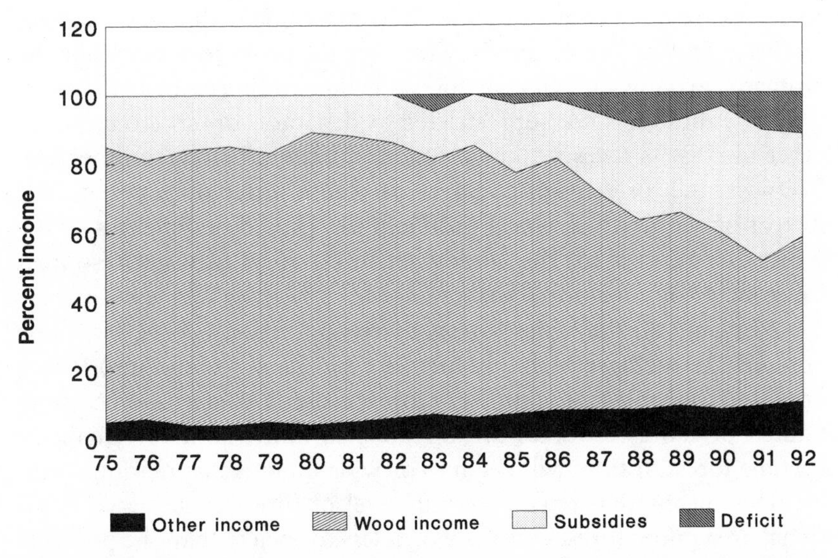

Figure 4.3. Income sources from the public forest enterprise in Switzerland. (*Source:* Rauch-Schwegler 1994.)

than CHF 10 million in 1987, it is now greater than CHF 44 million. Without subsidies from the Swiss Confederation and the cantons, who make available some CHF 150 million in support, the deficit would approach CHF 200 million (OFEFP 1993a, 1995). It is estimated that approximately half of the public forestry enterprises in Switzerland are relying on subsidies, 35 percent are showing a profit, and the remaining 15 percent are breaking even (OFEFP 1993a). With the addition of subsidies for forest conservation activities, the pressure on an ever-shrinking pot of funds is even greater. It is conceivable that with fewer funds available for forest production support, and with zoning laws limiting conversion of land under forest, owners may decide to leave forest land in a more "natural" state, thereby avoiding the losses being experienced in the forest industry, and possibly still benefit from government support for their activities. This action would effectively preserve the forest for future, as yet unknown, benefits and keep use options open. However, it would also incur opportunity costs.

Costs Incurred by Forest Owners

The central economic problem concerns how to allocate and distribute scarce resources between competing demands. This implies that resources are finite because they are in competition with other resources (cf. Price 1990). This is certainly true for the Swiss forests. While the ultimate carrying capacity for forests is limited by ecological factors, their distribution is further restricted by demands for land to be put to other use. Swiss forests take up space that could be put to other use. Likewise, the forest itself could be used in a different way (e.g., for recreation rather than wood production). Thus different uses of the forest carry a cost for the landowner in the form of other resources foregone or opportunity costs.

Not only do the Swiss forests carry opportunity costs, but these are likely to increase with any further conversion of land into forest, or with stronger laws to limit how forest is used. As use becomes more limited or more resources are foregone to allow land to remain or become forest, forests will begin to provide diminishing marginal utility. This means that once a certain level of forest exists, each additional area under forest cover is worth less to society than the previous one. Based on these assertions, economists might argue that there is a "socially optimal" amount of forest or mix of forest uses. Beyond these levels, the cost of providing additional uses or forest cover to society exceeds the benefits they provide.

This implies that forests are not free resources. They carry costs, and these costs have to be justified by the benefits provided to society. If forests provided no benefits to society at all, or if some of the benefits are not recognized (erosion and avalanche protection in the early years and biodiversity conservation today), they would likely be eliminated and replaced with uses of greater perceived value such as agriculture or housing developments. Under this analysis, those who care about the continued existence of forests should seek to maximize their value to society (Table 4.6).

Discussion

Under the new FFL of 1993, equal emphasis is given to the forests' multiple functions. The concept is an ecosystem approach that includes not only the commercial aspects but those of biological diversity, ecosystem function, and vitality. While this is an important step forward in policy and legislation, Swiss forests continue to be judged on their ability to produce timber. The concept of sustained yield is assumed to be an adequate measure of sustainability. However, it is becoming increasingly clear that financial considerations alone measure only a fraction of all the important forest values (cf. Franklin 1993). While the area of Switzerland under forest cover has remained relatively stable, the question of forest quality is more directly related to biological diversity. Specifically, if biodiversity is to be maintained, the Swiss forests must be viewed from a broader perspective than simply timber production.

The Fallacy of Sustained-Yield Forestry

Traditional forestry, as has been practiced in Switzerland, is based on managing for an oversimplified state (stand structure, stocking, and so on). In fact, much of the current ecological/forest information is based on such conditions (cf. OFEFP 1989, 1993a; Office fédéral des questions conjoncturelles 1991). The concept of sustained-yield forestry was developed under a clearly defined set of conditions. These conditions included that all areas of forest could be managed, that a demand for wood existed, and that there would be an adequate workforce to supply this demand.

Because of the rough terrain throughout much of Switzerland, it has not been possible to manage all areas of the forest. However, in those areas most accessible—parts of the Jura, the Plateau, and the Prealps—stand structure and stocking have been radically changed from natural conditions, resulting in extensive impacts on the biological diversity of those areas.

While the demand for local wood still exists, it has decreased or been removed by technological change. For example, the introduction of more efficient transportation networks meant that wood needed in areas where supplies were expensive or difficult to obtain, such as in the mountains, could be supplied more cheaply from areas where trees grew faster and could be harvested at lower cost, such as on the Plateau. Additionally, the introduction of new fuel sources, such as fuel oil and electricity, minimized the demand for fuelwood.

The question of an adequate workforce is closely linked to the issue of demand. If the demand for wood remains high, prices will be high enough to pay people to harvest it. Conversely, when demand is low, the cost of employing people becomes excessive, and the harvest will decrease. This has been the case in Switzerland since the 1950s when other industries came to dominate the economy. Today almost two-thirds of the country's gross national product is generated by service industries, about one-third by other industry and trade, and less than 5 percent through agriculture and forestry. In Switzerland as a whole, wage levels in forestry have increased far faster than wood prices since 1950, particularly since 1970; wood prices actually decreased in the 1980s (Affolter 1985). In addition, forestry is not perceived as a particularly desirable profession, since wages are relatively low, accident rates are high, and many positions are part-time. The number of full-time forestry jobs decreased by about 40 percent from the 1950s to the 1980s (Butora 1984; Schwingruber 1985).

Since many of the conditions assumed for sustained-yield forestry based principally on wood production are no longer valid, policies based on this philosophy are inappropriate for managing much of the forest. Yet such policies are still at the core of forestry as practiced in Switzerland today (cf. OFEFP 1989, 1993a, 1995). As a consequence, the structural and functional diversity of the forests has been degraded to the point that the provision of public goods has been negatively affected.

A Different Approach to Forestry in Switzerland

Given that the provision of public goods is a primary objective of the FFL, then a different, more holistic philosophy of forestry is necessary if management is to be effective. Dudley (1992) has put forward four broad criteria by which a more holistic view of forest management might be developed.

Authenticity measures how closely a forest mirrors the natural forest of the area and, in ecological terms, defines optimal conditions for maintaining biodiversity. In countries such as Switzerland, most, if not all, reference forests have long disappeared, making it difficult to know how "authentic" a forest is.

Forest health assesses the forest with respect to disease or pollution damage, and health of forest flora and fauna. These criteria include factors outside the direct control of the owner or manager of the forest, such as air pollution and climate change. Other aspects of forest health are more directly impacted by management practices such as the use of chemicals and the choice of tree species. In Switzerland, forest health has been steadily decreasing, most notably through needle loss and tree mortality believed to be linked to air pollution and insect infestation.

Environmental benefits include benefits that extend beyond the boundaries of the forest such as biodiversity and genetic resource conservation, soil and watershed protection, local climate benefits, carbon sequestration, and climate stabilization. All of these are essentially public goods. Treatment of these issues is now required in forest plans under the new FFL if owners are to benefit from federal compensation. Specific proposals to change forest management practices to more holistic approaches are now being put forward in Switzerland (LSPN 1989).

Value to humans is a measurement of the economic and social value of the forest area including wood products such as timber, pulp, and fuelwood; nontimber products such as fruits, nuts, forage, game animals, medicines; recreational activities; and esthetic and cultural qualities. Implicit is that forest *value* includes both the market value of, for example, timber, and the nonmarket value of the existing forest to different user groups (hiker, mushroom collector, hunter, and so on). The importance of these will differ considerably in different forest areas and among different users. Efforts are now being made in Switzerland to value all of the forest's services realistically (Rauch-Schwegler 1994).

These criteria can be difficult to reconcile with each other. For example, as has been demonstrated in this chapter, there are clear conflicts between achieving a high level of wood production and maintaining the forest ecosystem in a natural state. A nature reserve and an exotic conifer plantation may both have high value to people in terms of providing recreational resources, but a reserve will score high on biodiversity protection and low on economic value of timber production, while the reverse will be true for the plantation. Properly designed forest policy must make room in the forest estate for all of these, sometimes conflicting, benefits.

The ecological systems upon which Swiss forestry depends have limits. By neglecting certain aspects of these systems in pursuit of commodity extraction as the primary output, many values of the forest ecosystem have been compromised or lost. It is important that the move beyond simple commodity management of forests continues in Switzerland, to the management of ecological systems (cf. LSPN 1989). This will require recognizing them as the dynamic units that they are, functioning at a multitude of spatial and temporal scales, with multiple values for different sectors of society.

Acknowledgments

This chapter was funded as part of a wider project supported by WWF–World Wide Fund for Nature under project 9Z0534.04. A number of staff at WWF provided guidance and input to this review including Chris Elliott, 'Wale Adeleke, Maria Boulos, and Monika Borner. Curtis Freese, project manager, provided direction throughout the process. Support was received from a number of individuals, organizations, and institutions. Those providing information or comments on the document included Martin Gerig, Bibliothek, Eidgenössische Forschungsanstalt für Wald, Schnee und Landschaft, Birmensdorf; Christian Marti, Schweizerische Vogelwarte, Sempach; Urs Breitenmoser, Swiss Lynx Project, University of Bern; Willi Zimmermann, Professur für Forstpolitik, Eidgenössische Technische Hochschule (ETH), Zurich; Martin Spinnler, assistant chair of Forest Policy and Forest Economics, ETH, Zurich; Urs Tester, Ligue Suisse pour la Protection de la Nature (LSPN); Phillipe Domont, Découvrir la Forêt, Yverdon-les-Bains; Eric Treboux, inspecteur des forêts, District de Nyon; Office fédéral de l'environnement, des forêts et du paysage (OFEFP), Centre suisse de documentation pour la recherche sur la biologie de la faune sauvage, Zurich and Bern; and Eidgenössische Drucksachen- und Materialzentrale, Bern. We are grateful to all of these individuals, organizations, and institutions for their time and help.

References

Affolter, E. 1985. Zunehmende Zwangsnutzungen: Holzmarkt und Holzverwendung aus der Sicht der Waldwirtschaft. *Schweizerische Zeitschrift für Forstwesen* 136:805–818.

Altweg, D. 1988. *Scénarios des conséquences du dépérissement des forêts—les avalanches pourraient nous coûter des milliards. Rapport Sanasilva sur les dégâts aux forêts 1988.* Birminsdorf: Office fédéral de l'environnement, des forêts et du paysage.

Amann, F. 1994. Der Brutvogelbestand im Allschwilerwald 1948/9 und 1992/3. *Der Ornithologische Beobachter* 91:1–23.

Auer, C. 1956. Die volkswirtschaftliche Bedeutung des Gebirgwaldes. *Schweizerische Zeitschrift für Forstwesen* 107:319–326.

Barth, W.-E. 1987. *Praktischer Umwelt- und Naturschutz.* Hamburg: Verlag Paul Parey.

Bowles, J. B. 1963. Ornithology of changing forest stands on the western slope of the Cascade Mountains in Central Washington. M.Sc. thesis, University of Washington, Seattle.

Broggi, M. F., and G. Willi. 1993. *Réserves forestières et protection de la nature.* Contributions à la protection de la nature en Suisse, no. 14. Basel: Ligue Suisse pour la protection de la Nature (LSPN).

Brown, B., J. Henderson, T. McShane, R. Stephens, and G. van der Ray. 1980. *Riparian vegetation/streambank stability inventory.* Olympia, Washington: USDA Forest Service, Olympic National Forest.

Bruderer, B., and R. Luder. 1982. Die "Rote Liste" als Instrument des Vogelschutzes—Erste Revision der Roten Liste der gefährdeten und seltenen Brutvogelarten der Schweiz, 1982. *Der Ornithologische Beobachter,* suppl. to vol. 79.

Bruderer, B., and W. Thoenen. 1977. *Liste rouge des espèces d'oiseaux menacées et rares en Suisse.* Sempach: Comité suisse pour la protection des oiseaux.

Butora, V. 1984. Programme zur Steigerung der Arbeitssicherheit in der schweizerischen Forstwirtschaft. *Schweizerische Zeitschrift für Forstwesen* 135:785–792.

Cornes, R., and T. Sandler. 1986. *The theory of externalities, public goods and club goods.* Cambridge, England: Cambridge University Press.

Covington, R. 1993. Worth saving: The Alps. *Travel & Leisure* October: 68–71.

Debrot, S., and D. Meyer. 1989. Dégradation de l'écosystème forestier: Analyse et ébauches de solutions. *Schweizerische Zeitschrift für Forstwesen* 140:965–976.

Dudley, N. 1992. *Forests in trouble: A review of the status of temperate forests worldwide.* Gland, Switzerland: World Wide Fund for Nature.

Ellenberg, H., and F. Kloetzli. 1982. Waldgesellschaften und Waldstandorte der Schweiz. *Mémoires de l'Institut fédéral des recherches forestières.* 48: 588–930.

Ellenberg, H., R. Mayer, and J. Schauermann. 1986. *Ökosystemforschung—Ergebnisse des Sollingsprojektes.* Stuttgart: Verlag Eugen Ulmer.

Franklin, J. F. 1993. The fundamentals of ecosystem management with applications in the Pacific Northwest. In *Defining sustainable forestry,* ed. G. H. Applet, N. Johnson, J. T. Olson, and V. A. Sample, 127–144. Washington, D.C.: Island Press.

Franklin, J. F., K. Cromack, W. Denison, A. McKee, C. Maser, J. Sedell, F. Swanson, and G. Juday. 1981. *Ecological characteristics of old-growth Douglas-fir forests.* USDA Forest Service General Technical Report PNW-118. Portland, Oregon: Forest Service, Pacific Northwest Forest and Range Experiment Station, USDA.

Franklin, J. F., T. Spies, D. Perry, M. E. Harmon, and A. McKee. 1986. Modifying Douglas-fir management regimes for nontimber objectives. In *Douglas-fir stand management for the future,* ed. C. D. Oliver, D. P. Hanley, and J. A. Johnson, 373–379. Seattle: College of Forest Resources, University of Washington.

Gebhard, J. 1985. *Nos chauves-souris.* Basel: Ligue Suisse pour la Protection de la Nature.

Gonseth, Y. 1987. Atlas de distribution des papillons diurnes de Suisse (Lepidoptera, Rhopalocera), avec liste rouge. *Documenta faunistica helveticae* 6, Neuchâtel: Centre suisse de cartographie de la faune.

Grossenbacher, K. 1988. Atlas de distribution des amphibiens de Suisse. *Documenta faunistica helveticae* 7:1–207.

Harmon, M. E., J. F. Franklin, F. J. Swanson, P. Sollins, S. V. Gregory, J. D. Lattin, N. H. Anderson, S. P. Cline, N. G. Aumen, J. R. Sedell, G. W. Lienkaemper, K. Cromack, and K. W. Cummins. 1986. Ecology of coarse woody debris in temperate ecosystems. In *Advances in ecological research* 15:133–302. New York: Academic Press.

Hauselmann, P. 1993. *Forest Stewardship Council: Swiss consultation report.* Granges-près-Marnand: FSC.

Heberlein, T. A. 1977. Density, crowding and satisfaction. In *Proceedings, River recreation management and research symposium,* ed. D. W. Lime and C. A. Fasick, 67–76. St. Paul, Minnesota: U.S. Forest Service General Technical Report NC-28.

IFN. 1991. *Inventaire forestier national suisse.* Birmensdorf: Institut fédéral de recherches sur la forêt, la neige et le paysage, section inventaire forestier national.

Krutilla, J. V., and A. C. Fisher. 1985. *The economics of natural environments.* Washington, D.C.: Resources for the Future.

Langenegger, H. 1984. Mountain forests: Dynamics and stability. In *The transformation of Swiss mountain regions,* ed. E. A. Brugger, G. Furrer, B. Messerli, and P. Haupt. Bern: Messerli.

Leibundgut, H. 1984. *Unsere Waldbäume.* Stuttgart: Huber.

LSPN. 1989. *Thèses pour d'avantange de nature en forêt.* Contributions à la protection de la nature en Suisse, no. 12. Basel: Ligue suisse pour la protection de la nature.

MacArthur, R. H. 1984. *Geographical ecology: Patterns in the distribution of species.* Princeton: Princeton University Press.

Mader, H.-J. 1979. Die Isolationswirkung von Verkehrsstrassen auf Tierpopulationen untersucht am Beispiel von Arthropoden und Kleinsäugern der Waldbiozönose. *Schriftenreihe für Landschaftspflege und Naturschutz* 19:1–127.

Mader, H.-J. 1981. Der Konflikt Strasse—Tierwelt aus ökologischer Sicht. *Schriftenreihe für Landschaftspflege und Naturschutz* 22:1–99.

Mannan, R. W. 1982. *Bird populations and vegetation characteristics in managed and old-growth forests, northeastern Oregon.* Doctoral dissertation, Oregon State University, Corvallis.

Marti, C. 1992. Vom Inventar zum Schutz. *Ornis* December: 6:31–33.

Marti, C. 1993. *Aide-mémoire: Sylviculture et Grand Tétras.* Berne and Sempach: Office fédéral de l'environnement, des forêts et du paysage (OFEFP) and Station ornithologique suisse.

Marti, C., P. Meile, and U. Bühler. 1988. Sylviculture et grand tétras. *Forestier Suisse* 124(7–8):36–37.

Mayer, H. 1976. *Gebirgswaldbau-Schutzwaldpflege.* Stuttgart: Fischer.

Mehr, C. 1989. Are the Swiss forests in peril? *National Geographic Magazine* 175:637–652.

Office fédéral de l'environnement, des forêts et du paysage (OFEFP). 1989. *La forêt suisse aujourd'hui: Une interprétation de politique forestière de l'inventaire forestier national suisse (IFN).* Berne: Office fédéral central des imprimés et du matériel.

OFEFP. 1993a. *La forêt suisse: Un portrait.* Documents environnement no. 3, forêt. Berne: OFEFP.

OFEFP. 1993b. *Zum Verhältnis zwischen Forstwirtschaft und Natur- und Landschaftschutz.* Berne: OFEFP.

OFEFP 1995. *La développement durable des forêts suisses.* Berne: OFEFP.

Office fédéral des questions conjoncturelles. 1991. *Prospectives d'approvisionnement en bois ronds provenant de la forêt suisse.* Berne: FNP (IFN), ASEF, OFEFP.

Perrig, C.-A., and A. Fux. 1945. *Recueil des lois, décrets, arretês et instructions du Canton du Valais concernant l'économie forestière.* Sion: Département Forestier du Canton du Valais.

Price, M. F. 1990. *Mountain forests as common-property resources: Management policies and their outcomes in the Colorado Rockies and the Swiss Alps.* Forstwissenschaftliche Beiträge, 9. Zurich: ETH.

Rauch-Schwegler, T. 1994. *La forêt, capital et intérêts: Combien vaut la forêt suisse?* Yverdon-les-Bains: Découvrir la forêt.

Samuelson, P. A. 1954. The pure theory of public expenditure. *Review of Economics and Statistics* 36:387–389.

Schelbert-Syfrig, H. 1988. *Wertvolle Umwelt. Ein wirtschaftswissenschaftlicher Beitrag zur Umwelteinschätzung in Stadt und Agglomeration Zürich.* Zurich: Züricher Kantonalbank.

Schütz, J.-P. 1982. La sylviculture et l'écologie se rejoignent dans le traitement régulier de nos forêts. *Schweizerische Zeitschrift für Forstwesen* 133:5–18.

Schwingruber, C. 1985. Ergebnisse 1984 der Lohnerhebung in der schweizerischen Forstwirtschaft. *Wald und Holz* 66:811–819.

Speight, M. 1989. Les invertébrés saproxyliques et leur protection. Conseil de l'Europe, Collection Sauvegarde de la nature, no. 42. Strasbourg: Conseil de l'Europe.

Tromp, H. 1980a. Hundert Jahre forstliche Planung in der Schweiz. *Mitteilungen Eidgenössische Anstalt für das forstliche Versuchswesen.* 56:253–267.

Tromp, H. 1980b. Bannwälder. *Mitteilungen Eidgenössische Anstalt für das forstliche Versuchswesen.* 56:324–328.

Vester, F. 1986. *Ein Baum ist mehr als ein Baum.* Stuttgart: Kösel Verlag.

von Fellenberg, G. 1981. *Der Wald in der Natur. Vol. I, Berner illustrierte Enzyklopaedie,* 88–105. Bern: Büchler and Co.

Wilhelm, M. B. 1988. Routes forestières: Le dialogue est possible. Communiqué de presse du Service d'informations forestières, Solure.

FIVE

Management of Living Resources in the Matang Mangrove Reserve, Perak, Malaysia

N. Gopinath and P. Gabriel

Malaysian mangrove forestry is exemplified by the Matang Mangrove Forest. Stretching over an area of 40,151 hectares (ha) in the state of Perak, some 300 kilometers (km) north of the national capital, Kuala Lumpur, Matang's management regime boasts of a genealogy that goes back 90 years. The wealth of forest records and historical data has enabled successive generations of forestry authorities to maintain Matang as a source of quality timber for the production of charcoal, poles, and firewood. This has given Matang the distinction of being described as one of the best-managed mangrove forests in the world (McNicoll 1992).

While Matang's value in the nearshore marine ecosystem and in coastal erosion control has been recognized throughout its history, the reserve's management plans have never addressed these functions. In addition, Matang supports a small but growing visitor traffic based on ecotourism. In the first holistic review of Matang's monetary and nonmonetary values, Cheng (1994a) found little information on the Matang mangroves apart from their value as a forestry resource. The sustainability of other resource uses that are dependent on the Matang mangroves has received little attention.

We attempt to evaluate comprehensively Matang's myriad resources and the extent to which the existing management regime is socio-economically and ecologically sustainable. The main objectives of the study are (1) to provide a comprehensive overview of the biological and economic resources of the Matang mangroves; (2) to examine how well the present management regime addresses the diverse resource values, including biodiversity conservation, of the Matang Reserve; and (3) to explore how the present management regime can be modified so as to take a more holistic approach in the management of the mangrove.

Mangrove Forests in Malaysia

Extent and Type

Of around 653,400 ha of mangrove forests in Malaysia, Peninsular Malaysia has about 120,000 ha, mainly along the sheltered west coast that borders the Straits of Malacca. Malaysia is probably the geographical center for several genera of mangrove, especially *Rhizophora, Bruguiera, Sonneratia, Avicennia, Ceriops,* and *Lumnitzera* (Chapman 1975).

The extent and exact limits of these forests are continually changing (Wyatt-Smith 1963). In estuaries, where rivers meet with tides, there is extensive silt deposition (Chapman 1975). When sufficient silt has accumulated, the estuary bed becomes gradually colonized by pioneer mangrove species such as *Avicennia* spp. and *Sonneratia* spp. Such an accreting mangrove shoreline is often characterized by a dense low crop of trees advancing seawards (Carter 1959). On the landward side, continuous accretion and deposition lead to a progressive change into dryland forest known locally as *hutan darat.* The hutan darat is considered an inferior forest and is characterized by a heterogeneous crop of poorly formed trees.

Watson (1928) divided Malaysia's mangrove forests into five distinct vegetation types based on species composition and dominance:

1. The Api-api-Perapat (*Avicennia-Sonneratia*) type is divided into two subtypes: (a) the shoal type, dominated by *Avicennia alba,* which is formed offshore by deposition of sand and silt; (b) the accretion type, dominated by *Avicennia marina,* which is formed by the combined actions of tidal currents and silt brought down by rivers.

2. The Berus (*Bruguiera cylindrica*) type, dominated by almost pure stands of *B. cylindrica,* occurs primarily behind the accretion type of *A. marina.*

3. The Lenggadai (*Bruguiera parviflora*) type occurs near riverbanks and is associated with *Rhizophora* forest. In poorly managed *Rhizophora* stands, *B. parviflora* predominate.

4. The Bakau (*Rhizophora*) type, dominated by *Rhizophora apiculata,* occurs inland behind the *B. cylindrica* type. *Rhizophora mucronata* is the dominant species along riverbanks.

5. The Tumu (*Bruguiera gymnorhiza*) type marks the final stage of mangrove development and the onset of transition to dryland forest. Natural regeneration of mangrove is often limited because of abundant *Acrostichum* ferns that infest all forest floor gaps. In addition, the mud crabs build extensive mounds that are often above the reach of tidal influence and lead to formation of dryland forest.

Environmental and Economic Significance

Mangrove forests in Malaysia have been managed and preserved because of their ecological and economic significance since the turn of the century. Most (468,437 ha, or 72%) of the mangrove forests in the country have been designated reserves and are managed on a 20- to 30-year cutting cycle. However, the declining significance of wood products from mangrove forests led to a cessation of commercial exploitation of mangroves in 1986, except in areas designated for local production. In addition to forest products, the mangrove forests maintain shorelines and contain erosion, and play a critical role in the marine ecosystem and associated fisheries. A detailed review of these roles is outside the scope of this chapter, but they are summarized below.

Forestry

Mangroves have traditionally been exploited for a variety of forestry products (NATMANCOM 1986). In economic value per hectare, they rank among the most valuable of forest types. Tang et al. (1984) estimated the economic value of forest products from mangrove forests of Peninsular Malaysia at RM 28,300/ha (US$1 = RM 2.50) compared with inland dipterocarp forests at RM 11,500/ha. While significant at a local level, mangrove timber constitutes less that 0.02 percent of the overall export value of Malaysian timber, which amounted to RM 12 billion in 1993 (Dept. of Forestry 1994; Perak State Forestry Dept. 1994). The major mangrove forest products include charcoal, poles, and fuelwood. Mangrove poles are used extensively as piling material due to their resilience and ability to resist decay, especially in anaerobic conditions.

Less significant, but still important at a local level, are mangrove bark, thatch, palm sugar, and wood chips. The leaves of the palm, *Nypa fruticans*, are used extensively in rural communities as roofing material (Fong 1984), while in Sabah and Sarawak, mangroves were once a major source of wood chips for the manufacture of rayon (Liew et al. 1975; Chai 1977; Liew 1977; Chai and Lai 1984; Phillips 1984).

Maintenance of the Nearshore Marine Ecosystem

Mangroves play a critical role in the maintenance of the nearshore environment and its productivity. They act as nutrient traps, sequester heavy metals, and break down organic pollutants, thereby protecting the fragile and commercially valuable seagrass and coral reef ecosystems from siltation and contaminants from land-based activities. Leaf litter from mangrove trees is exported as detritus to coastal waters. During the process of decomposition, the protein content of the decomposing material is enhanced by microbial activity, making it an attractive food source in coastal waters. Gong and Ong (1990) estimated the biomass released by the Matang Reserve to be about 1,015,980 metric tons (MT) annually. This huge export of food material supports commercially valuable species of fish and shrimp (MacNae 1974; Leh and Sasekumar 1984).

Mangroves are also vitally important in providing nursery grounds for commercially important fish and shrimp. The use of mangroves by the fry through subadult stages of milkfish (*Chanos chanos*) and some penaeid shrimp is well known. However, other valuable species such as mullets (*Mugil* sp.), siganids, carangids, and serranids, together with other small fish that serve as food for more valuable species, also depend on mangroves as nursery grounds. Leh and Sasekumar (1984) and Thong and Sasekumar (1984) reported the presence of significant amounts of mangrove detritus in the guts of many mangrove-associated species. Mangrove forests and their associated waterways have been shown to harbor four to ten times the densities of juvenile fish and shrimp compared to adjacent mudflat and seagrass habitats (Robertson 1991).

The role of mangroves is particularly crucial in Malaysia, since most of the country's fish landings come from nearshore and coastal waters. Jothy (1984) estimated that 32 percent of fisheries landings in the country are associated with mangroves. In 1993, there were more than 41,479 artisanal fishermen in the country, most of whom fished within three nautical miles of the coastline (Dept. of Fisheries 1994). Their combined landings amounted to 280,776 MT, or 27 percent of the total catch of 1,047,350 MT. The contribution of coastal fishing grounds goes beyond just catch volume. Most of the higher valued fisheries are located in coastal waters. This is exemplified by its wholesale catch value of RM 524,317,175, which amounted to 31 percent of the value of the total fish landings in 1993.

Shrimp are particularly reliant on mangroves (D' Cruz 1991). The dependence of many species of shrimp on mangrove forests in Malaysia has been reported by Chong and Sasekumar (1990) and Sasekumar (1980). In 1993, nationwide landings of shrimp alone amounted to 81,694 MT valued at RM 450,461,790. The largest shrimp grounds in the country are off the states of Perak (which accounted for 37% of the landings) and Selangor (23%), both on the west coast of Peninsular Malaysia, and Sabah (16%) and Sarawak (12%). These states possess some of the largest tracts of mangrove in the country.

Along with capture fisheries, the farming of the blood clam (*Anadara granosa*) is undertaken extensively on the mangrove mudflats off the west coast of Peninsular Malaysia. In 1993, more than 5,040 ha of mudflats were utilized for this purpose, producing more than 77,755 MT of clams valued at RM 26.8 million, mainly for domestic consumption. Malaysia is considered one of the world's largest producers of the commodity.

Prevention of Shore Erosion and Storm Damage

Mangrove areas act as a buffer on the landward side, protecting the shoreline and preventing damage to coastal dwellings and infrastructure (Snedaker and Getter 1985; PHILNATMANCOM 1987). On the seaward side, mangroves act as barriers against coastal and riverbank erosion, dampen storm surges, and aid in stabilizing the coastline and riverbanks (Carlton 1974). Erosion problems have frequently arisen because of unplanned mangrove clearing (Chan 1984).

General Description of the Matang Mangrove Forest

The Matang Mangrove Forest Reserve was established in 1902, and the first plan for forest exploitation was developed in 1904 (Noakes 1952). The reserve lies on the Straits of Malacca, at approximately 4° 45′ N and 100° 35′ E. Its southern and northern boundaries are delineated by the Kuala Pancor and Kuala Gula estuaries, respectively. The reserve is crescent-shaped, about 52 km in length and 13 km wide in the middle. It covers 40,151 ha (B. K. Gan, pers. comm. 1995), accounting for 40 percent of the total mangrove area in Peninsular Malaysia (Fig. 5.1).

Six major river estuaries intersect the reserve (Kuala Jarum Mas, Kuala Pasir Hitam, Kuala Jaha, Kuala Larut, Kuala Selinsing, and Kuala Kalumpang), besides the Kuala Gula and Kuala Pancor rivers at the boundaries. Except for some scattered areas of dryland forest, practically the whole reserve (95%) is tidal swamp. While the tidal range in the area is about 3 meters (m), the topography allows for a wide range of inundation patterns ranging from land flooded by every tide to land that is inundated only during the highest spring tides. Most of the reserve (about 35,000 ha, or 87%) is considered productive forest, while the balance is considered unproductive from a forestry standpoint, consisting of newly accreted forest and successional dryland forest (Gan 1991; Cheng 1994a).

Seaward, outside the administrative boundary of the reserve, are wide expanses of mudflat that extend several hundred meters from the shore. These mudflats, which represent accreting coastlines, are not vegetated and are largely inundated during high tides.

The settlements in the Matang mangrove have been described by Chan (1986). The main communities in the Matang mangrove areas are the Chinese fishing villages of Kuala Sepetang, Bagan Pancor, Kampung Pasir Hitam, and Kuala Sangga Besar. The majority of the male population living in these villages are full-time commercial fishermen. In fact, all the villages are major trawler bases (Choy 1991). Satellite settlements are often found scattered along the banks of the mainland or island mangrove estuaries not far from these villages. Many are

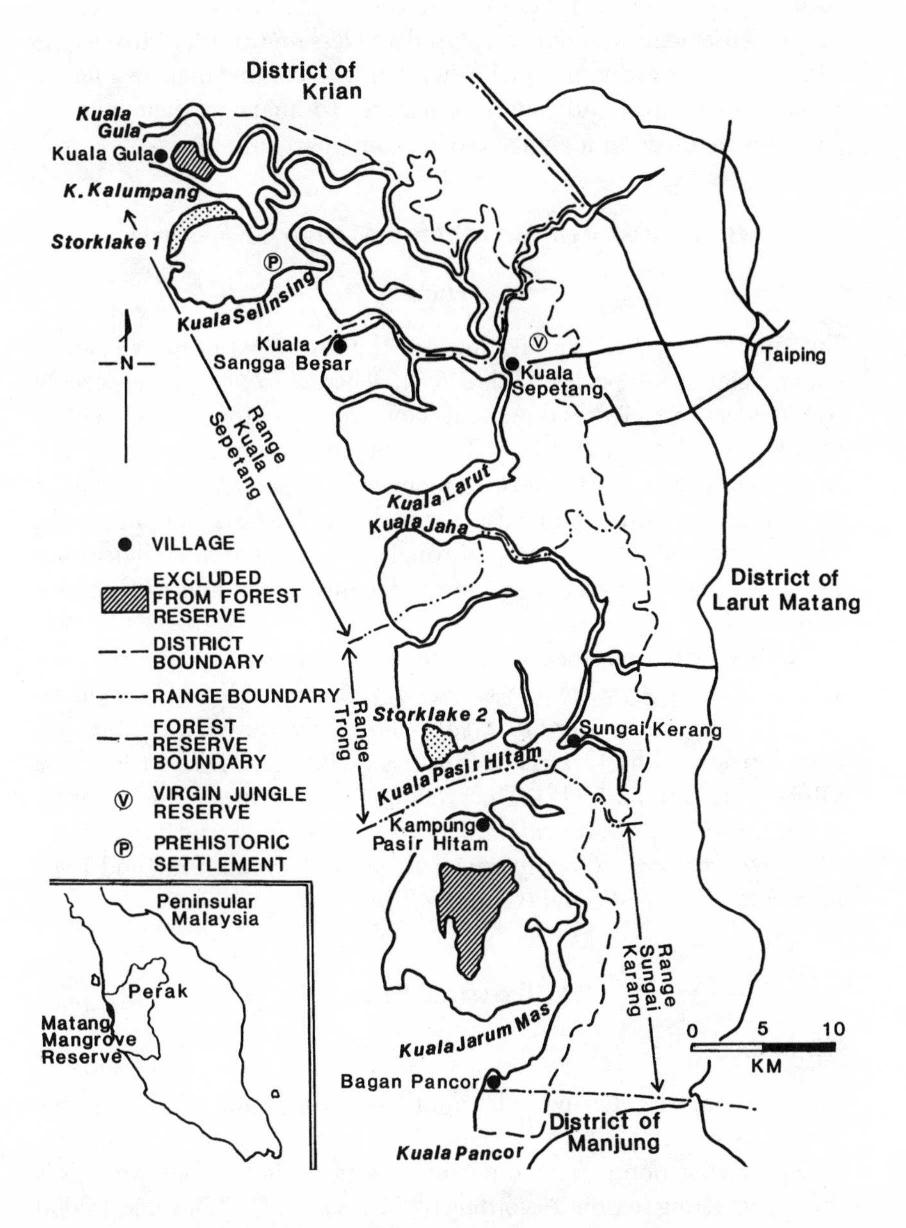

Figure 5.1. Map of Matang Mangrove Forest Reserve.

isolated and are accessible only by boat. There are no accurate population census data for the villages or satellite settlements.

Smaller settlements of Malays are usually found along the upstream banks of the mangrove tidal rivers. Unlike the Chinese villages, the Malay settlements are more dispersed and less mercantile. Most of the Malay villagers are traditional fishermen who confine their fishing activity to nearshore and estuarine waters. To augment their income, some are involved in logging activities on a part-time basis.

Biodiversity of the Matang Mangrove Reserve

Flora

The only comprehensive inventory of the reserve's flora has been of commercial tree species (Table 5.1). The *Rhizophoras,* especially *R. apiculata,* are clearly dominant, covering more than 40 percent of the reserve (Silvius et al. 1987). This dominance is not entirely natural. *Rhizophora* spp. are the most economically important tree species in the mangrove, forming the basis for both the charcoal and pole trade. The management of the forest through reforestation and silviculture has thus been oriented toward sustaining and expanding *Rhizophora* output.

Other mangrove species have more restricted distributions, with species such as *A. marina, Sonneratia alba,* and *B. cylindrica* dominant in some areas. *Ceriops tagal* is not native to Matang, but has been introduced by forestry authorities (I. Bukhari, District Forestry Office, pers. comm. 1995). The predominant tree species of Matang's dryland forest are *R. apiculata, Heritiera littoralis, Ficus microcarpa, Flacourtia jangomas, Oncosperma tigillarium, B. gymnorhiza,* and *Teijsmanniodendron hollrungi* (Chan 1989).

Terrestrial Fauna

Avifauna

The presence of extensive tidal mudflats for foraging and mangrove forests as roosting sites has made the Matang Mangrove Reserve a major congregation point for avifauna, particularly for migratory birds during wintering season. According to Silvius et al. (1987), some 45,000 shorebirds have been counted during peak passage along the west coast

Table 5.1. Commercially Important Trees of the Matang Mangrove Forest

Family	Scientific name	Common name
Avicenniaceae	*Avicennia alba*	Api-api putih
	Avicennia officinalis	Api-api ludat
Palmae	*Nypa fruticans*	Nipah
Rhizophoraceae	*Rhizophora apiculata*	Bakau minyak
	Rhizophora mucronata	Bakau kurap
	Bruguiera gymnorhiza	Tumu merah
	Bruguiera cylindrica	Berus
	Bruguiera parviflora	Lenggadai
	Bruguiera holinesii	Berus mata buaya
	Ceriops tagal	Tengar
Sonneratiaceae	*Sonneratia alba*	Perepat
	Sonneratia caseolaris	Berembang
	Sonneratia ovata	Gedabu
Meliaceae	*Xylocarpus granatum*	Nyireh bunga
	Xylocarpus gangeticus	Nyireh batu
Pteridaceae	*Acrostichum aureum*	Piai raya
	Acrostichum speciosum	Piai lasa
Luguminosae	*Intsia retusa*	Merbau ipil

Source: Department of Forestry, Peninsular Malaysia 1994.

of Peninsular Malaysia, of which the Matang area forms the most significant part. At an annual turnover rate of three to six times a year, it is estimated that between 200,000 and 400,000 shorebirds may use these areas during migration (Parrish and Wells 1984). Surveys of migratory birds in Matang estuaries during the period 1985–92 revealed that some 24 species of migratory birds use the reserve during autumn (Aug.–Nov.), winter (Dec.–Feb.), and spring (March–May) (Thompson 1994).

The Matang Mangrove Reserve also provides habitat for populations of rare birds such as the milky stork (*Mycteria cinera*) and lesser adjutant stork (*Leptoptilus javanicus*). It is the only place in Peninsular Malaysia to support a viable population of the endangered milky stork with an estimated population of about one hundred individuals (Silvius et al. 1987). Significant breeding populations of another rare bird, the black-crowned night heron (*Nycticorax nycticorax*), have been found just outside the reserve (Ratnam et al. 1987). In total, 154 bird species from 41 families have been observed in the Matang Mangrove Reserve (Siti Hawa 1994).

Mammalian Fauna

There is no comprehensive study of mammals in the reserve. Shariff (1984) reported the presence of significant populations of the endangered smooth otter (*Lutrogale perspicillata*) in the Kuala Gula estuary. A survey by Burhanuddin et al. (1994) found wild pig (*Sus scrofa*), smooth otter, small-clawed otter (*Aonyx cinera*), long-tailed macaque (*Macaca fascicularis*), and leopard cat (*Felis bengalensis*) to be common. As noted later, the cave bat (*Eonyeteris spelea*) and flying fox (*Pteropus vampyrus*), both of economic importance, use the forest of the reserve.

Cheng (1994a) reported that the tiger (*Panthera tigris*) once inhabited the Matang mangroves. Interviews with long-time residents of the area indicated that tigers tended to inhabit the dryland forest and that they had not been sighted in recent years. The tiger is probably restricted to inland forests due to the removal of the corridor that originally linked it with the mangrove forest.

Reptilian Fauna

Reptiles of the Matang Mangrove Reserve include the monitor lizard (*Varanus salvator*) and reticulated python (*Python reticulatus*). The estuarine crocodile (*Crocodilus porosus*) may still inhabit small creeks in the reserve, but this remains to be confirmed (Cheng 1994a).

Aquatic Fauna

Preliminary surveys of the aquatic macrofauna have been undertaken by a number of authorities (Lui 1991; Othman and Arshad 1991; Chong et al. 1994; Sasekumar et al. 1994; Singh and Sasekumar 1994; Yap et al. 1994; Zainal and Sasekumar 1994). Sasekumar et al. (1994) found 117 species of fish from 49 families in the channels, mudflats, and inshore waters of the reserve. Ambassids and sciaenids were the most abundant. Commercially valuable food fishes such as sea bass (*Lates calcarifer*), mullets, plotosid catfish (*Plotosus canius*), and mackerel (*Rastrelliger kanagurata*) also inhabit the reserve. Fish diversity was highest in the mangrove channels and adjacent mudflats. Fifteen species of penaeids (mainly *Penaeus*) and five species of palaemonid (mostly *Macrobrachium*) have been reported (Chong et al. 1994). The role of mangroves as a nursery ground for the penaeid shrimp fishery in the area has been well documented (Lui 1991; Sasekumar 1991), and there is little doubt that the reserve plays a vital role in the maintenance of the shrimp fisheries that operate in the nearshore area.

Fifty-two families of macrobenthic fauna, dominated by polychaetes, molluscs, crustaceans, and echinoderms, have been recorded in the reserve (Othman and Arshad 1991). Other common invertebrates include crabs of the genera *Sesarma* and *Uca*, as well as the commercially important *Scylla serrata*.

Current Consumptive and Socioeconomic Value of the Matang Mangrove Reserve

Forestry

The primary output of the Matang mangrove is billets for the production of charcoal and fuelwood, and poles for construction. Forest extraction in Matang is based on an areal concession called a *coupe* that is allocated on a renewable basis to contractors for one year. The concession areas are clearly demarcated, and heavy penalties are imposed for violations. License conditions currently require concessionaires to replant after clear felling. Logging is undertaken by chainsaws and axes. The logs are hauled by hand and wheelbarrow to the nearest waterway, where they are loaded on barges for transport to any one of numerous landing points within the forest. Apart from the use of chainsaws, there is very little mechanization, and almost all the work is undertaken using manual labor. There are no timber roads that allow forest harvesting and transportation equipment to move freely. Transportation is entirely by barge.

Fuelwood

Originally, fuelwood was the single most important product coming from the forest. In fact, at one time the Malaysian railway system was entirely reliant on fuelwood from Matang (Noakes 1951). However, the use of fuelwood has significantly declined over the years with the widespread use of fossil fuels, electricity, and liquefied petroleum gas. Contributing to this change has been the poor return from the sale of fuelwood that currently sells for only RM 62.50 per MT compared with RM 410 for charcoal. With the decline in fuelwood extraction, no fuelwood coupe is currently allocated to contractors.

Among the species logged for fuelwood, *R. apiculata* was preferred because it was heavier, harder, ignited easily, and burned with an even flame with little smoke. Though fuelwood production no longer represents a viable economic activity, it is still important at a local level. The firing of the charcoal kilns, for instance, is entirely with fuelwood. No gas or electricity is employed. Splitting of the fuelwood billets is still undertaken in the traditional manner by casual laborers who are paid RM 0.70 per billet.

Charcoal

The charcoal industry, which began in Matang around 1924, is still thriving. Thai immigrants introduced the highly efficient Siamese-type kiln in 1930 (Robertson 1940; Noakes 1952), and with these beginnings Matang became the base for the largest charcoal industry in the country, a preeminence it maintains to this day. In 1992, for example, more than 75 contractors operated 336 charcoal kilns based on the output from the reserve. The contractors are allocated tracts of mangrove to clear based on the number of kilns they own. On the average, each contractor ends up with about 13 ha of forest concession to log annually. The preferred tree for charcoal is *R. apiculata,* which produces premium grade charcoal.

The igloolike charcoal kilns have a capacity of 40 MT per charge. Workers are paid RM 35–RM 70 per filling, depending on their job. The work is backbreaking and, in the stuffy, unventilated conditions that prevail in the kilns, extremely uncomfortable. The fire schedule usually comprises two main stages, "big burn" and "small burn." The process requires approximately 28 days to convert the greenwood into charcoal. The production of charcoal per kiln is 10.5 MT, or a conversion rate of 26 percent. Cheng (1994a) found that the price of each metric ton of charcoal fetched RM 438, or RM 4,599 per burn. Given that each kiln can be used for an average of nine burns per year (Harun

1981), total earnings per kiln equate to RM 41,391 annually. Assuming all 336 kilns are operating at full capacity, annual earnings derived from the charcoal industry would be about RM 13.9 million.

In spite of the fact that the use of charcoal as a domestic and industrial fuel has been largely supplanted by other energy sources, demand still outstrips supply. The charcoal industry has been saved from joining the fuelwood sector by the demand for activated charcoal and charcoal briquettes from overseas markets. At present, most of the charcoal is sent to factories in Perak for reprocessing into briquettes, after which they are exported, mainly to Japan.

Poles

The second most important forest product after charcoal is poles, primarily from *R. apiculata*. The poles are supplied during thinning or intermediate felling operations that are carried out during years 15–19 and 20–24 of the 30-year cycle. These operations form part of the silvicultural regime employed in the management of the forest. In the 1990–99 working plan, areas earmarked for the first thinning and second thinning operations are estimated to total around 2,145 ha per year. There are some 75 licensed pole contractors, and each of them is allotted 25–32 ha per year. According to Cheng (1994a), the price of poles ranges from RM 1.70 to RM 3.75 each, depending on the length, straightness, and diameter. Like charcoal, concessions for poles are awarded on a yearly basis.

Thinning operations are usually supervised strictly so that there is no overexploitation that would jeopardize greenwood for future charcoal production. The length of the poles is usually 6 m taken from a single stem in the first thinning. In the second thinning, however, each stem can sometimes yield two 6-m poles. Pole diameter varies from 7 to 13 centimeters (cm). Poles are used primarily as piling and scaffolding material in the construction industry.

Traditional Material

Many coastal communities traditionally harvested a variety of mangrove forest products, such as *Acrostichum* fronds and *Nypa* thatch. Despite being a weed in the *Rhizophora* stand, the *Acrostichum* frond stalk is currently exploited for use in supporting vegetable plants (Chan and Salleh 1986). The defoliated stalks are sold at RM 0.10 per hundred pieces (Cheng 1994a). Forest products such as thatch from the *Nypa fruticans* are also important traditional building materials, though their use has lost out to corrugated galvanized sheets in recent years. No permits are required for harvesting these two products, and no data are available on their total commercial value in the reserve.

State Revenues and Employment

The Matang mangroves have been an important source of income for the state government of Perak, particularly through fees and royalties imposed on logging permits and forest products. Minor sources include permit fees for the kilns and fines for violation of forestry regulations. Royalties and premiums are periodically updated. The latest revision, undertaken in 1989, is given in Table 5.2. In 1992, gross revenues to the state from the reserve amounted to RM 1.23 million (Perak State Forestry Dept. 1994).

In addition to its financial contribution to the state and national economy, the Matang Mangrove Reserve is a source of substantial em-

Table 5.2. Revised Premiums and Royalties Charged by the Government of Perak on Forest Products from the Matang Mangrove Forest Reserve, 1989

Premiums	
Charcoal	RM 340.00 per ha of greenwood
Firewood	RM 1.00 per MT of greenwood
Poles	RM 13.00 per ha
Royalties	
Charcoal	RM 180.00 per kiln or RM 17.15 per MT
Firewood	RM 2.00 per MT
Poles	
60–90 cm girth	RM 4.50 per 30 m
30–60 cm girth	RM 3.00 per 30 m
10–30 cm girth	RM 2.00 per 30 m
< 10 cm girth	RM 0.05 per piece
Other fees	
Permit for temporary use of Forest Reserve for charcoal kilns	RM 12.00 per annum
Fee for operation of charcoal kiln	RM 10 per kiln per year

Source: Perak State Forestry Department 1994.

ployment. Ong (1978) estimated that the forest industry provided direct employment for 1,400 people and indirect employment for a further 1,000. Current figures are unavailable. However, all the contractors interviewed complained of a severe shortage of labor, especially skilled labor, to operate the kilns and extract timber.

Wildlife

Wildlife use in the Matang Mangrove Forest is not well documented. Hunting is undertaken mainly by amateur hunters and for recreational purposes. Professional hunting is almost nonexistent. Most hunters operate in groups and share their hunt, primarily for personal consumption. Excess wildmeat is then sold to the licensed animal dealers who in turn sell it to restaurants. The flying fox, in particular, is much sought after by local restaurants. A sizable population of the bat inhabits the reserve, feeding mainly off the nectar of *Durio* sp. and *Sonneratia* sp. It is difficult to estimate the hunting effort in the Matang forests. Though hunting licenses are issued by the Perak State Wildlife Office, these licenses enable hunting anywhere within the state. The situation is further complicated by the fact that hunters from other states also frequent the Matang mangroves.

The reserve's mangroves and the cave fruit bats that depend on them probably play an important role in maintaining production of the highly prized durian fruit (*Durian zibethinus*) elsewhere in the state. Durian flowers are pollinated almost solely by the bat, which eats the flower's nectar and pollen. Before the durian comes into flower, however, nectar from *S. alba* is a preferred food of the bats. The economic value of this functional role of the bats and mangroves is impossible to calculate, but must be substantial (Lee 1980, cited by Cheng 1994a).

Capture Fisheries

The waters off the state of Perak support one of the largest fisheries industries in the country. In 1993, Perak's fishermen landed 179,406 MT of fish, or 23 percent of the total catch of Peninsular Malaysia, valued at RM 420 million (Dept. of Fisheries 1994). There are nine major fishing bases in the state: Bagan Datoh, Pangkor, Pantai Remis, Bagan Pancor, Kampung Pasir Hitam, Sungai Kerang, Kuala Sepetang, Kuala Sangga Besar, and Kuala Kurau. Five of these—Bagan Pancor,

Kampung Pasir Hitam, Sungai Kerang, Kuala Sangga Besar, and Kuala Sepetang—are within the reserve, while Pantai Remis is immediately south and Kuala Kurau is just north of the reserve (Choy 1991).

Though the importance of mangroves in nearshore and coastal fisheries has been recognized in the reserve's working plans, there has been little examination of the industry in the region. Choy (1991) provided the first comprehensive review of the fisheries industry in the Larut Matang District, wherein most of the reserve lies. However, in considering the impact of the Matang Reserve mangroves, it would be inadequate to confine our discussions to the Larut Matang District alone, as adjacent fishing grounds are no doubt also highly dependent on the reserve.

There were 1,853 inboard fishing boats operating off the Pantai Remis-Matang-Kuala Kurau coastline corridor in 1993. Both commercial trawlers, as well as artisanal gear such as gill netting, hook and line, and fishing traps, were employed by the fishing population (K. Siva, pers. comm. 1995). In addition, Choy (1991) reported that 200 unlicensed mechanized push nets operated in the Larut Matang District, and the collection of crabs remained a major traditional activity. The capture of fish for personal consumption by villagers in the area was also a significant activity. Though small in terms of total output, this subsistence catch probably plays an important role in the nutrition of the villagers.

At least 3,903 fishermen operated from the seven bases in the corridor in 1993. Of this number, 1,954 were in Larut Matang District. At least twice as many were employed in associated trades such as boat building, marketing, and other services. Approximately half of the fishermen were employed onboard the trawlers, while the rest were involved in the operation of the artisanal gears (K. Siva, pers. comm. 1995). Some 91,642 MT (amounting to 51% of the state's catch) valued at RM 214,340,236 were landed at these seven bases in 1993. Of this, 50,073 MT (47%) came from the five ports in the Matang Reserve (Dept. of Fisheries 1994).

Aquaculture

Perak is the country's leading producer of the blood clam. In 1993, the blood clam culture beds in the state accounted for 47,916 MT or 63 percent of the national production of 77,755 MT. The most important culture grounds in the state are the extensive mudflats fringing the seaward edge just outside the Matang Reserve. The culture of the blood clam is, in fact, thought to have originated in Bagan Pancor in 1948 (Pathansali and Soong 1958).

The biology and culture of the blood clam have been extensively studied by a number of authorities (Pathansali and Soong 1958; Pathansali 1977; Devakie 1986; Ng 1994). The industry is entirely reliant on naturally produced seed (spat). Blood clam spatfalls are largely confined to specific areas in Peninsular Malaysia. The Matang mangroves represent one of the most important spatfall grounds in the country, contributing more than half of the state's spat production (Choy 1991). While the clam breeds throughout the year, there are two peaks in spat production, December–January and April–July. There is at present no hatchery production of the spat, although laboratory production has proved successful (Kamal 1986).

Spats of 6.4–10 millimeters (mm) are removed from the spatfall grounds and sown in sheltered mudflat areas along the coastline (Ng 1994). The spats are harvested by hand using wire scoops with a 6.4-mm mesh. The spats are then transported to the culture grounds where they are broadcast directly onto mudflats, particularly in sheltered bays and inlets, where currents are too weak to carry the spat away. Stocking densities are highly variable, depending on the experience of the culturists and the size of the spat. In Matang, the sowing rate is 50–120 tins/ha. One tin contains 18 kilograms (kg) of spat with 4,000–6,000 spat/kg (Ng 1994). Thinning and redistribution may be carried out to ensure optimal growth.

Though many of the permits are held under the local fishermen's association, the collection of the spat and the management and operation of the farms are by individuals. Farmers must also hold a Temporary Occupation License that allocates them land rights on a year-to-year basis. Each lot allocated for blood clam farming is identified on the license and marked off by pegs and poles.

No feeding is involved. Blood clams, like all bivalves, are filter feeders, and are thus highly dependent on microorganisms present in the water as a source of food. The density of microorganisms relates closely to nutrient levels in the water, which is linked to detrital output from the mangroves. After a culture period of about 16 months, the clams are harvested on a staggered basis using wire scoops at the end of long poles.

In 1994, blood clam culture in the Larut Matang District occupied an estimated 1,889 ha of mudflats, while production amounted to 33,463 MT valued at RM 8.7 million. This is equivalent to a production of 17.7 MT per ha and a value of RM 4,606 per ha. A total of 190 culturists were involved, most of them based in Kuala Sepetang and Kuala Gula estuaries (K. Siva, pers. comm. 1995). Apart from nominal fees for the permits and licenses, the state does not accrue any direct revenue from blood clam farming.

Cage culture of finfish is also carried out, particularly in the Kuala Sangga Besar estuary. In 1993, 6,000 such cages were operated by more than 70 culturists in the Matang Reserve area. Production amounted to 720 MT valued at RM 4.32 million. The culture revolves basically around the sea bass, though other species such as the greasy grouper, *Epinephelus salmoides,* and mangrove jack, *Lutjanus argentimaculatus,* are also farmed.

Unlike blood clam culture, cage culture is an intensive culture activity, and there is little, if any, direct reliance on natural productivity. The cultured fish are fed the bycatch from trawlers that operate from nearby fishing bases. The role of the mangrove in this situation is more to ensure the stability of the estuarine environment, as well as to sequester deleterious pollutants from the water. The cages in Kuala Gula, for instance, have experienced losses due to pollution on at least two occasions (Choy Siew Kong, pers. comm. 1995).

Tourism

Tourism is the latest economic activity in Matang and the one for which there are the fewest data. The tourism industry in the country is highly developed, and, in terms of arrivals, Malaysia rates as the third most important destination in Asia (Cheah 1995). Though tourist infrastructure in the Matang Reserve is still limited and constraining, the Perak State Government actively promotes Matang as a tourist destination, catering particularly to the rapidly growing international ecotourism market. There are several major attractions in the reserve that serve as a focus for tourist arrivals (Fig. 5.1).

Kuala Gula Bird Sanctuary. Storklake 1, often better known as the Kuala Gula Bird Sanctuary, has become a focal point for bird watchers throughout the country, and even internationally. The local wildlife office estimates that at least eighty to one hundred foreigners and two hundred local tourists visit the sanctuary during the nesting period between September and March. The visitor flow at other times is small to negligible.

Kuala Sepetang Mangrove Boardwalk. Located at the Forestry Office in Kuala Sepetang, the mangrove boardwalk is a popular destination as well. The Forestry Office estimates visitor flow is about five hundred to eight hundred visitors per month.

Historical Sites. Kuala Sepetang has historical value as the terminus of the first railway line in the country. Within the reserve itself, there are two prehistoric settlements that date back to 2200 B.C. to A.D. 1000 (Nik Hassan Shuhami 1991) (Fig. 5.1). Though the site has been extensively surveyed, it is not significantly important as a tourist draw.

From anecdotal information, it appears that as many as ten thousand visitors per year visit the various attractions in Matang. Studies of revenues from tourism in the reserve have not been made. However, if we assume that the average tourist spends two days and one night visiting the reserve, and that each spends RM 140 for food and accommodations and RM 30 for travel, the total annual revenue flow would be RM 1,700,000. The growth of tourism has been particularly visible in the small village of Kuala Gula (Fig. 5.1). Dilapidated village dwellings have been demolished and villagers moved into new housing. Chalets have sprung up to cater to visitors, while some villagers provide bed and breakfast accommodations. A boardwalk reaching out to the major bird-watching grounds is also under construction. Many complain of the lack of seed capital to purchase boats to cater to the growing crowd of visitors.

It is unknown to what extent tourism has improved job opportunities at the various sites. Given the somewhat seasonal flow of visitors, however, it is likely that tourism so far has succeeded more in providing additional income opportunities for the area's residents than in creating full-time employment.

Coastal Protection

The Matang mangroves buffer coastal communities and agriculture from the erosive and other potentially destructive effects of storms and wave action. Cheng (1994a) cites a case in Bagan Datoh where clearing of mangroves led to extensive damage of agricultural lands by sea erosion. In response, the government planned to construct a 6-km-long rock bund at a cost of RM 3.33 million per km. He extrapolates that if the Matang mangroves were to be cleared, the cost of constructing similar rock bunds would total RM 170 million.

Management of the Matang Mangrove Reserve

Forestry

The management of the Matang Mangrove Reserve comes under the purview of the Forest Department of the State of Perak. For administrative purposes, the reserve is divided into three ranges: Kuala Sepetang (20,864 ha), Kuala Trong (10,818 ha), and Sungai Kerang (8,469 ha) (Fig. 5.1). The overall management of the reserve is undertaken by the district forestry officer and the assistant district forestry officer based in Taiping, the administrative capital of the district. Each range is covered by a ranger and a team of support staff including foresters, laborers, and boatmen.

When first established, the objective of the reserve's management plan was to ensure the sustainable production of high-quality greenwood for charcoal, fuelwood, poles for construction, and fishing stakes. Though this basic tenet has remained throughout, the management plan has acquired a number of additional objectives. The general management objectives, as described by B. K. Gan (1991), are as follows:

—Maximize production, exploitation, and utilization of greenwood for the charcoal industry, to meet local demand as well as for export.

—Make available cheap fuelwood for local requirements.

—Maximize production of prime poles to maintain a steady supply of quality and cheap construction material, especially for the housing industry and other construction needs.

—Provide livelihood, employment, and cheap structural materials for local communities.

—Rehabilitate unproductive and degraded forests through guiding succession in accordance with sound ecological practices.

—Diversify into the production of other wood and nonwood resources.

—Encourage and maintain the ecosystem for nondestructive aquaculture activities.

—Preserve a highly productive mangrove forest ecosystem as a feeding and breeding ground for marine life.

—Preserve sufficient areas for conservation, research, seed source, recreation, and training in mangrove management.

—Conserve and protect the foreshore and riverbanks from erosion and damage by strong winds and tides.

Outside of forestry, no quantitative values have been ascribed to the objectives, either in terms of output or the size of the industry it is expected to support. Two basic strategies for meeting these objectives have been employed throughout: (1) a planned cutting and silvicultural cycle to ensure the sustainability of forestry products, and (2) establishment of virgin jungle reserves for conservation as well as coastal reserves to sustain the nearshore marine environment and to mitigate the possibility of coastal erosion. We focus first on the cutting and silvicultural system.

The regimes employed in forest management have changed considerably since the reserve was first established in 1902 (Noakes 1952; Gan 1991). The first working plan covered only the insular part of the reserve and involved a cutting rotation of 20 years. This regime was extended to the entire reserve in 1908. In 1914, the rotation period was increased to 25 years, with thinnings on the twelfth and twentieth year of the cycle. In 1924, a minimum girth system of 30 cm was tried. In 1925, a rotation of 40 years was introduced. Meanwhile, a "retention of standards" (or mother trees) system required that, initially, 25 seed trees per ha, and then from 1925 to 1940, 50 seed trees per ha, be retained during cutting, particularly in areas lacking regeneration. The "retention of standards" was eventually abandoned in the 1940s because better regeneration could be attained through plantings. Shortly after 1950, the rotation age was reduced to 30 years based on the first 10-year working plan by Noakes (1952). The rotation of 30 years was based on data obtained from sample plots which showed that the mean annual volume increment for *Rhizophora* spp. culminated at about 23 years. A further provision of seven years took into account potential constraints in the growth of the trees, thus providing ample time for maturation. The smaller, more consistent tree size also facilitated manual extraction of the timber. Subsequent 10-year working plans for the reserve have been prepared by Dixon (1959) for the 1960–69 period, by Darus (1969) for 1970–79, and by Harun (1981) for 1980–89.

The operational sequence of the current 30-year rotation cycle is provided in Table 5.3. In the 1990–99 working plan, the total allotment area for charcoal production is proposed at 7,980.3 ha (25%), first thinning at 10,244.1 ha (31%), and second thinning at 11,214.8 ha (34%), while the remaining 3,147.7 ha (10%) are designated as reserve areas (B. K. Gan, pers. comm. 1995). At one time, a third thinning was undertaken, but it was abandoned when it was discovered that second thinning plots had better regeneration and growth than expected (Serivastava and Bal 1984).

Thinning has a duel function. The first is the production of poles for the construction industry. Second, thinned stands have a greater mean annual diameter increment and better natural regeneration (Noakes 1952). Chan et al. (1982) also observed a higher recruitment in thinned stands. Some of this new growth will reach pole size by the time of final felling.

The productive areas are divided into three periodic blocks, each to be worked within 10 years in succession. The periodic blocks are distributed among all three ranges in the reserve. Yield regulation is on an areal basis. In the current (1990–99) plan, the yield is estimated at 170 MT/ha. The volume of greenwood in the blocks allocated for the period is estimated to be sufficient to support 336 charcoal kilns for the 10-year period. Though the plan does not specifically call for concessions to be allocated to each individual charcoal kiln, in practice, this is what happens. In the present plan, allocation per kiln is 2.9 ha as opposed to 2.8 ha in the previous plan.

It is significant to note that the yield for the last two working plans has been entirely from second generation crops, except for 2 percent of the forest that is a holdover from previous cycles. This has led to a consistency in terms of yield, billet size, and species composition compared with previous crops. The entire productive forest can now be categorized into three periods: crops of 1–10 years (Period I), 11–20 years (Period II), and 21–30 years (Period III) (Ashaari and Hadi 1977).

The present silviculture system is not without problems and constraints. These constitute both lapses in implementation as well as natural biological and physical stressors (Salleh and Chan 1986; Tang et al. 1984). *R. apiculata* that are marked for retention for the second

Table 5.3. The 30-year Rotation Cycle of Matang Mangrove Forest Reserve

Year	Type of operation
–1	Boundary operation, blocking, and preliminary assessment of stocking —Surveying, boundary cutting, and demarcation of individual felling subcoupes. —Determination of the extent of inundation classes, dryland, and disturbed forest.
0	Final felling —Trees of 8-cm diameter and above are clear felled for charcoal and firewood. *Rhizophora mucronata, R. apiculata, Bruguiera gymnorhiza, B. cylindrica,* and *Ceriops tagal* are used mainly for charcoal, while *B. parviflora* are used mainly for firewood. —For coupes that border rivers and the sea, a 3-m buffer is left intact to obviate erosion. —In the process of felling, the charcoal/firewood contractors are required to girdle or fell all nonuseful species such as the *Sonneratia griffithii, S. caeolaris, S. ovata,,* and *B. cylindrica.* By the time the blocks are fully exploited, only *Rhizophora* spp. are left standing.
1	Estimation of areas that require planting —Invading ferns (*Acrostichum speciosum* and *A. aureum*) are eradicated by Velpar 90 (Hexazione) and manual means.
2	Enrichment planting with *Rhizophora apiculata* and *Rhizophora mucronata* —Spacing 1.2 m × 1.2 m for the former and 1.8 m × 1.8 m for the latter. Planting conducted only if natural regeneration is less than 75 percent of desired densities. —Seeds are supplied by charcoal contractors from areas slated to be logged. They pay casual laborers RM 9 for 1,000 seeds of *R. apiculata* and RM 12 for 1,000 seeds of *R. mucronata.* Sowing is done by the Forest Department.
3	Inspection of planted areas and refilling
15–19	First thinning —1.2-m stick used to measure between trees to reach the density desired. The one good tree to be retained is usually near the corner of the working block or beside a riverbank, and all trees within a radius of 1.2 m from this tree are to be felled. —Trees with good bole structures are extracted as standard poles, while the malformed ones are extracted for shorter length poles or firewood.
20–24	Second thinning —Done with 1.8-m stick.
30	Final felling —Monitoring by Forest Department.

Source: Gan 1991.

thinning and final felling, or are outside the productive forest area, are often cut. According to Harun (1981), the first could be controlled by close monitoring during thinning operations to ensure that all marked trees are retained. The only means of controlling the second form of overexploitation is by patrolling the reserve regularly and identifying vulnerable spots, for example, areas where *R. apiculata* occurs in high densities and near riverbanks where access is easy.

Various factors affect regeneration and seedling survival. This can have important economic consequences, since in 1992 the estimated cost of establishment of *R. apiculata* stands was RM 222 per ha (Perak State Forestry Dept. 1994). In the absence of seed trees, it is common for logged areas to be devoid of regeneration. The presence of slash materials after logging further impedes natural regeneration. The problem is compounded in areas of deep water, where the movement of slash material is likely to damage any seedling that has taken root.

Infestation by two species of ferns, *Acrostichum aereum* and *Acrostichum speciosum,* is another major problem. Responding to full light, they quickly colonize the ground after clear felling, forming dense thickets that effectively prevent mangrove regeneration. Artificial regeneration of mangrove seedlings thus calls for the extermination of the fern. This is usually accomplished through the use of an herbicide or by manual means.

Another weed species is *B. parviflora,* which regenerates very rapidly after clear felling. The tree is considered undesirable because of its limited economic value. Thus, before planting, all *B. parviflora* seedlings are slashed. Epiphytic plants can also be problematic. Serivastava et al. (1988) found the climber *Derris trifolia* was a primary cause of mortality in some areas, particularly in six to nine-year *Rhizophora* crops. Infestations of young mangrove seedlings by barnacles have also been reported (Zamora 1987; Ramly 1992).

Crabs and monkeys are also known to damage seedlings by feeding on them. Among the crabs, the *Sesarma* species of graspid crabs are the most menacing. The attacks vary in intensity and duration, with *R. apiculata* being particularly susceptible (Noakes 1952). The extent of damage can be so severe that planting becomes impractical. The use of potted seedlings of about three to four months old has overcome this problem somewhat since the plants acquire a woody stem by this age and are able to resist crab attacks (Gan 1991). Little is known of the extent of damage caused by monkeys. Preliminary indications are that *R. apiculata* is more susceptible than *R. mucronata* in this respect (Chan 1988). The problem, however, is not considered severe enough to warrant reduction of monkey populations in the forest (Salleh and Chan 1986).

Among the physical stressors, windstorms, called Sumatras, are a major cause of forest damage, especially to the mangroves along the shoreline. Salleh and Chan (1986) reported that between 1966 and 1980 these storms were responsible for the erosion loss of more than 658 ha of forest in the reserve.

The second strategy regarding forest and coastal reserves consists of the following components. A 3-m buffer zone is established along river channels and the seas as a matter of course during felling. Virgin jungle reserves have also been established covering all the different forest types in Matang. At present the virgin jungle reserves cover an area of 41.2 ha. Seed stands (i.e., stands of mother trees) are also being established to ensure supplies of seeds. Two areas, Storklake 1 and Storklake 2, have been set aside as bird sanctuaries (Fig. 5.1).

The accreting zone is left undisturbed as a reserve to prevent erosion as well as for maintenance of the marine ecosystem. Consisting mainly of *Avicennia* spp. and *Sonneratia* spp., this zone does not support *Rhizophora* growth, and attempts to develop it for forestry have not been successful (Chan et al. 1986).

Wildlife

Wildlife management in the reserve is the responsibility of the Department of Wildlife and National Parks (PERHILITAN). However, the Matang forest comes under the administrative purview of the Perak Forestry Department. Thus the wildlife authorities work closely with the forest authorities where Matang is concerned. PERHILITAN conducts research on migratory birds and mangrove mammals such as otters and enforces wildlife regulations. This includes enforcement of the ban against the hunting of otter, which is designated Totally Protected under the Wildlife Act of 1972, and the issuance of hunting licenses for some 14 species of animals listed as Protected under Schedule 2 and Schedule 4, which allow hunting subject to season and bag limit conditions. Huntable species include 10 species of birds (mostly shorebirds), the flying fox (*Pteropus vampyrus*), long-tailed macaque (*Macaca fascicularis*), silvered leaf monkey (*Presbytis cristatus*), wild pig (*Sus scrofa*), and reticulated python (*Python reticulatus*). Hunting restrictions are standard throughout the country. More specific regulations for the reserve would be impractical because of inadequate data about its wildlife populations.

Until 1990, there was no specific wildlife management plan operational in the Matang area, though there were general statements in the various working plans with respect to "maintaining and conserving wildlife," and the *Avicennia-Sonneratia* belt along the coastline and virgin forest reserves were important for wildlife conservation. The working plan instituted a more proactive role in wildlife conservation in 1990, specifically for shorebirds. The document clearly addresses the need for foraging and roosting sites for shorebirds. Two marsh areas that are major congregation points for shorebirds, Storklake 1 and Storklake 2, have been set aside as bird sanctuaries. PERHILITAN is establishing an interpretation center for these sanctuaries at Kuala Gula. Though these sanctuaries are not official in the legal sense (the Forestry Department does not have the legal authority to declare bird sanctuaries), they do represent an important management change in the reserve.

Capture Fisheries

Fisheries resources in the country are managed by the Federal Department of Fisheries, which is also responsible for development work, extension, and technical support services and research. The Fisheries and Forestry departments operate independently, and there is no formal mechanism that brings the two authorities together with respect to the Matang Reserve's management.

The working plan takes note of the vital role of the mangroves in the sustainability of the nearshore and coastal fisheries. In this respect, the *Avicennia-Sonneratia* belt has been identified as the primary marine sanctuary of the Matang forests. However, this belt has not been established with the concurrence or involvement of the fisheries authorities, nor has there been any work to indicate that the belt is sufficient to sustain the nearshore fisheries at original levels.

The Department of Fisheries has a countrywide national fisheries management policy, though it has no specific management plan with respect to the Matang waters. A strictly enforced licensing regime restricts inshore and coastal waters up to 5 km from the shoreline exclusively for artisanal fishermen, who operate traditional gears such as drift/gill nets, hook-and-line, and bag nets. Commercial fishermen, who work the trawlers and purse seiners, are not allowed to operate in this zone.

Smaller commercial boats (up to 40 gross tons) are allowed to operate 5–12 km offshore, while larger (40–70 GRT) boats are licensed to fish only 12–55 km offshore. Deep-sea fishing boats (above 70 GRT) are confined to seas beyond 55 km. This zoning system prevents conflicts between the different groups of fishermen. In addition, it also prevents the commercial fishermen from putting any further pressure on nearshore and coastal fisheries resources.

The coastal fisheries sector of the country, particularly of Peninsular Malaysia, has long been considered overfished (Khoo 1976). Currently, management policies are therefore directed toward reducing the number of inshore and nearshore fishermen. Outside of deep-sea fishing, no new licenses are issued for inshore fishing vessels, and artisanal fishermen are actively encouraged to leave the industry and seek alternative employment.

Aquaculture

Aquaculture comes under the administrative ambit of the Department of Fisheries. The Matang Mangrove Forest Working Plan lists as an objective the maintenance of the ecosystem for the practice of nondestructive aquaculture activities. However, beyond the creation of the *Avicennia-Sonneratia* belt, there are no specific measures for the management of aquaculture activities.

Under existing regulations, both spat collectors and culturists must obtain a license. As the mudflats on which the blood clam are cultured are outside the reserve's boundaries, there is no licensing requirement by the Department of Forestry. The only additional permit is a Temporary Occupation License from the Land Office in the district for culture sites. However, the length of the coastline and the difficulties of access enable many culturists to farm illegally a larger area than is approved by the Land Office. Anecdotal information from two culturists indicates that the area under culture may be at least 25 percent greater than that approved under the land licenses.

Size regulations prevent the collection of spat less than 6.4 mm and the harvest of clams less than 2.5 cm. The size regulations are directed at preventing harvest of undersized clams as well as allowing adult clams to spawn at least once before they are removed for marketing. Since the spat and adult clams are landed at nearby fishing bases and transported by road, these regulations are effectively enforced.

Cage culture operators also have to obtain a permit from the Fisheries Department which specifies both the location and number of cages that can be operated in any given area. Resource conflict is preempted by locating the cages in deeper lagoons and estuaries where clam culture is not practicable. In addition, the cages are sited such that they do not obstruct boat traffic or become a navigational hazard. There is no restriction, however, on the management regime or the species reared. Because cage culture is marginal to management questions within the reserve, we do not assess its sustainability in the subsequent section.

The Sustainability of the Matang Mangrove Resource

Despite the multiplicity of objectives cited in the working plan, the management of the Matang forest has been in the past directed almost exclusively toward the production of wood for the charcoal, pole, and fuelwood industries. In evaluating the strengths of the plan, therefore, it is important to keep these objectives in mind. The other objectives have remained secondary to forestry.

The basis of the sustainable use of any renewable natural resource is that the rate of offtake must be equal to, or less than, the rate of regeneration (Cheng 1994a). The reserve's working plans do not specify the norms by which sustainability can be measured. Sustainability can be measured in terms of financial output (i.e., it continues to sustain an industry in terms of output value), in terms of production (production volume of one or more commodities), or in terms of the overall integrity of the ecosystem. The three criteria are not necessarily interlinked and may even be counterproductive with respect to each other. We have employed these three criteria to evaluate the effectiveness of the reserve's management on a more holistic basis.

Forestry

Financial Sustainability

Financial sustainability relates to economic output based on the market prices of forest products and production volume. A related indica-

Table 5.4. Economic Output of Forest Products in the Matang Mangrove Forest Reserve

Year	Value of forest output (RM)				Total
	Charcoal	Fuelwood	Poles	Total	(1990 RM)
1980[a]	18,060,980	702,000	1,791,025	20,554,005	28,366,001
1990[b]	20,026,783	50,000	2,177,124	22,253,907	22,253,907
1992[c]	15,540,555	—	2,171,714	17,711,269	16,248,870

Sources: [a]Harun 1981; [b]B. K. Gan, pers. comm. 1995; [c]Perak State Forestry Department 1994.

tor is the revenue that accrues to the state from economic activities based in the reserve. To make meaningful comparisons we have corrected all figures and expressed them in 1990 Ringgits. This is based on official indices (Ministry of Finance 1986, 1992; N. Shyamala, pers. comm. 1995) as follows: 1992 = 109; 1990 = 100; 1980 = 72.46; 1949 = 32.05.

Limited data suggest an overall decline in the total financial output of the forest (Table 5.4), but are inadequate to analyze the importance of price versus production trends in contributing to this decline. In 1980, output value was more than RM 28 million per year, while by 1992 this had declined to RM 16 million per year. This represents a decline of RM 12 million over 12 years. Though part of this trend could be due to a decrease in the size of the coupes tendered out, output value (in 1990 Ringgits) per ha under concession also declined. Harun (1981) reported a figure of RM 8,445/ha for 1980. By 1990, this had declined to RM 6,994/ha (B. K. Gan, pers. comm. 1995) and by 1992 to RM 6,000/ha.

State revenue has also shown a considerable decline over the decades since the forest was first managed, despite periodic increases in premiums and fees. Net annual revenue accruing to the state (income less operating expenses and reforestation) declined in real terms (1990 Ringgits) from RM 858,115 in 1949 (Noakes 1951) to RM 764,134 in 1980 (Harun 1981), to RM 330,480 in 1992 (Perak State Forestry Dept. 1994). This decline is attributable to increasing management costs since 1949, as well as to decreasing revenues since 1980.

A corollary consideration in ensuring financial sustainability is the long-term viability of the charcoal and pole industry, toward which the existing management regime is basically directed. Though demand at present exceeds supply, this may not prevail in the long term. The production of charcoal is more an art form than a science. The industry is almost totally dependent on skilled labor, and there is very little mechanization employed. The process of charcoal production is undertaken by skilled operators who gauge charcoal quality through the color and smell of the smoke. Under these circumstances, the loss of skilled traditions and labor can effectively destroy the industry. Most of the workers employed in the industry now have been in it all their lives.

In the present context of the Malaysian economy, where export and manufacturing growth has led to labor shortages, it is likely that the charcoal industry will face intractable problems in the next few years as the present cohort of workers retires. The exacting nature of the work involved and the uncomfortable working environment are unlikely to attract new recruits. This is already being felt by kiln owners. The industry may be able to survive through mechanization, but shifting a traditional cottage activity into an organized manufacturing operation would require structural changes within the industry itself.

Malaysia would also have to contend with other low-cost producers from the region such as Burma, Thailand, and Indonesia. As their economies grow, and gas and other fuels supplant charcoal, these countries would have a surplus production capability that would allow them to cater to the same market Malaysia is currently servicing. The volume of production from these countries is not insignificant. In 1982, for instance, Burma produced nearly 350,000 MT of charcoal (U Than and U Saw 1984). Again, little information exists on the international charcoal trade to allow a more clear-sighted assessment of future potentials.

The long-term prospects of poles as construction material is also debatable. There has been a decided shift in the local construction industry toward modern building methods that preempt the use of poles. Environmental concerns have also come into play. In 1994, the Malaysian government prohibited the use of mangrove poles for piling and construction purposes by government agencies, specifically the Public Works Department and the Drainage and Irrigation Department. The restriction is bound to be expanded to the whole construction industry in the not-too-distant future.

Production Sustainability

Questions have been raised with respect to the sustainability of timber output in the reserve, but data are inconclusive. Saenger et al. (1981)

noted that timber production appeared to decline with each cycle. This trend was also observed by Tang et al. (1984), who found that charcoal kiln licensees had to request that more areas be logged in order to feed charcoal kilns. Salleh and Chan (1986) reported that the timber outturn from second generation stands in Matang yielded only 158 and 136 MT/ha, for the periods 1967–69 and 1970–79, respectively. This was lower than the expected yields of 197 MT/ha and 165 MT/ha, respectively, and far lower than the 299 MT/ha from the first generation and virgin stands (Tang et al. 1984). However, B. K. Gan (pers. comm. 1995) estimated the average output in 1980–89 to be 174 MT/ha, and his estimate for 1990–99 is 171 MT/ha.

Evaluation of forest productivity is complicated by the fact that only 0.8 percent, 4.2 percent, and 60 percent of the areas felled in the periods 1950–59, 1960–69, and 1970–79, respectively, were actually 30 years of age, whereas the other areas felled consisted of 31–40 and 40-and-above age classes. Further, prime areas with large trees that were earmarked for logging in the first felling periods have undergone successional degradation into dryland forest (Gan 1991).

The comparison of output figures is also hampered because the plan does not stipulate a standardized means of assessing yields and each author has applied different means of calculating output. B. K. Gan (pers. comm. 1995) notes that inadequate monitoring of actual harvest makes accurate assessments of yields difficult. The absence of a standardized protocol is a serious shortcoming of the plan because it does not enable managers to monitor real output of the reserve, and renders the question of sustainability unanswerable.

An indirect indicator of productivity is coupe size, which has steadily increased from 1.6 ha/kiln in the 1950–59 plan (Noakes 1951) to 2.9 ha/kiln in the 1990–99 plan (B. K. Gan, pers. comm. 1995). The fact that coupe sizes need to be increased indicates that unit productivity may be dropping. It is unclear if the present coupe size is sufficient, as labor shortages have made it difficult to harvest the coupes completely.

Environmental Sustainability

The impact of reserve management on ecosystem processes and biotic diversity is difficult to assess. A preliminary assessment of a virgin forest reserve near Kuala Sepetang indicates that the original forest was multicanopied, dominated by 30–33-m *R. apiculata* trees. *B. gym-*

norhiza and *B. parviflora* occurred only as small trees and only in areas where the canopy had been opened up by the death of the larger trees (Putz 1978). Putz and Chan (1986), however, established that if succession were permitted without interference, both *B. gymnorhiza* and *B. parviflora* would come to occupy a significant portion of the forest profile, reducing the relative abundance of *R. apiculata*.

The deliberate removal of most mangrove species and the exclusive planting of *R. apiculata* in much of the reserve has thus had the effect of changing the fundamental structure of the forest. There are virtually no areas of old primary forest left. The management of the reserve has led to a forest type that is structurally different, being more dense and more uniform. The effect of growing trees of the same age and height has destroyed the multicanopied structure of the primary forest, and with it, presumably, the complex food web and ecological niches that this structure engenders. As Saenger et al. (1981) noted, while sustained yield management augured well for charcoal, fuelwood, and pole production, it has shifted the natural mangrove ecosystem to that of a monoculture plantation with *Rhizophora* species.

Although small areas have been set aside for biodiversity conservation and sources of seed, the vast majority of Matang is now basically a plantation forest. It is difficult to assess the environmental consequences of this shift, but there has probably been an overall loss in biodiversity. Given the antiquity of the mangrove ecosystem, it is likely that complex host-specific relationships have evolved between various tree species and mangrove fauna. For instance, gut analysis of the tree-climbing crab, *Episesarma versicolor,* in Singapore mangroves indicates that the animal has a distinct preference for *A. alba* (Sivasothy et al. 1991). The potential loss of specific feeding niches provided by different mangrove species would have an impact down the food chain. Many valuable commercial species of fish predate actively on adult mangrove invertebrates (Sasekumar 1984; Sasekumar et al. 1984). The larvae of many invertebrates, especially of *Uca* spp., also serve as a valuable food source for planktophagous fish and other animals (Macintosh 1979).

The shift in species diversity may also affect detrital flow to the nearshore marine environment. Detritus, specifically leaf litter, is the major contributor to mangrove productivity (Gong 1984). The amount of leaf litter exported varies considerably depending on the site and the dominant species in the mangrove. Gong et al. (1984) reported that the gross litterfall in the Matang Reserve varied from 11.4 to 6.96 MT/ha per year for the planted forest compared with the 7.63 MT/ha per year for the virgin jungle reserve. The nutrient value of the litterfall also varied considerably. Litterfall in the virgin jungle reserve was substantially higher in phosphates as compared with that in the planted forests. How this may affect floral and faunal diversity is uncertain.

Wildlife

The effects of management on terrestrial wildlife, without detailed historical data, are also largely unknown. Though the tiger is no longer found in the reserve, its absence cannot be directly attributable to forest management under the working plan. Without exception, senior villagers in Kuala Gula claimed that a far wider variety of birdlife existed in the 1960s. Officers at the local wildlife office noted a reduction in recruitment in stork populations, especially the milky stork and lesser adjunct (Y. Alang, pers. comm. 1995). While a reduced juvenile population in the milky stork was reported by Silvius et al. (1987), there is no published information to confirm that the lesser adjunct suffers the same fate. Both birds nest in the highest trees in the natural multistoried mangrove forest, building solitary nests which they maintain for two years or more before moving on (Medway and Wells 1976; Howes and NPWO 1986). The present management regime has largely eliminated the multistoried characteristic of the forest, which may affect stork reproduction by reducing potential nest sites.

Fisheries and Aquaculture

Among the economic activities that depend on Matang, it is probably fisheries that have the most widespread impact, be it in terms of volume, employment, or species directly affected. We examine the value of the reserve for fisheries by reviewing socioeconomic and landings data.

A review of the fishing population in the Larut Matang District shows a general increase in the number of fishermen and fishing vessels from 1973 to the early 1980s, and a decline in both since then (Fig. 5.2). A review of the total landings of the district over the same time (Fig. 5.3) shows a highly variable pattern of production, with a general increase from 1973 to 1990, but tapering off since then.

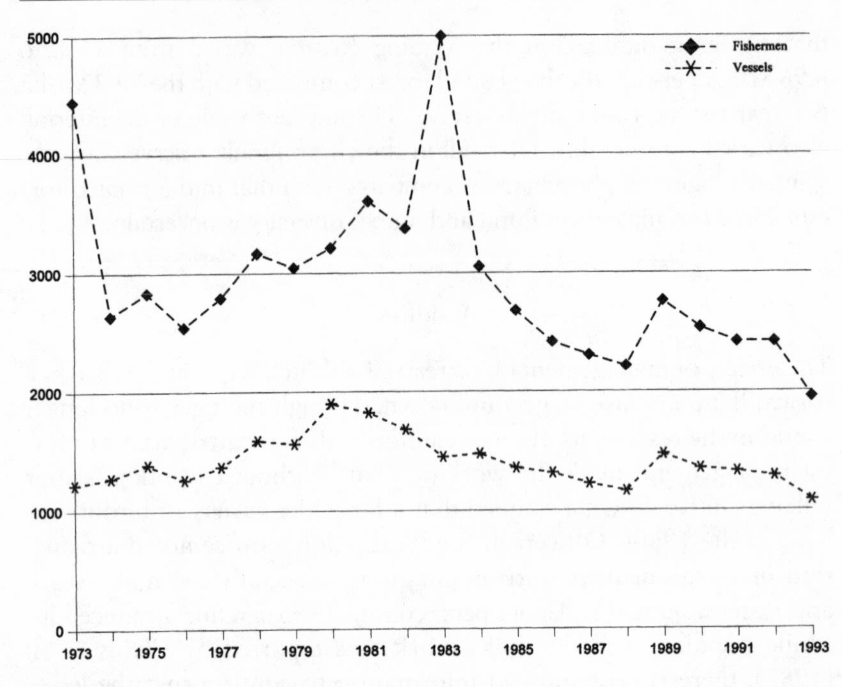

Figure 5.2. Number of fishermen and fishing vessels in Larut Matang District, 1973–93. (*Source:* Department of Fisheries 1974–94.)

Total landings, however, do not correlate clearly with mangrove productivity. This is because a significant proportion of the landings come from offshore areas where the trawlers are restricted. To obtain a clearer idea of mangrove fisheries, we reviewed landings of sergesid shrimp from 1973 to 1988, the last year for which data are available. Shrimp of the genus *Acetes,* which are the most important in these landings, are closely linked to the mangrove environment (Mariana 1991). Sergesid shrimp are found in nearshore waters, estuarine waterways, and mudflats. They are harvested by push nets, a traditional gear operated by artisanal fishermen. Most sergesid shrimp landed in Perak come from the Larut Matang and the adjacent district of Krian. Shrimp are used in the preparation of a highly valued shrimp paste, the production of which is centered around Kuala Gula. Data indicate that landings increased sharply during the 1980s, and then declined from a peak of more than 5,000 MT in 1980 to 1,225 MT in 1988 (Fig. 5.3).

A major problem facing fisheries authorities is the *sorong,* a mechanized push net used to capture shrimp that churns the sea bottom, creating extremely turbid conditions. Sorong operators accounted for much of the increase in shrimp landings in the 1970s, but the gear may have been equally responsible for the subsequent decimation of stocks. Mechanized sorongs were banned in 1971, but enforcement has been problematic. Hand-operated push nets are still allowed.

The working plan's objective to introduce only compatible aquaculture activities has succeeded in preventing the proliferation of shrimp farms in the forest area. Shrimp farming involves totally clearing the forest and excavating ponds for culture purposes. The decimation of the forest is total because any remaining mangrove vegetation is likely to interfere with farming operations.

Shrimp farming has, in fact, been a factor in the destruction of mangrove forests in Thailand, Malaysia, and Indonesia (Phillips et al. 1993). In the Philippines, aquaculture activities were primarily responsible for the clearing of more than 338,000 ha of mangrove forest since 1968 and seriously affected the coastal fisheries catch (Primavera 1989; Camacho and Bagrinao 1986). In Thailand, more than 38,000

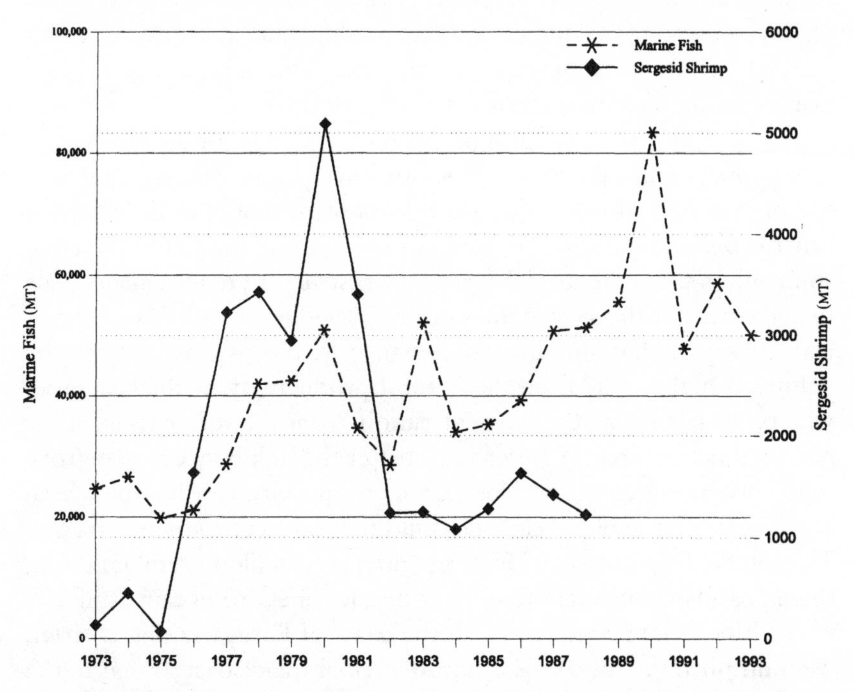

Figure 5.3. Landings of marine fish, 1973–93, and Sergesid shrimp, 1973–88, in Larut Matang District. (*Source:* Department of Fisheries 1974–94.)

ha of mangrove, accounting for 13 percent of the country's mangroves, were cleared for shrimp farming between 1979 and 1986 (Phillips et al. 1993). In Malaysia, 20 percent of the available mangrove resource has been slated for aquaculture development (Tengku 1985), and the development of aquaculture in mangroves has been cited as a major cause of concern by Ong (1982). More specifically, the number of shrimp aquaculture ponds in Perak jumped from 10 in 1983 to 214 in 1992. While this growth has been concentrated in the Sitiawan region some distance south of Matang, some aquaculture farms now fringe the reserve, and there have been requests for aquaculture concessions within the reserve (Cheng 1994b). It seems clear that the value of forest products from Matang over the decades has played an important role in averting the conversion of the Matang mangroves into shrimp ponds.

Effluent discharge from fish farming areas has been known to cause problems of pollution in some river systems (Beveridge and Phillips 1993). The clearing of freshwater swamp forests for pond construction has also led to flooding in Malaysia (Consumer Assoc. 1992). In Khulna, Bangladesh, shrimp farms have been blamed for salinization of the soil leading to diminished rice and coconut production. Commercial shrimp farming has also displaced the indigenous populations, forcing them to migrate (Hidden 1993).

In contrast to shrimp farming, on-bottom culture of the blood clam represents perhaps the ideal aquaculture system for Matang. It makes use of that part of the mangrove environment that is of little use for other economic purposes. It does not involve any trauma to the environment as there are no deleterious infrastructure requirements. The management of the system does not involve the use of feeds or chemicals that could have an impact on mangrove ecosystems. Though the culture is highly reliant on the natural productivity of the nearshore waters, in particular the nutrient runoff from the forest itself, there has been no research to understand better the link between mangrove functions/management and blood clam productivity. Nor has there been any monitoring of blood clam landings to establish production trends. Though the Department of Fisheries does record blood clam landings, these are based on state rather than district lines. To obtain estimates of the blood clam production off the coast of Larut Matang District, we multiplied the ratio of the culture area in the district to that of the state with the overall state production. In 1994, for instance, 1,511 ha of the Larut Matang mudflats were licensed for clam culture. Assum-

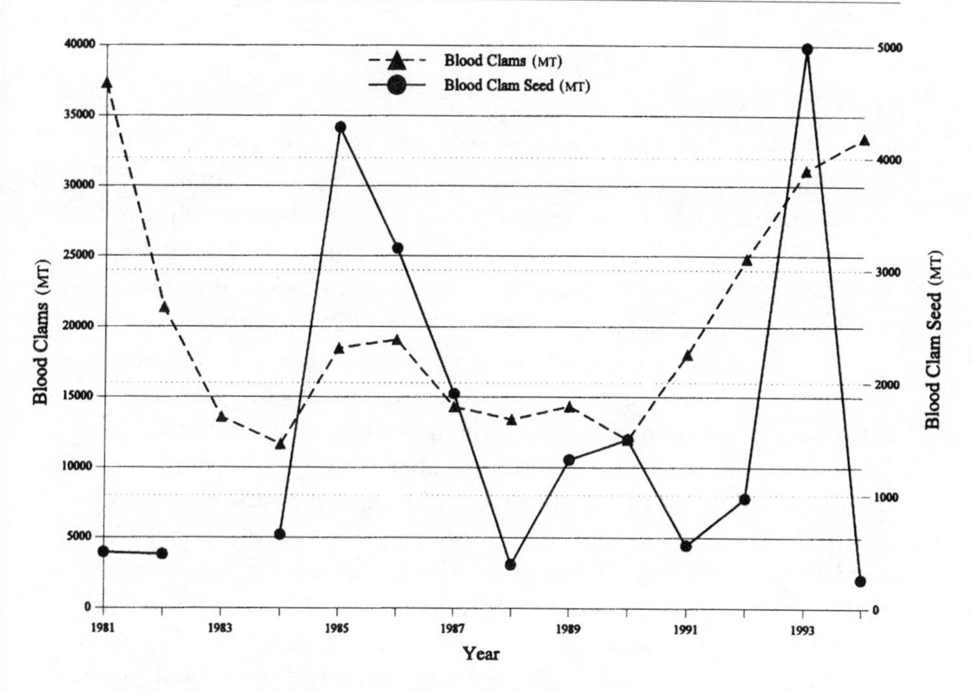

Figure 5.4. Estimated blood clam and blood clam seed production in Larut Matang District, 1981–94. (*Sources:* Ng 1994; Department of Fisheries, Perak [unpublished].)

ing that an additional 25 percent was utilized illegally, the total area under culture in the district would be 1,889 ha, compared with the state total of 2,883 ha.

Our estimate of landings for clam seeds is similarly based on Choy's (1991) estimation that the spatfall grounds in Matang accounted for about half the state's production. The results are charted in Figure 5.4. The figures suggest substantial variation in clam production in the Larut Matang District over the period 1981–94. In the absence of accurate and long-term production data, it is impossible to interpret or assign any cause to the drop in production from 1982 to 1988.

Spatfall over the same period was also variable. Such variability has been noted since the very onset of the industry. Pathansali and Soong (1958) suggested spatfall declines could have been associated with changes in the substratum, while others have reported loss in fecundity due to adverse environmental conditions (Broom 1983). Research is required to understand better the link between the reserve's management and blood clam production.

Conclusions

The Matang Mangrove Forest Reserve is valuable for its timber products, which in 1992 amounted to 451,196 MT with a value of RM 17.7 million. However, the reserve's natural heritage also contains other socioeconomic and environmental values, such as fisheries, tourism, and wildlife, that go far beyond charcoal and poles. The present management strategies, however, with a focus on extraction of timber, do not address management of the reserve's resources in toto.

The absence of a comprehensive strategic plan is a serious shortcoming, even where forestry resources are concerned. The working plans have never set quantitative productive targets for sustained forest production. Targets were set on an areal basis or according to the size of the industry to be supported, rather than in terms of the inherent productive capability of the forest. This has skewed the plan toward simply setting areal quotas (coupes), with less emphasis on improving forest production or evaluating industry skills and potentials. The earlier plans did not institute a standardized methodology for establishing yields. As such, yields were not seriously monitored as an indicator of management success, and though some authorities suggest there is a decline in forest productivity, data are inadequate for a rigorous assessment.

The financial sustainability of the present management regime is also doubtful. All indicators point to a declining economic output, as well as declining revenues to the state. Such a decline was, in fact, alluded to by several of the kiln owners we spoke to, who maintained that while demand for charcoal was good, they had made "better" money in the 1960s.

The social value of mangrove-based forestry is also declining. The need for charcoal and firewood as domestic fuel has all but ended. Charcoal is in demand for the manufacture of briquettes and activated carbon. While the former is for domestic use, it does not represent an indispensable part of an average household's energy requirements. In the context of the current economic growth of the country, which emphasizes manufacturing and service-related industries, the labor-intensive nature of charcoal production is untenable. Though the people that operate the charcoal kilns have been doing so for decades, and the workers have been at it all their lives and are unlikely to be able to find alternative employment elsewhere, this in itself would be insufficient justification for maintaining the kilns. If the anecdotal information is correct and the kilns are seriously bereft of young workers, then it may not be too far in the future when the kilns will either have to close down or upgrade by mechanizing their manufacturing operations. Mangrove products such as thatch are also of little use and of local importance only, having largely been replaced by galvanized iron, asbestos, and tiles. The production of poles is still of major importance, although that market is also likely to decline in the future.

The development of alternative forestry activities would blunt the diminution of the charcoal and pole trade. Chan (1989) discovered that the dryland forest in Matang had substantial wood resources that could be exploited. Gan (1991) recommended the establishment of indigenous plantations of *Oncosperma tigillarium, Xylocarpus* spp., *Intsia bijuga,* and rattan in unproductive dryland forest. While some of these are not commercially exploited at the present time, they have long-term potential in the furniture and construction industries. He also recommended apiculture in the *Avicennia* forest.

Fisheries are yet another industry that relies on the reserve for its sustenance. In 1993, the waters off the reserve yielded 50,073 MT of fish valued at RM 117.1 million (Dept. of Fisheries 1994). The industry provided direct employment to 1,954 people, and at least twice as many were involved in associated occupations. Though there has been a decline in the coastal fishing population of the district, a move that is openly encouraged by the authorities, total landings of marine fish have not declined. The role of the Matang mangrove ecosystem in maintaining the coastal fisheries has not been elucidated, but prudence dictates that management of the reserve take into account maintenance of these productive fisheries.

Aquaculture, specifically the farming of the blood clam, is probably the only activity that does not face an uncertain future. In 1993, an estimated 34,183 MT of blood clam valued at RM 13 million were produced in Matang. The mudflats off the reserve are not open to any other kind of economic activity. Blood clam farming is not particularly labor intensive, and markets are not limiting. Under these circumstances, the long-term prospects of the industry appear bright.

Tourism is the latest development in Matang. Estimates from anecdotal information indicate that tourist arrivals may contribute at least RM 1.7 million annually to the local economy. Tourist infrastructure must be upgraded if this sector is to go further. Tourism, if prop-

Table 5.5. Economic and Social Values from Biodiversity Uses Associated with the Matang Mangrove Forest Reserve

Activity	Output (MT)	Value (RM)	Direct employment	Indirect employment
Forestry[a]	451,196	17,711,269	1,400	1,000
Fisheries[b]	50,073	117,114,510	1,954	3,908
Aquaculture[b]	34,183	13,020,000	260	N/A
Tourism[c]	N/A	3,930,000	N/A	N/A
Total		151,775,779	4,082	8,266

Sources: [a]1992 output figures based on Perak State Foresty Department 1994, and employment figures based on Ong 1978; [b]1993 data for Larut Matang District, Department of Fisheries 1994; [c]estimates for 1993.
N/A, not applicable.

erly planned, would allow the reserve to be exploited on a nonconsumptive basis, without economic loss to either the state or the reserve's current users.

The socioeconomic value of the forest is suggested in Table 5.5, which summarizes the known economic output that is wholly or partially dependent on the mangrove ecosystem of the reserve. Of the estimated RM 151 million annual output during 1992–93, 77 percent came from fisheries, 9 percent from aquaculture, and 12 percent from forestry. (Fisheries and aquaculture are for Larut Matang District, which should closely approximate those for the Matang Reserve.) There are other benefits and services from the forest whose value and social impact have not been adequately documented. Even with the limited information at hand, it is obvious that forestry represents only a small proportion of the economic value of the forest. Forestry output is far surpassed by fisheries landings and almost matched by aquaculture. Tourism is still small, but has a promising future.

Management should give increased attention to biodiversity conservation. The virgin jungle reserves, the *Avicennia-Sonneratia* belt, and the bird sanctuaries are important beginnings, but fall far short of what is needed. The virgin jungle reserves amount to 41 ha of forest, or just 0.1 percent of the reserve. The *Avicennia-Sonneratia* belt has been set aside mainly because it is of no use from a forestry standpoint. Further research is required to establish its contribution to the nearshore marine environment and if the present extent is sufficient. More generally, we have not been able to assess the environmental impact of the working plans and forest management regime because of the lack of baseline data and the absence of monitoring key components of the Matang mangrove ecosystem in general. The charcoal industry for which Matang is being largely managed is a tradition-bound industry that, we believe, will eventually be phased out. This would effectively undermine the present rationale for its management, and open it to other, more destructive options, such as pond-based shrimp farming. The full range of socioeconomic benefits of the reserve must therefore be considered in setting management goals.

While the present management regime has its weaknesses, it does not mean that it is ineffective. Nevertheless, a fundamental review of the plan and its objectives is needed so that it is relevant to contemporary Malaysian needs.

Given the enormous significance of the Matang Mangrove Reserve for both its biodiversity and diverse economic values, it may be worthwhile to revoke its status as a forest reserve, including some of the foreshore areas, and redeclare it a national park. Though Malaysia has several parks, administered both federally as well as by the states, none of them epitomizes a mangrove environment. Matang could thus stand out as the first national mangrove park. Designation as a park would need to include provisions for resource use by local communities, and may need to be implemented incrementally in step with the declining needs of the charcoal industry. Such an approach would enable the holistic management of Matang's resources as well as facilitate the gradual phasing out of economically untenable activities. In this manner, the Matang Mangrove Reserve can continue to serve the nation, and indeed, the world, as a showcase of mangrove wealth.

References

Ashaari, M., and Y. Hadi. 1977. Management of mangrove forest in Peninsular Malaysia. In *Workshop on mangrove and estuarine vegetation,* ed. P.B.L. Serivastava and R. Abdul Kader, 39–48. Serdang, Malaysia: Faculty of Forestry, University of Agriculture.

Beveridge, M.C.M., and M. J. Phillips. 1993. Environmental impact of tropical inland fisheries. In *Environment and aquaculture in developing countries. International Centre for Living Aquatic Resources Management (ICLARM) Conference Proceedings,* ed. R.S.V. Pullin, H. Rosenthal, and J. L. MacLean, 213–237. Manda, Phillipines: ICLARM.

Broom, M. J. 1983. Gonadal development and spawning in the *Anadara granosa* L., Bivalvia, Arcidae. *Aquaculture* 30:211–219.

Burhanuddin, M. N., A. Norizan, S. Sukigara. 1994. Estimation on the density of common palm civet (*Paradoxurus hermaphroditus*). In *An oil palm plantation at Kuala Gula, Perak, Malaysia. A field study report in Matang Mangrove Forest.* Kuala Lumpur: Department of Wildlife and National Parks and Japan Wildlife Research Center.

Camacho, A. S., and T. Bagrinao. 1986. Impact of fish pond development on the mangrove ecosystem in the Philippines. In *Mangroves of Asia and the Pacific: Status and management,* ed. R. M. Umali, P. M. Zamora, R. R. Golera, R. S. Jara, and A. S. Camacho, 263–405. Technical Report of the UNDP/UNESCO Research and Training Pilot Programme on Mangrove Ecosystems in Asia and the Pacific, RAS/79/002. Manila: UNDP/UNESCO.

Carlton, J. M. 1974. Land building and stabilisation by mangroves. *Environmental Conservation* 1:285–294.

Carter, J. 1959. Mangrove succession and coastal changes in south-west Malaya. *Institute of British Geography Publication* 26:79–88.

Chai, P. K. 1977. Mangrove forests of Sarawak. In *Proceedings of workshop on mangrove and estuarine vegetation in Malaysia,* ed. P.B.L. Serivastava and R. Abdul Kader, 1–5. Serdang, Malaysia: Faculty of Forestry, University of Agriculture.

Chai, P. K., and K. K. Lai. 1984. Management and utilisation of mangrove forests of Sarawak. In *Proceedings of the Asian Symposium on Mangrove Environment—Resources and Management,* ed. E. Soepadmo, A. N. Rao, and D. J. Macintosh, 785–795. Kuala Lumpur: University of Malaya.

Chan, H. T. 1984. Coastal and river bank erosion of mangroves in Malaysia. Paper presented at the NARU Conference and Workshop on Coasts and Tidal Wetlands of the Australian Monsoon Region. November 4–11, 1984, Darwin, Australia.

Chan, H. T. 1986. Human habitation and traditional uses in Matang mangrove. In *Mangroves of Asia and the Pacific: Status and management,* ed. R. M. Umali, P. M. Zamora, R. R. Golera, R. S. Jana, and A. S. Camacho, 313–317. Technical report of the UNDP/UNESCO Research and Training Pilot Programme on Mangrove Ecosystem, Project RAS/70/002. Manila: Natural Resources Management Center of the Philippines.

Chan, H. T. 1988. Rehabilitation of logged over mangrove areas using wildlings of *Rhizophora apiculata. Journal of Tropical Forest Science* 1:187–188.

Chan, H. T. 1989. A note on the tree species and productivity of a natural dryland mangrove forest in Matang, Peninsular Malaysia. *Journal of Tropical Forest Science* 1:399–400.

Chan, H. T., U. Razani, and F. E. Putz. 1982. Growth and natural regeneration of a stand of *Rhizophora apiculata* trees at the Matang mangroves following thinning. In *Symposium on mangrove forest ecosystem productivity in South East Asia,* ed. A. Y. Kostermans and S. S. Sastroutomo. Biotrop Special Publication no. 17. Bogor, Indonesia: Biotrop.

Chan, H. T., U. Razani, and F. E. Putz. 1986. A preliminary study on planting of *Rhizophora* species in an *Avicennia* forest at the Matang mangroves. In *Prosiding Seminat II Ekosistem Mangrove,* ed. S. Soemodohardjo. Bogot, Indonesia: Lembaga Ilmu Pengetahuan.

Chan, H. T., and M. N. Salleh. 1986. Traditional uses of the mangrove system in Malaysia. In *Mangroves of Asia and the Pacific: Status and management,* ed. R. M. Umali, P. M. Zamora, R. R. Golera, R. S. Jana, and A. S. Camacho. Technical Report of the UNDP/UNESCO Research and Training Pilot Programme on Mangrove Ecosystem in Asia and the Pacific, RAS/79/002. Manila: Ministry of Natural Resources.

Chapman, V. J. 1975. *Mangrove vegetation.* Vaduz, Germany: J. Cramer.

Cheah, B. K. 1995. Ranked no. 3 in Asia. *Malaysia Tourism* January/February. Kuala Lumpur: Tourist Development Corporation of Malaysia.

Cheng, H. K. 1994a. *The biodiversity value, sustainability and management of the Matang mangroves.* WWF Malaysia Project Report MYS 274/93. Kuala Lumpur: WWF Malaysia.

Cheng, H. K. 1994b. *Mangroves in jeopardy: The economics of prawn aquaculture and its implications on mangrove management in Perak.* WWF Malaysia Presentation Paper.

Chong, V. C., and A. Sasekumar. 1990. The mangrove-coastal fisheries connection: The Selangor case. In *Proceedings of the Workshop on Mangrove Fisheries and Connections,* ed. A. Sasekumar, 51. ASEAN–Australian Marine Science Project. Kuala Lumpur: Ministry of Science, Technology and the Environment.

Chong, V. C., A. Sasekumar, and K. H. Lim. 1994. Distribution and abundance of prawns in a Malaysian mangrove system. In *Proceedings, Third ASEAN-Australian Symposium on Living Coastal Resources,* Vol. 2: *Research Papers,* ed. S. Sundara, C. R. Wilkinson, and L. M. Chou, 437–445. Bangkok, Thailand: Chulalongkorn University.

Choy, S. K. 1991. The commercial and artisanal fisheries of the Larut Matang District of Perak. In *Proceedings of the Workshop on Mangrove Fisheries and Connections,* ed. A. Sasekumar, 27–40. ASEAN–Australian Marine Science Project. Kuala Lumpur: Ministry of Science, Technology and the Environment.

Consumer Association of Penang. 1992. *5,000 acres of peat forests destroyed.* Utusan Konsumer no. 263. Penang, Malaysia.

Darus, M. H. 1969. *Rancangan kerja bagi hutan simpan paya Larut Matang.* Perak, Malaysia: Perak State Forestry Department.

D' Cruz, R. 1991. The role of mangrove inlets as a habitat for juvenile prawns. In *Proceedings of the Workshop on Mangrove Fisheries and Connections,* ed. A. Sasekumar, 53. ASEAN–Australian Marine Science Project. Kuala Lumpur: Ministry of Science Technology and the Environment.

Department of Fisheries. 1974–94. *Annual fisheries statistics [1973–93].* Kuala Lumpur: Ministry of Agriculture.

Department of Forestry, Peninsular Malaysia. 1994. *Forestry statistics: Peninsular Malaysia.* Kuala Lumpur: Ministry of Primary Industries.

Devakie, N. 1986. Observation on the current status and potential of cockle culture in Malaysia. Paper presented at the Workshop on the Biology of *Anadara granosa* in Malaysia, January 22–23, 1986, Penang, Malaysia.

Dixon, R. G. 1959. Working plan for the Matang Mangrove Forest Reserve, Perak. Department of Forestry, Perak, Malaysia.

Fong, F. W. 1984. Nypa palms: A neglected mangrove resource. In *Proceedings of the Asian Symposium on Mangrove Environmental Resources and Management,* ed. E. Soepadmo, A. N. Rao, and D. J. Macintosh, 663–671. Kuala Lumpur: University of Malaya.

Gan, B. K. 1991. Forest management in Matang. In *Proceedings of the Workshop on Mangrove Fisheries and Connections,* ed. A. Sasekumar, 15–26. ASEAN–Australian Marine Science Project. Kuala Lumpur: Ministry of Science, Technology and the Environment.

Gong, W. K. 1984. Mangrove primary productivity. In *Productivity of the mangrove ecosystem: Management implications,* ed. J. E. Ong and W. K. Gong, 10–19. Penang, Malaysia: Universiti Sains Malaysia.

Gong, W. K., and J. E. Ong. 1990. Plant biomass and nutrient flux in a managed mangrove forest in Malaysia. *Estuarine and Coastal Shelf Science* 31:519–530.

Gong W. K., J. E. Ong, C. H. Wong, and G. Dhanarajan. 1984. Productivity of mangrove trees and its significance in a managed mangrove ecosystem in Malaysia. In *Proceedings of the Asian Symposium on Mangrove Environment—Resources and Management,* ed. E. Soepadmo, A. N. Rao, and D. J. Macintosh, 216–225. Kuala Lumpur: University of Malaya.

Harun, H. Abu Hassan. 1981. *A working plan for the Matang mangroves, Perak, 1980–89.* Perak, Malaysia: Perak State Forestry Department.

Hidden costs of shrimp farming. 1993. *Infofish International.* 6/93:37. Kuala Lumpur.

Howes, J. R., and National Parks and Wildlife Office. 1986. Evaluation of Sarawak wetlands and their importance to waterbirds. Report no. 3: Pulau Bruit. *Interwader* no. 10. Kuala Lumpur.

Jothy, A. A. 1984. Capture fisheries and the mangrove ecosystem. In *Productivity of the mangrove ecosystem: Management implications,* ed. J. E. Ong and W. K. Gong, 129–141. Penang, Malaysia: Universiti Sains Malaysia.

Kamal, Z. M. 1986. Notes on the maturation and spawning of the cockle (*Anadara granosa* L.) under culture conditions, its induced spawning and larval rearing. Paper presented at the Workshop on the Biology of *Anadara granosa* in Malaysia, January 22–23, 1986, Penang, Malaysia.

Khoo, K. H. 1976. Optimal utilisation and management of fisheries resources. *Journal of the Malaysian Economics Association* 13(1–2):40–51. University of Malaya, Kuala Lumpur.

Lee, D. 1980. *The sinking ark: Environmental problems in Malaysia and Southeast Asia.* Kuala Lumpur: Heineman.

Leh, M. U., and A. Sasekumar. 1984. Feeding ecology of prawns in shallow water adjoining mangrove shores. In *Proceedings of the Asian Symposium on Mangrove Environment—Resources and Management,* ed. E. Soepadmo, A. N. Rao, and D. J. Macintosh, 331–353. Kuala Lumpur: University of Malaya.

Liew, T. C. 1977. Mangrove forests of Sabah. In *Proceedings of the Workshop on Mangrove and Estuarine Vegetation in Malaysia,* ed. P.B.L. Serivastava and R. Abdul Kader. Kuala Lumpur: University of Agriculture.

Liew, T. C., M. N. Diah, and W. Y. Chui. 1975. Mangrove exploitation and regeneration in Sabah. *Malayan Forester* 38:260–270.

Lui, C. L. 1991. Prawn recruitment in Matang, Perak, Malaysia. In *Proceedings of the Workshop on Mangrove Fisheries and Connections,* ed. A. Sasekumar, 41–46. ASEAN–Australian Marine Science Project. Kuala Lumpur: Ministry of Science, Technology and the Environment.

Macintosh, D. J. 1979. Predation of fiddler crabs (*Uca* spp.). In *Estuarine Mangroves. Proceedings of the Symposium on Mangrove Estuarine Vegetation in South East Asia,* ed. P.B.L. Serivastava, A. Manap Ahmad, G. Dhanarajan, and I. Hamzah, 101–110. Biotrop Special Publication no. 10. Bogor, Indonesia: Biotrop.

MacNae, W. 1974. *Mangrove forests and fisheries.* FAO/UNDP Indian Ocean Program (Indian Ocean Fishery Commission), IOFC/DEV/74/34. Rome: United Nations Food and Agricultural Organization.

McNicoll, A. 1992. The Matang mangrove: A showcase in forest management. *Seed Centre News* 8:13.

Malaysian Timber Council. 1994. *Profile of the Malaysian timber industry.* Kuala Lumpur: Malaysian Timber Council.

Mariana, A. 1991. The occurrence of *Acetes* (Sergesid Shrimp) in the mangrove environment. In *Proceedings of the Workshop on Mangrove Fisheries and Connections,* ed. A. Sasekumar, 54. ASEAN–Australian Marine Science Project. Kuala Lumpur: Ministry of Science, Technology and the Environment.

Medway, L., and D. R. Wells. 1976. *The birds of the Malay Peninsula.* Kuala Lumpur: H. Witherly & Sons and University of Malaya.

Ministry of Finance. 1986. *Economic report 1986/87.* Kuala Lumpur.

Ministry of Finance. 1992. *Economic report 1992/93.* Kuala Lumpur.

NATMANCOM. 1986. *Guidelines on the use of the mangrove ecosystem for brackish-water aquaculture in Malaysia.* A working group of the Malaysian National Mangrove Committee (NATMANCOM), Kuala Lumpur.

Ng, F. O. 1994. Current status, constraints and future development of cockle farming in Malaysia. Paper presented at the Seminar on Aquaculture Practices, March 29–31, 1994, Universiti Pertanian Malaysia, Selangor, Malaysia.

Nik Hassan Shuhami, N.A.R. 1991. Recent research at Kula Selinsing, Perak. *Indo-Pacific Prehistory Association Bulletin* 11:141–152.

Noakes, D.S.P. 1951. Notes on the silviculture of the mangroves forest of Matang, Perak. *Malayan Forester* 14:183–196.

Noakes, D.S.P. 1952. *A working plan for the Matang Mangrove Forest Reserve, Perak.* Kuala Lumpur: Caxton Press.

Ong, J. E. 1978. The Malaysian mangrove environment. Paper presented at the Regional Seminar on Human Use of the Mangrove Environment and Management Implications, Bangladesh.

Ong, J. E. 1982. Aquaculture, forestry and conservation in Malaysian mangroves. *Ambio* 11:252–257.

Othman, B.H.R., and A. Arshad. 1991. The macrobenthos community of the Matang mangrove channels and inshore waters. In *Proceedings of the Workshop on Mangrove Fisheries and Connections,* ed. A. Sasekumar, 56–69. ASEAN–Australian Marine Science Project. Kuala Lumpur: Ministry of Science, Technology and the Environment.

Parrish, D., and D. R. Wells. 1984. Wader survey report: Aeriel surveys. In *Interwader 1983 Report,* ed. D. Parrrish and D. R. Wells. Kuala Lumpur: INTERWADER.

Pathansali, D. 1977. Culture of cockles and other molluscs in the coastal waters of Malaysia. In *Proceedings of the Asian 1st Meeting of Experts on Aquaculture,* 149–155. Jakarta, Indonesia. ASEAN 77/FAEgA/DOC WP21.

Pathansali, D., and M. K. Soong. 1958. Some aspects of cockle (*Anadara granosa*) culture in Malaya. *Proceedings of the Indo-Pacific Fish Council* 8(2):26–31.

Perak State Forestry Department. 1994. *Management, utilisation and development of the Matang Mangrove Forest Reserve.* Perak, Malaysia: Ipoh.

Phillips, C. R. 1984. Conservation and rehabilitation of mangroves in Sabah, E. Malaysia. In *Proceedings of the Asian Symposium on Mangrove Environment—Resources and Management,* ed. E. Soepadmo, A. N. Rao, and D. J. Macintosh, 809–820. Kuala Lumpur: University of Malaya.

Phillips, M. J., C. K. Lin, and M.C.M. Beveridge. 1993. Shrimp culture and environment: Lessons from the world's most rapidly expanding warmwater aquaculture sector. In *Environment and aquaculture in developing countries,* ed. R.S.V. Pullin, H. Rosenthal, and J. L. Mclean, 171–197. ICLARM Conference Proceedings no. 31. Manila: International Centre for Living Aquatic Resources.

PHILNATMANCOM. 1987. Country paper: Philippines. In *Mangroves of Asia and the Pacific: Status and management,* ed. R. M. Umali, P. M. Zamora, R. R. Golera, R. S. Jara, and A. S. Camacho, 175–210. Technical Report of the UNDP/UNESCO Research and Training Pilot Programme on Mangrove Ecosystems in Asia and the Pacific, RAS/79/002. Manila: UNDP/UNESCO.

Primavera, J. 1989. The social, ecological and economic implications of intensive prawn farming. *SEAFDEC Asian Aquaculture* 11(1):1–6.

Putz, F. E. 1978. *A survey of virgin jungle reserve in Peninsular Malaysia.* Kuala Lumpur: Forestry Department of Peninsular Malaysia.

Putz, F. E., and H. T. Chan. 1986. Tree growth, dynamics and production in a mature mangrove forest in Malaysia. *Forest Ecology and Management* 17(2–3):211–230.

Ramly, A. 1992. *National conservation strategy for mangrove genetic resources in Malaysia.* Madras, India: M. S. Swaminathan Research Foundation (ITTO-CRSARD Project).

Ratnam, L., A. Jasmi, F. Gombek, and S. Hawa Yatim. 1987. Aspects of the demography and reproduction of the night heron at Sungei Burung. Paper presented at the International Conference on Wetland and Waterfowl Conservation in Asia.

Robertson, A. I. 1991. Fish, prawns and mangroves: Patterns and processes. In *Proceedings of the Workshop on Mangrove Fisheries and Connections,* ed. A. Sasekumar, 114–130. ASEAN–Australian Marine Science Project. Kuala Lumpur: Ministry of Science, Technology and the Environment.

Robertson, E. D. 1940. Charcoal kilns in the Matang mangrove forests. *Malayan Forester* 9:178–183.

Saenger, P., E. L. Hegerl, and J.D.S. Davie. 1981. *First report on the global status of mangrove ecosystems.* Commission of Ecology, Working Group on Mangrove Ecosystem, International Union for Conservation of Nature and Natural Resources (IUCN). Gland, Switzerland: IUCN.

Salleh, M. N., and H. T. Chan. 1986. Sustained yield management of the Matang mangrove. In *Mangroves of Asia and the Pacific: Status and management,* Technical report of the UNDP/UNESCO Research and Training Pilot Programme on Mangrove Ecosystems, ed. R. M. Umali, P. M. Zamora, R. R. Golera, R. S. Jana, and A. S. Camacho, 319–324. RAS/70/002. Manila: Natural Resources Management Center of the Philippines.

Sasekumar, A. 1980. *The present state of the mangrove ecosystem in S.E. Asia and the impact of pollution: Malaysia.* South China Sea Fisheries Development and Coordinating Programme, SCS/80/WP/94b (Rev). Manila: UNDP.

Sasekumar, A. 1984. Mangrove secondary productivity. In *Productivity of mangrove ecosystems: Management implications,* ed. J. E. Ong and W. K. Gong, 20–28. Penang, Malaysia: Universiti Sains Malaysia.

Sasekumar A. 1991. A review of mangrove fisheries connections. In *Proceedings of the Workshop on Mangrove Fisheries and Connections,* ed. A. Sasekumar, 47–50. ASEAN–Australian Marine Science Project. Kuala Lumpur: Ministry of Science, Technology and the Environment.

Sasekumar, A., V. C. Chong, and H. Singh. 1994. The fish community of Matang mangrove waters. In *Proceedings of the Third ASEAN-Australian Symposium on Living Coastal Resources.* Vol. 2: *Research Papers,* ed. S. Sundara, C. R. Wilkinson, and L. M. Chou, 457–464. Bangkok, Thailand: Chulalongkorn University.

Sasekumar, A., T. L. Ong, and K. L. Thong. 1984. Predation of mangrove fauna by marine fishes. In *Proceedings of the Asian Symposium on Mangrove Environment—Resources and Management,* ed. E. Soepadmo, A. N. Rao, and D. J. Macintosh, 378–384. Kuala Lumpur: University of Malaya.

Serivastava, P.B.L., and H. S. Bal. 1984. Composition and distribution pattern of natural regeneration after second thinning in Matang Mangrove Reserve, Perak, Malaysia. In *Proceedings of the Asian Symposium on Mangrove Environment—Resources and Management,* ed. E. Soepadmo, A. N. Rao, and D. J. Macintosh, 761–784. Kuala Lumpur: University of Malaya.

Serivastava, P.B.L., S. L Guan, and A. Muktar. 1988. Progress of crop in some *Rhizophora* stands before first thinning in Matang Mangrove Reserve of Peninsular Malaysia. *Pertanika* 11(3):365–374.

Shariff, S. 1984. Some observations on otters at Kuala Gula and National Park, Pahang. *Journal of Wildlife and National Parks* 3:75–88. Department of Wildlife and National Parks, Malaysia, Kuala Lumpur.

Silvius, M., H. T. Chan, and I. Shamsuddin. 1987. *Evaluation of wetlands of the west coast of Peninsular Malaysia and their importance for natural resource conservation.* Kuala Lumpur: World Wide Fund for Nature–Malaysia.

Singh, H. R., and A. Sasekumar. 1994. Distribution and abundance of marine catfish (Fam. *Ariidae*) in the Matang mangrove waters. In *Proceedings of the Third ASEAN-Australian Symposium on Living Coastal Resources.* Vol. 2: *Research Papers,* ed. S. Sundara, C. R. Wilkinson, and L. M. Chou, 471–477. Bangkok, Thailand: Chulalongkorn University.

Siti Hawa, Y. 1994. *Avifauna of Matang mangrove forest: A field study report in Matang mangrove forest.* Kuala Lumpur: Department of Wildlife and National Parks and Japan Wildlife Research Center.

Sivasothy, N., D. H. Murphy, and Peter K. L. Ng. 1991. In *Proceedings of the Workshop on Mangrove Fisheries and Connections,* ed. A. Sasekumar, 220–237. ASEAN-Australian Marine Science Project. Kuala Lumpur: Ministry of Science, Technology and the Environment.

Snedaker, S. C., and C. D. Getter. 1985. *Coastal resources management guidelines.* Renewable Resources Information Series. Coastal Management Publication no. 2. Columbia, South Carolina: Research Planning Institute.

Tang, H. T., H. Abu Hassan Harun, and E. K. Cheah. 1984. Mangrove forests of Peninsular Malaysia: A review of management research objectives and priorities. In *Proceedings of the Asian Symposium on Mangrove Environment—Resources and Management,* ed. E. Soepadmo, A. N. Rao, and D. J. Macintosh, 796–808. Kuala Lumpur: University of Malaya.

Tengku Ubaidillah, A. K. 1985. *Aquaculture development strategies and programmes.* Report of the Aquaculture Conference. Kuala Lumpur: Department of Fisheries Malaysia with International Convention Secretariat, Ministry of Agriculture.

Thompson, J. 1994. *Review of shorebird distribution in the Matang forest area, Malaysia. A field study report in Matang mangrove forest.* Kuala Lumpur: Department of Wildlife and National Parks and Japan Wildlife Research Center.

Thong, K. L., and A. Sasekumar. 1984. The trophic relationship of the fish community of the Angsa Bank, Selangor, Malaysia. In *Proceedings of the Asian Symposium on Mangrove Environment—Resources and Management,* ed. E. Soepadmo, A. N. Rao, and D. J. Macintosh, 385–399. Kuala Lumpur: University of Malaya.

U Than, H., and H. U Saw. 1984. Mangrove forests in Burma. In *Proceedings of the Asian Symposium on Mangrove Environment—Resources and Management,* ed. E. Soepadmo, A. N. Rao, and D. J. Macintosh, 82–85. Kuala Lumpur: University of Malaya.

Watson, J. G. 1928. *Mangrove forests of the Malay Peninsula.* Malayan Forest Records. Kuala Lumpur: Forest Department, Malaysia.

Wyatt-Smith, J. 1963. *Manual of Malayan silviculture for inland forests.* Malayan Forest Records no. 23. Kuala Lumpur: Department of Forestry, Peninsular Malaysia.

Yap, Y. N., A. Sasekumar, and V. C. Chong. 1994. Scianid fishes of the Matang mangrove waters. In *Proceedings of the Third ASEAN-Australian Symposium on Living Coastal Resources. Vol. 2: Research Papers,* ed. S. Sundara, C. R. Wilkinson, and L. M. Chou, 491–498. Bangkok, Thailand: Chulalongkorn University.

Zainal, Z. A., and A. Sasekumar. 1994. The macroinvertebrates in intact and cleared mangrove forests in Malaysia. In *Proceedings of the Third ASEAN-Australian Symposium on Living Coastal Resources.* Vol. 2: *Research Papers,* ed. S. Sundara, C. R. Wilkinson, and L. M. Chou, 433–436. Bangkok, Thailand: Chulalongkorn University.

Zamora, P. M. 1987. Protection, conservation, and rehabilitation of Philippine mangrove areas. Paper presented at the UNEP/ROPME Workshop on Coastal Area Development, Kuwait.

SIX

Sustainable Forestry in Concept and Reality

Andrew J. Hansen

In this twenty-seventh year since Earth Day, 1970, resource professionals, if not the public, now recognize that the human condition is intimately tied to ecological systems. A core principle of the emerging field of ecological economics (Costanza 1991), for example, is that human capital is fueled by natural capital (e.g., ecological productivity, biodiversity) and that human economies can only be understood in the context of ecological systems. These linkages are graphically evident in West Africa where Kaplan (1994) attributes the current spiral of social upheaval, crime, and warfare ultimately to depletion of natural resources. Such observations lead inevitably to the conclusion that humanity will benefit by managing ecological systems sustainably. In essence, this means that we must manage ecological systems to sustain their capacity to remain productive in perpetuity, with resource extraction levels being set within this context.

Approaches for sustaining resources have sprung up somewhat independently in the fields of forestry, agriculture, fisheries, and nature reserve management. However, the core questions about sustainability are similar in all these fields. What attributes of ecological systems promote the long-term productivity of plants and animals? What management strategies will, in reality, sustain these ecological systems? Will these strategies reduce resource outputs in the near term? Can society afford these short-term costs to gain long-term benefit? These questions have been most fully explored in relation to forestry, driven in large measure by controversy over harvest of natural forests in the tropics and the Pacific Northwest United States (PNW). New forestry practices based on ecological theory have been developed to better sustain forest productivity and biodiversity (Hunter 1990; Franklin 1992). The long-term effectiveness of these new practices in maintaining productivity and biodiversity, however, is not well tested. Similarly, relatively little is known about their short-term economic costs and benefits.

This chapter explores the conceptual basis of sustainable forestry and evaluates the state of knowledge on its effectiveness. First, the concept of sustainability is examined. Next I consider the properties of a forest that are thought to confer sustainability. Management approaches for maintaining or restoring these properties while also producing wood products are then examined. The last section summarizes studies of how well these approaches are sustaining forest biodiversity and productivity in the PNW and the consequences for short-term wood production. The major conclusion is that we do not yet know how effective sustainable forestry is and that much work is needed on how to measure and achieve sustainable resource management.

The Concept of a Sustainable Forest

Sustainable forestry is a substantial departure from the maximum sustained-yield paradigm that guided forestry during most of the twentieth century (SAF 1993). Traditional sustained-yield management involved "continuing the flow of one or more products within constraints imposed by environmental and economic factors" (SAF 1993, p. xv). Timber and livestock forage were often the focus of management, and other resources such as endangered species or water quality were considered only to the extent that they constrained commodity production.

This paradigm has fallen from favor on public lands for at least two reasons. First, society has placed increased value on forest resources in addition to wood and forage including recreation, wilderness, biodiversity, phytochemical medicines, and so on. Sustained-yield management was too narrowly focused on traditional forest resources to maintain adequately these other commodity and noncommodity values. Second, ecological studies revealed that traditional sustained-yield management often did not sustain long-term timber and forage production. As we came to understand better how forest ecosystems work, it became apparent that intensive plantation forestry removes some elements of forest structure, composition, and function that are essential for long-term primary and secondary forest productivity (hereafter termed forest productivity) (Perry et al. 1989; Hansen et al. 1991; Franklin 1992; Lanksy 1992). Similarly, the very notion of an even flow of forest products over time ignores the temporal dynamics of ecological systems (Mladenoff and Pastor 1993). We now know that the carrying capacity of populations within forests (including commercial tree species) varies over time as a function of variations in climate, natural disturbance, and other environmental factors.

The concept of sustainable forestry, in contrast, uses knowledge of how ecosystems function as the basis for management. The goal is to keep the ecosystem working properly and to harvest resources only up to a level that does not disrupt the ecosystem. According to Franklin (1993, p. 127), "sustainability refers to maintenance of the potential for our land and water ecosystems to produce the same quantity and quality of goods and services in perpetuity." The states of resources derived from an ecosystem are typically linked, so that managing for one resource requires consideration of other components of the ecosystem. Wood production in a forest, for example, is influenced by the diversity of soil organisms that process nutrients required by trees. Similarly, primary productivity of plants is essential for maintaining populations of endangered species that are higher on the food chain.

Sustainable forest management aims to maintain the key components and linkages of an ecosystem so that its ability to provide goods and services will not be impaired over time. The wholeness of an ecosystem is best described by the term *biological integrity.* Karr and Dudley (1981) defined biological integrity as "the capability of supporting and maintaining a balanced, integrated, adaptive community of organisms having a species composition, diversity, and functional organization comparable to that of natural habitat of the region." Hence sustainable management primarily aims at maintaining the biological integrity of the forest. Natural resources, whether they be wood, forage, endangered species, or medicinal compounds, are then utilized or maintained at levels that do not reduce the biological intregity of the forest.

Primary productivity, for example, is the result of complex interactions between the physical environment and organisms. Hence sustainable forestry seeks to prevent the degradation of soils, nutrients, and key ecological processes (e.g., nutrient cycling, succession, trophic interactions) that drive primary productivity. It also seeks to prevent the loss of species diversity, as a myriad of organisms are involved in primary productivity (Franklin 1993; Mladenoff and Pastor 1993).

The concept of sustainable forestry is closely related to another popular term, *ecosystem management*. Ecosystem management is the process by which entire ecosystems are managed to achieve desired objectives. For many public land management agencies in the United States, ecosystem management is the means used to strive toward sustainability (Grumbine 1994). It "attempts to maintain the complex processes, pathways and interdependencies of forest ecosystems and keep them functioning well over long periods of time, in order to provide resilience to short-term stress and adaptation to long-term change. Thus, the condition of the forest landscape is the dominant focus, and the sustained yield of products and services is provided within this context" (SAF 1993, p. xxi).

Moreover, ecosystem management "recognizes that natural disturbance regimes have provided the basic blueprint for sustaining ecological pattern and process on the landscape, and emphasizes management which emulates—not duplicates—many of the characteristics of these regimes" (SAF 1993, pp. 14–15).

If sustainable forestry is appealing in concept, it is difficult in reality. A challenge of sustainable forestry is to understand forest ecosystems well enough to maintain biological integrity even while extracting resources. Among other things, this means that we must be able to quantify biological integrity in order to measure our success. A second challenge to sustainable forestry involves the will of society. If traditional sustained-yield management extracted commodities at levels that impaired the forest ecosystem, it is likely that short-term resource production levels will need to be reduced for long-term sustainability. Is society willing or able to forgo resource production in the near term in favor of a lower but sustainable level of resources over the long term? Or alternatively, can society afford not to sustain forest productivity over the long term? These questions focus attention on the forest conditions, states, and properties that promote biological integrity and how they are influenced by alternative management strategies.

Elements of the Sustainable Forest

The plethora of ecological studies of natural forests in the last decade (e.g., Franklin et al. 1981; Harris 1984; Ruggiero et al. 1991), and more recently of managed forests (Hunter 1990; Saunders et al. 1991; McComb et al. 1993), provide a basis for delineating the ecological characteristics of forests that sustain forest productivity and biodiversity. These characteristics were summarized by Hopwood and Island (1991), Franklin (1992, 1993), Seymour and Hunter (1992), and Swanson and Franklin (1992). The key forest elements involve forest structure, species composition, and ecological processes. These elements are reviewed below with reference to three spatial scales: stand, landscape, and region.

Stand Level

Forest structure denotes the physical arrangement of biotic material in the ecosystem in both the vertical and horizontal dimensions. It is usually quantified in terms of mean and variation in live tree density and size class, canopy layering, standing dead tree density and size class, volume of woody debris on the forest floor, and understory abundance. Forests with high levels of variation in these factors are termed "structurally complex." Structural complexity is positively associated with long-term forest productivity and biodiversity (Franklin et al. 1981).

Forest structure influences the distribution of local environmental conditions (e.g., microclimate) and resources (e.g., nutrients, soil moisture, habitats). These factors in turn influence the numbers and types of niches available to organisms and the rates of processes such as decomposition and primary productivity. Structurally complex forests provide many different types of niches and, consequently, relatively high levels of species diversity. Some of these species play key roles in ecosystem processes such as nutrient cycling. For example, some lichens and bacteria are relatively abundant in structurally complex forests because of the large live trees and downed woody debris. These organisms fix nitrogen from the atmosphere, facilitating increased nitrogen availability to plants and enhanced primary productivity (Harmon et al. 1986).

Traditional forest management practices were designed to reduce structural complexity to favor crop species. This often reduced species diversity (Hunter 1990; Hansen et al. 1991) and likely reduced long-term forest productivity (Perry et al. 1989; Lansky 1992).

The numbers and types of organisms within the forest is termed *species composition.* Forest managers often seek to maintain native species because of their direct value for food, medicines, and esthetics. Species are also valued for the ecological services they provide. These include decomposition, nitrogen fixation, maintenance of soil structure, reducing leaching of nutrients, dispersing propagules of other species, and controlling pest populations. Hence foresters also manage for native species composition to maintain long-term forest productivity. Traditional forest management selects against some forest organisms through direct mortality (e.g., using herbicides to inhibit species competing with crop species), truncating early and late seral stages, or altering forest structure and niche diversity (Franklin 1992).

Among the ecological processes that influence biodiversity and long-term forest productivity is natural disturbance. Events such as wildfire, windthrow, and insect outbreaks kill some organisms, favor growth of others already established at the site, and allow colonization by new emigrants to the site (Pickett and White 1985). Natural disturbance seldom removes all biological material from a site. Rather, a rich "biological legacy" of living organisms, propagules, or woody debris commonly persist through disturbance (Franklin 1992). Such legacies foster relatively rapid recovery of ecological processes and biodiversity to disturbed patches. Natural ecosystems experience characteristic disturbance regimes where disturbance type, frequency, size, and intensity occur within a certain range of variability. Native organisms are adapted to these disturbance regimes, and in many cases are dependent on them (Swanson et al. 1993). Disturbance, and the ecological succession that follows, thus enhances structural complexity and species diversity and may contribute to long-term forest productivity.

Traditional silviculture involves the suppression of natural disturbances (e.g., wildfire) and the imposition of novel disturbances (e.g., clear cutting) (Urban et al. 1987). This commonly reduces forest structural complexity and species diversity. A well-documented example comes from the Inland West of North America where serious forest health problems resulted primarily from fire suppression and selective removal of fire-resistant tree species (Gast et al. 1991).

Landscape Level

Landscapes are mosaics of stands and other patch types. The spatial arrangement and interaction among patches across the landscape strongly influence ecological processes and species habitats. Thus analysis of landscape patterns is increasingly recognized as an important aspect of forest management (Harris 1984; Franklin and Forman 1987).

Ecologists initially chose to work in the centers of ecosystems and ignore influences from adjacent ecosystems. Recently it has become apparent that the flows of materials, energy, and organisms among ecosystems are key determinants of ecosystem function (Wiens et al. 1985; Hansen and di Castri 1992). The landscape surrounding a stand, for example, exerts control over the microclimate within the stand, the recruitment of propagules and organisms into the stand, and the disturbance regime in the stand (Franklin and Forman 1987). Also, many organisms operate over spatial scales larger than individual stands, and their population dynamics reflect the types, sizes, and juxtapositioning of patches across the landscape (Hunter 1990; Pulliam et al. 1992).

A widely known example of the consequences of ignoring landscape patterning involves the process of forest fragmentation. In several regions of the world, individual forest stands were converted to agricultural fields or other nonforest patches without regard for the broader landscape (Burgess and Sharpe 1981). This resulted in a reduction in total forest area, a reduction in forest patch density and size, and an increase in the distance between forest patches. Edge effects such as increased predation levels and more extreme microclimates penetrated to the centers of the smaller forest patches, leading to the local extinction of forest interior species (Terborgh 1989). Dispersal of forest organisms among forest patches was also reduced, inhibiting recolonization of the suitable patches.

Elsewhere in the world, changing land-use policies have led to the abandonment of agricultural lands, massive afforestation, and major shifts in forest species composition and structure (Lepart and Debussche 1992; McKibben 1995). Such "unplanned" changes in landscape patterns demonstrate the value of managing the spatial structure of landscapes. Many ecosystem management efforts are now carefully designing the spatial and temporal patterns of landscapes to best accomplish management objectives, often with reference to the patterns created by environmental gradients and natural disturbance regimes (e.g., Cissel et al. 1994).

The main elements of landscape structure that are relevant to forest sustainability involve patch type, patch size and shape, patch edge characteristics, the spatial arrangement of patches relative to one another and relative to underlying environmental gradients, and change in these elements over time (Forman and Godron 1985). Patch types in forest management are typically defined by vegetation type, seral stage, site quality, or habitat suitability. These factors obviously are strong drivers of species diversity and long-term site productivity. By altering the range of seral stages across the landscape and the abundance of each, for example, forest managers may strongly influence the species diversity and rates of ecological processes. Traditional forest management has typically reduced the variety of patch types, favoring commercially valuable vegetation types and younger, faster-growing seral stages.

The size and shape of a patch mediate the strength of interaction with adjacent patches. Patches that are small or have a high edge to interior ratio have relatively more edge and relatively less core area. Hence such patches will be more influenced by the surrounding landscape than large patches with relatively large interior areas. For example, forest fragmentation often leads to a loss of forest interior species because predators and brood parasites are better able to access their hosts in the small forest patches resulting from fragmentation (Payton 1994). The effects of patch size are somewhat mediated by patch edge structure. Narrow, abrupt edges, such as those produced by clear cutting, are more likely to expose a forest patch to surrounding patches than are wider, less abrupt edges (such as "feathered" edges around a harvest unit) (Hansen et al. 1992).

Managing stand juxtapositioning allows foresters to facilitate or inhibit exchanges among patches. Dispersing harvest units evenly over the forest, for example, facilitates seed recruitment from adjacent forests and provides habitats that require both early- and late-seral vegetation. This design, however, also enhances the spread of disturbances (e.g., windthrow and pathogens), extreme microclimates, and predation into the remaining forest patches (Franklin and Forman 1987).

Another important element of landscape function is the location of patch types relative to topographical, climatic, and edaphic gradients. These gradients may strongly influence habitat suitability and rates of ecological processes. For example, human settlement tends to occur in places in the landscape that are most favorable in climate, soils, and primary productivity. By default, nature reserves were often placed in the least productive places. The productive settings that remain undeveloped (e.g., lowlands and riparian areas in mountainous landscapes) may act as source areas for native species and contribute emigrants to the surrounding less productive habitats, and thus may have high conservation value. The goals of forest sustainability can best be achieved by "tailoring" land-use practices to the landscape by taking into account these environmental gradients.

Finally, the timing of change in the spatial structure of landscapes is critical to management. For example, early seral stages in the PNW support nitrogen-fixing plants that may significantly enrich soil productivity. Traditional forest management in the PNW has often shortened the duration of the open-canopy stage relative to natural succession and reduced natural nitrogen inputs (Franklin 1992).

These complexities of landscapes offer substantial challenges to ecosystem management. How can land managers identify an appropriate future landscape design without complete knowledge of how organisms and ecological processes will respond to the design? Some managers advocate maintaining modern landscapes within the natural range of variability from presettlement times (USFS 1993). The rationale is that native organisms and ecological processes are adapted to these landscape patterns and have the best chance to persist under them in the future (Swanson et al. 1993). This approach presents problems because it is difficult to identify the natural range of variability (NRV), and modern changes (roads, exotic species, alteration of disturbance regimes, climate change) may not allow landscapes within NRV to maintain biodiversity. An alternative approach is to design landscapes from "first principles" based on the best current knowledge (Hansen et al. 1993). For example, Murphy and Noon (1992) recommend a regional design for maintaining viable populations of the northern spotted owl that is based on the life-history attributes of the species. Most ecosystem management efforts use combinations of both approaches.

Regional Level

Many of the factors that influence the sustainability of landscapes also apply to regional scales. The abundance of ecosystem types and their spatial distributions across regions affect the presence and persistence of regional populations of native organisms. Humans are integral parts of most regions, and planning at this scale requires difficult decisions on allocation of land use among nature reserves, multiple-use, agricultural, and urban areas (Hunter 1990). There has been considerable analysis and debate on the size and connectivity of nature reserves (Meffe and Carroll 1994). In some cases, careful analyses were performed across regions on the spatial patterning of reserves and multiple-use forests for maintaining endangered species (FEMAT 1993). Less work has been done to incorporate broad environmental gradients into regional design. Multiple ownerships and agency jurisdictions present a special challenge to regional management (Slocombe 1993). Some of the more innovative efforts have attempted to manage the full gradient of human impact from nature reserves to urban areas (Slocombe 1993; Noss and Cooperrider 1994). This requires considerable cooperation among the diverse stakeholders.

In sum, ecological studies make clear that forest productivity is a complex product of interactions between forest structure, biodiversity, ecological processes, and environmental gradients. These interactions vary among the stand, landscape, and regional spatial scales. Managing forests sustainably is clearly not a simple process. Franklin (1992, pp. 42–43) warns that "organisms, structures, or processes cannot be discarded without ecological consequences. Hence forest simplification—the basis for most current commodity management—needs be approached with caution and humility." This realization has led to a growing body of knowledge on management strategies and prescriptions for sustaining forests.

Strategies for Sustaining Forests

Much of the thinking about ecological approaches to forest management originated in the PNW, guided by the influential writings of Franklin et al. (1981), Harris (1984), Harmon et al. (1986), Franklin and Forman (1987), Ruggiero et al. (1991), Swanson and Franklin (1992), and Franklin (1993). Largely independent work from other regions by Hunter (1990), Saunders et al. (1991), Lenz and Haber (1992), Seymour and Hunter (1992), and Mladenoff and Pastor (1993) has been convergent with that from the PNW. Collectively, these reviews suggest strategies for sustaining the long-term productivity and biodiversity of forests while also providing resources for human use. These strategies involve attention to the interactions among forest structure, biodiversity, and ecological processes (Table 6.1).

Stand Level

Strategies to achieve structural complexity within stands focus on the retention of live and dead plants at harvest and on encouraging their development over the rotation. Canopy structure has been called "the master regulator of the forest" (Horne and Hickey 1991, p. 121), and retention of live trees is a cornerstone of sustainable forestry. Most authors suggest leaving a range of tree and shrub size classes typical of natural forests. Methods to achieve retention include the marking of individual trees not to be cut, selection of trees to be cut, and gap cutting. Considerable uncertainty remains on the benefits and costs of the densities of plants retained and their spatial distributions (dispersed or clumped). Other strategies to promote structural complexity of live plants are to use variable densities for precommercial and commercial thinning and longer rotations to grow larger trees. Similarly, snags and downed woody debris can be maintained by retention at harvest, by recruitment via longer rotations, and by selective killing of live trees.

Attention to the species composition of retained plants during harvest and thinning also provides for the maintenance of plant diversity. Native species can be recruited into depauperate forests by maintaining seed-bearing individuals across the landscape, mixed species plantings, maintenance of a canopy structure to favor poorly represented species, and use of longer rotations to allow establishment of late-seral species.

Providing appropriate levels of structural complexity and plant species diversity will help ensure the ecological processes and biodiversity necessary for long-term forest productivity. Beyond this, atten-

Table 6.1. Means for Managing Forests Sustainably

Objective	Strategies	Management prescriptions	Possible ecological benefits
		Stand Level	
Maintain/restore structural complexity	Retain live trees in harvest units	Individual tree selection Group selection Live plant retention by size class	Provide habitats for canopy species Maintain mycorrhizal community Recruit snags and woody debris Reduce nutrient leaching
	Retain snags in harvest units	Snag retention by size class	Provide habitats for cavity-nesting, and dead-wood–dependent species Recruit woody debris
	Retain woody debris in harvest units	Woody debris retention by size class Minimize whole-tree harvesting Minimize postharvest burning	Provide habitat for dead-wood–dependent species Reduce erosion Maintain soil moisture and soil structure
	Recruit large trees, large snags, woody debris	Longer rotations Carry retained live trees over full rotation Kill trees to recruit snags	As above
	Create spatial heterogeneity in regenerating cohort	Variable-density thinning	Maintain early-seral and nitrogen-fixing species
Maintain/restore plant diversity	Retain native species in harvest units	Live plant retention by species	Maintain ecological services provided by species including: nitrogen fixation, seed dispersal, soil structuring, pollination, predation, and parasitism to control pests

	Recruit native species	Maintain seed-bearing individuals in/near harvest units Mixed species planting, Vary overstory spatial heterogeneity to favor shade-tolerant and shade-intolerant species Retain native species when thinning Longer rotations to recruit late-seral species	Maintain ecological services provided by species including: nitrogen fixation, seed dispersal, soil structuring, pollination, predation, and parasitism to control pests
Maintain/restore long-term site productivity	Maintain/restore structural complexity and plant diversity	As above	As above
	Minimize soil disturbance	Aerial suspension of logs via helicopter, balloon, high-lead logging	Reduce erosion and nutrient leaching
		Domestic animal logging Nonpermanent roads Minimize scarification for site preparation	Maintain soil structure Reduce runoff Maintain soil organisms
Maintain/restore disturbance processes	Maintain natural disturbances in natural range of variability	Do not suppress all wildfires, windthrows, pest outbreaks, landsliding	Maintain natural stand structure, function, and composition
	Augment natural disturbance	Prescribed burning	Recreate natural stand structure, function, and composition
		Silvicultural practices as above Controlled herbivory	

(*continued*)

Table 6.1. (*continued*)

Objective	Strategies	Management prescriptions	
Landscape and Regional Levels			
Establish land use objectives	Designate land allocations (e.g., reserves, multiple-use, high commodity production)	Landscape and regional assessment of ecological and human processes and needs	Achieve both ecological and human benefits
Maintain/restore desired diversity of ecosystems and seral stages	Retain existing native vegetation types, ecosystems, and seral stages	Identify important native vegetation types and ecosystems based on location of environmental gradients, species composition, structure, and function Maintain via land-use allocation, management prescriptions (e.g., silviculture, prescribed fire), or maintenance of natural disturbance Seek cooperative agreements among land owners	Retain key habitats and functions across landscape and region

	Restore desired ecosystems and seral stages that were lost	As above Selective plantings Species reintroductions	Retain key habitats and functions across landscape and region
Maintain/restore desired spatial patterning of ecosystems and seral stages	Maintain/restore patch size, shape, and connectivity to accomplish objectives	Identify desired spatial patterning based on natural range of variability or ecological principles Maintain/restore desired spatial patterns via spatial patterning of management prescriptions and natural disturbance	Minimize loss of species or processes due to landscape and regional scale phenomena (e.g., fragmentation, metapopulation dynamics)

Sources: Hopwood and Island 1991; Seymour and Hunter 1992; Swanson and Franklin 1992; Franklin 1993; and author's own work.

tion to soil disturbance and natural disturbance is recommended. Mechanized forest management is often associated with soil scarification, compaction, and erosion that alter soil structure and soil nutrient content, and reduce productivity. Such soil disturbance can be reduced via aerial suspension of logs, use of nonpermanent roads, yarding logs with domestic animals, and thoughtful site preparation techniques.

Like human management activities, natural disturbance strongly influences forest ecology. The challenge to managers is to maintain or restore the desirable aspects of natural disturbance in managed forests. Rather than suppressing natural disturbance like wildfire, stands and landscapes can be managed to encourage appropriate levels of natural disturbance. In some forests, prescribed burning, domestic animal herbivory, and other management activities are needed to provide the role once played by natural disturbance.

Landscape and Regional Levels

New approaches to forest management place much greater emphasis on landscape and regional factors than does traditional forestry. While ecological theory provides guidance on how to manage complex landscapes, empirical tests are few and much is yet to be learned on how to sustain biodiversity and productivity across large areas.

A first step is to set specific objectives for the management of landscapes and regions. This involves carefully identifying the suite of species and processes that are to be sustained over the planning area and the types of resources to be made available for human use. Such objectives provide a basis for specifying desired future landscape conditions (DFC). This is the arrangement of stand types or ecosystems (depending on the spatial scale) over the landscape and through time that are thought to accomplish landscape objectives.

At the regional scale, DFC involves allocating broad land uses such as for nature reserves, multiple-use, agricultural, and urban lands. This allocation has often been done by default in the past with settlements and agriculture claiming the most productive sites and nature reserves being placed in locations not suitable for other purposes. Regional planning is now shifting toward a more rigorous approach based on careful ecological and socioeconomic assessment and evaluation of alternative designs. For example, the Gap Analysis Project (Scott et al. 1987) is identifying ecosystem types across the United States that are now poorly represented in nature reserves, in order that they may be granted greater protection. Similarly, the Man in the Biosphere Program advocates organizing regional management along gradients in land use, including nature reserves, multiple-use buffer areas, commodity-production zones, and urban areas (Noss and Cooperrider 1994). Careful assessment of socioeconomic and ecological consequences of alternative regional management strategies was recently completed in the PNW (FEMAT 1993).

At the landscape level, DFC is often set using a combination of the so-called coarse-filter and fine-filter approaches (Hunter 1990). The coarse filter focuses on the maintenance of ecosystem types across the landscape under the assumption that most species and ecological processes will be maintained within those ecosystems. The difficult question remains of what types and spatial patterns of ecosystems will best accomplish the management objectives. One option is to maintain the landscape within the range of variation of presettlement times. A second option is to maintain representation of each ecosystem type across the landscape.

The fine-filter approach, in contrast, focuses on the species and processes that are unlikely to be perpetuated under the coarse filter. Detailed studies of species habitat requirements and demography are often used to derive landscape designs to maintain the viability of the species (Murphy and Noon 1992). Considerable work is still needed on how to integrate the coarse- and fine-filter approaches in ways that are most effective and cost efficient (Hansen et al. 1993).

Once the desired arrangement of landscape elements is specified, the next challenge is to devise a management scheme to achieve these patterns. Foresters have considerable experience with scheduling silvicultural manipulations and roads over the landscape. Little is known, however, of the socioeconomic consequences of nontraditional harvest unit sizes, or of the ecosystem sizes, shapes, and dispersions required for ecological objectives. Current strategies in multiple-use forests focus on managing patch sizes and edge characteristics to provide the desired mix of interior and edge habitats (Franklin and Forman 1987). In regions like the PNW where many species require forest interior habitats, harvest units may be aggregated in some portions of the landscape with large forest patches maintained in other places in the landscape. Also, relatively high levels of live trees may be retained around the outer portions of harvest units to buffer edge effects into the surrounding forest. Special attention is often paid to riparian zones in forest management. Streamside habitats are often rich in species diversity and productivity and are consequently protected from timber harvest. The resulting riparian buffers represent linear habitat networks that may facilitate movement of organisms across the landscape (Harris 1984).

Landscape-scale experiments are a means of reducing uncertainty about the consequences of alternative spatial designs. Unfortunately, relatively few such experiments have been performed because of difficulty in getting land allocated for this purpose and the high cost of implementing and monitoring the experiment over appropriately long time scales. Instead, many scientists and managers now advocate adaptive management (Walters 1986) where management implementations are designed and monitored to provide knowledge on their effectiveness.

Adaptive Management

Adaptive management offers a logical approach for sustaining forest stands, landscapes, and regions. The approach integrates science and management to improve understanding with each management implementation. Several models of adaptive management have emerged (Walters and Holling 1990; Everett et al. 1993; FEMAT 1993). The scheme of Hansen et al. (1993, 1994a) (Fig. 6.1) contains the core

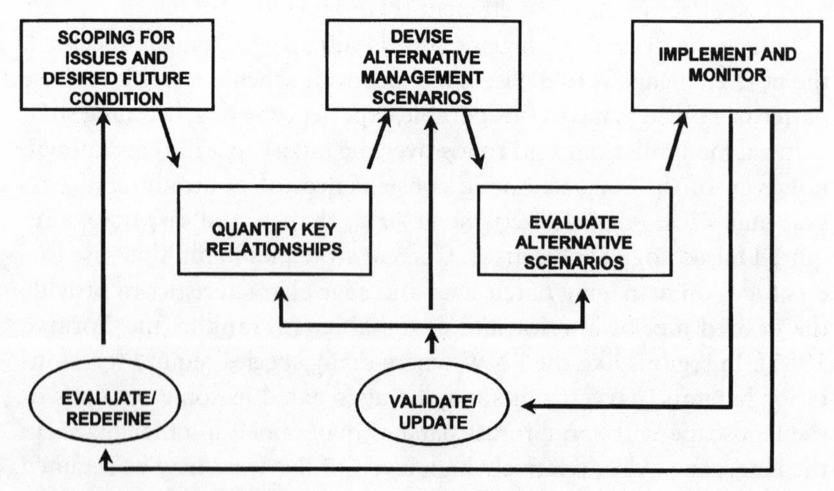

Figure 6.1. A conceptual model of adaptive management useful for forest management. (*Source:* Hansen et al. 1993, 1994a.)

elements found in most adaptive management models. After the DFC is specified, key ecological relationships (e.g., forest succession patterns, animal habitat relationships) are quantified using inventory data, results of scientific studies, and other information. Alternative scenarios for achieving the DFC are then devised and evaluated, often using decision support systems such as landscape simulation models (e.g., Pulliam et al. 1992; Hansen et al. 1994a). These projections help managers to decide which scenario is best able to achieve the DFC. Monitoring the results of the implementation reveals how well in fact the scenario worked and provides information for updating model parameters. Iteratively cycling through these steps offers the opportunity to learn from each management action and, ultimately, to make more informed management decisions.

The Effectiveness of Sustainable Forestry

As the preceding pages reveal, the concept of sustainable forestry has stimulated a great deal of thinking on the goals of forest management, forest ecology, and alternative silvicultural approaches. The ultimate question is how well can we sustain biodiversity and productivity in forests also used for resource extraction. Unfortunately, the answer is not yet in hand. The ecologically based approaches described above have been widely implemented only in the last five years or so. The major land management agencies in the United States and Canada have not yet enacted broadscale monitoring efforts, so ecological responses to these initial implementations are not yet quantified. Moreover, biodiversity and ecological productivity are difficult and expensive to measure, especially at landscape and larger scales. And finally, benchmark data (e.g., from natural systems) are not widely available for evaluating current trends in biodiversity and productivity.

The poor state of knowledge was revealed in the literature searches I did for this chapter. The databases Agricola and Biosis revealed only a handful of papers published in the last 10 years that did empirical evaluations of ecological forestry approaches.

Some of the earliest insights for the stand level may come from field studies of green-tree retention in harvest units, particularly in the PNW. Natural disturbance in this region typically leaves substantial levels of biological legacy which is thought to strongly influence biodiversity and productivity (Franklin 1992). Since the late 1980s, scientists and managers have worked together to design and implement harvest designs that retain various densities of trees and shrubs in a range of size classes. Several studies of these stands are under way (Fischer et al. 1994; McComb et al. 1994; Acker 1995; Schowalter 1995; Griffiths et al. 1996; Rose and Muir in press). Similar work has been initiated in the south-central and eastern United States (Nichols and Wood 1993; Petit et al. 1993).

Work to date in the PNW suggests that green-tree retention and longer rotations support levels of structural complexity and native species diversity closer to those of natural forests. However, these strategies have lower wood production than short-rotation clear cutting. For example, the densities of large trees retained in 22 harvest units in western Oregon (mean 13.0/hectare [ha], range 1.3–56/ha) were similar to those of large residual trees in 14 natural stands that had undergone burning by wildfire 60–130 years ago (mean 21.4/ha, range 3–57/ha) (Acker 1995; Hansen et al. 1995a). Similarly, a simulation experiment (Hansen et al. 1995b) found that structural complexity (stand density, mean tree diameter, and variation in tree diameter) in retention stands was more similar to old-growth stands than were stands that had been clearcut. These differences in structural complexity between retention and clearcut stands persisted for more than one hundred years after harvest.

Species composition and diversity are also influenced by green-tree retention. Species richness of regenerating trees was higher under even low levels of canopy-tree retention (5/ha) than following clear cutting, with shade-tolerant species being better represented with retention (Hansen et al. 1995b). Similarly, Schowalter (1995) found that canopy arthropod diversity was greater in retention stands than in clearcuts, but did not differ between retention and old-growth stands. Trophic diversity of arthropods was also higher in retention stands than clearcuts, suggesting that processes like predation are better maintained with canopy-tree retention. Bird diversity was predicted to be higher in retention stands than clearcuts in the computer simulation study (Hansen et al. 1995b). This finding is supported by field data from Hansen and Hounihan (1996) and Nichols and Wood (1993), but not by Vega (1993). All these studies found that some species of birds reach peak abundance in retention units and that many closed-canopy specialists were also found in retention units. Not all species of birds benefit from green-tree retention, however. A few open-canopy species avoided stands with even low densities of canopy trees, probably because of elevated predation rates in these stands.

Relatively few studies have examined rates of ecological processes as a function of canopy density. Griffiths et al. (1996) found that mats of ectomycorrhizal fungi decreased in abundance with distance from canopy trees in retention units. Many of these fungi are symbiotic with trees, providing nutrients that enhance tree growth and possibly survival. Hence retained trees may serve as refugia for ectomycorrhizal fungi following harvest and may increase seedling survival and growth rates (Griffiths et al. 1996). Similarly, retained trees act as refugia for arthropod predators and parasites, which may enhance the control of herbivorous arthropods by these predators and parasites (Schowalter 1995). Interestingly, predation on artificial bird nests was found to be higher in retention stands than in clearcuts (Vega 1993). Retained trees may serve as perches for avian predators, allowing them to locate the nests of prey species better. Nichols and Wood (1993) drew the same conclusion.

Foresters are concerned that canopy-tree retention reduces primary productivity and output of wood products. Unfortunately, existing studies do not allow this concern to be addressed sufficiently. A field study found that net primary productivity was substantially lower in retention stands than in stands that had been clearcut within the previous 10 years (Hansen et al. 1994b). The benefits of retention for primary productivity, however, may be expressed in later seral stages. Computer simulations over full rotations by Birch and Johnson (1992) and Hansen et al. (1995b) predicted reductions in wood production of 18–30 percent under retention of five trees per ha as compared to clear cutting. Wood production was further reduced with higher levels of retention, though the effect was less pronounced. Neither simulation model considered the possible loss of nutrients and reduction in production under clear cutting. Data from 60- to 130-year-old natural stands established after wildfire indicated that growth rates of regenerating trees were inversely related to density of residual trees (Acker 1995). In contrast to the simulation studies, however, Acker found that total tree growth (residual and regenerating trees) rates in this study did not differ as a function of residual tree density.

The consequences of green-tree retention for tree growth have important implications for forest economics. Hansen et al. (1995b) predicted that the cumulative value of wood products over multiple rotations would decrease with green-tree retention density, but at a lower rate than does wood volume. The higher value of the larger trees produced by retention partially compensated for the reduction in cumulative wood volume. For the same reason, cumulative value of wood products did not decrease under longer rotations.

In sum, these initial studies from the Pacific Northwest indicate that canopy-tree retention and longer rotations will better sustain forest biodiversity than traditional plantation forestry. The maintenance of forest structural complexity, species diversity, trophic diversity, and mycorrhizal communities may also help to sustain long-term ecological productivity. The effects of these factors on forest growth were not considered in the computer simulation studies. It is possible that mycorrhizal communities and nutrient cycles remained intact in the natural stands studied by Acker (1995) and explain why total tree growth was not reduced by the presence of residual canopy trees. Clearly, more research is needed to determine how well these ecologically based management strategies will maintain long-term forest productivity. Ecological theory suggests that the near-term reduction in wood production under these ecological approaches will be compensated by long-term sustainability, especially in the face of uncertainties such as global climate change. More empirical work is needed to determine if this hypothesis is correct.

Conclusion

Resource sustainability is like world peace: everyone agrees it is a good idea, but few people know how to maintain it and many are unwilling to pay the near-term costs. Now that the concept has been clearly elucidated by scientists and widely embraced by policymakers, the challenge is to make sustainability a reality. I suggest three key questions that need more attention from forest scientists and managers. How can the general strategies for sustaining forests best be adapted to local conditions? To what extent are these strategies being employed in forests around the world? How effective are these strategies in sustaining forest productivity and biodiversity, and what are the short-term costs to society?

Relatively few places in the world have had extensive research and adaptive management directed at forest sustainability. While the general principles that have emerged from these places (e.g., structural complexity confers biodiversity) are likely to apply to many ecosystems, there is broad agreement that the principles need to be adapted to each local environment. For example, forest fragmentation may have strong negative effects in some regions, but be the natural condition in others (Hansen et al. 1992; Hansen and Urban 1992). Similarly, salvage logging may improve forest health in some situations in eastern Oregon, but reduce forest health in other parts of the Inland West (Della Salla et al. 1995). In the words of Gordon (1993, p. 242), "Manage where you are." Development of research and adaptive management protocols would help more scientists and managers to conduct these types of studies in their local forests. Among the factors that need to be considered are the life histories of the species in the system, the role of natural disturbance, similarities between current management approaches and natural disturbance, the distribution of local environmental gradients, and their effects on ecological productivity and biodiversity.

We would also benefit from taking stock of the types and frequencies of forest management practices that are being implemented in each region. The rapid shift from maximum sustained-yield management to ecosystem management has stimulated many new practices. However, there are few data on what practices have been implemented where and how well they achieved the desired forest structure and composition. For example, the U.S. Department of Agriculture Forest Service Northern Region has a goal to maintain modern stands and landscapes within the natural range of variability of presettlement times (USFS 1993). But no summary exists of the types and frequencies of current stand and landscape designs. Inventories of stand and landscape conditions would provide a basis for evaluating which designs are most likely to achieve management goals.

Perhaps the greatest need, however, is to determine which management strategies sustain biodiversity and forest productivity most effectively and what levels of commodity extraction are compatible with these strategies. While it is widely assumed that sustainable resource management is possible, it is plausible that the short-term costs are greater than some societies will be able to bear. Many examples exist of humans harvesting a natural resource to total depletion and then suffering the consequences (Perlin 1989). The most important finding of this review is that we do not know how well current strategies for sustaining forests are working. While some research is under way, these studies are specific to only a few regions of the world. Moreover, the appealing concept of adaptive management has not yet been widely implemented for forest management. This knowledge shortfall can be remedied by a greater commitment to adaptive management, large-scale experiments, retrospective studies of natural systems, and simulation modeling. It is important that these studies consider several types of response variables (ecological and socioeconomic) and gradients in treatment intensity. This will allow for trade-off analyses to determine the relative costs and benefits of each strategy. Considering several treatment intensities will help to identify thresholds where response variables may change abruptly (Hansen et al. 1995b). Such knowledge gives managers an objective basis for evaluating which strategy will best optimize the management objectives. Let us hope that these studies will guide us toward the ability to sustain both ecological systems and human societies.

References

Acker, S. 1995. Structure, composition, and dynamics of two-aged forest stands in the Willamette National Forest: A retrospective study concerning green-tree retention. Unpublished report. Forest Science Department, Oregon State University, Corvallis, Oregon.

Birch, K., and K. Johnson. 1992. Stand-level wood-production costs of leaving live, mature trees at regeneration harvest in coastal Douglas-fir stands. *Western Journal of Applied Forestry* 3:65–68.

Burgess, R., and D. Sharpe, eds. 1981. *Forest island dynamics in man-dominated landscapes.* New York: Springer-Verlag.

Cissel, J., F. Swanson, W. McKee, and A. Burditt. 1994. Using the past to plan the future in the Pacific Northwest. *Journal of Forestry* August:30–46.

Constanza, R., ed. 1991. *Ecological economics: The science and management of sustainability.* New York: Columbia University Press.

Della Salla, D., D. Olson, S. Barth, S. Crane, and S. Primm. 1995. Forest health: Moving beyond the rhetoric to restore healthy landscapes in the inland northwest. *Wildlife Society Bulletin* 23(3):346–356.

Everett, R., C. Oliver, J. Saveland, P. Hessburg, N. Diaz, and L. Erwin. 1993. Adaptive ecosystem management. In *Eastside forest ecosystem health assessment,* vol. 2 of *Ecosystem management: Principles and applications,* ed. M. E. Jensen and P. S. Bourgeron, 351–364. Portland, Oregon: United States Department of Agriculture Forest Service.

FEMAT. 1993. *Forest ecosystem management: An ecological, economic, and social assessment.* Portland, Oregon: United States Department of Agriculture Forest Service.

Fischer, C., J. Smith, and R. Molina. 1994. Mycorrhyzal diversity within a green-tree retention unit: A preliminary report. Unpublished report. U.S. Department of Agriculture Forest Service Pacific Northwest Research Station, Corvallis, Oregon.

Forman, R.T.T., and M. Godron. 1985. *Landscape ecology.* New York: John Wiley and Sons.

Franklin, J. F. 1992. Scientific basis for new perspectives in forests and streams. In *Watershed management: Balancing sustainability and environmental change,* ed. R. Naiman, 5–72. New York: Springer-Verlag.

Franklin, J. F. 1993. The fundamentals of ecosystem management with applications in the Pacific Northwest. In *Defining sustainable forestry,* ed. G. Aplet, N. Johnson, J. Olson, and V. Sample, 127–144. Washington, D.C.: Island Press.

Franklin, J. F., K. Cromark, W. Denison, A. Mc Kee, C. Maser, J. R. Sedell, F. J. Swanson, and G. Juday. 1981. *Ecological characteristics of old-growth Douglas-fir forests.* General Technical Report PNW-118, Pacific Northwest Forest and Range Experiment Station. Portland, Oregon: United States Department of Agriculture Forest Service.

Franklin, J. F., and R.T.T. Forman. 1987. Creating landscape patterns by forest cutting: Ecological consequences and principles. *Landscape Ecology* 1:5–18.

Gast, W. R., Jr., D. W. Scott, C. Schmitt, D. Clemens, S. Howes, C. G. Johnson, Jr., R. Mason, and F. Mohr. 1991. *Blue Mountains forest health report.* Pacific Northwest Region. Portland, Oregon: United States Department of Agriculture Forest Service.

Gordon, J. 1993. Ecosystem management: An idiosyncratic overview. In *Defining sustainable forestry,* ed. G. Aplet, N. Johnson, J. Olson, and V. Sample, 240–244. Washington, D.C.: Island Press.

Griffiths, R., G. Bradshaw, B. Marks, and G. Lienkaemper. 1996. Factors influencing the spatial distribution of ectomycorrhizal mats in coniferous forests of the Pacific Northwest, USA. *Plant and Soil* 180:147–153.

Grumbine, E. 1994. What is ecosystem management? *Conservation Biology* 8:27–38.

Hansen, A. J., and F. di Castri, eds. 1992. *Landscape boundaries: Consequences for biotic diversity and ecological flows.* New York: Springer-Verlag.

Hansen, A. J., and P. Hounihan. 1996. A test of ecological forestry: Canopy retention and avian diversity in the Oregon Cascades. In *Biodiversity in managed landscapes: Theory and practice,* ed. R. Szaro, 401–424. London: Oxford University Press.

Hansen, A. J., and D. L. Urban. 1992. Avian response to landscape pattern: The role of species life histories. *Landscape Ecology* 7:163–180.

Hansen, A. J., T. A. Spies, F. J. Swanson, and J. L. Ohmann. 1991. Lessons from natural forests: Implications for conserving biodiversity in natural forests. *BioScience* 41:382–392.

Hansen, A. J., D. L. Urban, and B. Marks. 1992. Avian community dynamics: The interplay of human landscape trajectories and species life histories. In *Landscape boundaries: Consequences for biological diversity and ecological flows,* ed. A. J. Hansen and F. di Castri, 170–195. New York: Springer-Verlag.

Hansen, A. J., S. L. Garman, B. Marks, and D. L. Urban. 1993. An approach for managing vertebrate diversity across multiple-use landscapes. *Ecological Applications* 3:481–496.

Hansen, A. J., R. Patten, and G. DeGayner. 1994a. Demonstration of landscape modeling and analysis: Mitkof Island, Tongass National Forest. Bozeman, Montana: Montana State University.

Hansen, A. J., R. Spencer, A. Moldenke, and A. McKee. 1994b. Ecological processes linking forest structure and avian diversity in western Oregon: A test of hypotheses. In *Biodiversity, temperate ecosystems, and global change,* ed. T. Boyle and C. Boyle, 217–247. New York: Springer-Verlag.

Hansen, A. J., W. McComb, M. Raphael, M. Hunter, and R. Vega. 1995a. Bird habitat relationships across a gradient of stand structure and ages in natural and managed forests in western Oregon. *Ecological Applications* 5:555–569.

Hansen, A. J., S. Garman, J. Weigand, D. Urban, W. McComb, and M. Raphael. 1995b. Ecological and economic effects of alternative silvicultural regimes in the Pacific Northwest. *Ecological Applications* 5:535–554.

Harmon, M., J. F. Franklin, F. Swanson, P. Sollins, S. Gregory, J. Luttin, W. Anderson, S. Cline, N. Aumnen, J. Sedell, G. Lienkaemper, K. Cromack, and K. Cummins. 1986. Ecology of coarse woody debris in temperate ecosystems. *Advances in Ecological Research* 15:133–302.

Harris, L. 1984. *The fragmented forest.* Chicago: University of Chicago Press.

Hopwood, D., and L. Island. 1991. *Principles and practices of new forestry: A guide for British Columbians.* Victoria, British Columbia: British Columbia Ministry of Forests.

Horne, R., and J. Hickey. 1991. Ecological sensitivity of Australian rain forests to selective logging. *Australian Journal of Ecology* 16:119–129.

Hunter, M. 1990. *Wildlife, forests, and forestry: Principles of managing forests for biological diversity.* Englewood Cliffs, N.J.: Prentice-Hall.

Kaplan, R. D. 1994. The coming anarchy. *Atlantic Monthly* 274(2):44–76.

Karr, J. R., and D. R. Dudley. 1981. Ecological perspective on water quality goals. *Environmental Management* 5:55–68.

Lansky, M. 1992. *Beyond the beauty strip*. Gardiner, Maine: Tilbury House.

Lenz, R., and W. Haber. 1992. Approaches for the restoration of forest ecosystems in northeastern Bavaria. *Ecological Modeling* 63:299–317.

Lepart, J., and M. Debussche. 1992. Human impact on landscape patterning: Mediterranean examples. In *Landscape boundaries: Consequences for biotic diversity and ecological flows,* ed. A. J. Hansen and F. di Castri, 76–106. New York: Springer-Verlag.

McComb, W., T. Spies, and W. Emmingham. 1993. Douglas-fir forests: Managing for timber and mature-forest habitat. *Journal of Forestry* December: 31–41.

McComb, W., J. Tappiener, L. Kellogg, C. Chambers, and R. Johnson. 1994. Stand management alternatives for multiple resources: Integrated management experiments. In *Expanding horizons of forest ecosystem management: Proceedings of the 3rd Habitat Futures Workshop,* ed. M. Huff. General Technical Report PNW-GTR-336. Portland, Oregon: United States Department of Agriculture.

McKibben, B. 1995. An explosion of green. *Atlantic Monthly* 275(4):61–83.

Meffe, G., and C. Carroll. 1994. *Principles of conservation biology.* Sunderland, Massachusetts: Sinauer Associates.

Mladenoff, D., and J. Pastor. 1993. Sustainable forest ecosystems in the northern hardwood and conifer forest region: Concepts and management. In *Defining sustainable forestry,* ed. G. Aplet, N. Johnson, J. Olson, and V. Sample, 145–180. Washington, D.C.: Island Press.

Murphy, D. D., and B. R. Noon. 1992. Integrating scientific methods with habitat conservation planning: Reserve design for northern spotted owls. *Ecological Applications* 2:3–17.

Nichols, J., and P. Wood. 1993. Effects of two-aged timber management on neotropical migrant songbird density and reproductive success. Unpublished report. West Virginia Cooperative Fish and Wildlife Resources Unit, Morgantown, West Virginia.

Noss, R., and A. Y. Cooperrider. 1994. *Saving nature's legacy.* Washington, D.C.: Island Press.

Payton, P. 1994. The effect of edge on avian nest success: How strong is the evidence? *Conservation Biology* 8:17–26.

Perlin, J. 1989. *A forest journey.* Cambridge, Massachusetts: Harvard University Press.

Perry, D. A., M. P. Amaranthus, J. G. Borchers, and S. L. Brainard. 1989. Bootstrapping in ecosystems. *BioScience* 39:230–237.

Petit, R., D. Petit, T. Martin, R. Thill, and J. Taulman. 1993. Predicting the effects of ecosystem management harvesting treatments on breeding birds in pine-hardwood forests. Paper presented at the Symposium on Ecosystem Management Research in the Ouachita Mountains: Pretreatment Conditions and Preliminary Findings. Hot Springs, Arkansas, October 26–27, 1993.

Pickett, S.T.A., and P. S. White, eds. 1985. *The ecology of natural disturbance and patch dynamics.* New York: Academic Press.

Pulliam, H. R., J. B. Dunning, and J. Liu. 1992. Population dynamics in complex landscapes: A case study. *Ecological Applications* 2:165–177.

Rose, C., and P. S. Muir. n.d. Relationships of green-tree retention following timber harvest to forest growth and species composition in the western Cascade Mountains, U.S.A. *Canadian Journal of Forest Research.*

Ruggiero, L., K. Aubry, A. Carey, and M. Huff, eds. 1991. *Wildlife and vegetation of unmanaged Douglas-fir forests.* General Technical Report PNW-GTR-285, Pacific Northwest Forest and Range Experiment Station. Portland, Oregon: United States Department of Agriculture Forest Service.

SAF. 1993. *Task force report on sustaining long-term forest health and productivity.* Bethesda, Maryland: Society of American Foresters.

Saunders, D., R. Hobbs, and C. Margules. 1991. Biological consequences of ecosystem fragmentation: A review. *Conservation Biology* 5:18–32.

Schowalter, T. 1995. Canopy arthropod communities in relation to forest age and alternative harvest practices in western Oregon. *Forest Ecology and Management* 78(1/3):115–122.

Scott, J. M., B. Csuti, J. D. Jacobi, and J. E. Estes. 1987. Species richness: A geographic approach to protecting future biological diversity. *BioScience* 37(11):782–788.

Seymour, R., and M. Hunter. 1992. *New forestry in eastern spruce-fir forests: Principles and applications to Maine.* Miscellaneous Publication 716, Maine Agricultural Experiment Station. Orono, Maine: University of Maine.

Slocombe, D. S. 1993. Implementing ecosystem-based management. *BioScience* 43:612–622.

Swanson, F. J., and J. F. Franklin. 1992. New forestry principles from ecosystem analysis of Pacific Northwest forests. *Ecological Applications* 2:262–274.

Swanson, F. J., J. A. Jones, D. A. Wallin, and J. Cissel. 1993. Natural variability: Implications for ecosystem management. In *Eastside forest ecosystem health assessment,* vol. 2 of *Ecosystem management: Principles and applications,* ed. M. E. Jensen and P. S. Bourgeron, 85–100. Portland, Oregon: United States Department of Agriculture Forest Service.

Terborgh, J. 1989. *Where have all the birds gone?* Princeton, N.J.: Princeton University Press.

Urban, D. L., R. V. O'Neill, and H. H. Shugart. 1987. Landscape ecology. *BioScience* 37:119–127.

USFS. 1993. *Our approach to sustaining ecological systems.* Missoula, Montana: United States Department of Agriculture Forest Service, Northern Region.

Vega, R. 1993. Bird communities in managed conifer stands in the Oregon Cascades: Habitat associations and nest predation. Thesis, Oregon State University, Corvallis, Oregon.

Walters, C. J. 1986. *Adaptive management of renewable resources.* New York: McGraw-Hill.

Walters, C. J., and C. S. Holling. 1990. Large-scale management experiments and learning by doing. *Ecology* 71:2060–2068.

Wiens, J. A., C. S. Crawford, and J. R. Gosz. 1985. Boundary dynamics: A conceptual framework for studying landscape ecosystems. *Oikos* 45:421–427.

SEVEN

Brazil Nuts

The Use of a Keystone Species for Conservation and Development

Jason W. Clay

Brazil nut trees (*Bertholletia excelsa*) are generally acknowledged as a keystone species (if not the keystone botanical species) in Amazonian forests. The harvest of their seed has generally been acknowledged as the largest extractive income generator, following rubber, in the region for a century. More recently, Brazil nuts have become the cornerstone of production and marketing strategies aimed at both conservation and income generation in the Amazon region. I summarize here the current understanding of Brazil nut distribution and ecology; collection methods; uses and economic potential; historic, current, and potential markets; and the overall impact of Brazil nut harvest on the species. In addition, I identify and recommend further research and development activities which could improve the conservation of the species as well as income derived from it.

An average of 42,000 metric tons (MT) of Brazil nuts are harvested annually from wild trees in Amazon forests and sold into the nut market (Gill and Duftus 1988a,b, 1991). An estimated two hundred thousand people take part in the commercial nut harvest. In addition, an untold number of nuts are harvested in the wild by an even larger number of people for subsistence use. The harvesters of nuts destined both for subsistence and commercial ends range from tiny, isolated indigenous groups to long-standing peasant communities and more recent colonists.

Brazil nuts comprise 1.5 percent of the international edible nut trade, which is estimated at US$2.3 billion annually.* The relative share of Brazil nuts in the edible nut trade has fallen over the last two decades because of a decline in total Brazil nut production, a drop in the price of other substitutable nuts, and an increase in the overall world nut market.

There are approximately 90 million hectares (ha) of forests used for extractivism in the Amazon, with about 20 million ha containing Brazil nut trees. About half of this area (10 million ha) is mixed with rubber and other extractive trees (LaFleur 1991c). Given an average of at least one adult Brazil nut tree per ha, there is a minimum total stand of at least 20 million Brazil nut trees in full production (LaFleur 1991c). During the past five years, subsidies for rubber in Brazil have been eliminated, and this has led Brazil nut gatherers to harvest and sell larger quantities of nuts since they are now, in many areas, the largest single source of income.

In Bolivia, more than a dozen new factories were established in the 1990s to shell nuts for export. These factories are fed by Brazil nut collectors who scour the countryside harvesting all available nuts. In addition, Bolivian factories (due to lower energy and labor costs as well as lower local and national taxes) are able to pay higher prices for raw nuts than Brazilian counterparts across the border (Amazonia Trading Company 1991). This has made it possible, for the first time, to pay high transportation costs from remote, isolated areas in both Bolivia and Brazil and still remain competitive in the market. Consequently the nut offtake is increasing, and there is some concern that the increased harvests will have an adverse effect on the populations of Brazil nut trees.

Brazil Nuts (Bertholletia excelsa): *The Plant and Related Species*

The Brazil nut (*Bertholletia excelsa* Humb. & Bonpl.) is the most economically important species in the family Lecythidaceae. Its edible seeds

*The term *Brazil nuts* is used to designate both shelled and in-shell Brazil nuts, unless otherwise specified.

are collected entirely from stands of trees within primary forests. While more than 95 percent of those seeds collected for sale or barter are sold into an international marketing system, Brazil nuts continue to be extremely important sources of food for local populations. There is considerable evidence that many contemporary Brazil nut stands were created through the efforts of Native American populations in the Amazon to create food supplies for themselves and future generations (D. A. Posey, pers. comm. 1990; C. R. Clement, pers. comm. 1991).

Because of its economic importance, the Brazil nut has been the subject of nearly three hundred works (Vaz Pereira and Costa 1981; Mori and Prance 1990a). Even so, we know very little about its overall distribution, reproduction, productivity, and economic and subsistence importance throughout the entire region.

Related Species

There are eleven genera of Lecythidaceae in South America, with approximately 199 known species (Mori and Prance 1990b), a few of which produce edible nuts. The family is essentially tropical and the subfamily, Lecythidoideae, is found exclusively on the South American continent. *Excelsa* is the sole species in the genus *Bertholletia* (de Souza 1963).

The closely related genus *Lecythis* contains several edible species, notably the sapucaia or paradise nut group of species, the most widespread of which is *L. pisonis* = *L. usitata* (Mori and Prance 1990a,b). The sapucaia is a large tree, although smaller in stature than the Brazil nut, and native to the *terra firma* (dry uplands) and *várzeas* (whitewater floodplain terraces) of Amazonia, eastern Brazil, and the southeastern transition zones to *cerrado* (upland, drier scrub areas). Sapucaia fruit vary enormously in size, from about the size of a Brazil nut to five or more times that, although they contain about the same number of seeds (10–25) (Mori and Prance 1990a; Clement 1993). Although the seeds are generally reputed to have a better flavor than the Brazil nut, sapucaia is rarely found at market because the fruit capsule opens while on the tree, providing a feast for bats, parrots, and monkeys, but leaving little for humans to collect (Clement 1993). Attempts to graft sapucaia onto Brazil nut and vice versa have failed (Clement 1993).

Description and Phenology

Brazil nuts are large trees, with canopy emergents frequently attaining 50 meters (m) in height, and straight cylindrical unbranched trunks attaining 1–2.5 m in diameter at breast height (dbh). The openly branched crown occurs at or above canopy level and may have a diameter of 20–30 m as an emergent (Clement 1993).

The Brazil nut is an allogamous species, possibly presenting very small levels of autogamy (Buckley et al. 1988; O'Malley et al. 1988). Seeds are contained in fruits with an extremely hard outer shell, the mesocarp, locally called the *ouriço.*" The fruits are large (10–12 centimeters [cm] in diameter), weighing 0.2–1.5 kilogram (kg) and containing about 18 seeds (from 10 to 25). The seeds are about 3.5–7 cm long by 2 cm wide, and weigh 4–10 grams (g). The unshelled seeds represent about 25 percent and shelled seeds about 13 percent of the weight of the fruit (Müller et al. 1994). The Brazil nut is, in fact, a seed rather than a nut, but popular usage continues to prevail.

Flowers are large and aromatic. In eastern Amazonia, Brazil nut flowering starts at the end of the rainy season (September) and extends through the dry season to February, with the greatest intensity in October–December (Moritz 1984). Flowering can extend into the rainy season (S. A. Mori, pers. comm. 1994). In the Manaus and the western Amazon, flowering starts earlier because the dry season starts in June, and extends to August or September. The Brazil nut fruit takes about 15 months to develop to maturity (Moritz 1984), and thus trees are not annual producers (de Souza 1984). Fruit fall starts in the wettest part of the rainy season. This is February to April in eastern Amazonia and November to March in the western Amazon.

Brazil nuts, protected by the hard ouriço in addition to their individual shells, are resistant to insects and animals. While in the protective ouriço and hanging on the tree, the nuts do well without ventilation, but are plagued by rot caused by a fungus of the genus *Aspergillus* (de Souza 1984) as soon as the stems dry and they fall to the ground.

A. J. Sampaio (pers. comm. 1944, as cited in Mori and Prance 1990a) reports that seed size and the number of seeds per fruit pod vary considerably. B. W. Nelson (pers. comm. 1990) reports that seed size may vary within the same population and that intrapopulational variations may be as great as interpopulational variations. There are, however, insufficient data on the wide range of Brazil nut trees throughout the Amazon to support such conclusions. In the past, preference was given to the larger seeds both for shelled and in-shell nuts. Today, smaller nuts are generally preferred in the confection industry and command a higher price.

Distribution, Abundance, and Ecology

The Brazil nut tree is found principally on nutrient-poor, well-structured, and well-drained oxisols and utisols (Clement 1993). It is not found in areas with poor drainage or on excessively compacted soils. Diniz and Bastos (1974) report that the Brazil nut is found in areas with a mean annual rainfall of 1,400–2,800 millimeters (mm), a mean annual temperature of 24°–27°C, and a mean annual relative humidity of 79–86 percent. These ranges cover the Klöppen "Ami," "Afi," and "Awi" climate regimes, giving it an extremely wide climatic tolerance.

The seed is most commonly dispersed by agoutis (*Dasyprocta* spp.), but squirrels and other animals also extract the seeds from the hard fruit pod (Viana et al. 1994b). Some seeds are consumed immediately, others are stored for later consumption or are forgotten and germinate (Mori and Prance 1990b). This seed dispersal mechanism must have severely limited the Brazil nut's original range, because viable seeds would have great difficulty in crossing major river tributaries (Clement 1993). The arrival of humans in Amazonia, however, probably changed the Brazil nut's distribution dramatically. Its evident food value, due to oils and protein, and flavor must have made it a preferred extractivist resource very early in the human occupation of Amazonia. Later, seeds and perhaps seedlings were planted by Amerindians into their agroecosystems. This practice continues among the Amerindians in southeastern Amazonia today (Posey 1985) and elsewhere in Amazonia among Amerindians, *caboclos* (mixed blood long-term natives), and modern colonists. Müller et al. (1980) suggest that most "natural" stands of the Brazil nut were created through Amerindian intervention. Agouti dispersal may then have completed the species' occupation of a new area (Clement 1993).

Today Brazil nut trees are found in most of Amazonia, the adjacent Guyana highlands and forested lowlands, and the upper Orinoco River basin. In Amazonia it is found in Brazil, French Guiana, Surinam, Guyana, southern Venezuela, southeastern Colombia, eastern Peru, and northern Bolivia. There are some curious irregularities in its distribution that give support to Müller et al.'s (1980) hypothesis of Amerindian intervention as a means of distribution. The area around Manaus, although well populated at contact, was free of the Brazil nut until it was planted in historic times (B. W. Nelson, pers. comm. 1990). Perhaps the Amerindian groups that occupied these areas did not consider the Brazil nut to be as delectable as other groups obviously did (Clement 1993).

The Brazil nut tree occurs as a rare emergent or upper canopy component of the terra firma forest throughout much of its range, at an average abundance of 1 tree or less per ha, although occasionally occurring in stands of 15 to 20 trees per ha (Clement 1993). In natural stands, Brazil nut trees grow slowly and produce late. The tree begins to bear fruit at 20 years and only at about 24 years does it reach a mature production. In small open areas within forests such as those cleared for subsistence farming or spaces where large trees have fallen, Brazil nut trees grow much faster, bearing fruit in 12 years and reaching commercial production by 16 years. Cleared and gap areas have more seedlings which, according to forest dwellers, is because agoutis are responsible for burying nuts in these areas. Grafted Brazil nut trees reach commercial production between seven and eight years after planting (de Souza 1963; Müller et al. 1980). All stands increase their production as they get older and larger. It is not known when or why most Brazil nut trees die. Because of the extensive primary and secondary root systems of the Brazil nut tree, rarely is the tree torn up by storms or strong winds. The oldest known producing Brazil nut trees in the Amazon are in Pará and are estimated to be 1,600 years old with a circumference of more than 16 m (C. H. Müller, pers. comm. 1991). Most native stands include trees that are hundreds of years old.

Brazil nuts are gap dependent (Mori and Prance 1990b), which means that they attain reproductive size only when growing in a light gap created in the forest by the fall of a large tree or other similar disturbances. Viana et al. (1994b) found that 83.7 percent of juvenile trees (less than 40 cm dbh) had either small or very small canopy openings and concluded that "the most rapid and inexpensive way to increase productivity of Brazil nut [trees] in the study area seems to be liberation thinning." As the study pointed out, "liberation thinning of suppressed individuals is relatively inexpensive (US$0.08/plant) and can increase the size of the adult population through increased recruitment from the large pool of juveniles and increase the production of

Brazil nuts through more favorable light regimes for small adult plants."

Yields, Collection Methods, and Processing

Estimations of the yields of Brazil nut trees vary widely. A natural stand of three to four Brazil nut trees per ha is reported to produce an average of about 36 liters or 20 kg of raw nuts (Müller et al. 1980). According to Brazil nut harvesters, the average production of an adult Brazil nut tree is between 24 and 36 kg of raw nuts per season (2,700 to 4,050 raw nuts or 135 to 203 ouriços) per tree. A tree with an average yearly production of 30 kg has an annual income value of US$6 to the *castanheiro* (Brazil nut gatherer) and US$24 annually at the FOB (free on board) export level. Castanheiros have claimed that a fully developed healthy adult tree can produce 89 kg of raw nuts (LaFleur 1991c). Viana et al. (1994b) found unshelled nut production per tree to vary from 1.5 to 105 kg in the vicinity of Xapuri, a town in Acre, Brazil.

Rosengarten (1984) reports that a good production year is generally followed by a poor one, as the tree uses most of its accumulated reserves and takes more than a year to accumulate more. Alternate-year bearing is common in undomesticated species and will be one of the first traits modified by selection in the ongoing Brazil nut improvement program executed by the Agricultural Research Center for the Humid Tropics (CPATU) (Clement 1993).

A conservative estimate based on these numbers, assuming commercial production to be 50,000 MT per year, indicates that nuts are being harvested from 2–3 million Brazil nut trees in any one year. This means that an area of approximately 2–3 million ha (Mori and Prance 1990a), or 20–30 percent of the potential area, is being exploited today. Jim LaFleur and Wim Groeneveld (pers. comm. 1993) believe from their work with Brazil nuts in both Brazil and Bolivia that the actual area that is exploited is less than 10 percent, possibly only 5 percent, of the potential. There is concern, however, about the effect of harvesting on Brazil nut populations because, where they are harvested, collectors gather as many as possible. Despite its economic and social importance, there is very little quantitative information on the pollination, dispersal, regeneration, and growth of Brazil nut trees in old-growth forests in Amazonia (Mori and Prance 1990a), and evidence for any downward population trend is lacking. This is complicated by the fact that, with increasing deforestation in the eastern region of the Amazon, the major areas of Brazil nut production are shifting away from the state of Pará in Brazil to the western Amazon, for example, the Brazilian states of Rondônia, Acre, and Amazonas, as well as Bolivia.

Woodroof (1979), Rosengarten (1984), Moritz and Ludders (1985), and Mori and Prance (1990b) discuss collection and processing of the Brazil nut and describe the life and hard times of the castanheiros, so only a brief outline will be presented here. In the western Amazon the Brazil nut crop is harvested from December through March, during the rainy season. The Brazil nut harvest (but not the commercial shelling of the nuts) complements that of latex (raw rubber), which is tapped during the dry season, thus allowing the same collectors to be involved in both activities. In some cases, gatherers clear undergrowth from beneath the trees to facilitate collection. Most often, however, collection takes place year after year under the same tree so annual clearing is minimal. Harvesting is conducted individually or in small groups; sometimes families harvest nuts together.

When mature, Brazil nuts fall to the ground. Fruits are usually collected in the morning when it is drier and when there is less chance of nuts falling from the tree. Because of competition from agoutis, insects, and fungi, the castanheiros visit the trees regularly. A good day of collecting will yield 700–800 fruits, containing 7,000–20,000 seeds (the latter figure is extremely unlikely, however, and was derived from the multiplication of the highest collection figure by the highest seed number figure given previously) (Clement 1993).

Collecting can be dangerous in areas where land rights are contested or where colonists have moved into a region traditionally used by long-term residents. In 1985, six would-be collectors were killed for attempting to collect Brazil nuts from underneath trees that others considered their own.

The ouriços are generally collected and put into piles to the side of the tree. In this way, the collector can take the nuts out of the fruit without fear of being hit by a falling fruit. The empty fruit shells weigh more than the seeds they contain, so most collectors do not want to carry unnecessary weight back to their houses. Instead, they remove the seeds from the pod in the forest, generally an afternoon activity. Collectors use either a machete or an ax to open a round hole at the top of the fruit similar to the way a green coconut is opened. The seeds are then removed. They are generally washed near the house or storage shed. During this process any seeds that float are thrown away because they are bad. Only the seeds that sink are saved. The seeds are then left to dry for a few hours and eventually stored out of the rain.

A conscientious harvester (or one who has a buyer that will pay a higher price for better-quality nuts) will sun-dry his seeds for a few days and store them in a dry environment. With a relative humidity of 80 percent during the rainy season, however, nuts only partially dry. The lack of adequate storage conditions available to forest residents is the major reason that seed quality is not very high when the harvester finally gets his harvest to market.

Most harvesters sell their nuts to intermediaries who travel throughout the interior during the harvest season. These merchants then sell the nuts to other intermediaries or directly to a processing plant, generally located in a major urban center. Traditionally Manaus and Belém have been the major processing centers, although Riberalto in Bolivia and Xapurí in Acre, Brazil have become important recently (Clement 1993).

At the processing plant, the nuts are sorted again, first by floating, then by visual inspection. They may be graded at this stage or after shelling. In the traditional and smaller processing plants, if the nuts are to be shelled, they will be dried to shrink the kernel from the shell, then soaked for 8 to 24 hours in room temperature water. The nuts are soaked after drying to make them pliable so that they do not shatter when shelled. In larger factories the nuts are steamed in autoclaves to make them pliable, though the steam is now suspected of causing the oil in the nut to spoil more quickly, thereby shortening the nut's shelf life. Almost all Brazil nuts are shelled by hand. (Some processing plants are experimenting with putting unshelled nuts in liquid nitrogen which shatters the shell, while others are trying to develop mechanical shellers, but this is difficult due to the irregular shape of Brazil nuts.) The nuts are then sorted by size (tiny, midget, medium, large, chipped, and broken) and checked for quality and shell fragments. They are then dried to less than 4 percent moisture to reduce the chance of spoilage (La Fleur and Groenveld 1990; Clement 1993).

Modern packaging involves placing graded, high-quality, shelled nuts into laminated plastic bags, which are vacuum-sealed. Nitrogen gas or carbon dioxide is then introduced (flushed) into the package to eliminate oxygen and further conserve quality. Bag sizes are a standard 20 kg. These can be stored and shipped conveniently, although most are shipped shortly after packaging to avoid having the nuts in the hot tropical climate any longer than necessary. Inventories are kept at low temperatures (Clement 1993). Processed nuts can be kept in cold storage (2°–10°C) for more than a year without noticeable loss of quality.

The Brazil nut production alternative to collection in the wild is plantations. While such systems have received considerable attention, and plantings of orchards have increased in recent years, they are still largely theoretical and no significant production exists. Furthermore, the advent of plantations, if they ever become productive and common, would reduce the price paid for Brazil nuts and likely force a number of people who currently support themselves through the harvest and sale of Brazil nuts either to turn to other sources of income such as logging, cattle ranching, or agriculture, or abandon the forest to others who would undertake one of those more destructive activities.

Viana et al. (1994b) argue that "productivity of extractive systems is low because there has been basically no research to develop technologies to improve productivity through natural forest management in ways that are accessible to traditional forest communities. Furthermore, the pervasive commercial structure and lack of assurance over land tenure have not given the incentive to extractivists to manage their resources sustainably. Now that land tenure is not a problem in many extractive reserves, productivity cannot be increased because conventional forestry and agricultural research does not provide the answers to the questions faced by extractivists."

As a somewhat intermediate system between natural forest production and plantations, Padoch et al. (1987) report that the Brazil nut is being planted in a market-oriented agroforestry system near Iquitos, Peru. Although Brazil nut is not the major species in that system, this does show that Amazonian agroforest farmers, without government-sponsored extension support, are following Amerindian traditions (Clement 1993).

To support rainforest peoples and conserve significant areas of tropical forest, however, an extractivist system of forest management appears, on paper at least, to be able to compete with plantation Brazil nut production so long as it is combined with one or more of the following activities: enrichment programs, on-site processing, and direct marketing. These factors eliminate most of the middlemen currently involved and allow the harvesters to receive a greater percentage of the final value of their nuts. Without the combination of these activities, the harvesters have little or no incentive to expand, or sometimes even to continue, harvesting Brazil nuts, much less compete with more capital-intensive operations. When this happens the fate of the forest is left to the next colonist, which generally means chainsaws and fire.

Though it is against the law to cut Brazil nut trees in Brazil, the trees generally cease to produce when the forest around them is removed. Viana et al. (1994a) have conducted the most extensive research on the decline of Brazil nut production due to forest clearing. Their study found high Brazil nut tree mortality when pasture was established in forest areas. They observed that in 20-year-old pastures all trees were dead; in 9-year-old pastures all trees were dead or dying; and even in 3-year-old pastures, 28.6 percent of Brazil nut trees had died and 35.7 percent were dying. No regeneration of Brazil nut trees was found in any of the pastures sampled. They also found that fruit production was almost nonexistent. Thus, as Viana et al. (1994a) conclude, "current Brazilian legislation that protects Brazil nut production in areas converted to pastures has been ineffective in protecting the resource base. It has not prevented mortality and consequent local extinction of Brazil nut in areas deforested for pasture establishment."

Deforestation clearly affects overall Brazil nut production. Production data, for example, show that in Pará where deforestation is most severe, Brazil nut production has declined from a high of more than 37,000 MT in 1973 to 18,000 MT in 1987. Ironically, production increased in the area until 1973, because the same colonists who were clearing forests for pasture also provided the largest labor force ever in the area to collect nuts from areas that previously had been unharvested.

The future of Brazil nut as a crop can be threefold: as an extractivist product; as an agroforestry/forest management component; and as a modern monoculture plantation crop. As already mentioned, the first option will help conserve tropical forests and the cultures of numerous tropical-forest peoples, both Amerindian and caboclo. The second option is the preferred route to enrich already deforested areas with Brazil nut and provide for the long-term capitalization of the Amazonian farmer. The final option is for the already capitalized investor and may be limited due to pest and disease infestations which commonly attack dense plantings of most tropical plants (Freire and Ponte 1976, as cited in Clement 1993).

The Selenium-Sulfur Connection

Some Brazil nuts can cause hair loss if consumed in large quantities because of high concentrations of selenium. Thorn et al. (1978) report on the presence of selenium in Brazil nuts imported into England. One shipment of Brazil nuts was rejected for this reason in Germany, where acceptable levels of selenium are lower than in most other importing countries. Palmer et al. (1982) and Chansler et al. (1986) report on experimentally induced selenium toxicity in laboratory animals.

Brazil nuts appear to concentrate selenium because it is structurally similar to sulfur, and the plant is fooled. Sulfur is an essential nutrient in seed protein, but sulfur is frequently deficient in Amazonian soils, especially after decades or centuries of Brazil nut harvesting (C. R. Clement, pers. comm. 1990). If the soil contains significant amounts of selenium, it may be used by the plant instead of sulfur. If selenium is not present, there appears to be no problem, other than reduced yields, as sulfur becomes limiting (Clement 1993).

A simple solution to the selenium problem is to fertilize the affected area with sulfur. This is commonly practiced by the caboclos along the Madeira River to improve yields (V. Cruz Alves, pers. comm., as cited in Clement 1993), although there are no reports of selenium problems there. The caboclos use pharmaceutical sulfur, the only form available to them, and apply it by drilling a small hole in the bark of the tree as far as the cambium layer. They then place 1–2 g of sulfur into this hole and close it with pitch or *jatobá* (*Hymenaea courbaril*) resin. If the sulfur is applied after the flowering season, the following flowering season will result, according to caboclos, in significantly improved yields (C. R. Clement, pers. comm. 1990). This practice points to a solution to the selenium and low yield problems in older Brazil nut tree groves throughout Amazonia, but no experimental work has been done to test the validity of this practice (Clement 1993).

Uses and Economic Potential

The Principal Use

Brazil nuts are consumed raw, roasted, salted, and used in ice creams or as prepared confectionery items (Rosengarten 1984; Mori and Prance 1990b). They are an important ingredient in shelled nut mixtures (Rosengarten 1984). Woodroof (1979) presents 50 recipes for the Brazil nut, mostly for confectionery uses. Clark and Nursten (1976) analyzed the seed flavor components (Clement 1993). I have developed and introduced more than one hundred products into North American and European markets that use Brazil nuts or Brazil nut-derived raw materials (e.g., oil, flour, paste/butter) (see Clay 1989, 1990, 1992; Clay and Clements 1993).

Mori and Prance (1990b), citing Zucas et al. (1975), report that Brazil nut protein contains all the essential amino acids, although without reporting which may fall below FAO/WHO (1973) recommended limits. They do report that Brazil nut protein has a lower nutritional value than casein, suggesting that some of the essential amino acids are limited (Clement 1993). Brazil nuts contain minerals such as phosphorus (693 milligrams [mg]100 g dry weight), potassium (715 mg), iron (3.4 mg), and sodium (1.0 mg) in addition to vitamins such as thiamine (0.96 mg), riboflavin (0.12 mg), niacin (1.6 mg), and traces of vitamin A (Clement 1993).

Brazil nuts require considerable care in handling. The largest single cause of damage to harvested Brazil nuts probably occurs during the multiweek, open-barge journey from forest to factory. As a result of rot and mold, many accumulate aflatoxins, which can cause the rejection of whole batches of nuts exported in-shell (FAO 1986). Because the nuts are very rich in oils, they rancify easily and absorb foreign flavors which overpower their own (Woodroof 1979, as cited in Clement 1993).

Over the last decade, Brazil nut production for export has hovered around the 40,000 MT level. In fact, given that, on average, 25–30 percent of all harvested nuts spoil in transit and storage, more than 50,000 MT are currently being harvested for commercial markets. One simple way to increase production (and the price paid to harvesters) would be to reduce losses. Production could also be expanded in other ways if demand increases or if production and transportation costs are reduced. Additional production could come from the millions of forest trees that are currently not harvested because of the low prices paid to most collectors. The low prices are due partly to slack international demand and partly to the number of middlemen involved in getting the nuts from the forest to the processing plants. Residents of the extractivist reserves now being set up in Acre, Amapa, Pará, and Rondônia are trying to organize local processing and cooperative commercialization programs so that a greater proportion of the end value of the nuts can be paid to the collectors. This should encourage greater harvests while improving collector incomes. A number of international nongovernmental organizations (NGOs), as well as North American and European businesses, are cooperating with extractivists in these attempts (Clement 1993).

Secondary Uses

The Brazil nut tree produces one of the finest timbers of Amazonia, as it has a straight grain, is easy to work, takes a finish readily, has a pleasing appearance, and is very durable (Loureiro et al. 1979). Although the felling of Brazil nut trees is prohibited by law in Brazil, there is thought to be a considerable black market for its wood (Mori and Prance 1990b), and illegal logging occurs (Kitamura and Müller 1984).

Large numbers of Brazil nut trees are left to die and decay in pastures. The magnitude of the waste of the resource is suggested by Viana et al. (1994a): "There were an estimated 800,000 ha of pastures in Acre as of 1987 . . . , most of which [is] in prime Brazil nut habitat. If the timber volumes surveyed in our old-growth forests are representative of the area deforested, an estimated 18,888,000 m^3 of Brazil nut timber has been lost. Considering stumpage value of US$20/ m^3, a total of US$377,760,000 was lost up to 1987. Considering a processed value of US$100/$m^3$ and 50% loss in processing, a total of US$944,400,000 was lost up to 1987 by the state economy."

Even if the calculations are off, the value of timber wasted is phenomenal. Compounding this loss is the fact that most deforestation in the state has occurred since 1987. In addition, two other states (Pará and Rondônia) have greater natural stands of Brazil nut trees and greater deforestation.

SUDAM (1979) concluded that Brazil nut trees have superior silvicultural characteristics, including fast growth (more than 1 m/year in the first decade), straight trunk, and tolerance or resistance to pests and disease in plantation, at least when cultivated for timber. G. Hartshorn (pers. comm. 1994) has observed that Brazil nut can even be abandoned in second growth and still achieve excellent growth and trunk shape. This suggests that Brazil nut could be used to help restore degraded sites (100,000 square kilometers [km^2] in Amazonia) as a multipurpose species, yielding nuts after 15–20 years and timber after 50–100 years (Clement 1993).

As with most nuts, the Brazil nut is rich in oils, variously reported at 65–70 percent of dry seed weight (Woodroof 1979; Pesce 1985). This oil is rich in unsaturated fatty acids (75%) and may be attractive for various culinary uses (Woodroof 1979). Assunção et al. (1984) report on the stability of this oil. Nuts that are rejected for export could be pressed for oil, if a market is found. This would encourage quality control, now rather precarious in some processing plants, and increase the value of the nuts to the collectors (Clement 1993).

One of the reasons frequently cited for conserving the Amazonian rainforest is to conserve the genes that make up its enormous species and intraspecies diversity, since these are the raw materials for the biotechnology revolution now under way in the developed world. Brazil nut exemplifies this concept. The nut is rich in sulfur amino acids (methionine, cysteine), which are deficient in the seed of the common bean (*Phaseolus vulgaris*), for example (Clement 1993). Since the common bean is a major source of protein in the third world, two laboratories (one in California, one in Brasilia) are racing to put Brazil nut amino acids into the bean (Gander 1986). The relevant gene has been isolated, inserted into the bean genome, and found to be expressed in bean callous tissue. Both laboratories are having some trouble regenerating the transgenic plants from callous and they must then determine that the amino acid genes are expressed in the seed (E. S. Gander, pers. comm., as cited in Clement 1993). When the transgenic plants are finally grown out, they must then be bred conventionally to obtain cultivars suitable for distribution to farmers. Once this process is completed, the common bean will have nutritional characteristics much superior to those it now has and will become an even more important staple than it currently is (Clement 1993). The steps followed in this process to date can be traced in Kamiya et al. (1983), Altenbach et al. (1984), Ampe et al. (1986), Castro et al. (1987), Sun et al. (1987a,b), Plietz et al. (1988), and Guerche et al. (1990).

Historical Production Data

Brazil nuts entered world commerce in the late eighteenth century, introduced by Dutch traders during the period that they attempted to colonize eastern Amazonia. A prosperous trade developed soon after Brazil opened its ports to world trade in 1866, and the Brazil nut has been an important item of trade since that period (Mori and Prance

1990b). When shipments began leaving from the port of Belém in Pará, they became known as Pará nuts.

During the nineteenth century, the most important economic event in the Brazil nut region was the rubber boom. The search for rubber resulted in waves of migration into the Amazon by people hoping to make fortunes in the rubber industry (Clement 1993). By the end of the century, rubber seeds taken from the Amazon had been successfully cultivated in Southeast Asia, most notably in Malaysia. These new areas increased world production considerably and resulted in the decline of rubber prices and, eventually, the collapse of the market from 1911 to 1932. The rubber boom was over, and the rubber tappers were left out in the forest with a considerably lower income. Despite the decline in rubber prices, most of the immigrants remained and continued to tap rubber, but they augmented this income with the indigenous activities of hunting and gathering (LaFleur 1991c). The gathering of Brazil nuts became an important new source of income. Cities in the Amazonian interior such as Guajará-Mirim in Rondonia and Xapuri in Acre constructed Brazil nut processing plants to prepare nuts for sale and transport to exporters. Nuts were processed and exported either in-shell or shelled (LaFleur 1991c).

During World War II, the demand for rubber increased again when the Japanese occupied the rubber plantations in Malaysia, and forest dwellers were given economic incentives to devote most of their time to tapping. During the 1940s, there was a new wave of immigrants to help meet this increase in demand, again primarily from the northeast of Brazil. Other extractive products became less important or were abandoned, and most of the existing Brazil nut processing plants closed down (Hemming 1987, as cited in LaFleur 1991c).

Priorities reversed again after the war, when rubber prices fell as a result of increased production from the then-liberated Malaysia, and Brazil nuts reemerged as an important source of income for forest dwellers. The trade was concentrated in the hands of a few surviving trading companies located principally in Belém, Manaus, and Obidos (LaFleur 1991c).

The Brazil Nut Economy

Brazil nuts are a commodity and as such have prices that are set by world trade. These prices at any point in time are a function of world supply and demand and world stocks (or expectations of same). An important variable affecting Brazil nut prices is the cost of substitutes (i.e., other nuts).

The price that an importer will pay is based on international prices. All importing costs, including transportation from source, insurance, and port costs, are subtracted from the base international price to give an FOB exporter's price. The exporter's price is, therefore, given or fixed, and all of the exporter's costs must be subtracted from this sale price (FOB) (LaFleur 1991b,c).

An individual exporter can mainly affect his profit through changing costs rather than the selling price, as the latter is more or less fixed for all exporters relative to each one's distance from the market and cannot generally be directly influenced by the seller (although "green" or "organic" classifications are attempts by producers, processors, or importers to differentiate their products in the marketplace and thereby receive higher prices). The price that an exporter will pay for raw material is consequently adjusted according to the distance, time, and therefore the cost of getting it to the processing plant and port of embarkation, including spoilage and financing. If the FOB exporting price (selling price) falls, the exporter will lower his price for raw material, that is, international price changes will be passed on to the gatherer, especially when prices decline (LaFleur 1991c).

There are few possibilities to achieve economies of scale in the processing of Brazil nuts. The most important part of the work is individual nut shelling by hand-operated tools. What economies of scale there are include central ovens for drying, autoclaves for steam flashing, and vacuum packers. All these processes can also be done with simpler and cheaper technology that has similar productivity and the possibility of lower costs. The majority of processing in Brazil is highly centralized with high costs for transportation, drying, storing, and processing. The raw nuts are carried thousands of kilometers by people and animals and on small boats and barges to be processed in the major cities of the Amazon. As a result, up to 30 percent of the nuts are lost to spoilage, adding to the overall cost (LaFleur 1991c).

One economy of scale, however, is the minimum requirement for international lots. In general, nuts are sold by the container. A short container holds about 14.5 MT. If producers cannot fill and sell a container in a timely way, importers will not be interested in their product. Labor costs for unloading smaller amounts would be prohibitive in places such as New York or Rotterdam. For that reason, containers that can be unloaded by cranes are required. Some small producers have been known to pool their output to achieve this minimum requirement.

Historically, however, the reason for centralized processing is not due to economies of scale so much as to the concentration of capital. Centralization is mainly a result of greater distribution of available government and private capital, both subsidized and nonsubsidized, in urban areas. Wealthy investors with connections, business experience, and good credit standing are able to finance and run large factories. As a result, extractivists (rural, not wealthy, with little formal business experience, no access to credit, and few assets to use as collateral) are largely limited to gathering. In mature economies, businesses run the gamut from tiny family-run businesses to large centralized industries. Much of the middle part of this spectrum is lacking in less developed economies like that of the Amazon. Gatherers are poor and undercapitalized and unable to undertake more processing locally, while centralized industries are often overcapitalized and inefficient.

Theoretically, if there is a high-cost, competitive industry, competition will transfer any excessive industrial profits to the producers via higher prices. If or when a lower-cost processing system is introduced, it will drive out the high-cost one by offering a higher price to the producers, thereby depriving the high-cost system of raw material. This means that an increase in efficiency will directly benefit the producers. However, if the original industry has a monopoly on investment and working capital (in this case because of intrinsic government policies), the new low-cost industry will be thwarted. In a high-cost, noncompetitive industry the price to the producers/gatherers will be only as high as needed to garner an adequate supply and could mean excessive profits to the processors/exporters (LaFleur 1991c).

Decentralization of Brazil nut processing existed prior to World War II, until the war demand caused a shift back to rubber production under rigidly organized government support. Today, decentralization of processing is taking place in Bolivia where plants are being set up near raw material sources. The number of small processing plants is increasing, prices paid to collectors are increasing, and the nuts are being exported at lower prices than those from Brazil. Given that both Bolivian and Brazilian processors buy their raw material from the same border region and pay the same price to the extractivists/gatherers, the

Brazilian industry has higher costs and lower profits while the Bolivian system has lower production costs and higher profits. There is at present only one small processing plant in Brazil which is owned and operated by the collectors, the Cooperativa Agro-Extractivista de Xapuri (LaFleur 1991c).

With local processing there is a potential to lower costs, including transportation, and to reduce spoilage, thereby reducing the financial cost of forward payments. Also, family labor can be utilized, and payment to family members can be based on production or piecework as opposed to fixed salaries. An additional benefit to local economies in processing regions is the transfer of added value to these regions by generating local employment and tax revenues (LaFleur 1991c).

Current and Potential Markets

The Brazil Nut Market and Salient Features of the Edible Nut Market

The following discussion is taken largely from the ground-breaking work of LaFleur (1991c). World production of raw edible nuts is presently about 2.5 million MT annually, with an estimated international trade value of US$2.3 billion. The most notable market shares are held by groundnuts (peanuts) and hazelnuts, with cashew nuts, desiccated coconut, almonds, walnuts, and Brazil nuts of lesser importance. Groundnuts and almonds have had the most significant average annual growth rate in production from 1970 to the present, increasing 8.6 percent and 4.0 percent, respectively. Interestingly enough, Brazil nuts and cashew nuts are the only nuts that show a negative annual growth rate over this same period, which indicates that both are losing their share in the world edible nut market. Brazil nuts have gone from occupying 3.5 percent of the world market share in 1970 to 1.7 percent in 1988. This is an 18 percent loss of volume of market share, and per capita consumption of Brazil nuts actually declined (LaFleur 1991c).

In 1987, the average price of cashew nuts per ton was US$7,016, while hazelnuts were selling at US$4,040/ton. During the same period, Brazil nuts were being sold at US$2,398/ton, which represents the third lowest price among the principal edible nuts. Only groundnuts and desiccated coconut were priced lower at US$707/ton and US$796/ton, respectively. An extrapolation for the 1987 market value (average world imports multiplied by the average price/ton) for edible nuts resulted in values of US$548 million for hazelnuts and US$489 million for almonds, compared with only US$33 million for Brazil nuts. In terms of market share, hazelnuts are responsible for 25 percent, while 22.6 percent is held by almonds and 19.1 percent by cashew nuts (LaFleur 1991c). Brazil nut total market value is the lowest among the principal edible nuts due to a relatively low product price combined with low production rates.

The supply of edible tree nuts as a group is highly price-inelastic in the short term because of the lag between planting and production, as well as the economic lifespan of the trees. Most perennial edible nut trees remain in production for more than 30 years. Brazil nuts, on the other hand, are gathered almost exclusively from natural stands and can produce for more than five hundred years.

The world demand for edible nuts is characterized by low price and income elasticities. This means that as prices fall or incomes rise, the increase in demand will be less than proportional. This is true of most nonessential, "luxury" food commodities where consumption is a function of habit and taste rather than of necessity. Consumers respond little to price or income changes. The opposite is true for basic food staples in developing or low-income countries, where demand responds proportionally to changes in prices and income.

There is also a high degree of price correlation between different types of nuts because of the degree of substitution among them. This is especially true of Brazil nuts. Approximately 80 percent of shelled Brazil nuts shipped to the United States are sold to baggers and salters for use in mixed nut products (Karas 1991). The exact proportion of Brazil nuts in such mixes is small but depends on both availability and price.

Low price and income elasticities of demand for edible nuts result in a high price flexibility of supply (LaFleur 1991c), that is, if there is an increase in the supply of edible nuts, the price will fall by the inverse of their collective price elasticity (flexibility). Given a price elasticity of demand for the nut industry of approximately –0.25 (coffee, cocoa, sugar, and pepper are similar), the price flexibility in relation to supply (the change in price in response to changes in supply) will be –4 percent. This means that for every 1 percent increase in world supply, the price will fall about 4 percent (the inverse of 0.25), all other variables remaining equal (LaFleur 1991c).

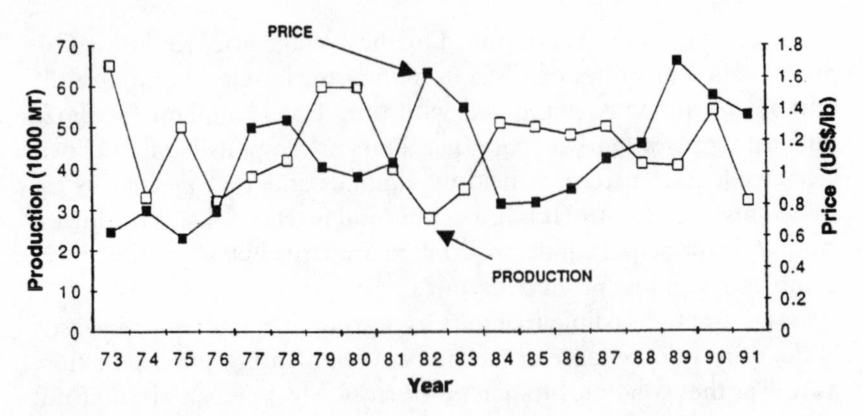

Figure 7.1. World Brazil nut production and price, 1973–91. (*Source:* LaFleur 1991c.)

Most nonessential food products normally have very high price flexibility in relation to changes in world supply. This means that any change in supply causes a relatively large change in price (LaFleur and Jones 1984). This is not the case for Brazil nuts because of their high degree of substitution in the edible nut market (mixed nuts and fillers) and the fact that Brazil nuts hold a very low percentage of this market (only about 2%). If Brazil nuts were to increase in supply by 10 percent, there would be a mere 0.2 percent increase in the total nut market (LaFleur 1991c).

From 1970 to 1991, the yearly average change in Brazil nut supply has been 30.15 percent, while the price, which changes in response to variations in supply, fluctuates on average 21.53 percent annually, resulting in a market flexibility of Brazil nuts (average price change in relation to production changes) of approximately –0.71 percent (Fig. 7.1). In other words, for the 21-year period, for each 1 percent change in supply, there has been an average 0.71 percent short-term price change in the opposite direction (LaFleur 1991c).

While the estimated price change in the short term is approximately –0.71, in the long term this price will be determined by the price flexibility of supply for all edible nuts as substitution takes place over time. This is caused by a "ratchet effect," that is, as changes in consumption habits take place they tend to become permanent over time. This means that people are no longer consuming a new product because it is cheaper (the reason for the initial change)—just the opposite—they will revert back to their original consumption patterns only if the price of the original product goes down (LaFleur 1991c).

Brazil nuts are an extractive product and nut gatherers are responsive to price changes, but there is a limit to increases in supply. This limit is determined by the actual production of any season. Other factors that affect supply are the time and costs involved in transporting nuts from the gathering site to a place where buyers will take delivery. Deforestation of Brazil nut producing regions has cut down on production with the result that Brazil nuts are not available in the quantity sought by buyers during certain times of the year. This, coupled with an increased production of other nuts, has contributed to loss of market share. The opportunity for extractivists to earn more money in activities such as prospecting for gold or hiring out their labor for urban and farm work has also brought about a decrease in gathering. The low market share is even more significant when one considers that the price of the nuts does not react proportionally to the decrease in supply, but is insulated because of the degree of product substitution in the edible nut market. The danger is obvious: as deforestation takes place and production of Brazil nuts declines, other nuts will take over the market share lost by Brazil nuts. As a result, long-term demand for Brazil nuts will be replaced by demand for other nuts (LaFleur 1991c).

The high degree of product substitution and the relative scarcity of Brazil nuts in the marketplace offer opportunities to expand the market share through increased production without a proportional decrease in price. The opportunity to expand market share and total income by increasing production provides a potent argument against deforestation and for better management of existing trees as well as the planting of new trees.

Supply

Commercially produced Brazil nuts come mainly from the Amazon forests of Brazil, Bolivia, and Peru. Production has varied from highs of 60,000 MT/year to lows of about 30,000 MT/year (LaFleur 1991c). Historically, world production of Brazil nuts has been dominated by Brazil. With overall Brazil nut production fluctuating widely, landed CIF (the value of a good that has cleared customs and is free of any liens) New York prices paid per pound have changed inversely. Thus, in 1979 and 1980 when production was at a peak (60,000 MT), prices were relatively low (US$1.04/lb in 1979 and US$0.98/lb in 1980). Prices of US$1.63 in 1982 and US$1.70 in 1989 were at their highest when world production was at its lowest (Fig. 7.1) (LaFleur 1991c).

Bolivia's export figures have been lower than actual production, as previously a large percentage of their nuts were sold into Brazil for processing and shipped as Brazilian product. More recently, Bolivia has increased its processing capacity and now processes almost all of its crop and some of Brazil's. In past years, Bolivia's nut industry, which is centered in the towns of Cobija and Riberalta, was characterized by consistently poor quality nuts and unreliability of shipments. More recently, factories have been plagued by financial mismanagement and problems associated with forward contracting. Three companies currently dominate the industry, but new companies are constructing plants in the producing areas of Riberalta and Cobija. The supply of nuts gathered still remains limited by the lack of roads into the richer Brazil nut forests. New plants will increase competition for the current nut supply or expand collection into new areas. It is becoming apparent, however, that not all the companies in Bolivia will survive even the next few years.

Current Markets

Almost all Brazil nuts are destined for international markets. Price, and therefore cost, are the most important considerations in determining what type of nuts industries, traders, and brokers will buy. The market is divided into shelled and in-shell nuts. In-shell nuts are sold predominantly during October, November, and December for Christmas and Thanksgiving holidays in the United States and Europe, particularly Germany, Italy, and the United Kingdom. Shelled nuts are sold mainly to roasters for mixed nut snack items, with the United Kingdom accounting for more than half of the market, followed by the United States at 25 percent. Some Brazil nuts are also used in chocolate products (LaFleur 1991c).

In the early 1980s, the government of Brazil began studies in processing Brazil nuts for consumption in alternative products. These experiments included toasting nuts in microwave ovens, testing milk and flour derived from nuts, and carrying out studies of other alternatives for marketing the highly nutritious nut (de Souza 1984). To date, this work has not had an impact on the Brazil nut market.

The domestic market for Brazil nuts is a fraction of the export market and is largely determined by international prices and local income levels. Historically, between 3 percent and 5 percent of exporter-processed Brazil nuts have gone into local markets. This is roughly the equivalent of 500 to 900 tons of shelled nuts. In addition to these processed nuts, a small quantity of in-shell nuts reach the domestic market. These generally come from local and cottage industries. Shelled and in-shell nuts are sold in some major supermarket chains, particularly in Brazil (LaFleur 1991c).

The internal market has a high income elasticity for demand, with increased income causing a more than proportional positive change in demand. By contrast, a small increase in price will give rise to a more than proportional negative change in demand. This is the opposite of the price and income elasticities of Brazil nuts in importing nations (LaFleur 1991c).

Market Potential

Profits in the Brazil nut industry are divided between manufacturers and sellers of final consumer products; traders or importers, exporters, brokers, and other intermediaries; the processors of raw nuts; regional traders; and finally, the gatherers (Fig. 7.2). The processors make their profit from processing as well as trading for raw nuts, but the selling or trading of goods for nuts is, in most cases, a more important revenue generator than the processing of nuts themselves (LaFleur 1991c).

In order to obtain a greater market share, production must increase, and the most efficient way to increase production and to lower production costs is to cultivate. Even enrichment planting of Brazil nut trees more densely in the forest and closer to the gatherers and processors will lower the time and distances of gathering and therefore the production costs. Processing the nuts closer to the harvest site can reduce losses, lower transport costs, and increase profitability (LaFleur 1991c).

There is a real opportunity for gatherers to become producers (i.e., those who process nuts for sale). The present processors are mostly located in areas where the cost of land and labor are high as compared

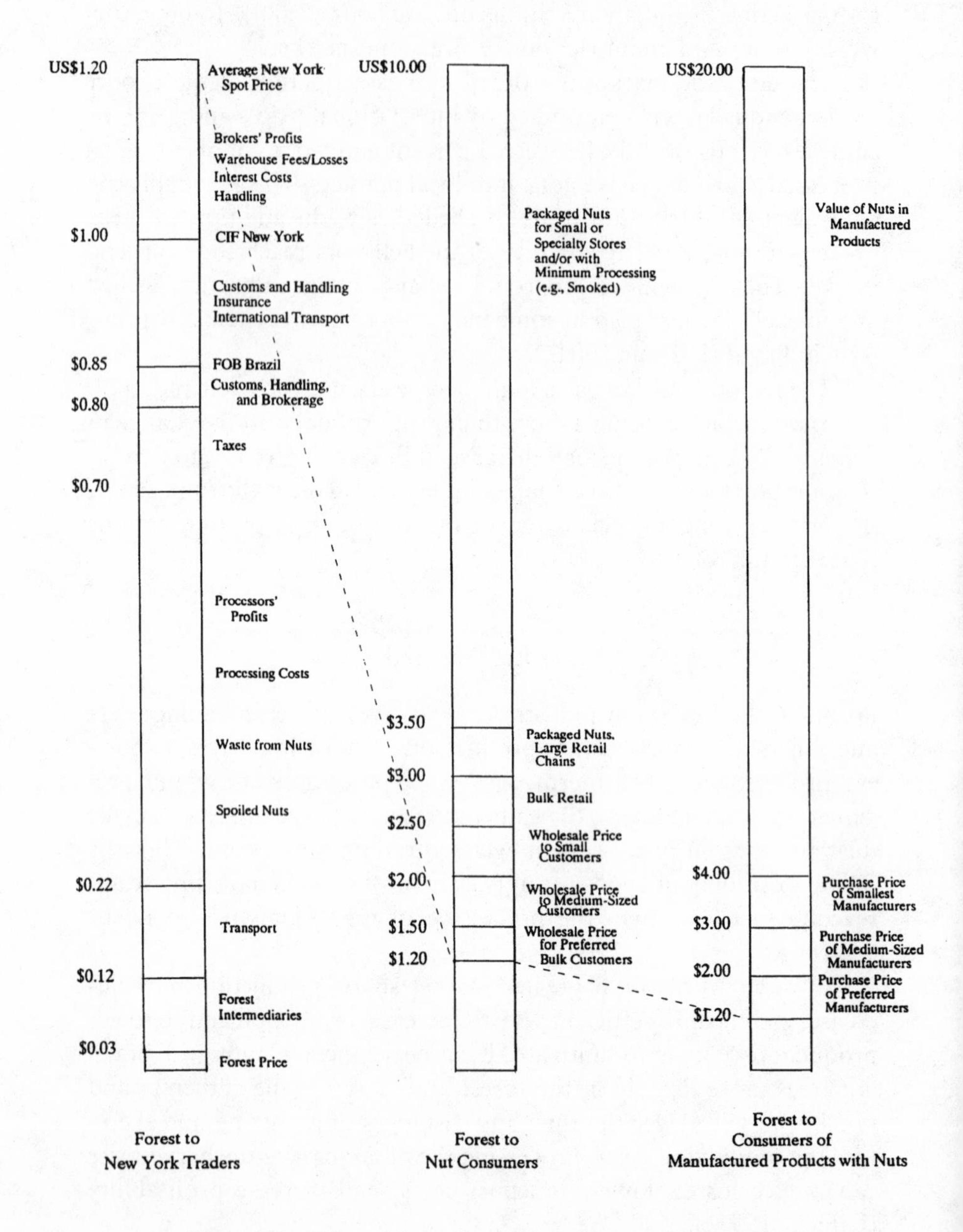

Figure 7.2. Value added to Brazil nuts. (*Source:* Clay 1996.)

to the extractivists, who are in areas of low economic value and make use of low-cost family labor (LaFleur 1991c).

The real market potential for Brazil nuts lies in increasing supply to the importing or high-income countries and lowering the price for the developing countries' markets. Increasing the supply will take market share away from other edible nuts in importing or high-income countries while lowering price in the world market, thus encouraging an increase in demand in developing producer countries (LaFleur 1991c).

Product Marketing and Distribution

Marketing Channels

Gatherers, normally men, go into the forest, sometimes several days' walk from their homes, to gather fallen nuts. This work requires carrying and managing heavy loads and can mean days or even weeks in the forest. Gatherers sell the in-shell nuts to middlemen (*marreteiros* and *regatões*) who trade mainly in merchandise or forward goods in exchange for a guarantee of future delivery of products. The middlemen transport the nuts by mule, ox, or small boats to more central collection points where buyers or buyers' agents take delivery. From regional cities located on the edge of forests and navigable rivers, the nuts are shipped by barge (before the end of May when river levels are high enough to allow large barges to be towed into the area) in lots of 250–1,000 MT of in-shell nuts. A price differential for nuts exists based on the distance between the processing plant and the place where nuts are bought (LaFleur 1991c). Acre and Rondônia receive less than two-thirds of the price paid for nuts sold in Belém.

After processing and packaging, the nuts are ready for shipment. The majority of sales are on an FOB basis where the exporter is responsible for all costs until the cargo is loaded onto the ship and physically and legally ready for shipment.

Within the present system, local populations control only the Brazil nuts they gather in the first two stages of the marketing chain: the gathering of the nuts and their sale to middlemen. In this system, women have a minor role at best. By contrast, under a value-added cooperative model of transporting or processing nuts, extractivists have the opportunity of participating in the whole marketing chain including stocking, processing, selling, and exporting. In this alternative system, women occupy a significant role in the shelling of the nuts (LaFleur 1991c).

Price Formation

Prices paid to the harvester in Brazil result from the following costs which are deducted from the export price. The shipper pays a currency exchange broker's fee of 0.375 percent on the FOB price of the nuts for preparing and processing his exchange contract. In addition, there is an exchange contract tax of 0.2 percent. The value-added tax on commerce, merchandise, and services is 12 percent of the FOB value. There are port and handling charges of about 2 percent and administration and warehousing costs of approximately 2 percent on the FOB value. The margin for the exporter is estimated at 3 percent. This leaves an "ex-factory" price from which the processor has to subtract all his costs. The financial cost of forwarding monies to buy the nuts is calculated and discounted back to the price paid to extractivists. Production costs for processing are subtracted, and the plant's processing yields are estimated to arrive at a price for nuts as raw material. This price does not reflect transportation costs but is the price that would be paid if the nuts were purchased at the factory door. The cost of transporting raw nuts from the place of purchase to the processing plant, and other costs involved in shipping, storing, or handling, figure in calculations of the price to be paid to the harvesters (LaFleur 1991c). The "value-added" journey of the Brazil nut from the forest, via the local buyer, processor, exporter, importer, distributor, and manufacturer to the final consumer is traced in Figure 7.2 (Clay 1996).

Technical Assistance and Information on Markets

Brazil nut gatherers receive virtually no extension services or technical assistance regarding how to evaluate the impact of harvesting on natural stands of Brazil nuts. Methods to enrich natural stands in forests, ways to reduce postharvest losses, and marketing information are practically nonexistent for Brazil nut gatherers. What the extractivist knows about the Brazil nut market, he learns from the buyers of the nuts. This source is obviously not without bias. What the extractivist knows about tree management or any technical skills for collecting or preprocessing nuts, he has learned through his own experience as a nut gatherer and from the collective experience of communities of nut gatherers.

Extractivists sell their Brazil nuts to middlemen with the price being based on an order from the middlemen's buyers/processors. The value is calculated based on the international market and the conditions of the sale, that is, whether it is a forward or a spot transaction. In a forward sale, the cost of financing the merchandise over the period of time until the nuts are delivered is discounted from the purchase price.

The real need at this level of the marketing chain is for more competition or more buyers (particularly environmental "green" buyers or those interested in fair trade) for the isolated seller. Ideally, additional buyers could appear in the market in the form of extractivists' commercial associations and cooperatives or international NGOs that have a proven record with local groups.

In general, the Brazil nut market lacks standardized reporting and information from the producer countries, and what little exists is not readily available to gatherers or their organizations. Most market information comes from importing countries. Access to systematic reports and statistics would help local producer groups obtain a fairer price for their goods and would increase the options available to producers and sellers. Ideally, research should be expanded to analyze the economic feasibility and social acceptance of decentralized processing (whether urban or rural) and to monitor existing projects in detail.

Conclusions and Recommendations

Conclusions

The single most serious threat to Brazil nut trees in the Amazon is the chainsaw. Clearing land for pasture has destroyed Brazil nut stands in Pará, Rondônia, Acre, and Bolivia. The continued conversion of forest to pasture will do more to eliminate Brazil nut trees, as well as the ecosystem that they dominate, than the collection and sale of Brazil nuts ever will.

Much land in the Amazon (up to 20%) has a higher long-term economic worth when harvested for renewable products such as fruits, resin, and nuts (especially Brazil nuts) than when the land is cleared for unsustainable agriculture, cattle ranching, or timber production. Nevertheless, to date, economic forces have continued to promote clearing rather than preservation of the forest. Until 1989, land clearing had been encouraged through subsidization and Brazilian government policies, especially for the building of roads. This has opened up new areas of the forest, increasing the supply of land and depreciating its value to the point where it has become profitable to exploit land as a short-term nonrenewable resource (cut, burn, use, and move on) (LaFleur 1991c).

The Brazil nut industry is highly concentrated with three of the major processor/exporters owned by members of the same family and controlling more than half of the world market (LaFleur 1991c). These three and the other major Brazil nut processors are all located in urban areas with a significantly higher labor cost than in the forests or nearby communities where the nuts originate. Raw in-shell nuts have to be transported thousands of kilometers from source to processing plant with significant losses due to spoilage. Two-thirds of the gross weight of the raw material entering the processing plants is discarded via shelling. There are few economies of scale in current processing plant locations that can justify the transportation costs and losses to spoilage for a product that is still hand-shelled and manually classified and selected (LaFleur 1991c).

However, while decentralizing the shelling of Brazil nuts by locating shelling factories closer to where the nuts are harvested will certainly become more common because of the economics of the trade, it will also tend to increase the number of areas from which Brazil nut sales will be profitable, often for the first time. This means that now, more than ever, detailed, site-specific studies, as well as more general regional analyses, must be undertaken if we are to understand the impact of the harvest and sale of Brazil nuts on ecosystem function, as well as the impact on Brazil nut populations and productivity.

There are little data, to date, concerning replacement costs for Brazil nut trees and consequently what the cost of Brazil nuts should be from a replacement or an environmental point of view. As it is, "markets" determine prices based more on the cost of mining resources (e.g., labor and capital) than replacing them. In most instances, if a substitute is available, it will greatly influence the price of a commodity such as Brazil nuts. This is not an environmentally friendly way to set prices. Environmental premiums or green fees are ways to place a higher value on natural resources, but, to date, markets still ignore replacement costs. Studies on the pricing of Brazil nuts, as well as many other commodities, versus the environmental costs to produce them are needed to shed more light on this problem.

There is also a lack of standardized reporting and information from the producer countries, and what little exists is not readily available to Brazil nut harvesters or their organizations (LaFleur 1991c). Without access to reliable market information, collectors are left to the mercy of monopoly traders and processors. No formal credit is available to extractivists, who receive advances of funds and merchandise in exchange for their guarantee of future delivery of Brazil nuts. Without access to credit, there is little chance that gatherers' organizations will ever become processors or exporters. Finally, little research has been done on the extractive economics and management of natural stands of Brazil nut trees and other forest products; research and technical development are biased toward plantation activities and centralized processing. What all this means is that if collectors are not able to make better livings from the harvest and sale of products such as Brazil nuts, they will use the forests in other ways (e.g., logging or mining) or clear the forests entirely for pasture or crops.

It is ironic that, although the economic history of the Amazon has been based almost exclusively on extractivism and these extractivist activities have supported the region's population for centuries without destroying the ecosystems that so many are now trying to "preserve," the information on and experience of extractivism were not utilized in the formation of economic development models for the region (LaFleur 1991c). Conservationists should avoid making the same mistake.

The greatest threat to Brazil nuts in the Amazon is the same as the greatest threat to biodiversity—the conversion of forests to agriculture or pasture land. As more land is cleared of natural forest, the number of Brazil nut trees diminishes and future production potential decreases. This results in giving up the market share within the edible nut trade without the compensation of higher prices for the remainder of the "on-tree" stock. This promises a bleak future for the Brazil nut industry as a whole. It also means, incidently, the loss of vast amounts of biodiversity.

Recommendations

The following recommendations (taken from LaFleur 1991c and my own observations) focus on socioeconomic and market research and action that would help Brazil nut harvesters receive higher incomes for their work so that they remain in the forests as the first line of defense against deforestation and other destructive land uses.

1. *Taxation policies should encourage the rational use of forest resources and generate income and employment in the area.* Specifically, Brazil should eliminate or limit the value-added tax (ICMS) on Brazil nuts. This tax on shelled nuts, which amounts to 12 percent of the export (FOB) price, should be applied to unshelled nuts. In the state of Acre, which recently adopted this policy, the state has provided the opportunity to increase statewide income from US$1 million on the sale of unshelled nuts to more than US$6 million for the sale of shelled nuts. This tax is paid, in effect, by the extractivists. A reduction or elimination of the tax would increase the price paid to extractivists, resulting in more nuts being collected and an expansion into more isolated areas with higher transportation costs. Assuming that an increase in the price paid to extractivists would cause a proportional increase in supply, the increase in supply would in turn cause a drop in FOB prices. The fall in FOB prices would be less than proportional to the increase in supply (0.71%) and would stimulate the demand for Brazil nuts and accommodate the increment. Gatherers would have higher income through higher prices or more production (their prices could fall back as FOB prices drop, but not to original levels). State tax revenue loss would be offset by an increase in export income and environmental benefits intrinsic to Brazil nut gathering, and increased taxes on income and consumer goods associated with a potential sixfold local revenue increase as in Acre. In addition, by increasing production of Brazil nuts, Brazil has an opportunity to earn more foreign exchange. This is because prices decrease only 0.71 percent for every 1 percent increase in production. A 10 percent increase in shipments would generate a 2.7 percent increase in total export earnings.

2. *Secure better land or land-use rights for producers so that they can make long-term investments in crop management, enrichment programs, warehousing, and decentralized shelling operations.* At present, the great majority of extractivists rent or work on lands controlled by others. These extractivists have the legal status of *posseiros,* or squatters, with rights of permanency but without title to the land. Long-term investments such as planting new stocks of Brazil nut trees will

not take place without a clear definition of land rights. It is debatable whether wealthy landowners can afford to hire the labor required for enrichment programs.

3. *Create a system of rural extension and technical assistance supported by research in natural resource development.* At the extractivist level, services could concentrate on better management of natural stands, the planting of new trees, and techniques for lowering costs and increasing productivity of the harvest, transport, and storage processes. Technical information is also needed concerning local, decentralized processing systems including mechanization and other site-appropriate technologies. Institutions and individual researchers should be encouraged to exchange production and management information. Research must be expanded to analyze the economic viability and social acceptance of decentralized processing and to monitor existing projects in detail.

4. *Change the official regulations and unofficial attitudes concerning extractivism and local community-based commercial organizations so that these smaller local groups have access to the investment and working capital necessary for processing nuts.* There is a need for financial institutions that deal directly with extractivists and their organizations to reassess guarantees for loans. Once extractivist organizations are recognized as legitimate debt holders, physical stocks of nuts and a percentage of the economic returns from local processing can be used as collateral for working capital loans. The history of individual or group production levels in relation to production forecasts can be used to determine working capital requirements.

5. *Develop a systematic reporting and information distribution system.* Systematic reporting and information distribution in producer countries is essential for an effective market intelligence system. A comprehensive system should include a monthly periodical to provide information on crop forecasting and development, producer prices, FOB prices for different qualities of nuts, shipping costs and shipping schedules, as well as other technical information. Also needed is a complete list of all producers, buyers, sellers, importers, exporters, brokers and agencies, freight forwarders, shipping agencies, and others included in the commercial chain in order to provide transparency and competition in the system. Communications media to make this information available to extractivists and their organizations are also essential.

6. *Develop a specific program for decentralized processing to increase the income of forest dwellers and residents of nearby towns, and lower processing and transportation costs.* Such a program would include a research component to analyze socioeconomic impacts and site-appropriate technology involved in decentralized processing, as well as an extension and training component to assist extractivist organizations in developing the processing plants at the local level. In addition, a credit program is needed to support community associations and gatherers in activities essential to decentralized processing.

References

Altenbach, S. B., K. W. Pearson, F. W. Leung, and S.S.M. Sun. 1984. Molecular cloning and characterization of the Brazil nut sulfur-rich protein CDNA. *Plant Physiology* 75(1)(suppl.):65.

Amazonia Trading Company. 1991. *Brazil nut market report.* Liverpool, England: Amazonia Trading Company.

Ampe, C., J. Van-Damme, L.A.B. Castro, M.J.A.M. Sampaio, M. Van Montagu, and J. Vandekerchkhove. 1986. The amino-acid sequence of the 2S sulphur-rich proteins from seeds of Brazil nut (*Bertholletia excelsa* H.B.K.). *European Journal of Biochemistry* 159(3):597–604.

Andrade, F. A. 1968. *Conjuntura da castanha-do-Pará; relatório preliminar.* Belém: Superintendência de Desenvolvimento da Amazônia.

Assunção, F. P., M.H.S. Bentes, and H. Serruya. 1984. A comparison of the stability of oils from Brazil nut, Pará rubber and passion fruit seeds [*Bertholletia excelsa, Hevea brasiliensis, Passiflora edulis*]. *Journal of the American Oil Chemists' Society* 61(6):1031–1036.

Buckley, D. P., D. M. O'Malley, V. Apsit, G. T. Prance, and K. S. Bawa. 1988. Genetics of Brazil "nut" (*Bertholletia excelsa* Humb. & Bonpl.: Lecythidaceae): I. Genetic variation in natural populations. *Theoretical and Applied Genetics* 76:923–928.

Castro, L.A.B., Z. Lacerda, R. A. Aramayo, M.J.A.M. Sampaio, and E. S. Gander. 1987. Evidence for a precursor molecule of Brazil nut 2S seed proteins from biosynthesis and CDNA analysis. *Molecular and General Genetics* 206(2):338–343.

Chansler, M. W., M. Mutanen, V. C. Morris, and O. A. Levander. 1986. Nutritional bioavailability to rats of selenium in Brazil nuts and mushrooms. *Nutrition Research* 6(12):1419–1428.

Clark, R. G., and H. E. Nursten. 1976. Volatile flavour components of Brazil nuts, *Bertholletia excelsa* (Humb. and Bonpl.). *Journal of the Science of Food and Agriculture* 27(8):713–720.

Clay, J. 1989. How reserves can work. *Garden Magazine* October:13(5):2–4.

Clay, J. 1990. A rain forest emporium. *Garden Magazine* January/February: 14(1):2–7.

Clay, J. 1992. *Report on funding and investment opportunities for income generating activities that could complement strategies to halt environmental degradation in the Greater Amazon Basin.* Washington, D.C.: Biodiversity Support Program/USAID.

Clay, J. 1996. *Generating income and conserving resources: Twenty lessons from the field.* Washington, D.C.: World Wildlife Fund.

Clay, J., and C. R. Clement. 1993. *Selected species and strategies to enhance income generation from Amazonian forests.* Rome: U.N. Food and Agricultural Organization.

Clement, C. R. 1993. Brazil nut. In *Selected species and strategies to enhance income generation from Amazonian forests,* ed. J. Clay and C. R. Clement. Rome: U.N. Food and Agricultural Organization.

de Souza, A. H. 1963. *Castanha do Pará: Estudo botânico.* Rio de Janeiro, Brazil: Ministério da Agricultura.

de Souza, M. L. 1984. *Estudo de processos tecnológicos para a obtenção de produtos da castanha-do-Brasil.* Fortaleza, Brazil.

Diniz, T. D. de A. S., and T. X. Bastos. 1974. Contribuição ao conhecimento do clima tipico da castanha do Brasil. Boletim tecnico. *IPEAN* 64:59–71.

FAO. 1986. *Food and fruit-bearing forest species: Three examples from Latin America.* FAO Forestry Paper 44/3. Rome: U.N. Food and Agricultural Organization.

Freire, F.C.O., and F. F. Ponte. 1976. A meloidoginose da castanha do Pará, *Bertholletia excelsa* H.B.K. Bol. *Cearense Agronômica* 17:57–60.

Gander, E. S. 1986. Tecnologia de DNA recombinante em plantas. *Ciência e Cultura* 38(7):1178–1185.

Gill and Duffus. 1988a. *Edible nut market.* Report Nos. 102–127, February 1981–May 1988. London: Gill and Duffus.

Gill and Duffus. 1988b. *Edible nut statistics.* June. London: Gill and Duffus.

Gill and Duffus. 1991. *Edible nut market.* Report No. 131 (August). London: Gill and Duffus.

Guerche, P., E.R.P. Almeida, M. A. Schwarztein, E. S. Gander, E. Krebbers, and G. Pelletier. 1990. Expression of the 2S albumin from *Bertholletia excelsa* in *Brassica napus. Molecular and General Genetics* 221(3):306–314.

Hemming, J. 1987. *Amazon frontier: The defeat of the Brazilian Indians.* London: Macmillan.

Kamiya, N., K. Sakabe, N. Sakabe, K. Sasaki, M. Sakakibara, and H. Noguchi. 1983. Structural properties of Brazil nut 11s globulin, excelsin. *Agricultural and Biological Chemistry* 47(9):2091–2098.

Karas, J. 1991. *Brazil nut market research.* Boston: Cultural Survival.

Kitamura, P. C., and C. H. Müller. 1984. *Castanhais nativos de Marabá-PA: Fatores de depredação e bases para a sua preservação.* Documentos 30:1–32. Belém, Brazil: Empresa Brasileira de Pesquisa Agropecuária. Centro de Pesquisa Agropecuária do Trópico Umido.

LaFleur, J. 1991a. Beneficios econômicos de descentralização. Paper given at the 4th Congresso Nordestino de Ecologia, Recife, Brazil.

LaFleur, J. 1991b. Trip report: Research on market channels for Brazil nuts. Recife, Brazil: ECOTEC.

LaFleur, J. 1991c. *The Brazil nut market.* Recife, Brazil: Sociedade para Desenvolvimento Tecno-Ecológico (ECOTEC).

LaFleur, J., and J. Bryon. 1988. *Pepper study.* São Paulo, Brazil. Unpublished.

LaFleur, J., and W. Groeneveld. 1990. Avaliação da Usina de beneficiamento de castanha-do-Pará da Cooperativa Agro-Extrativista de Xapuri-Acre. Xapuri, Brazil. Unpublished report.

LaFleur, J., and G. Jones. 1984. A review of some econometric work on cocoa prices and production with extended comment on the long cocoa cycle. *Oxford Agrarian Studies,* 13:67–89. Oxford University Institute of Agricultural Economics.

Loureiro, A. A., M. F. Silva, and J. C. Alencar. 1979. *Essências madeireiras da Amazônia,* vol. 2. Manaus, Brazil: Instituto de Pesquisas da Amazônia.

Mahar, D. 1988. *Government policies and deforestation in Brazil's Amazon region.* Washington, D.C.: World Bank.

Margrave, G. 1648. Historiae rerum naturalium brasiliae. In *Historiae naturalis brasiliae, auspicio et beneficio illustris,* G. Piso and G. Margrave, 128–293. Leiden: I. Mauriti Com. Nassau.

Miers, J. 1874. On the Lecythidaceae. *Linnean Society of London* 30(2): 157–318.

Miller, C. 1990. Natural history, economic botany, and germplasm conservation of the Brazil nut tree. Thesis, University of Florida, Gainesville.

Mori, S. A. 1992. The Brazil nut industry: Past, present, and future. In *Sustainable harvest and marketing of rain forest products,* ed. M. Plotkin and L. Famolare, 241–251. Washington, D.C.: Island Press.

Mori, S. A., and G. T. Prance. 1990a. Taxonomy, ecology, and economic botany of the Brazil nut (*Bertholletia excelsa* Humb. & Bonpl.; Lecythidaceae). *Advances in Economic Botany* 8:130–150.

Mori, S. A., and G. T. Prance. 1990b. Lecythidaceae—Part II. The zygomorphic-flowered New World genera (*Bertholletia, Corythophora, Coura-*

tari, Couroupita, Eschweilera, and *Lecythis*). *Flora Neotropica Monographs* 21(2):1–376.

Moritz, A. 1984. *Estudos biológicos da castanha do Brasil* (*Bertholletia excelsa* H.B.K.). Documentos 29:1–82. Belém: Centro de Pesquisas Agropecuários do Tropico Umido-Empresa Brasileira de Pesquisa Agropecuária.

Moritz, A., and P. Ludders. 1985. Present situation and possibilities for the development of the Pará nuts in Brazil (Stand und Entwicklungsmöglichkeiten des Paranussanbaues in Brasilien). *Erwerbsobstbau* 27(12):296–299. Berlin.

Müller, C. H., F. J. Câmara Figueirêdo, A. K. Kato, J. E. Urano de Carvalho, R. L. Benchimal Stein, and A. de Brito Silva. 1994. *A cultura das castanheiras-do-Brazil.* Centro de Pesquisa Agroflorestal da Amazônia Oriental-CPATU. Brasília: Serviço de Promulgação de Informação (SPI).

Müller, C. H., I. A. Rodrigues, A. A. Müller, and N.R.M. Müller. 1980. *Castanha do Brasil: Resultados de pesquisas.* Miscelânea 2:1–25. Belém: CPATU-EMBRAPA.

O'Malley, D. M., D. P. Buckley, G. T. Pramel, and K. S. Bawa. 1988. Genetics of Brazil nut (*Bertholletia excelsa* Humb. and Bonpl.: Lecythidaceae). *Theoretical and Applied Genetics* 76(6):929–932.

Padoch, C., J. Chota Inuma, W. de Jong, and J. Unruh. 1987. Market-oriented agroforestry at Tamshiyacu. *Advances in Economic Botany* 5:90–96.

Palmer, I. S., A. Herr, and T. Nelson. 1982. Toxicity of selenium in Brazil nuts to rats. *Journal of Food Science* 47(5):1595–1597.

Pesce, C. 1985. *Oil palms and other oilseeds of the Amazon* (trans. D. V. Johnson). Algonac, Michigan: Reference Publications.

Plietz, P., B. Drescher, and G. Djamaschun. 1988. Structure and evolution of the 11S globulins: Conclusions from comparative evaluation of amino acid sequences and x-ray scattering data. *Biochemie und Physiologie der Pflanzen* 183(2–3):199–203.

Posey, D. A. 1985. Indigenous management of tropical forest ecosystems: The case of the Kayapó Indians of the Brazilian Amazon. *Agroforestry Systems* 3:139–158.

Rosengarten, Jr., F. 1984. *The book of edible nuts.* New York: Walker.

Sampaio, A. J. 1994. A flora amazônica. In *Excerptos da revista Brasileira de geografia: Amazônia Brasileira.* Instituto Brasileira de Geografia e Estatística, Rio de Janeiro.

SUDAM. 1979. *Características silviculturais de espécies nativas e exóticas dos plantios do Centro de Tecnologia Madeireira.* Estação Experimental de Curua-Una. Belém, Pará: Superintendência de Desenvolvimento da Amazônia.

Sun, S.S.M., S. B. Altenbach, and F. W. Leung. 1987a. Properties, biosynthesis and processing of a sulfur-rich protein in Brazil nut (*Bertholletia excelsa* H.B.K.). *European Journal of Biochemistry* 162(3):477–483.

Sun, S.S.M., F. W. Leung, and J. C. Tomic. 1987b. Brazil nut (*Bertholletia excelsa* H.B.K.) proteins: Fractionation, composition, and identification of a sulfur-rich protein. *Journal of Agricultural and Food Chemistry* 35(2):232–235.

Thorn, J., J. Robertson, D. H. Buss, and N. G. Bunton. 1978. Trace nutrients: Selenium in British food. *British Food Nutrition* 32(2):391–396.

Vaz Pereira, L. C., and S.L.L. Costa. 1981. *Bibliografía de castanha-do-Pará* (*Bertholletia excelsa* H.B.K). Belém, Pará: Empresa Brasileira de Pesquisa Agropecuária, Centro de Pesquisas Agropecuário do Trópico Umido.

Viana, V., R. A. Mello, L. M. de Morais, and N. T. Mendes. 1994a. Deforestation, decay of Brazil nut populations in pastures, and forest policies in the Amazon: The case of Xapuri. Unpublished manuscript.

Viana, V., R. A. Mello, L. M. de Morais, and N. T. Mendes. 1994b. Ecology and management of Brazil nut populations in extractive reserves in Xapuri, Acre. Unpublished manuscript.

Woodroof, J. G. 1979. *Tree nuts: production, processing, products,* 2d ed. Westport, Connecticut: AVI Publishing.

Zucas, S. M., E.C.V. Silva, and M. I. Fernandes. 1975. Farinha de castanha-do-Pará: Valor de sua proteína. *Revista Farmacêutica e Bioquimica da Universidade de São Paulo* 13:133–143.

EIGHT

The Impact of Palm Heart Harvesting in the Amazon Estuary

Jason W. Clay

Palm heart harvested from açaí (*Euterpe oleracea* Mart.) in the Amazon estuary provides nearly 30,000 jobs and generates some US$300 million in annual sales (*Diário do Pará* 1993). Yet, short of clearing land for permanent pasture, the current commercial harvest of açaí for palm heart is one of the more destructive practices in the region today (P. Ramos, pers. comm. 1991; Clay 1992; Bovi and de Castro 1993; Clay and Clement 1993). Clearcut harvesting of palm hearts year after year is not sustainable. Within as little as five years, such harvesting practices lead to noticeable declines in production (Pollak et al. 1993). Spot checks of export-quality palm heart from the Belém area indicate that not even 5 percent of the cans contain palm stems that are of a size that is legal to harvest in Brazil.

Though many palm spears are harvested for palm heart, the multistemmed *E. oleracea* is the most commonly harvested. To date, none of the vast natural stands of this palm are sustainably managed for commercial palm heart harvests. Yet studies have shown that the palm can readily be managed by selectively harvesting stems every three to four years. Management can be made even simpler by cutting only large stems. Research has suggested that annual production under a controlled management system is greater than the first harvest from naturally occurring stands of the palm and can continue indefinitely (Pollak et al. 1993).

One incentive for sustainable management systems for harvesting *E. oleracea* is to help harvesters earn more income by adding more value to what they harvest and by making the sale of the product conditional on sustainable management practices. In this way they would be compensated for sustainable management. Furthermore, they would earn more income but harvest fewer palm stems. Harvesters could capture some of the value that is added to the product that they have harvested for years but sold only in the form of raw, unprocessed stems (Clay 1996).

Most of the value added to palm heart is captured far from the forest (Clay 1996). Thus extractors earn the least and foreign buyers and retailers capture the most value added to palm heart, with a progressively larger share earned by those intermediaries closest to the consumer. If harvesters are to earn enough income to cover their costs of sustainably harvesting palm heart, this picture needs to change. If consumers are not going to pay more, then each level of profit taking between the producer and the consumer will have to be reduced so that harvesters receive a greater share.

On the face of it, extractors, palm heart processors, distributors, wholesalers, retailers, and even consumers all stand to lose from the unsustainable management of native stands of açaí. Increasingly, there are few other vast natural stands to exploit. The cost of planting even the fastest-growing, multistemmed palm species would be far more expensive than the implementation of limited impact palm management/harvest systems. All sectors of the industry stand to gain from the implementation of sustainable management systems, but none of them will invest the required funds in such programs unless all are required to do so.

I review here the causes, particularly the market incentives, that fuel the unsustainable harvest of an important palm heart species, *E. oleracea,* and the environmental impacts of this harvest. I then review management alternatives for improving sustainability and reducing impacts.

*Açaí (*Euterpe oleracea *Mart.): The Plant and Related Species*

There are at least two different palm species known as açaí in the Brazilian Amazon. *E. oleracea,* called açaí-do-Pará, is an extremely abundant, multistemmed palm that occurs in the *várzeas* (annually flooded areas along the banks of whitewater rivers) and in the Amazon River estuary, where it frequently forms monospecific populations. Within the classification *oleracea* there are a number of variations. For example, açaí-do-igapó and the açaí-da-várzea, while both *E. oleracea,* have a number of marked differences. *Euterpe precatoria* Mart., called açaí-da-terra-firma, is a single-stemmed palm (solitary), common in central and western Amazonia, that occurs along the valley edges on the *terra firma* (nonflooded, upland plateaus), in areas without annual flooding (Bovi and de Castro 1993).

In other parts of Brazil and South America, other *Euterpe* species are used as sources of palm heart, especially *Euterpe edulis.* This single-stemmed species, in the south and southeast of Brazil, was very important as a source of palm hearts until the 1960s when production throughout the region declined sharply due to unsustainable harvesting.

Description and Phenology

Açaí (henceforth refers to *E. oleracea*) is a slender, multistemmed, monoecious palm that can attain more than 24 meters (m) in height. It can have more than 45 stems in different stages of growth and fructification, depending on insolation. In natural stands, four to eight well-developed stems per mature plant are common (Bovi and de Castro 1993), with up to 600 plant clumps per hectare (ha). The plant is well adapted to high light levels and periodically waterlogged soils. In low-lying areas, soil and organic particles trapped on the root system slowly construct a mound around the plant (Bovi and de Castro 1993).

Açaí is predominantly allogamous (outbreeding). Small bees and flies are the principal pollinators. Flowering can start as early as four years if the palm grows in full sunlight (Bovi and Castro 1993). Flowering and fruiting occur throughout the year.

Seed dispersal of açcaí over short distances is by rats and other rodents. Long-distance dispersal is by birds (Zimmerman 1991) and by floods along stream banks. Germination begins 3–11 months after seeds fall, with the bulk of seeds germinating in the first 30–60 days. There is no long-term dormancy. The fruit epicarp is readily eliminated by natural decomposition, aided by microorganisms, insects, or passing through the digestive tract of some birds. Although the germination percentage is high in laboratory conditions, reaching 90 percent (Bovi et al. 1989), in nature it may attain only 50–60 percent (Bovi and de Castro 1993). Seedling survival is low, especially from the one-leaf stage until the plants are about 50 centimeters (cm) in height. Competition for light is the main factor limiting survival and seedling growth (Bovi and de Castro 1993).

Plant growth rate is slow during the first three years, especially compared to some of its forest competitors. Small seedlings (one-leaf stage) are able to survive without much growth, awaiting more favorable conditions, especially light. Stem growth usually occurs two to three years from seedling stage, when the plant is at about 1 m in height. After that, growth in height increases rapidly. Normally it takes four to six years for the main stem to reach the minimum size for economically viable palm heart harvest. Once the palm reaches the canopy, stem growth is slower and constant (Bovi and de Castro 1993).

Distribution, Abundance, and Ecology

Açaí is widely distributed in northern South America but attains its greatest coverage and economic importance in Pará state, Brazil, where it is found throughout the state. The major occurrence is in the Amazon River estuary, in an area estimated at 25,000 square kilometers (km^2) (Lima 1956). Much of this area (ca. 88% according to Lima 1956) is frequently inundated, and açaí is the most ecologically important tree species (Anderson 1988). Calzavara (1972) conservatively estimated the coverage of açaí in the Amazon estuary to be 10,000 km^2. It is especially common on varzeas, slightly less so in the *igapós* (perennially flooded, black- or clear-water forest swamps), and even less so on the terra firma (Bovi and de Castro 1993). Plants from these three edaphic conditions look quite different. Frequently the palms are found in almost pure stands (*açaizais*), representing, along with burití (*Mauritia flexuosa*), the most prominent feature of the vegetation landscape (Peters et al. 1989a). It is more common, however, to find the palm dominating as one of the most abundant species in a highly diverse forest.

Population density of the species in the Amazon River estuary depends mostly on soil conditions. On average, the plant population is between 230 and 600 clumps per ha, considering only clumps with stems higher than 2 m (Jardim 1991 and pers. comm. 1992). Total population ranges from 2,500 to 7,500 plants per ha, with most of the population (50%) in the first seedling stage (1–2 leaves and about 20–25 cm tall) (Bovi and de Castro 1993).

Uses and Economic Potential of Açaí: A General Overview

Principal Uses

Açaí produces a wide variety of market and subsistence products (Strudwick 1990; Nascimento 1991, 1992a–c, 1993; Urpi et al. 1991). Anderson (1988) listed 22 uses for all plant parts, from the leaves to the roots. Local people use açaí principally for the preparation of a thick, dark purple liquid obtained by maceration of the pulp of the ripe fruits. The liquid, locally called *açaí* or *vinho de açaí* (although it is not a fermented or distilled beverage), is not particularly nutritious. The nutrient content reported by Mota (1946), Campos (1951), and Altman (1956) is as follows: 1.25–4.34 percent (dry weight) protein; 7.6–11 percent fat; 1–25 percent sugar; 0.050 percent calcium; 0.033 percent phosphorus; and 0.0009 percent iron. It also has some sulfur, traces of vitamin B_1, and some vitamin A. Caloric content ranges from 88 to 265 calories per 100 grams (g), depending on the dilution and on the complement. Yet the liquid is extremely filling, especially when mixed with manioc (*Manihot esculenta*) flour. It is usually not drunk but eaten with a spoon, and forms a major and basic part of the diet of most of the inhabitants of the lower Amazon River. Individual daily consumption of up to 2 liters has been reported. It has a metallic, somewhat nutty flavor, with a creamy purple texture and slightly oily taste and appearance (Bovi and de Castro 1993).

The açaí liquid is so popular that there are establishments in small and large towns throughout the region that make and sell it in half-liter plastic bags. Although a basic part of the diet of the poor, açaí liquid is popular throughout all socioeconomic levels. In Belém alone, as much as 50,000 liters of the fruit are sold each night for the early morning market. Juice is now shipped to and consumed in Rio de Janeiro, São Paulo, and other major cities in the south of Brazil. Details of açaí liquid making, consumption, and marketing are described by Strudwick and Sobel (1988).

Açaí liquid is extremely perishable, so its consumption has been restricted to a purely regional level for the past century. However, freezing the pulp allows it to be easily transported, even if it is somewhat expensive. Attempts have also been made to dehydrate the liquid to preserve it (Melo et al. 1988). The dehydrated product is suitable for consumption up to 115 days after packing. With this product, açaí could be made available throughout the year without the expense of cold storage or the weight of shipping water, and could be exported to other national and international markets. However, the energy and technology costs of dehydration are quite high.

Açaí can also be preserved by adding sugar, but Amazonian residents do not prefer it sweet. Consequently this form of preservation is not practiced. If markets are developed for the fruit in the south of Brazil or North America or Europe, it is possible that making a jam would be one way to prepare the product for export from the region.

This report, however, is focused on the other major product from açaí: the palm heart. It consists of the tender, whitish, immature leaves of the palm, found just above the growing point on each stem. Once removed, it is a flexible cylinder about 45 cm long and 2–3 cm wide. Because açaí is a multistemmed palm, selective harvesting of individual stems does not kill the plant, allowing the removal of palm heart year after year. Palm heart has almost no nutritional value (Ferreira and Yokomizo 1978; Ferreira et al. 1982), but it is widely appreciated in a variety of dishes. Although regional (Amazonian) consumption of palm heart is minimal, there is a large internal Brazilian market for it, especially in the southeastern states, where it is a common appetizer and where, in addition, a wide variety of dishes made with it can be found in local restaurants (Bovi and de Castro 1993).

Commercial extraction of açaí palm heart began in the Amazon River estuary in the 1960s as a consequence of the decimation caused by heavy exploitation of native stands of *E. edulis* in southeastern Brazil (Renesto and Vieira 1977). Açaí is currently the world's main source of palm heart, with the Amazon River estuary being the principal producing region. By 1992 there were 120 registered processing plants in the region, most of them situated at the edges of rivers, particularly on Marajó Island. Numerous smaller factories exist based on family labor and affiliated with or selling their product to the larger ones. The degree of sophistication of these plants varies, but most of them are very precarious and process a low-quality product (Bovi and de Castro 1993).

Canned heart of palm production per processing plant is variable, ranging from 6 to 30 metric tons (MT) per month. There is already a shortage of the raw material in many locations in the Amazon River estuary due to overharvesting and lack of management of the native stands. Some processing plants run only two to three days per week. This shortage has also spurred the migration of some floating processing plants from the estuary in Pará to varzea areas upriver in Amazonas state, where they exploit (again destructively) both local populations of açaí and previously unexploited (for palm heart) populations of *E. precatoria,* thus denying an important food resource to local residents (C. R. Clement, pers. comm. 1991; Bovi and de Castro 1993; ECOTEC 1993; Pollak et al. 1993).

In addition to the palm heart, the soft inner core of the palm stem, including the growing point and the pulp below it, also has considerable market potential. Having slightly more weight and volume than the heart, this part of the plant can be processed the same way and later marinated and used in salads. Since this part of the plant is currently discarded, any net income derived from it would tend to increase the economic viability of palm heart extraction.

Finally, there are a number of new, or secondary, uses that are currently being considered for açaí. A pigment extract from the fruit (anthocyanin) can provide natural red and purple dyes for the food industry (Iaderoza et al. 1992). While there is increasing interest internationally in natural, vegetable-based dyes both for food and printing, it is not clear that açaí will ever compete with beets, which produce roughly the same color more cheaply, and closer to North American and European markets (Lepree 1994). Most of the other uses of açaí (e.g., feed for livestock, thatch, or construction materials) are merely by-products of the harvest of fruit or palm heart.

Palm Heart Production and Sales: A Historical Overview

Palm hearts are a delicacy throughout the world. Brazil has been the primary source of palm heart since the 1950s, when they were first introduced on international markets. Historically, however, Brazil has always consumed more palm heart than it has exported.

Initially, palm hearts were extracted for local and national consumption from the forests of southern and southeastern Brazil from the species *E. edulis*. This species, however, is single-stemmed, so the plant does not survive palm heart extraction. Intensive harvesting of the species both for increased domestic consumption and, after 1950, for export, led to the collapse of the palm heart industry in southern and eastern Brazil by the end of the 1960s (Ferreira and Paschoalino 1987).

In the early 1970s, palm heart companies, faced with depleted reserves of natural stands of *E. edulis* in the south and southeast of Brazil, moved their extractive industry to the estuary of the Amazon River, where vast, nearly monocrop, natural stands of açaí palm are found (Calzavara 1972). In addition, due to the importance of açaí fruit to the local diet, many residents of the region cultivate the palm for subsistence or for sales into local markets (Anderson and Ioris 1992).

By 1975, Pará state, which encompasses much of the estuary of the Amazon, accounted for 96 percent of Brazil's total palm heart production. By 1993, it was estimated that the industry generated nearly 30,000 jobs in the region as well as gross revenues of some US$300 million per year (*Diário do Pará* 1993).

Economic Summary

Anderson (1988) studied the commercial products sold by a single family living on the Ilha das Onças, near Belém, Pará, during the course of one calendar year. Though it is not clear how "representative" the family was, he found that some 35,000 hearts of açaí were extracted during this period, representing an income of US$2,916 (or an average of US$0.083/stem). At that time, the family harvested US$15,532 worth of fruits (78,885 kilograms [kg] of product). Together these two products accounted for 75 percent of the forest products sold by the family.

Anderson (1988) also pointed out the fragility of the system. For example, an increase in demand and in the number of palm heart processing plants was already seen to pose a threat to the economic and ecological balance of the estuary. A rise in prices for the palm heart or a fall in the price for açaí fruit because of the improved processing and

Table 8.1. Brazilian Production and Consumption of Palm Heart (All Species), 1973–1980 and 1986–1992

Year	Brazilian production (MT)	Estimated percentage consumed in Brazil	Brazil FOB[a] price (US$/kg)
1973	35,986	87.7	
1974	34,273	75.2	
1975	200,154	96.5	
1976	203,948	51.1	
1977	35,123	66.4	
1978	24,625	77.3	
1979	31,358	78.2	
1980	114,408	91.2	
1986	131,013	93.5	2.82
1987	142,060	93.2	3.59
1988	190,314	95.9	3.87
1989	202,439	97.1	3.60
1990	182,140	95.9	3.63
1991[b]	200,263	96.5	3.79
1992[b]	222,425	96.1	3.66

Sources: ECOTEC 1994; FIBGE (Anuário Estatistico do Brasil); Banco do Brasil; CACEX (Cateìra de Comércio Exterior); Inter-American Development Bank as cited in ECOTEC 1993.

[a]Free on board.

[b]ECOTEC estimates.

transport technology could result in the cutting of stems traditionally used for fruit production and result in general scarcity of fruit for at least a decade or more.

World trade in palm hearts in 1990 exceeded US$65 million, but total production in Brazil is valued at US$300 million. Some 85 percent or more of world exports come from the exploitation of native forest stands, principally of açaí in Brazil. Brazil is both the major world producer (more than 200,000 MT) and the largest consumer of palm hearts. Brazil is responsible for 70 percent of the world trade in palm hearts, exporting more than 10,000 MT, but consumes more than 10 times the amount exported (Tables 8.1 and 8.2).

Owing to the highly perishable character of palm heart derived from *Euterpe* species, especially discoloration due to enzyme-mediated oxidation, at least 85 percent of the total internal consumption of palm heart is in a processed form (cans or jars) and retail prices are nearly

Table 8.2. Brazilian Palm Heart Exports (Metric Tons; All Species) by Continent, 1986–1990

Year	Europe	Africa	Oceania	Asia	The Americas	Total
1986	5,984	11	23	52	2,387	8,458
1987	7,262	15	22	70	2,247	9,615
1988	6,159	11	4	123	1,476	7,773
1989	4,630	7	9	69	1,268	5,982
1990	5,620	3	7	111	1,660	7,401

Sources: ECOTEC 1994; Banco do Brasil; CACEX.

the same as those for export. Domestic palm heart is a classification of quality and size, rather than the destination of the product. Domestic palm heart is less choice, more fibrous, and cannot exceed 60 percent the volume of palm heart per 1-kg container. About 80 percent of Brazilian export-quality palm heart is consumed in Brazil.

Current Collection Methods and Yields

To harvest the palm heart, the entire stem must be cut down with a bill hook or an ax, or the harvester may climb the stem with the aid of a device made with palm leaf (called a *peconha*) and cut through the base of the crown shaft with a machete in order to bring down the entire crown. A man (only men harvest the palm heart commercially) can harvest as many as 300 palm hearts per day and is paid US$0.03–$0.10 for each palm stem harvested and delivered to the factory. Prices vary considerably, however. It takes two to four palm stems to make one pound of palm heart, or a total of US$0.06–$0.20 paid to the collector per pound. This contrasts with a final consumer price of US$5–$6 per pound in New York.

Uncontrolled commercial extraction of palm heart has been practiced in the Amazon River estuary since large-scale exploitation began in the late 1960s and early 1970s. It continues in many areas, with negative consequences for the regeneration of the natural açaí stands and also for the large segment of the rural population that depends on the harvest of açaí fruits for subsistence and sale (Bovi and de Castro 1993). Açaí cannot be managed to maximize the production of either fruit or palm heart without hurting production of the other. In the case of palm heart, collectors frequently cut every stem of sufficient size to yield a salable heart. Currently a plantation or dense natural stand produces about 3 MT of rough palm hearts per ha, only 20 percent of which is of sufficient size for export quality (Bovi and de Castro 1993).

There are several indications, some beginning 10–15 years ago, that the palm heart extraction industry in the Amazon estuary is not sustainable. Pollak et al. (1993) state that "the average palm heart today is smaller than it was 20 years ago, meaning that smaller, younger stems are being cut today. There are also few factories operating today in areas where palm hearts were exploited intensively in the past, suggesting that palm heart supplies have declined in these areas. While there is no evidence that the açaí palm is endangered as a species, these, and other signs of pressure, point to potential difficulties in supplying the local industry with palm hearts in the future."

Management Challenges

To maintain a steady supply of palm heart from one area, selective cutting of only a certain number of stems is possible. Residents of the Amazon River estuary have long managed natural stands of palms in this way and often enrich forests near their home with the açaí palm. They developed a management system that allowed for the sustainable harvest of palm heart in this region. By practicing "optimal" management—a combination of selective harvesting of açaí suckers from each clump, with selective thinning of forest competitors—the small holder can not only make cash from the direct sale of the palm heart but can also enhance fruit production for home consumption or sale from the remaining stems (Bovi and de Castro 1993).

This alternative management practice, that permits both fruit harvest and palm heart extraction, is increasingly common within relatively well populated peasant communities in the Amazon River estuary. Although enriching stands of açaí may, as explained later, reduce estuarine biodiversity in the long run, this practice can sustainably supply income to its practitioners while conserving vital ecosystem functions in the area.

This mixed management system, where açaí is managed for both fruit and palm heart, is most feasible when it is undertaken near a dwelling where the residents are not attempting to maximize either fruit or palm heart, but rather to maintain a steady supply of both. It could also be undertaken successfully near thriving markets for both products. Furthermore, it is more easily undertaken when labor in the area is relatively abundant. At this time, the natural stands of açaí that are being harvested for palm heart are, for the most part, not near populations where labor is relatively abundant nor are they generally utilized for more than the extraction of palm heart.

For such natural stands, there is a need to develop systems for large-scale sustainable exploitation. In addition, there is a large potential for the recuperation of degraded stands or the planting and use of palm stands in already degraded areas. In the Amazon estuary, agroforestry systems are feasible (indeed, they are practiced in many parts of the region). It is also possible that, in time, açaí plantations will become common and thereby make various açaí products more affordable while supporting increased demand.

To date, however, almost all açaí palm heart production comes from natural forests. Good data are lacking regarding whether the higher yields from a plantation or a managed forest can compensate for the extra costs (e.g., labor, management, fertilizing, and so on). Considering all the costs involved in the processing of palm heart (raw material, cans, processing, overhead and marketing, amortization and taxes), the raw material represents only 5 percent of the final export value of the product. Nonetheless, if current exploitation practices of natural stands continue without even minimal attempts to manage the stands for the mid- to long-term, evidence suggests that they will be decimated (Bovi and de Castro 1993). In the Amazon, agroforestry systems that include açaí, combined with the sustainable management of existing stands of açaí, are a major need.

Açaí and other "new" palms, such as pupunha, or peach palm (*Bactris gasipaes*) and açaí hybrids, are being cultivated in disturbed areas (e.g., in parts of the Brazilian states of Acre and Rondonia). The results of research in plant breeding and cultural management will soon be available for each palm species and for distinct edaphic and climatic conditions. With the new technologies being developed, palm heart plantations could be more productive and lucrative and the price of the product reduced (Bovi and de Castro 1993).

The Palm Heart Industry in the Amazon Estuary

The palm heart industry in the Amazon estuary is divided into three sectors: extraction, processing, and distribution. Extraction is conducted by estuary inhabitants who harvest wild palm hearts and sell them to processing plants. Processing is undertaken in small factories along riverbanks located throughout the estuary. Distribution or marketing of the finished product is undertaken by firms located in Belém that have connections to firms in the south of Brazil, primarily in São Paulo.

The costs of managing açaí in natural forest stands have an impact on each of the three activities identified above. Any sustainable (or at least limited impact) palm heart management system must not only identify the costs for each activity, but also address them within the structure of the business itself. Each of the three activities is discussed below as well as the costs associated with making palm heart harvesting and processing sustainable. The following discussion is drawn largely from Pollak et al. (1993).

Extractors

Although a few extractors are urban-based wage laborers, most are primarily residents of the Amazon estuary. Little is known about the land tenure of palm heart extractors. Extractors have generally harvested palm heart from their own areas, but increasingly they extract it from more distant sites. In general, however, they harvest from forests within 5–10 km of their homes. Almost all palm heart is harvested by men. Some work only a few days each year. Others work 50–100 days per year. Some are full-time employees of processing factories and harvest palm heart on a continuous basis.

Extractors are usually paid per açaí stem harvested rather than by the hour or the day. On Marajó Island, extractors earn as much as two and a half times the daily minimum wage equivalent (the minimum wage, normally calculated on a monthly basis, is considered a very good salary in the area). Since extractors are paid by the stems harvested and transported to processing factories, their earnings are affected not only by the number of stems that they can find and cut each day, but also by how long it takes to carry the hearts to the canoe, load them, transport them to the processing factory, and unload them.

Most extractors are subsistence-based forest dwellers. For cash income they sell rubber, Brazil nuts, açaí fruit, and in some cases manioc flour. Because of the long-term decline in rubber prices and the subsequent removal of rubber subsidies in Brazil during the past five years, many forest residents have had to look for other sources of cash income to make up for the lost income from rubber (P. Ramos, pers. comm. 1991). It is likely that increased palm heart demand in Brazil and for export has also stimulated palm heart harvesting. In the Amazon estuary, many extractors have stepped up their harvest of palm heart.

Pollak et al. (1993) describe the work as follows. "A typical palm heart extractor begins his day by sharpening his machete or ax (depending on the size trees to be harvested) and paddling to the area where he will harvest palm hearts. He moves through the forest from one açaí clump to another, cutting one or more açaí stems per clump, and lopping off the 50 to 100 cm with the ensheathed palm heart. He then removes the first two or three leaf-sheaths, leaving the inner sheaths to protect the palm heart during transport to the factory. The cutting done, the extractor returns by the same path to pick up the palm hearts and stacks them in his canoe. He then delivers the palm hearts to a factory or takes them to his house, where they are picked up for delivery to the factory."

Pollak et al. (1993) indicate that slightly more than half of an extractor's time is spent locating and cutting the palm and stripping the outer leaf-sheaths from the palm heart. About a quarter of the time is spent gathering, carrying, and loading and unloading the harvested palm hearts. About 10 percent of the time is spent in canoe travel either to and from the area to be harvested or to and from the processing factory. These figures, of course, vary tremendously depending on the density of açaí, and the proximity of the stand to the extractor's house, a navigable river, and the processing factory.

Productivity varies greatly depending on the size and abundance of açaí in the forest. These factors, in turn, vary tremendously depending on the interval between periods of extraction at the site. In general, Pollak et al. (1993) found that an extractor working a seven-hour work day can cut, on average, 200 small palm hearts, 175 medium hearts, or 150 large hearts per day. When the palm hearts are sold directly to a factory (which usually happens if the extractor lives within two hours distance), he receives approximately US$0.039 per small palm heart, US$0.052 per medium, and US$0.065 per large. He earns US$8–$10 per day. When the extractor sells his harvest to an intermediary, who in turn transports it to the factory, he gets about half that amount, but the time available to harvest the palm heart increases, so income may not be as adversely affected as it appears. Intermediaries are usually employed by the processing plant, which advances goods or money to pay the harvesters. The use of intermediaries may have as much to do with quality control as reducing the price a factory pays for the raw materials.

Processing Factories

Located throughout the estuary, processing factories are situated at the confluence of rivers or streams to facilitate the transport of the cut stems by extractors. The factories employ 15–30 workers who peel, cut, and can the palm hearts. When the palm hearts arrive at the factory, the remaining leaf-sheaths are removed and the hearts are cut into pieces for either export or domestic markets, according to their size and texture (the most tender are exported). The pieces are then placed either in cans for the export market or jars for the domestic market, covered with a water-based preservative solution of water, salt, and citric acid, cooked in a pressure cooker, and sealed (Brabo 1979).

Pollak et al. (1993) found that the average processing factory produced 29 MT/month (including the weight of the brine). It takes about 60,000 to 122,000 palm hearts, depending on their size, for that level of production (or about 400 to 610 man-days to harvest the palm hearts in the forest). They also found that a palm heart, independent of its size, yields an average of 1.8 pieces of export quality (36% of total palm heart's weight) and 0.9 pieces of domestic quality (23% of total weight). About half the palm heart bought by the processing factories is thrown away as waste, causing a large environmental problem in the vicinity of the factory.

There are 500 g (drained weight) of palm heart in a 1-kg can of export quality and 300 g in a 1-kg domestic jar. Approximately 48 percent (14.4 MT) of a typical factory's total monthly production (30 MT) goes to export-quality cans and 52 percent for domestic-quality jars (Pollak et al. 1993).

In July 1992, processing factories were paid US$0.44 for a 1-kg can of export-quality palm heart and US$0.28 for a 1-kg jar of domestic-quality palm heart, making gross monthly income for a factory producing 30 MT/month about US$10,700. Production costs at that time were about US$7,400/month, with 58 percent of the costs going toward the purchase of palm heart from extractors. This left the average monthly net income at US$3,315. Working exclusively with large, medium, or small palm hearts, average profits would be US$3,745, $3,343, or $2,858/month, respectively, with profit margins at 35 percent, 31 percent, or 27 percent, depending on the size of stems processed. Profits for the processor are lower for the smaller palm hearts due to the increased cost of processing by hand and the larger number of stems required to equal the same weight as larger palm hearts.

Most factories can be easily moved as native stands of açaí are depleted in an area. The processing factories employ workers, mostly women, from the surrounding area. While workers are supposed to be paid the legal minimum wage (plus benefits), very few processing facilities do so.

Most palm heart processing factories are financially indebted to specific distribution firms. The owners of the processing factories are responsible for maintaining the physical plant of the factory, procuring the raw material (açaí stems) from the region, and organizing the labor needed to process the palm heart. The distributor generally helps set up processing factories by providing steam furnaces and canning machines, and by supplying boxes, cans (for exports), jars (for domestic sales), salt, citric acid, and other inputs. This is in exchange for exclusive purchasing rights to a factory's production. The distributor also arranges for the collection of the canned product from the processing plant and its transportation to Belém.

Distributors

After processing, cans and jars of palm hearts are packed into boxes and sent by boat to a distribution firm in Belém. Most distributors handle 50–450 MT of processed palm heart each month, or about 150 MT on average. Large distributors handle the product of up to 15 different processing plants. Most of the distributors are, in turn, affiliated with large food businesses in the south of the country, principally in São Paulo.

Distributors purchase processed palm heart for US$0.44/kg can and US$0.28/kg jar. The distributor, in turn, sells the former for US$1.94/can and the latter for US$1.11/jar, for mean monthly gross returns of about US$230,000. The average firm's monthly profit is nearly US$60,000, or 28 percent of the gross returns. While returns are good, enterprises require some US$160,000 in working capital each month (Oliviera and Nascimento 1991).

Management of Açaí for Palm Heart in the Amazon Estuary

Problems with Existing Palm Heart Extraction Systems

Recent studies indicate that palm heart yields diminish significantly if extraction from the same area is too frequent. Areas harvested every one to two years yield fewer palm hearts over the course of four extraction cycles than stands harvested every four to five years. Frequent harvests appear to leave açaí stands with few mature stems and a relatively high proportion of dead clumps (Pollak et al. 1993). Without mature stems, little or no fruiting occurs to allow for natural reseeding of the overharvested areas. As shortened harvesting cycles occur over ever wider areas, wild palm hearts, particularly large ones, will be harder to find.

There are several indications that palm heart is being overexploited. The palm hearts processed today are generally smaller than those processed 20 years ago when palm heart processing was first relocated from the south of Brazil to the Amazon estuary (Bovi and de Castro 1993; Pollak et al. 1993). Brabo (1979) reported that in the 1970s, two palm hearts were sufficient to fill a 1-kg can. Today such large palm hearts are extremely rare. With the largest palm hearts available at this time, it now takes about four palm heart stems per can. In some cases, it can take as many as ten stems.

Another indication of declining production is found in the factories themselves. When factories were first opened in the region in the 1970s, they operated six or seven days per week, but today many operate only three or four days per week. According to Pollak et al. (1993), factory owners and managers attributed this situation to the shortage of palm heart. Inhabitants of the area also insist that overharvest of wild açaí for palm heart has depleted the population in the estuary.

Pressure on natural açaí stands appears to be greatest in counties (*municipios*) that are closest to Belém, where intensification of the palm heart extraction business occurred more quickly. Pollak et al. (1993) reported that on Marajó, the counties closest to Belém produced only 1 percent of the state's palm heart in 1989, compared to 64 percent in 1975 (Instituto Brasileia de Geografia y Estatística [IBGE]). During the same period, the more distant counties on Marajó increased their share of the state's palm heart production from 12 percent to 96 percent (IBGE). By 1992, the more distant counties contained 91 percent (88) of all registered factories on Marajó (Secretaria da Fazenda, unpublished document cited in Pollak et al. [1993]).

It is possible that the harvest of palm heart near large urban centers is declining due to the increase in lucrative urban markets for the fruit (A. B. Anderson, pers. comm. 1994). However, viable commercial fruit production must be no further than an overnight boat trip from an urban market, otherwise the fruit spoils. This hypothesis suggests that commercial palm heart production is not compatible with optimal fruit sales. In any case, outside of the urban market radius for açaí fruit, palm heart production is the dominant commercial use of açaí. It appears that factories are being relocated outside the fruit production radius, implying that palm heart production is not sustainable independent of urban-based fruit demand.

Disincentives to Sustainable Palm Heart Management

Periods of several years are necessary between harvests (Pollak et al. 1993) to maintain palm heart productivity, recovery, and growth. Frequent harvests result in smaller-sized hearts, sharp declines in palm heart production from wild stands, and reduced extractor incomes. To date, there has been little incentive at any level of the industry to avoid working with small palm hearts.

The disincentives to sustainable management of wild stands of palm heart start with the extractor. Once an extractor has waded through mud and hacked his way through undergrowth to reach the base of an açaí clump, he will be inclined to cut all stems, even the small ones, before trudging on to the next one. As Pollak et al. (1993) indicate, there are even advantages to cutting small stems: (1) they are easier to cut than thicker, bigger stems; (2) they have fewer leaf-sheaths surrounding the heart, so they are easier to prepare; (3) they require less work to harvest as well as to carry, load, and unload (28% of the work involved in palm heart extraction) when most of the leaf-sheaths have not been removed; and (4) more of them can be transported in the harvester's boat at once.

These factors translate into higher earnings from smaller hearts, too. In Marajó, an extractor earns US$12.21 for each 100 kg of small palm hearts that he carries out of the forest, compared to US$9.56 for each 100 kg of large palm hearts. Price incentives for large palm hearts still are not sufficient to overcome the return from smaller hearts. Thus harvesting smaller hearts is "rational" because it is lucrative. However, is it rational from the point of view of sustainability?

There are considerable financial incentives in the region to harvest palm hearts, whatever the size. Outside of palm heart harvesting, most individuals in the area do not even earn, on average, one minimum wage (US$67/month). Paid on a piece wage system, extractors' daily earnings from palm heart are two to three times the Brazilian minimum wage. In fact, from the point of view of income, and with the exception of areas close enough to Belém to sell açaí fruit, harvesting palm heart is better than any other legitimate activity in the region.

Uncertain land tenure is another reason that extractors may tend to overharvest açaí for palm hearts. Land in the region is frequently divided into large, unmarked plots. Many people live along rivers where rights are relatively clearly defined, but the ownership of the vast interiors behind the stream or riverbanks is less well defined. In many areas, families share access to the forest. Tenure issues are further complicated because outsiders often negotiate access to resources such as açaí with absentee landlords rather than with local residents. In general, the destructive harvest of palm heart is more highly concentrated on the areas of larger landowners or areas that are more sparsely populated.

Disincentives for sustainably managed natural stands of açaí also exist at the factory or processing level. Despite the fact that profits for processing plants decline by as much as 24 percent when working exclusively with small, illegal, and undersized palm hearts, factories continue to purchase small palm hearts. There are several reasons for this:

—Extractors bring all the palm hearts they harvest mixed together and will not separate them unless they can also sell the small ones. Instead, they would take all their harvested palm heart stems to another factory.

—Even small palm hearts are relatively lucrative for the factories. Profits are more than five times higher than profits from the other two predominant small industries in the Marajó regions—sawmills and brick factories (IMAZON study cited in Pollak et al. 1993).

—Many consumers prefer smaller palm hearts because they believe that they are more tender. This is particularly true in Brazil, but it is also true in many export markets as well.

As Pollak et al. (1993) state, "As long as consumers associate thick palm hearts with a fibrous product, instead of with a rationally harvested product, then small palm hearts will have more consumer appeal and firm owners will not discourage factories from processing them."

Distribution firms have also lacked effective incentives to insist on sustainably harvested large palm hearts from their suppliers. Depletion of wild palm heart has repeatedly forced distribution firms to shift their processing affiliates or suppliers to new areas over the past several decades. Yet the distribution firms continue to realize impressive profits (greater than US$500,000 per year for the larger distribution firms). A potential additional cost for using undersized palm stems is the slight risk of being fined by the Brazilian Environmental Agency (IBAMA) for selling undersized palm hearts. However, this is not a significant risk at this time and is not taken seriously.

For each of these reasons, managing palm heart harvests by increasing the time between harvests and only harvesting large stems does not necessarily result in higher incomes from palm heart for the individual, whether extractor, processor, or distributor. Furthermore, without secure land tenure, extractors are inclined to harvest palm heart as soon as possible, instead of waiting for it to reach a legal size. In this way, the extractor guarantees his own income, at least in the short term, while sharing the cost of palm stand degradation with other extractors or future generations who will depend on the same resource base (Pendleton 1992).

Attempts to Develop a Sustainable Palm Heart Extraction Industry

The Brazilian Environmental Agency has attempted to develop a sustainable palm heart industry in the Amazon by establishing minimum-size regulations for palm heart extraction and processing. By law, a palm heart must weigh at least 250 g after all leaf-sheaths are removed and must have a diameter of at least 2 cm. In fact, a palm heart that weighs at least 250 g has a diameter of about 2.5 cm. Diameter can be measured at any time, including after the product is cut and canned. But weighing can only take place when the palm heart is still in one piece. That is why diameter guidelines are so important for monitoring harvest when an inspector or regulator cannot be on site at all times. If 250-g stems are necessary to ensure long-term viable harvesting systems, as the research cited above and summarized in Pollak et al. (1993) suggests, then the diameter requirements should be changed to 2.5 cm.

In a recent random sampling of export-quality palm heart, the average diameter of all pieces measured was only 2.1 cm, meaning that they were derived from whole palm hearts that weighed, on average, only 180 g (Pollak et al. 1993). This means that the average palm heart being exported in the sampled cans was illegal and should not have been harvested, much less purchased, processed, and sold.

Recent samplings of palm heart from the Belém area sold at Safeway in the Washington, D.C., area indicate that 70 percent of the palm hearts in the cans had diameters less than 2 cm (Jason Clay, personal observation of 100 cans of palm heart from five different distributors). In fact, more than half of the 400-g cans contained 20 stems, an indication of severe overharvest of natural stands. By any standard, these stems would be classified as tiny and illegal for harvest.

Pollak et al. (1993) have developed equations that correlate the height and diameter at breast height (dbh) of an açaí stem with the weight and diameter of its palm heart. Their equations can be used to predict the size of a palm heart that would be obtained from a living stem:

$$\text{Palm heart weight (g)} = [27.552 \times \text{palm heart diam. (cm)}] + [35.517 \times \text{palm heart diam.}] - 16.603$$

$$\text{Palm heart diam. (cm)} = [0.295 \times \text{stem dbh (cm)}] - 0.513$$

$$\text{Stem dbh (cm)} = [3.390 \times \text{palm heart diam. (cm)}] + 1.739$$

$$\text{Stem height (m)} = [1.78 \times \text{stem dbh (cm)}] - 5.849$$

$$\text{Stem height (m)} = [6.599 \times \text{palm heart diam. (cm)}] - 3.905$$

Alternatively, knowing the size of a palm heart section in a can, the equations can be used to determine the size of the stem (diameter or weight of stem or height of tree) that was cut to produce that palm heart. For example, a 1.5-cm-diameter palm heart would come from an açaí tree about 7 cm in diameter and 6 m tall; whereas a 3.5-cm-diameter palm heart section would have come from an açaí tree that was 14 cm in dbh and 18 m tall (Table 8.3).

Using the guidelines from Table 8.4, it is possible to reconstruct recent harvests by surveying all açaí clumps in plots of known dimensions and measuring the diameters of all cut stems. Pollak et al. (1993) compared sites that had been subjected to regular harvests every four to five years with those harvested every one to two years over a 15-year period. The four- to five-year harvests yielded four to five times the produce of sites harvested every one to two years. This is more

Table 8.3. Açaí (*Euterpe oleracea* Mart.) Classification Categories: Stem Dimensions by Palm Heart Size and Weight

Dimension	Tiny	Small	Medium	Large	Extra large
Heart diameter (cm)	1.5	2.0	2.5	3.0	3.5
Heart weight (g)	99	165	244	338	445
Stem dbh (cm)	6.8	8.5	10.2	11.9	13.6
Stem height (m)	6.3	9.3	12.3	15.3	18.3
Pieces needed to fill 1-kg can	22	17	12	7	3

Source: Pollak et al. 1993.

Table 8.4. Açaí Palm Heart Extraction and Population Structure in the Estuary of the Amazon

	1- to 2-year harvest intervals	4- to 5-year harvest intervals	No harvest natural forest
Most recent harvest			
Number of stems harvested/clump	1.2	1.0	—
Diameter of harvested stems (cm)	6.2	10.6	—
Height of harvested stems (m)	5.2	13.0	—
Diameter of harvested palm hearts (cm)	1.3	2.6	—
Weight of harvested palm hearts (g)	76	262	—
Total harvest yield (g)	43,867	192,308	—
Postharvest açaí population structure			
Density of clumps/ha with stems greater than 2 m	481	734	620
Average dbh (cm) of stems greater than 2 m	4.9	6.3	8.5
Average height (m) of stems greater than 2 m	2.9	5.4	9.4
Percentage of stems with height:			
Less than 2 m	88	64	55
2 m to 4 m	8	14	13
Greater than 4 m	4	22	32
Percentage of dead clumps	25	11	12

Source: Pollak et al. 1993.

than 45 percent more palm heart per year. Furthermore, production levels from the shorter harvest cycles decline over time, even within the first four to five years.

Pollak et al. (1993) also found that there were significant differences in the açaí populations depending on whether they were harvested more or less frequently. Overall, the açaí population profile in the less frequently harvested areas was very similar to that found in natural stands, except that natural stands have an even higher level of dead clumps (Table 8.4).

Using the equations of Pollak et al. (1993), enforcement officials and consumers can determine if processed palm heart meets the legal size. For example, if a 1-kg can of palm heart contains 12 pieces, then, according to their equations, the average size is "medium": approximately six stems were used to fill the can, and the hearts were on average 2.5 cm in diameter. If the enforcement official found 15 or more pieces in the can, then at least eight açaí stems were used to fill the can, and it would be reasonable to assume, following these guidelines, that the harvesting that took place was not sustainable.

IBAMA's second strategy to protect natural stands of açaí is to require factories to establish management projects in which they demonstrate either that one stem is managed for each stem cut or one açaí seedling is planted for each stem harvested (Regulation no. 439, 1989). By 1990, there were 62 palm heart management projects (with an average size of 870 ha) registered with IBAMA. Some 80 percent were sponsored by Belém-based distribution firms (IBAMA, internal document cited by Pollak et al. 1993). However, the projects may often exist only on paper. In a 1992/93 survey, none of the factories visited purchased palm hearts that were obtained through managed production systems (Pollak et al. 1993).

Regardless of laws and regulations, palm heart extractors, processors, and distributors continue to depend on native açaí stands for raw materials. Factory owners or processors are reportedly wary of açaí management because of the relatively high initial investment. Pollak et al. (1993) report that net expenditures over the first six years to implement a low-intensity management project (including some clearing and thinning) are 15 percent of the factory's average profits during the same period. In addition, there is a six-year delay before the project reaches full production capacity.

Distribution firms do not invest in sustainable management practices either, even though such practices would represent only 3 percent of a distributor's profits during the first six years. More importantly, while firms provide a considerable amount of the capital investment for their supplying factories, they do not pay for the raw material. Most believe that the wild stands will not run out, and they see no reason to pay more in the short term for management to obtain lower-priced raw material in the long term if their competition is not required, by law, to do so as well.

In short, as of 1993, while many factories and distribution firms claimed to be conducting palm heart management projects, none thought that management would replace wild palm heart extraction. Nine out of ten distribution firms even admitted to investing profits from palm heart extraction in other activities such as cattle ranching, timber processing, or river transport (Pollak et al. 1993).

As the same study pointed out, "laws that mandate size regulations for palm hearts and require açaí palm management projects are examples of how, through a top-down approach, the government can impose standards on the palm heart industry geared toward making the industry sustainable. In the forest, however, it is the extractor who, upon deciding which stems to cut and which to leave for future harvests, exercises the ultimate control over whether or not the activity is sustainable. Conditions must be made favorable for the extractor to carry out that task in a way which, ultimately, protects the natural resource against over-extraction" (Pollak et al. 1993). The study goes on to recommend that "one way for extractors to receive a larger share of the benefits from working with palm hearts would be for them to operate their own management projects. This would make it easier to produce large palm hearts in a shorter amount of time, thereby increasing the amount of income an extractor could earn in a day. With clear title to land, access to investment capital, and an effective means of community organization, inhabitants would be well-positioned to responsibly manage the açaí palm. For example, twenty families, with 50 ha each (i.e., 1,000 ha total), could produce enough palm heart in managed forest tracts to supply a typical factory. If the family also ran the factory, they could receive an even greater portion of the profits" (Pollak et al. 1993).

Managing Açaí for Long-Term Palm Heart Extraction

From an environmental point of view, it is relatively easy to manage açaí for palm heart in the Amazon estuary region. The primary requirements for good growth are ample light and growing space, both of which are related factors. Measures to improve growing conditions include clearing undergrowth and ringing canopy trees with no market or subsistence value. Both activities adversely affect biodiversity. While this is a low-intensity management system, it still implies far more management and investment of time on the part of the extractor (10 days per ha in year one) than is common in the region today.

A high-intensity management system, as outlined by Pollak et al. (1993) and practiced by a large landowner or corporation, entails the enrichment of the number (and perhaps quality) of açaí stands in natural forests by planting seeds. Costs of such a system are quite high because the system requires outsiders to design and oversee the implementation plan with hired labor and high overheads.

Açaí seeds can be scattered on the forest floor to facilitate regeneration. Growth is improved if plots are thinned periodically of weeds, vines, and other unwanted species (reducing biodiversity considerably), and açaí clumps are groomed to contain approximately three large, three medium, and three small stems. According to Pollak et al. (1993), in an intensive management system, three large stems would be harvested for palm heart from each açaí clump every three years. It would take about six years of management to get to a three-year harvest cycle, with considerable labor invested in the initial year to clear undergrowth and plant and manage the açaí stands.

Pollak et al. (1993) compared the economic potential for both low- and high-intensity palm heart management, assuming an average clump density of 625/ha. Low-intensity management provides an average annual yield of 625 palm hearts/ha/yr (i.e., three stems/clump harvested every three years). High-intensity açaí management, as practiced by a Belém-based forest engineering firm that has designed most of the palm heart management projects in the estuary, could, theoretically at least, provide an annual production of 1,000 palm hearts/ha. In both cases, yields are estimated to average only 32 percent of full capacity for the first six years until a project reaches its full production potential (Lopes et al. 1982).

The low-intensity management system requires an area of 1,144 ha to supply a palm heart factory that produces 30 MT/month. A high-intensity açaí management program would require only 715 ha to supply such a factory. The cost of having an outside firm manage a high-intensity palm heart production system is much higher than having a system where individual owners or co-ops manage their own system. The management system run by an outside firm barely breaks even over a 20-year period. Most of the "profits" go to the managing firm rather than to the owners of the açaí stands or to the processing factory. While the high-intensity management scheme does offer a fair amount of employment, it is menial work with low pay and no chance of advancement. Furthermore, extractors are not directly involved in management or management decisions, even though their day-to-day labor either reinforces or undermines the viability of the system.

Pollak et al. (1993) have shown that açaí management for palm heart can result in significant long-term savings when compared to the cost of purchasing wild palm heart. Over a 20-year period, the average cost per palm heart for a factory using low-intensity management is US$0.001, compared to US$0.011 for high-intensity management or US$0.037 without any management. Thus it could be from 3 to 37 times as expensive to purchase wild harvested palm heart than from either low- or high-intensity management systems. One needs to be careful here, however. In fact, the study measures the cost of buying palm heart harvested by someone else on someone else's land. That is compared to the cost of hiring labor by the day to manage a system and harvest palm from one's own land. For owners/harvesters there are no opportunity costs calculated into the first model (e.g., how else could the labor or land or even the açaí be used). Harvesters harvesting from their own land or an extractive reserve need to be making a return not just on their labor but on the resources utilized as well. In short, resources have an inherent value which should be based on replacement costs or the increased costs of harvesting sustainably.

In the final analysis, it is obvious why, to date, management plans that have been designed by factories or distribution firms are unacceptable to either owners of açaí stands or extractors. Higher costs associated with managing palm heart must be borne by someone. Without an increase in the price paid to extractors, and by implication to processors as well, management costs are borne by the landowner or the extractor (through lower wages or piece rate paid per stem).

Considerable research is needed to understand the overall impact of the palm heart industry on the entire ecosystem of the Amazon estuary. Anderson (1988), for example, found that the number of tree species on managed açaí areas was reduced by nearly half from 52 to 28. Research is also needed on the types of incentive systems that would be effective with extractors, processors, and distributors.

Resource Management for a Floating Processing Plant

One of the easiest ways to prevent the overharvest of stems from a particular area is to move the factory—an option that is not readily available for a fixed factory. If rotating the factory's location is done each month, then the impact of the harvest could be negligible, particularly if the factory purchases only large stems (indicating sustainable harvest), whether they come from managed areas or from "virgin" stands. The impact of such harvests, however, will need to be monitored carefully for the entire duration of palm heart harvest to determine whether this hypothesis holds or not.

The main management issue from the point of view of the floating processing plant is where to tie up the factory and whether to return to the same areas year after year in order to let harvesters from those areas have a small, but guaranteed, source of income each year from the harvest and sale of palm heart. By putting the barge in ten locations per year, it might be possible to generate increased income for as many as 100 harvester families. There would also be similar increases in income to temporary workers in the processing plant and to those selling fuelwood used for processing. To distribute the income evenly in each area, each family could be given a quota for the number of stems it could sell, as well as a delivery schedule so that processing could go smoothly.

This system would not require the manipulation of the environment through radical thinning or clearing of competing tree species. Nor would it require either the full-time dedication of residents of any single area where harvesting and processing occurred or the importation of labor to the area for harvesting. The program would generate a smaller amount of income per person affected because harvesting would be undertaken over a much wider area used by many more families. However, it would spread income over a much larger population. Finally, the activity would tend not to diminish other economic or subsistence activities of individuals in the area, or change the composition of the ecosystem in such a way that would diminish an extractor's ability to abandon palm heart activities if for any reason that industry declined in importance.

In short, the mobile system is a more extensive use of the region's resources than the more intensive management associated with the fixed factory. Thus, while the income generated is less per ha, it is more per person-day invested. Furthermore, extensive management areas for palm heart can be more intensively managed as labor becomes less scarce or as other, more destructive uses of forests threaten local residents and forests alike.

Conclusions

Palm heart extraction, as currently practiced, has already destroyed the extensive stands of one palm species in southern Brazil and is rapidly depleting the native stands of another, related species, in the Amazon estuary. Several management/harvest plans that have been proposed appear to have the potential of limiting the environmental impact of palm harvests. Yet, while each company that processes palm heart is required to file a management plan with the government, none of those plans has been implemented.

The best strategy to reduce the environmental impact of palm heart harvesting is a carrot-and-stick approach. IBAMA, the Brazilian environmental regulatory agency, is charged with reviewing all palm heart management plans and certifying that all processed palm hearts are of legal size and weight and come from areas managed under approved plans. The stick aspect of the approach is to take IBAMA and the private palm heart processing and distribution companies to court for violating Brazilian laws which are designed to protect palm stands from degradation through overharvest.

With gross sales of US$300 million per year and profits at just under US$100 million per year, a legal suit against palm heart companies for present and past environmental damage could generate considerable fines. Under Brazilian law, individuals or organizations who sue for environmental damages to the country as a whole are not allowed to retain any money from settlements of the lawsuit. After legal fees have been paid, the remainder is put into an environmental fund. (A similar lawsuit recently netted US$10 million for the illegal cutting of mahogany on Indian lands in another part of Brazil.)

The carrot aspect of the approach is to provide sufficient economic/market incentives for local extractors, resource owners, processors, and distributors to stimulate the sustainable harvest of palm heart. Three alternatives are proposed here, which, with adequate monitoring, could be models of sustainable harvest of palm heart in the Amazon estuary. Each of the described management programs provides a viable alternative to the shutting down of the entire industry. At least some of the money generated by a settlement from palm heart processing and distribution companies could be invested in the establishment of viable, long-term management programs. However, even if that is not the case, the proposed management programs appear to be, on paper at least, economically viable in their own right.

The sustainable use of palm heart in the Amazon estuary highlights a number of issues common to the sustainable utilization of wild species in many parts of the world. First, resource ownership or tenure is central to any viable long-term management system. Second, the values generated by the harvesting, processing, and sale of wild species are traditionally captured far from the ecosystem that produced it. Green marketing is one way to return at least a portion of the value added to wild harvested products to the producers themselves. Green marketing is not usually possible except for exported products. Most third world consumers will not pay more for sustainably harvested green products. Third, management directed toward increased production of a single species of high economic value may involve tradeoffs for biodiversity. Fourth, sustainable management of wild (or even domestic) species has a cost. Currently these costs are borne (if they are covered at all) by producers. We must identify mechanisms that allow the costs of sustainable production to be passed on to consumers. By including the costs of palm heart management in the price of processed palm heart, we are attempting to include environmental "externalities" in the price consumers pay for goods. This approach needs to be attempted for the production of other wild and domestic species as well.

References

Altman, R.F.A. 1956. O caroço de açaí (*Euterpe oleracea* Mart.). *Boletim Técnico do Instituto Agronômico do Norte (IAN)* 31:109–111.

Anderson, A. B. 1988. Use and management of native forests dominated by açaí palm (*Euterpe oleracea* Mart.) in the Amazon estuary. *Advances in Economic Botany* 6:144–154.

Anderson, A. B., and E. M. Ioris. 1992. The logic of extraction: Resource management and income generation by extractive producers in the Amazon estuary. In *Conservation of neotropical forests: Working from traditional resource use,* ed. K. H. Redford and C. Padoch, 175–199. New York: Columbia University Press.

Bovi, M.L.A., and A. de Castro. 1993. Assau. In *Selected species and strategies to enhance income generation from Amazonian forests,* ed. J. Clay and C. Clement, 58–67. Rome: U.N. Food and Agricultural Organization.

Bovi, M.L.A., S. H. Spiering, and T. M. Melo. 1989. Temperaturas e substratos para germinação de sementes de palmiteiro e açaízeiro. *Anais do Segundo Congresso sobre Tecnologia de Sementes Florestais,* 43. Atibaia, São Paulo, Brazil.

Brabo, M.J.C. 1979. Palmiteiros de Muana-estudo sobre o processo de produção no benificiamento do açaízeiro. *Boletim do Museu Paraense Emilio Goeldi* 73:1–29.

Calzavara, B.B.G. 1972. As possibilidades do açaízeiro no estuario amazônico. Faculdade de Ciencias Agrarias do Pará, Belém, Brazil. *Boletim* 5:165–206.

Campos, F.A.M. 1951. Valor nutritivo de frutos brasileiros. *Instituto de Nutrição, Trabalhos e Pesquisas* 6:72–75.

Clay, J. W. 1992. Report on funding and investment opportunities for income-generating activities that could complement strategies to halt environmental degradation in the Greater Amazon Basin. Washington, D.C.: Biodiversity Support Program.

Clay, J. W. 1996. *Generating income and conserving resources: 20 lessons from the field.* Washington, D.C.: World Wildlife Fund.

Clay, J. W., and C. R. Clement, eds. 1993. *Selected species and strategies to enhance income generation from Amazonian forests.* Rome: U.N. Food and Agricultural Organization.

Diário do Pará. 1993. May 13:A-11.

ECOTEC (Sociedad para Desenvolvimento Techno-Ecolôgico). 1993. Produção e comercialização Pará as reservas extrativistas. Unpublished manuscript. Recife, Brazil.

ECOTEC. 1994. *Projeto e estudos preliminares de viabilidade econômico-financeira do sistema de processamento do palmito (Reserva Extrativista Rio Cajarí, Amapá). Unidade produtora de palmito em conserva (comparativo movel x fixa)*. Recife, Brazil: ECOTEC.

Ferreira, V.L.P., and J. E. Paschoalino. 1987. Pesquisa sobre palmito no Instituto de Tecnologia de Alimentos. In *Proceedings from the First National Conference of Researchers on Palm Hearts, May 26–28,* ed. Empresa Brasileira de Pesquisa Agropecuária (EMBRAPA), 45–62. Curitiba, Brazil: EMBRAPA.

Ferreira, V.L.P., and Y. Yokomizo. 1978. O aproveitamento da porção macia do estipe da palmeira juçara na alimentação humana. *Coletânea ITAL* 9:27–41. Campinas, São Paulo, Brazil.

Ferreira, V.L.P., M.L.A. Bovi, I. S. Draetta, J. E. Paschoalino, and I. Shirose. 1982. Estudo do palmito do híbrido das palmeiras *E. edulis* Mart., *E. oleracea* Mart. (açaí). I. Avaliaçães físicas, organolépticas e bioquímicas. *Coletânea ITAL* 12:27–42. Campinas, São Paulo, Brazil.

Iaderoza, M., V.L.S. Baldini, I. S. Draetta, and M.L.A. Bovi. 1992. Anthocyanins from fruits of açaí (*Euterpe oleracea* Mart.) and juçara (*Euterpe edulis* Mart.). *Tropical Science* 32:41–46.

Jardim, M.A.G. 1991. Aspectos da biologia reprodutiva de uma população natural de açaízeiro (*Euterpe oleracea* Mart.) no estuário Amazônico. Masters thesis, Escola Superior de Agricultura Luiz de Queiroz, Piracicaba, São Paulo, Brazil.

Lepree, J. 1994. Natural food colors flourish in wake of buyers demand. *Chemical Marketing Reporter* September 12:1994.

Lima, R. R. 1956. A agricultura nas várzeas do estuário do Amazonas. *Boletim Técnico do Instituto Agronômico do Norte* 33:1–164.

Lopes, A.V.F., J.M.S. Souza, and B.B.G. Calzavara. 1982. *Aspectos economicos do Açaízeiro*. Superintendência de Desenvolvimento da Amazônia, Belém, Brazil.

Melo, C.F.M., W. C. Barbosa, and S. M. Alves. 1988. Obtenção de açaí desidratado. *Boletim Pesquisa* 92:1–13. Centro de Pesquisas Agropecuárias do Trópico Umido, CPATU/EMBRAPA, Belém, Pará, Brazil.

Mota, S. 1946. Pesquisas sobre o valor alimentar do açaí. *Anais Associação Química Brasileira* 5(2):35–38.

Nascimento, M.J.M. 1991. *Produção e comercialização de palmito em conserva,* vol. 2. Belém, Brazil: Universidade Federal do Pará and World Wildlife Fund.

Nascimento, M.J.M. 1992a. *Mercado e comercialização de frutos de açaí,* vol. 3. Belém, Brazil: Universidade Federal do Pará and World Wildlife Fund.

Nascimento, M.J.M. 1992b. *Projeto de manejo florestal: Euterpe oleracea* Mart. Belem, Brazil: Universidade Federal do Pará and World Wildlife Fund.

Nascimento, M.J.M. 1992c. *Tecnologia para processamento do palmito.* Belem, Brazil: Universidade Federal do Pará and World Wildlife Fund.

Nascimento, M.J.M. 1993. *Palmito e açaí: organizacao empresarial e processo produtivo,* vol. 1. Belem, Brazil: Universidade Federal do Pará and World Wildlife Fund.

Oliveira, P.H.B. de, and M.J.M. Nascimento. 1991. Os trabalhadores rurais de Gurupa (PA) em busca de alternativas econômicas: Estratégias e o mercado do palmito em conserva Paraense. *Reforma Agraria* 21:91–120.

Pendleton, L. H. 1992. Trouble in paradise: Practical obstacles to non-timber forestry in Latin America. In *Sustainable harvest and marketing of rainforest products,* ed. M. Plotkin and L. Famolare, 252–262. Washington, D.C.: Conservation International and Island Press.

Peters, C. M., A. Gentry, and R. O. Mendelsohn. 1989a. Valuation of an Amazonian rainforest. *Nature* 339:656.

Peters, C. M., M. J. Balick, F. Kahn, and A. B. Anderson. 1989b. Oligarchic forests of economic plants in Amazonia: Utilization and conservation of an important tropical resource. *Conservation Biology* 3(4):341–349.

Pollak, H., M. Mattos, and C. Uhl. 1993. A profile of palm heart extraction in the Amazon estuary. Manuscript submitted to *Advances in Economic Botany.*

Renesto, O. V., and L. F. Vieira. 1977. *Análise econômica da produção e processamento do palmito em conserva nas regiões sudeste e sul do Brasíl.* Estudos Econômicos, Alimentos Processados 6, Instituto de Tecnologia de Alimentos, ITAL. Campinas, São Paulo, Brazil.

Strudwick, J. 1990. Commercial management for palm heart from *Euterpe oleracea* Mart. in the Amazon estuary and tropical forest conservation. *Advances in Economic Botany,* 8:241–248. New York: New York Botanical Garden.

Strudwick, J., and G. L. Sobel. 1988. Uses of *Euterpe oleracea* Mart. in the Amazon estuary, Brazil. *Advances in Economic Botany,* 6:225–253. New York: New York Botanical Garden.

Urpi, J. M., A. Bonilla, C. R. Clement, and D. V. Johnson. 1991. Mercado internacional de palmito y futuro de la exploitación salvaje vs. cultivado. *Boletim Informatívo* 3:6–27. Universidad de Costa Rica.

Zimmermann, C. E. 1991. A dispersão do palmiteiro por paseriformes. *Revista Ciencia Hoje* 12(72):18–19.

NINE

Linking Conservation and Local People through Sustainable Use of Natural Resources

Community-Based Management in the Peruvian Amazon

Richard E. Bodmer, James W. Penn, Pablo Puertas, Luis Moya I., and Tula G. Fang

Conservation programs in northeastern Peru are striving to conserve tropical forests through community-based approaches, implementing the concept that natural resource management can form a bridge between biodiversity conservation and the livelihoods of local people. This is critically important because forests outside of fully protected areas dominate the landscape in northeastern Peru, as in much of Amazonia, and these forests are used by local people for subsistence and market products. Such community-based approaches to conservation can work only if resources are not overexploited and if the economic, social, and political aspirations of the local people are included in the management programs (Ehrlich and Daily 1993; Holling 1993; Levin 1993). Therefore, natural resource management requires adequate information not only about species populations and ecosystems, but also about the people who most frequently use the resources (Browder 1992).

As a diverse array of natural resources is used by local people in Amazonia (wildlife, timber, nontimber plants, fish, among others), it is necessary to maintain intact natural habitat if these resources are to be used in the long term. This conserves not only the resources used by people but also the ecosystems in which these resources occur. It is also important that natural resource management in Amazonia include fully protected areas that act as controls for harvesting programs and reservoirs to replenish overexploited populations (McCullough 1987).

We use the case of the Reserva Comunal Tamshiyacu-Tahuayo (RCTT) in northeastern Peru to show how management programs are attempting to bridge local people and biodiversity conservation using a community-based approach. This community-based management of the RCTT is integrating different resource uses by establishing management programs on wildlife hunting, fishing, and forest plant extraction. These management programs are designed to (1) reduce nonsustainable uses of natural resources, (2) enhance sustainable uses, and (3) involve local people in data collection and analysis so that they may determine the status of resource populations and improve management. This community-based management should improve the economic and financial benefits of local people over the long term. Management programs of the RCTT do not pretend to attain theoretical sustainability; rather they are realistically trying to convert obviously nonsustainable uses of natural resources to more sustainable uses. In addition, management programs in the RCTT recognize that lack of natural resource management leads to losses in biodiversity and impoverishment of local people.

We analyze here information and events of the RCTT to see if community-based natural resource management can work as a method of conservation in Amazonia. First, we examine the events that led to the creation of the RCTT. Second, we look at the users of the RCTT, the different uses of wildlife, fish, nontimber plants, timber, and agricultural lands and the sustainability of these uses. Third, we analyze how resource use in the RCTT can become more sustainable. This will include analyses of how local people's economic and financial realities are affected by converting current resource uses to more sustainable uses. Finally, we examine whether the social and economic situation of the local people can incorporate more sustainable resource uses and fulfill the goals of biodiversity conservation of the RCTT.

The Reserva Comunal Tamshiyacu-Tahuayo (RCTT)

The Reserva Comunal Tamshiyacu-Tahuayo is located in the northeastern Peruvian Amazon, in the state of Loreto, extending over an

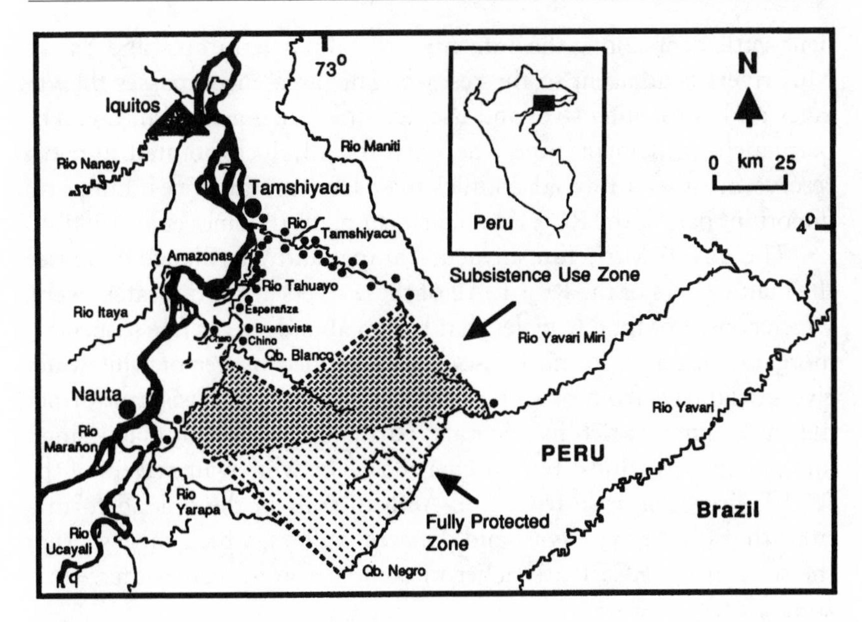

Figure 9.1. Location and zonation of the Reserva Comunal Tamshiyacu-Tahuayo and locations of surrounding communities.

area of 322,500 hectares (ha). The reserve is situated in the upland forests that divide the valley of the Amazon from the valley of the Yavari (Fig. 9.1). The city closest to the RCTT is Iquitos, located about 100 kilometers (km) northwest of the reserve with approximately 300,000 inhabitants. The reserve is bordered on the west by the upper Tahuayo and Quebrada (Qb.) Blanco rivers, the south by the upper Yarapa River, the east by the upper Yavari Miri River, and the north by the upper Tamshiyacu River.

The RCTT and adjacent lands are divided into three distinct land-use zones. These comprise (1) a fully protected core area of approximately 160,000 ha, (2) a buffer zone of subsistence use of approximately 160,000 ha, and (3) an area of permanent settlement which has no definite boundaries. The fully protected and subsistence areas fall within the official limits of the reserve and have no human settlements. The fully protected zone does not have extractive activities, whereas the subsistence zone is used by local residents of the permanent settlement zone for extraction of natural resources. Residents cannot set up permanent settlements or clear land for agricultural uses within the boundaries of the subsistence use or fully protected zones. The zone of permanent settlements along the Tamshiyacu, Tahuayo, Yarapa, and Yavari Miri rivers is adjacent to the reserve. This area encompasses the villages and is for intensive land-use activities such as agriculture. The permanent settlement zone was not officially incorporated into the reserve in order to avoid conflict over land-use practices, but is an important part of the RCTT management plans (Bodmer et al. 1990b).

The Yavari Miri, Tamshiyacu, Yarapa, and Qb. Blanco comprise the major rivers of the RCTT. All of these rivers are whitewater rivers, which consist of muddy water that has an abundance of fine suspended inorganic material. In most cases the suspended matter of whitewater rivers originates from erosion of Andean slopes and consists of former marine sediments rich in minerals, particularly calcium, magnesium, and phosphorus (Junk and Furch 1985). The whitewater rivers of the RCTT, however, arise from non-Andean soils in the upland formations that divide the Yavari and Amazon valleys, which suggests that the soils of the RCTT are richer in comparison to many other Amazonian sites.

The other major river in the RCTT is the Tahuayo, which runs parallel to the Amazon inside the *várzea* floodplain. Through most of the year this river has a blackish color, which results from organic breakdown of leaf litter that releases humic and fulvic acids (Junk and Furch 1985). The water chemistry of Tahuayo is not of a pure blackwater river but, like many várzea lakes and rivers, is a mixture of white and black waters (Coomes 1992b; Ayres 1993). The Tahuayo originates from a large lake situated on the edge of the várzea and *terra firme* forests and therefore does not have the high nutrient runoff from the upland forests.

There are many lakes along the Tahuayo, Yarapa, and Yavari Miri rivers within the RCTT. These Amazonian lakes differ from the continually renewed waters of rivers because they have stagnant water during periods of the year. Lakes are usually within floodplain systems and receive water during inundations, often through the groundwater. These lakes then partially release water when river levels fall, and become stagnant during low water periods.

Principal types of vegetation in the RCTT are whitewater floodplain vegetation of lakes, lagoons, islands, levees (*restingas*), backswamps, and channel bars (várzea) (Furch 1984; Pires and Prance 1985) and upland nonflooded terra firme vegetation. In addition, there are smaller specialized areas of plant communities within the RCTT such as nutrient-poor leached white sand *campinas,* and the relatively monospecific stands of *Mauritia flexuosa* (Palmae) known as *aguajals* (Kahn 1988).

The RCTT has a high diversity of faunal and floral groups (Castro 1991; Puertas and Bodmer 1993). This diversity is due in part to the RCTT having a combination of terra firme (with rich soil composition) and várzea habitats, and is part of the biogeographic pattern of high species diversity of western Amazonia (Gentry 1988). For example, at least 14 species of primates are found in the RCTT, the greatest diversity of primates reported for any protected area in Peru (Puertas and Bodmer 1993). The high diversity of anthropoid primates in the terra firme between the Yavari and Amazon probably arose from a combination of factors including Pleistocene refugia, river dynamics, and diversity of flora (Ayres and Clutton-Brock 1992; Puertas and Bodmer 1993).

People of the RCTT

The settlement zone of the RCTT is inhabited by nontribal people known in Loreto as *ribereños.* These rural folk commonly practice fishing, agricultural production, game hunting, small-scale lumber extraction, and collection of minor forest products (such as fruits, nuts, and fibers). Ribereños have diverse origins and include detribalized Indians, and varied mixtures of Indians, Europeans, and Africans (Lima 1991). The transformation of the Amazon from tribal groups to ribereños began with the earliest European immigrations and continued with detribalization imposed by missionaries, expansion of the slave trade, and influx of immigrants during the rubber epoch.

Ribereños, like Amazonian Indians, have a great knowledge of forest plants, agriculture techniques, and hunting and fishing methods. However, ribereños differ from most Indian groups because of their intricate involvement in the market economy on both regional and international levels (Padoch 1988). Indeed, products harvested by ribereños, such as spices, rubber, and furs, have traditionally been marketed in European countries and North America. Ribereños are renowned for their ability to switch product exploitation as markets change, which is one reason for their wide geographic mobility.

In Loreto, ribereños greatly exceed indigenous inhabitants (Amerindians). For example, data from the 1981 census reveal that 280,000 nontribal people live in the rural sector of Loreto, comprising 85 percent of the entire rural population. The remainder are indigenous people with a population of 50,000 (Egoavil 1992).

The first colonists entered the Tahuayo River basin shortly after the construction of a naval base in Iquitos in 1862 (Coomes 1992a). However, it was the rubber boom of 1880–1920 that brought a large influx of colonists to the area. With the crash of the rubber boom the area experienced net emigration. Communities of ribereños consolidated during the recession of the 1930s that saw an influx of people of Cocama/Cocamilla Indian origin. With the rise of market-oriented agriculture and an increase in extraction of forest resources after 1940, the Tahuayo River basin increased in population and continued to do so until the end of the 1980s. Indeed, the Tahuayo basin was considered to be rich in forest products and to have an abundance of agricultural land, which stimulated inmigration (Coomes 1992a).

Communities of ribereños in the RCTT are currently organized around political units, often with an elementary school and several health officials. Rules for land use and extraction of natural resources are determined by the consensus of inhabitants within each community. These rules not only govern titled land owned by community members and land officially recognized as being communal property, but also forest and fisheries resources of neighboring areas.

There are 32 villages in the Tahuayo, Tamshiyacu, Yarapa, and upper Yavari Miri river basins, with a total population of approximately six thousand inhabitants. These inhabitants use resources of the RCTT to varying degrees. Family income in the Tahuayo basin during 1988–89 ranged from US$0–$15,727, with a mean annual income of US$798 and a median of US$326. Incomes were skewed toward the lower end with 37 percent of households earning less than US$200/year, 68 percent earning less than US$600/year, and 89 percent earning less than US$1,600/year (Coomes 1992a). Virtually 100 percent of 541 households surveyed in the Tahuayo basin practiced some type of agricultural production, whereas 42 percent were involved in fishing as a major financial activity, 19 percent in wildlife hunting, 23 percent in the commercial extraction of nontimber plants, and 6 percent in the extraction of timber (calculated from Coomes 1992a).

Creation of the RCTT

A combination of factors led to the creation of the RCTT, including conservation initiatives of local communities in the permanent settlement zone, the outstanding biodiversity of the forests, the complementary nature of the area to the nearby Pacaya-Samiria National Reserve, and the extensive knowledge of the natural history of the area. The four groups that influenced the establishment of the RCTT included (1) the local communities, (2) government agencies, (3) nongovernmental organizations (NGOs), and (4) researchers. These groups coordinated their initiatives, but often had different approaches and reasons for creating the reserve.

The environmental actions taken by the communities of the upper Tahuayo were a major influence for the legal creation of the RCTT. These communities realized the extent of natural resource degradation occurring in the forests during the 1980s and began to take community initiatives to protect natural resources. Indeed, it was the open-access system that began with the abolishment of estates after the enactment of the agrarian law of 1969 that stimulated the uncontrolled extraction of natural resources. After 1970, the area now encompassing the RCTT was exploited extensively for timber, game animals, palm fruits, and fisheries by both local residents and small business operations from the city of Iquitos.

Natural resources were rapidly declining through the 1970s and early 1980s, and many were in short supply by the mid-1980s (Table 9.1). These natural resources fulfilled both financial requirements and subsistence needs of local inhabitants. The communities were particularly unhappy about the exploitation of fish by freezer vessels, the extraction of timber by city-based operators, and the hunting of meat by merchants from Iquitos. As a result, communities organized a system of controls that began to prohibit the extraction of natural resources by nonresidents.

During the 1980s, people living closest to the proposed reserve were seriously discussing the issue of fair natural resource use and continued to set community regulations among themselves. The villages of the upper Tahuayo began to manage five nearby lakes and levy taxes on game meat, fish, and floral resources sent to market by residents. Parties of resource extractors supported by entrepreneurs from Iquitos and the Amazon River area continued to enter the pro-

Table 9.1. Examples of Resources Depleted in Settled Areas of the Tamshiyacu and Tahuayo Rivers

Local name	Species	Use
Aguaje	*Mauritia flexuosa*	Fruit
Cedro	*Cedrela* spp.	Lumber
Choro	*Lagothrix lagothricha*	Meat
Coto	*Alouatta seniculus*	Meat
Huacrapona	*Iriartea* sp.	Flooring
Huasaí	*Euterpe precatoria*	Palm heart, building material
Lagarto negro	*Melanosuchus niger*	Hides, meat
Lagarto blanco	*Caiman crocodilus*	Hides, Meat
Leche huayo	*Couma macrocarpa*	Latex, fruit
Lobo del río	*Pteronura brasiliensis*	Fur
Lupuna	*Chorisia integrifolia*	Plywood
Maquisapa	*Ateles paniscus*	Meat
Naranjo podrido	*Parahancornia peruviana*	Fruit
Paiche	*Arapaima gigas*	Fish
Palo de rosa	*Aniba rosaedora*	Perfume
Sachavaca	*Tapirus terrestris*	Meat
Vaca marina	*Trichechus inungui*	Hides, meat

posed reserve, but were increasingly forced to deal directly with local inhabitants of the Tahuayo.

Community representatives approached the Ministry of Agriculture and the scientists working in the area to gain support for their community conservation initiatives. Acting together, they began the legal actions required to legislate a reserve. Coincidentally, the Peruvian government had recently enacted the new protected area category of "Community Reserve," which coincided well with the community's requirements and the conservation ambitions of the Regional Ministry of Agriculture.

Government officials held village meetings throughout the area between 1988 and 1990, with national and foreign scientists often present. The importance of socioeconomic surveys, scientific research, and the future creation of the Reserva Comunal Tamshiyacu-Tahuayo were major topics of discussion. While scientists and socioeconomic surveys were generally accepted, communities were wary of the involvement of government. Government officials were mistrusted and often poorly received by the villagers because of conflicts and inmigration stemming from government-sponsored land and credit programs of the mid-1980s. Poorly planned government announcements of the impending RCTT also caused some negative reaction. Most people in the area feared that the creation of a reserve would expel them from their lands because of experiences with the nearby Pacaya-Samiria National Reserve (Penn 1993).

People felt that if the government truly wanted to include local people in the establishment process, many new land concessions never would have been given and the actual geographic limits of the reserve would have better reflected local traditions of community resource use. This situation was partially resolved by the development of the Propuesta Técnica 1991 (Technical Proposal) compiled by the Gobierno Regional de Loreto for the reserve, which requires that authorities assist communities with their local vigilance systems (Moya et al. 1991).

Government agencies and nongovernment groups took particular interest in the area because of its unique biodiversity. For example, the RCTT is the only protected area in Peru that includes the red uakari monkey (*Cacajao calvus*), a species quite rare in Peru and considered vulnerable to extinction. The area was also of interest because it would complement the nearby Pacaya-Samiria Reserve. The RCTT, dominated by the upland terra firme forests, greatly contrasts with the flooded ecosystems of the Pacaya-Samiria Reserve and enhances the diversity of ecosystem types under protection in the northeastern Peruvian Amazon (Dourojeanni and Ponce 1978).

Another important consideration for creating the RCTT was the ever-increasing information on natural history and resource extraction in the area. The amount of published information on the RCTT includes more than fifty papers and reports and eleven theses.

Approaches to Community-Based Management

Once the RCTT was created (Resolución Ejecutiva Regional No. 080-91-CR-GRA-P) on June 19, 1991, debate began concerning who had access to the reserve and how much in the way of resources could be extracted. The four groups who are involved with management decisions in the RCTT include (1) the local communities, (2) government agencies, (3) NGOs, and (4) researchers. These groups coordinate many activities, but often have different approaches to community-based management of the reserve depending on the interests of the group and the resource of concern.

For example, after the creation of the reserve, extension work by NGOs involved both the regional government and local inhabitants to ensure that the number of people using the RCTT did not increase. Regional government officials, however, were promoting the reserve as something beneficial to the entire Tamshiyacu-Tahuayo River basin, including the town of Tamshiyacu (pop. 4,500) situated approximately 50 km from the reserve, which was considered by many communities to be outside the permanent settlement zone. Tamshiyacu authorities were anxious to secure benefits from the reserve, and authorities in Iquitos felt that Tamshiyacu approval was critical because it was the local government seat (Distrito de Fernando Lores) and regional policy was promoting rural development.

In December 1991 the government sent out copies of the reserve's legal decree, a map, and other information which included a vague letter stating the government's desire to implement a management plan. People living in villages closest to the reserve looked at the RCTT as a way to secure control for themselves, but they were uncertain as to the legality of the reserve and their role in protecting it. They wanted strong government assurances that they would have the authority to manage resource extraction in the area, something neither they nor NGO extension workers could procure from the government. Persistent political instability in Peru and the April 5, 1992 *autogolpe* by President Alberto Fujimori had kept most government personnel occupied with other concerns.

Beginning in 1989, extension workers have held many informal and formal meetings in villages of the upper Tahuayo to discuss the concept of the reserve and resource extraction. These meetings have evolved from a consciousness-raising experience to detailed discussions about the legalities of the reserve and the biology of sustainable resource use (Penn et al. 1993). Also, government officials from the Natural Resources, Agriculture, and Fisheries departments visit once or twice a year to learn of the people's needs and opinions and to help clarify the people's concerns over resource use. Meanwhile, researchers continued to collect information on resource use and its impact on species populations.

These efforts have resulted in the government's granting villages of the upper Tahuayo considerable authority in resource use decisions.

In general, final decisions on resource use and management in the upper Tahuayo River basin of the RCTT are voted on in a democratic manner free from the presence of government officials, NGO representatives, and researchers. For example, after several meetings, the villages of the upper Tahuayo drafted, voted on, and signed a resource use accord on October 20, 1992. This document includes a five-part agreement that defines those entitled to enter the reserve, rules on the quantity of resource extraction (game meat, palm hearts, and palm fruit), and regulations on fishing methods. It also suggests sanctions for those in violation.

These agreements have been amended frequently during community meetings. Amendments on resource management are often influenced by government programs, NGO activities, and researchers. For example, the planting of nontimber and timber species in agroforestry plots is promoted by NGO activities. Wildlife use has involved both researchers and an NGO-appointed wildlife extension officer. Fisheries management has been principally a community-run activity with some input from researchers and NGOs. Use of agricultural lands has been influenced by government programs, and illegal timber extraction has been promoted by the local police.

Uses of the RCTT

The most important economic activities for both financial income and subsistence of rural inhabitants of the RCTT are (1) agriculture/agroforestry, (2) fishing, (3) hunting, (4) harvesting timber, and (5) extraction of nontimber plant products (Table 9.2). As noted above, the three land-use zones of the RCTT form a continuum from intensive activities such as agriculture in the permanent settlement zone, to natural resource extraction in the subsistence zone, to no extraction in the fully protected area.

People living in the permanent settlement zone use the subsistence zone to varying degrees. For example, in the Tahuayo River basin of the RCTT, village representatives divided communities in the Tahuayo River basin into two groups: those upriver from the police post and those downriver. Villages in the upper Tahuayo (those upriver from the police post) extract the majority of game, fish, and plant products from the reserve. Those downriver generally extract resources outside the reserve. The villages of Esperanza, Charo, Buenavista, Cunshicu,

Table 9.2. Approximate Annual Revenues from Market Sales of Products from the Upper Tahuayo River Basin

Product	Economic value (US$)	Percentage economic value
Agriculture	71,000	44
Fish	43,000	26
Game meat	18,000	11
Timber	18,000	11
Nontimber plants	12,000	7

Sources: Bartecki et al. 1986; Coomes 1992a; Penn 1993; Bodmer et al. 1994a.

and Chino on the Tahuayo River are the five villages closest to the subsistence zone of the reserve and together have 171 resident households. The majority of these households seldom use the subsistence zone of the RCTT and rely mainly on resource extraction from the permanent settlement zone. Only around forty households from the Tahuayo River regularly use the reserve for extraction. However, almost all families use the subsistence zone of the reserve occasionally, especially in times of necessity. For example, during the flood season many residents of the upper Tahuayo basin extract resources from the subsistence zone for both cash and subsistence needs.

Use of the subsistence zone of the reserve can be illustrated by the extraction of resources by the village of Chino, the village closest to the reserve, and the people living in the surrounding forests of the Qb. Blanco. Wildlife is used by approximately 34 households from this area. These households use approximately 500 square kilometers (km^2) for hunting wildlife and collecting timber and nontimber plants. Of this 500 km^2, approximately 250 km^2 is situated within the subsistence zone of the reserve. The use of resources from the reserve varies greatly between households; for example, only two households gained more than US$2,000 annually in income from wildlife hunting, while the remainder gained less than US$1,500 (Fig. 9.2).

The permanent settlement zone is also used for natural resource extraction by people living around the RCTT. These areas consist of secondary forest, intact and disturbed primary forest, fallows, and wetlands. The majority of people living in the permanent settlement zone focus their extraction and agriculture in these areas outside the official boundaries of the RCTT.

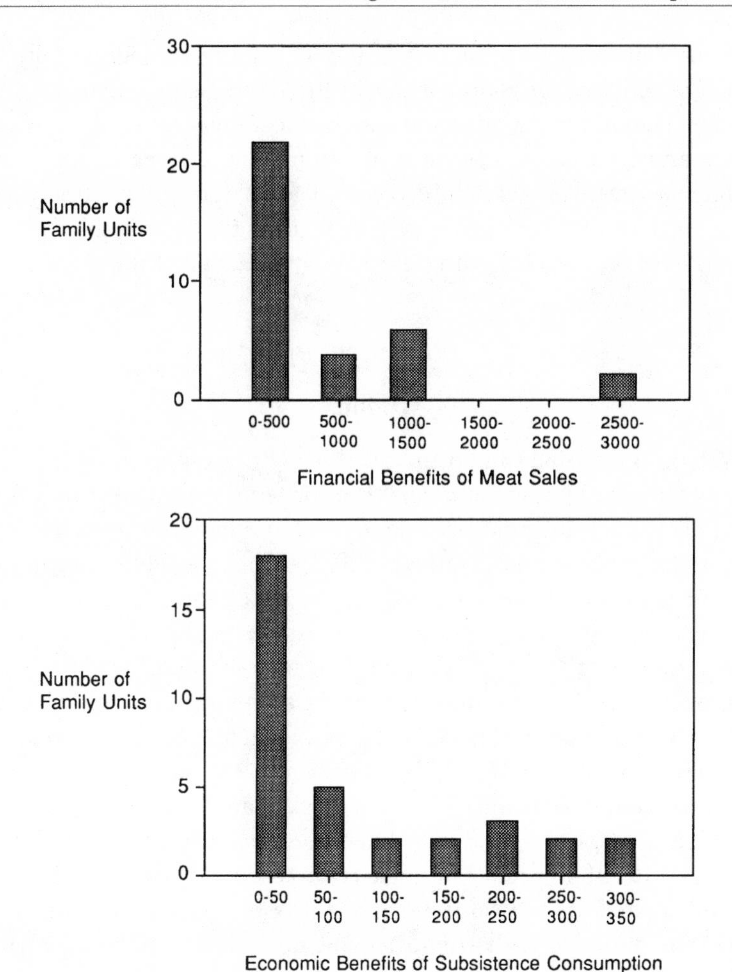

Figure 9.2. Distribution of annual financial benefits (in US$) from the sale of game meat and economic benefits from subsistence consumption of game meat in the upper Tahuayo River basin.

Most of the forests close to villages in the permanent settlement zone are managed by local people. Lakes, hunting zones, and areas with high concentrations of building materials (thatch and flooring) have been successfully managed by residents of the Tahuayo (Bodmer et al. 1990b; Penn and Alvarez 1990; Penn 1994). Villagers consider large parts of the permanent settlement zone as "private reserves" that belong to their respective communities or families. Indeed, villagers consider these to be both individual and communal property.

Thus management programs of the RCTT must integrate fully the extractive activities of people living in the permanent settlement zone. This area should be managed and enriched to supply residents with subsistence and cash needs. Programs of natural resource management and sustainable land use will help decrease deforestation in the permanent settlement zone, which in turn will strengthen the sustainable use and protection of biodiversity within the subsistence zone of the reserve.

Use and Management of Resources

Wildlife

Hunters in Amazonia obtain financial benefits from market sales and subsistence use of game mammals (Redford and Robinson 1987; Bodmer et al. 1990c). Commercial sale of meat and hides during a one-year study in the upper Tahuayo earned US$17,932 for the 34 hunting households of Chino and the surrounding forests of the Qb. Blanco, and made up 11 percent of the cash income for these inhabitants. Virtually all of the game animals were taken using 16-gauge shotguns. Ungulates and large rodents are the most important game animals sold for cash income (Table 9.3). For example, in the markets of Iquitos, 60.2 percent of the total volume of game meat sold between May 1986 and April 1987 was from ungulates, of which 64.8 percent was collared peccary (1,211 individuals) and 19.5 percent was white-lipped peccary (261 individuals) (Bendayán 1990). In Iquitos, a single collared peccary can fetch US$20–$30 dollars, making this trade a major source of revenue to hunters.

During the same study conducted in Chino and the surrounding forests of the Qb. Blanco, a total of 3,008 kilograms (kg) of game meat was not suitable for commercial use, and had a subsistence value of US$1/kg. These subsistence game species consisted mainly of primates, small rodents, edentates, marsupials, and carnivores, for which little demand exists in the markets of Iquitos. In total, mammals with commercial value were hunted in greater numbers than mammals used for subsistence.

Studies on wildlife use in the upper Tahuayo have shown that unmanaged hunting is overexploiting primates and tapirs, is probably overexploiting carnivores, edentates, and marsupials, and is not overexploiting deer, peccaries, and large rodents (Bodmer et al. 1994a,b). Estimates have been made of the sustainability of current offtake of deer and peccaries in the RCTT based on current population levels (Table 9.4). Results indicate that 15–35 percent of deer and peccary

Table 9.3. Annual Financial Benefits of Game Harvest in the Upper Tahuayo River Basin

Benefit	Price/animal (US$)	Number of animals	Total (US$)
Commercial meat value			
White-lipped peccary	30	166	4,980
Collared peccary	30	165	4,950
Red brocket deer	30	60	1,800
Gray brocket deer	20	28	560
Lowland tapir	80	38	3,040
Capybara	20	10	200
Paca	10	174	1,740
Total commercial meat value			17,270
Commercial pelt value	2	331	662
Total commercial value for hunters			17,932
Direct consumption value	1/kg	3,008 kg	3,008
Total direct benefits of mammalian harvests for hunters			20,940

Source: Bodmer et al. 1994a.

Table 9.4. Reproductive Productivity in Relation to Hunting Pressure of Amazonian Ungulates in Tahuayo Study Area

Parameter	Red brocket deer	Gray brocket deer	Collared peccary	White-lipped peccary	Lowland tapir
Litter size	1.0	1.0	1.7	1.6	1.0
Percentage of females reproductively active	45 (N = 40)	50 (N = 16)	44 (N = 62)	51 (N = 37)	50 (N = 8)
Number of gestations per year	1.5	1.5	1.5	1.5	0.5
Reproductive productivity of females: average number of young/female per year	0.67	0.75	1.11	1.23	0.25
Total reproductive productivity: average number of young/individual per year	0.33	0.37	0.55	0.61	0.12
Density (individuals/km^2)	1.8	0.8	3.3	1.3	0.4
Production (individuals/km^2)	0.6	0.3	1.83	0.80	0.05
Hunting pressure (individuals/km^2)	0.13	0.06	0.27	0.30	0.08
Percentage of production taken by hunters	22	20	15	38	160

Source: Bodmer 1994.

production is harvested. Such levels are apparently within a sustainable range (Bodmer 1994; Robinson and Redford 1994). Age structure curves of deer and peccaries in the RCTT suggest that current harvests of artiodactyls should not be increased, and it is assumed that harvests of large rodents also should not be increased (Bodmer et al. 1994a). A more sustainable hunt would therefore require cessation of hunting on primates, tapirs, marsupials, edentates, and carnivores, and set deer, peccary, and large rodent harvests at or below current levels.

A management program is being considered in the RCTT which allows hunters to cull a greater proportion of males of species that are not currently overharvested, and that prohibits hunting of overexploited mammals. A male-directed hunting program would support a harvest that does not degrade the game resources. This program would be used to manage wildlife in both the subsistence and permanent settlement zones. Hunters would set up the program with a wildlife extension worker. While communities have good intentions with resource management, they often require technical support from extension workers. In this case, plans were elaborated as a collaborative effort between the communities of the upper Tahuayo, extension workers, and researchers, thus taking into consideration the needs of the people and the biological limitations of the resources.

Male-directed hunting is commonly used for ungulate management in North America and Scandinavia to minimize the impact of harvests (Gill 1990). In many mammals, lifetime reproductive success of males is determined by access to females, but female reproductive success is limited more by resources than by the number of males (Clutton-Brock 1988). Recruitment in such mammalian populations is therefore relatively unaffected by a reduction in the proportion of adult males if the population is not near its carrying capacity, which is apparently the case for Amazonian ungulates and large rodents (J. Terborgh, pers. comm. 1989). An increase in the proportion of males taken is likely to increase recruitment only if it involves an actual reduction in females harvested. The management program should not permit the harvest of only males because this might have repercussions on recruitment by altering too drastically the ability of females to find mates (Ginsberg and Milner-Gulland 1994).

The 34 hunting households of Chino and the surrounding forests of the Qb. Blanco would incur a 26 percent reduction of total economic value and a 17 percent reduction of financial income if they ceased to harvest primates, tapirs, marsupials, edentates, and carnivores. This restricted hunt would reduce the extraction of mammalian biomass by 35 percent. The costs of setting up a more sustainable hunt would not be evenly distributed between family units. The costs would be greater for those households gaining more than US$1,500 in revenue from hunting, and thus programs replacing losses should be directed more to the households incurring greater costs. The labor cost of extracting meat for commercial sale would rise. Thus people who only occasionally hunt for commercial sale of meat would hunt less for marketable game and more for subsistence, while those hunters who rely on commercial sale of meat would require the programs mentioned above to overcome these added costs.

The communities of the upper Tahuayo could manage a male-directed hunt by requiring that market benefits be obtained only from male artiodactyls and large rodents, and subsistence benefits from female artiodactyls and large rodents. Interestingly, financial considerations of this management program would complement its biological justification. First, hunters would become more selective in harvesting male animals in order to maximize commercial benefits. Second, meat from females used for consumption would substitute for meat lost from prohibited species. Hunters would have less incentive to harvest primates, marsupials, edentates, or carnivores for consumption when meat from female artiodactyls and large rodents is available. Since the total number of peccaries harvested would remain the same, benefits of the peccary hide trade would not be affected by this management scheme.

A management program that permits only male artiodactyls and large rodents to be sold at market would decrease the total commercial meat benefits by 54 percent of the present hunt, if the sex ratio of the current harvest is used. Meat available for direct consumption would increase by 2.4 times the current level. However, by implementing a male-directed management program, the harvested sex ratio should become male biased as hunters attempt to maximize revenues.

Another advantage of this management program is that it should improve the nutritional condition of families. The nutritional condition of children in the permanent settlement zone of the RCTT is below the Peruvian standard. This is usually attributed to internal parasites (CETA 1986). However, high-quality protein sources are often unavailable to children because the heads of households sell high-quality foods at market for income. The proposed management plan for game hunting in the RCTT would require female deer, peccary, and large rodents to remain in households for subsistence, and not be sold at market. Thus children of the permanent settlement zone would have a greater opportunity to consume high-quality protein, which may help in raising their nutritional intake.

Fisheries

Fish caught in the permanent settlement zone of the RCTT are important both as subsistence food and as a commercial enterprise (Table 9.5). Indeed, fishing is the second most valuable economic activity (first is agriculture) for inhabitants of the upper Tahuayo, constituting 26 percent of income. Local habitat is an important factor in determining the

Table 9.5. Fish Commonly Used for Consumption and Sale from the Upper Tahuayo River and Neighboring Lakes

Common name	Species
Acarahuazú	*Astronotus ocellatus*
Arahuana	*Osteoglossum bicirrhosum*
Boquichico	*Prochilodus nigricans*
Bujurqui	*Cichlaururus* sp.
Carachama	*Pterygoplichthys multiradiatus*
Corvina	*Plagioscion* sp.
Chambira	*Rhaphiodon vulpinus*
Gamitana	*Colossoma macropomum*
Lisa	*Leporinus* sp.
Paco	*Piaractus brachpomun*
Paiche	*Arapaima gigas*
Palometa	*Mylossoma* sp.
Paña or Piraña	*Serrasalmus* sp.
Peje torre	*Practocephalus hemiliopterus*
Raya	*Potamotrygon histrix*
Sabalo	*Brycon* sp.
Shiruy	*Corydoras* sp.
Shuyo	*Hoplerythrinus unitaeniatus*
Tucunaré	*Cichla acellaris*
Yaraqui	*Prochilodus amazonensis*
Zungaro	*Pseudoplatistoma tigriinum*, *Brachiplatystoma filamentosum*

importance of fishing to local families of the RCTT. For example, fishing is often a major financial and subsistence activity of families that live in or near seasonally flooded várzea habitats of the Tahuayo basin. There are two complementary reasons for this. First, lowland várzea supports most of the Amazonian fisheries because these flooded forests are the principal zones for primary production of phytoplanktons and herbaceous floating plants, and the fruit production of these forests is specifically adapted for seed dispersal by fish. Most of the fish harvested from the upper Amazon come from várzea habitats (Bayley and Petrere 1989). Fishing is therefore a common and productive activity in these flooded regions, with an annual yield in Loreto of 15.2 kg/ha within the flooded forests (Bayley and Petrere 1989). Second, large game animals are often absent or their numbers greatly reduced in flooded habitats because they are unable to withstand the long periods of inundation (Bodmer 1990b). Since hunters' preferences are for large-bodied game (Bodmer 1995), it is often not worthwhile to hunt in seasonally flooded habitats because these larger game animals are absent or rare.

In contrast, nonflooded terra firme habitats have mainly small streams, often with fast-flowing currents. These water bodies are not productive fisheries since phytoplanktons and floating vegetation are often scarce. But terra firme is the favored habitat of large-bodied game. Thus hunting is a common and productive activity in nonflooded habitats, whereas fishing is rare.

Fishing techniques used in floodplains include the use of gill nets, seines, cast nets, harpoons, handlines, trotlines, and bows and arrows (Bayley and Petrere 1989). The most productive fishing occurs during dry seasons when river levels fall and fish become concentrated in shrinking lakes and channels. During high waters, fish are more dispersed and inhabit the large tracts of inundated forests, making harvests difficult. Amazonian fisheries are also dependent on migratory characin schools that roam larger rivers and are harvested with seines and gill nets (Goulding 1980).

Extraction of fish in the lakes of the upper Tahuayo was used for subsistence and income for local residents as early as the late nineteenth century. During the 1950s, fishing became a commercial activity for Iquitos-based operators who began sending freezer vessels to the upper Tahuayo. Commercial fishing steadily increased as the population of Iquitos grew and the demand for fish increased. By the early

1980s, the upper Tahuayo was so heavily fished that stocks began declining. According to local fishermen, the paiche (*Arapaima gigas*) became rare in the Tahuayo by the mid-1980s because of intense net fishing (McDaniel 1994).

The regional government did not have financial resources or trained personnel for effective fisheries management strategies, which might have included closed seasons, restricted areas, limitations on harvesting techniques, quotas, or prohibiting catches of certain species. However, communities were often sufficiently interested in managing fisheries around their villages to take action. Again, this was because of the importance of fish to families living in várzea habitat both for subsistence and cash income. The fear of losing a subsistence resource initiated the community-based management of fisheries in the upper Tahuayo (McDaniel 1994). One of the villages, Chino, formally organized its lake management and vigilance system during the early 1980s. Anyone from the village could enter, but the use of barbasco poison (*Lonchocarpus* sp.), large nets, and other devices was restricted or banned. Freezer boats and other commercial fishing vessels based in Iquitos were prohibited from entering any lakes near the village. Patrols were established to guard the lakes. The village of Chino organized, with community approval, a five-man rotating police force to guard five lakes near the village (Penn and Alvarez 1990), and was given usufruct rights by the regional ministry of fisheries.

The benefit of this community management was a secure supply of fish, and this was sufficient to keep the system in place. The fisheries management of the upper Tahuayo reestablished fish populations in both the permanent settlement and the subsistence zones. The local people currently appear to be harvesting fish at sustainable levels because size classes conform with sustainably harvested populations for the most commonly fished species (McDaniel 1994). However, these results also suggest that more intensive fishing would probably result in overexploitation, especially for gamitana (*Collosoma macroponum*), boquichico (*Prochilodus nigricans*), and fasaco (*Hoplias malabaricus*) (McDaniel 1994). In addition, paiche, which was harvested almost to local extinction before the community-based management, has repopulated the managed lakes to levels that can be harvested. Interestingly, this fish, which commonly weighs 125 kg (Goulding 1980), is captured only for community harvests. Revenue from these harvests is used for communal projects such as repairing the school buildings (McDaniel 1994).

Nontimber Plants

Nontimber plants are important resources for local inhabitants of the Peruvian Amazon and have a wide variety of uses for oils, fibers, foods, beverages, latex, medicines, toxins, and dyes, among others (Brack 1992). Nontimber plants can contribute to the economic value of natural forests. Peters et al. (1989) argue that this value exceeds that of clear cutting and timber extraction. For the most part, however, current uses of economically important species of nontimber plant products are not sustainable because many rural extractors in the Peruvian Amazon cut down individuals in the process of harvesting nontimber forest products (Vasquez and Gentry 1989; Bodmer et al. 1990a; Phillips 1993). In the RCTT, nontimber plant products are considered open-access resources to all community members in the subsistence and permanent settlement zones, resulting in destructive harvesting practices.

Palm products are currently one of the most economically important nontimber plant products of the Peruvian Amazon. For example, palm hearts (*Euterpe precatoria*) and palm fruits (*Mauritia flexuosa* and *Oenocarpus bataua*) for ice creams, fruit snacks, and drinks are used locally and occasionally exported (Kahn 1988) (Table 9.6). In fact, palm trees are the most important wild fruit resource in the Per-

Table 9.6. Palm Trees Commonly Used by Ribereños in the Peruvian Amazon

Common name	Species	Part	Product
Aguaje	*Mauritia flexuosa*	Fruit	Edible fruit, drink, ice cream
		Cortex	Starch
Aguajillo	*Mautitiella peruviana*	Fruit	Edible fruit
Chambira	*Astrocaryum* spp.	Fruit	Oil, edible fruit
		Fronds	Fiber
Huasaí	*Euterpe precatoria*	Apex	Palm heart
		Wood	Building material
		Fruit	Edible fruit
Irapay	*Lepidocaryum tessmannii*	Leaf	Roof material
Pijuayo	*Bactris gasipaes*	Fruit	Edible fruit, oil
Pona	*Iriartea* sp.	Wood	Building material
Sinamillo	*Oenocarpus mapora*	Fruit	Edible fruit, oil
Ungurahui	*Oenocarpus bataua*	Fruit	Edible fruit, drink, ice cream, oil
Yarina	*Phytelephas macrocarpa*	Leaf	Roof material, fruit

uvian Amazon and contribute 61 percent of the market value for wild fruit production in the Iquitos region (calculated from Peters et al. 1989). Palms provide significant income for both rural extractors and many urban families who work as middlemen and market vendors (Padoch 1987).

As with many nontimber forest products, palms are cut by rural extractors during harvests. This destructive practice appears to be related to several factors. Ribereños cut these palms partly because of the open-access system of the forests and also because of the difficulty of climbing these trees. The physical structure and height of the trees render climbing almost impossible and very dangerous. For example, *M. flexuosa* trees often reach 40 meters (m) in height and have a stegmata wood bark (containing silica bodies) that is extremely hard and slippery (Uhl and Dransfield 1987). Also, many rural extractors feel that the labor expended by climbing trees would be wasted, believing that other extractors from their community would cut the tree during the following harvest. Thus rural people feel they must maximize harvests by minimizing the time and labor expended, which is accomplished through cutting palms with axes.

One of the most important palms in the RCTT is the aguaje (*Mauritia flexuosa*), which provides financial income for many rural extractors and supports at least five hundred households in Iquitos (Padoch 1987). *M. flexuosa* occurs in almost monotypic stands in backswamp habitats that range in size from 1 to 10 ha or more (Uhl and Dransfield 1987; Kahn 1988). The patchy distribution of *M. flexuosa* helps ribereños in locating and collecting fruits. Trees are felled to harvest the fruits, and because of the dioecious reproductive system of *M. flexuosa,* only male trees remain in exploited swamps.

Harvesting of *M. flexuosa* has already destroyed many palm swamps close to villages in the permanent settlement zone of the RCTT. For example, in the Tahuayo River area most swamps closer than 10 km to the village of Chino were heavily damaged, and intact swamps of *M. flexuosa* usually could be found only at distances of more than 25 km. Thus there was a negative relationship between the degree of damage to *M. flexuosa* swamps and distance to the village (Bodmer 1994).

Other palms, such as the ungurahui (*Oenocarpus bataua*), also are being cut down for fruit harvests. Unlike *M. flexuosa, O. bataua* is evenly distributed throughout the forests (Kiltie and Terborgh 1983). Palm hearts (chonta) of *Euterpe precatoria,* the most common species

Table 9.7. Rank of Palm Fruits in Diets of Amazonian Ungulates

	Rank in diet[a]				
Palm species	Collared peccary	White-lipped peccary	Red brocket deer	Gray brocket deer	Lowland tapir
Astrocaryum sp.	3	4	28	Not eaten	10
Euterpe precatoria	16	10	1	1	Not eaten
Iriartea sp.	6	1	2	2	Not eaten
Oenocarpus bataua	2	2	30	Not eaten	2
Mauritia flexuosa	10	5	18	16	1

Source: Bodmer 1989.
[a]Rank of 1 indicates food eaten most frequently, with decreasing ranks representing decreased importance in diet.

used for palm heart in Loreto, are collected by cutting individual trees. Other species of palm heart, however, can be more sustainably harvested because they have multiple stalks that can be selectively cut without killing the individual (Anderson 1990). This also permits other stalks to grow for future harvests. But again, the open-access system makes a sustainable alternative unprofitable for the rural extractor because others are likely to cut the entire individual.

The extraction of fruit from natural habitats probably affects the nutritional condition of forest animals, which in turn affects their population growth by decreasing the potential carrying capacity. Ungulates and rodents are the most important game species and make up the majority of mammalian biomass extracted from the Peruvian Amazon. These large-bodied mammals are also primarily frugivores. For example, the diet of red brocket deer (*Mazama americana*) is 81 percent fruit, gray brocket (*M. gouazoubira*) 87 percent, collared peccary (*Tayassu tajacu*) 59 percent, white-lipped peccary (*T. pecari*) 66 percent, and lowland tapir (*Tapirus terrestris*) 34 percent (Bodmer 1989). The most important fruits for these ungulates and rodents are from palm species (Table 9.7), which are often very nutritious (Lopes et al. 1980). Thus a sustainable hunt could be increased if the palm resources were increased. However, if palm resources are being depleted, then the carrying capacity of ungulates and rodents will be lowered, which in turn will lower the number of individuals that could be sustainably harvested. Indeed, these density-dependent effects should be noticeable even at low populations of game.

The long-term benefits local people obtain from wildlife are thus directly linked to the palm fruit harvests. If people maintain high palm fruit production in natural areas and allow this fruit production to be consumed by wildlife, then game harvests can be increased. If, however, destructive harvests of palm fruits continue, then both the palm and game resources will be depleted and long-term benefits from these resources cannot be realized. Indeed, if no action is taken to manage the harvest of nontimber plants, the near, if not total, local extinction of *M. flexuosa, E. precatoria,* and *Jesenia* sp., among other species, is inevitable.

The RCTT requires management programs for forest fruit extraction that will maintain or possibly increase palm fruit and, as a consequence, game populations. To convert the nonsustainable use of palms to a more sustainable use, communities of the RCTT should (1) cease cutting palm trees, (2) have a substantial reduction in the collection of forest fruits in order to maintain the forest's carrying capacity for important game animals, and (3) plant palm trees whose fruits are currently collected in the forest, in privately owned agroforestry plots close to the villages (Bodmer et al. 1990a; Penn 1994).

Introducing nondestructive harvesting techniques will help maintain palm populations, but improving income for local people using nondestructive harvesting methods will depend on the proportion of fruit production that is harvested by the communities. Sufficient numbers of fruits must be left for dispersal and regeneration and for the forest animals that rely on palm fruit. The amount of fruit that should be left for regeneration and for consumption by game animals is an important management question that must consider ecological factors such as the density-dependent relationships of game animal production, the relationships between seed density and regeneration rates, and the socioeconomic choices between the importance of palm fruit harvests versus game harvests.

Ribereños will not destroy palm trees if they occur in small private plots but will collect the fruits for both household use and market sale. Interestingly, many palms grow considerably shorter when planted in open systems without competition and therefore do not require cutting or special climbing equipment. *M. flexuosa,* for example, will not grow to great heights when planted in open areas. Indeed, most *M. flexuosa* trees grown in open areas will reach only 6–15 m in height, which allows the fruit bunches to hang 2–10 m off the ground. In these cases, trees are not felled and the fruit may be easily harvested in a sustainable fashion.

The fruit production of planted palms will have several advantages over the cutting of wild trees. For one, inhabitants will have a renewable supply of palm fruits for market sale and subsistence consumption, and this will be a truly sustainable alternative. In addition, palm fruits in natural habitats of the RCTT will be mostly left for animal food, which in turn should strengthen game populations by increasing potential carrying capacity. While this strategy may decrease the financial income derived from the forest, it will increase the economic value of the forest for both local people and the broader society by having the forest continue its ecosystem functions. For local people this means an increase in game meat and the production of other important forest products.

The major disadvantage of this strategy is the time delay between planting and fruit production, which is around six to eight years. During this growing period the harvest of wild palms will continue, but once the agroforestry plots become productive, the degree of wild harvests should decrease dramatically because local extractors will no longer rely on wild fruit production.

Timber

Timber extraction was an important financial activity in the Tahuayo basin during the 1950s and 1960s, as in much of Loreto (Dourojeanni 1990). However, most of the economically valuable species in the upper Tahuayo were overharvested during this period (Coomes 1992a). Thus the current extraction is considerably reduced and, based on figures from the late 1980s, accounts for only 11 percent of the total financial income of the upper Tahuayo. In addition, the current use of timber does not appear to be sustainable because economically important species are being cut at greatly reduced sizes and are not being replanted. A more sustainable use of timber would be to (1) decrease the current level of extraction, (2) plant timber species in privately owned agroforestry plots close to villages (Penn 1994), and (3) introduce programs of natural forest management (Gorchov 1994).

The concentration of timber extraction on a few species has resulted in the depletion of large individuals of these species in the subsistence zone of the RCTT (Table 9.8). This overexploitation (known

Table 9.8. Timber Species Used in the Tahuayo River Basin and Their Status

Common name	Species names	Status
Caoba	*Swietenia macrophylla*	Overexploited
Capirona	*Calycophyllum spruceanum*	Unknown
Cedro	*Cedrela odorata*	Overexploited
	C. hissilits	Overexploited
Cumaceba	*Swartzia* spp.	Overexploited
Itahuba	*Aniba* spp.	Overexploited
	Pseudolmedia multinervis	Overexploited
Lagarto caspi	*Calophyllum brasiliense*	Overexploited
Lupuna	*Chorisia* spp.	Overexploited
Moena	*Endlicheria anomala*	Unknown
	Ocotea aciphylla	Unknown
	O. grandifolia	Unknown
	O. laxiflora	Unknown
	O. maynensis	Unknown
	Nectansra acutifolia	Unknown

as "high grading") might be converted to a more sustainable use through natural forest management or planting of these species in privately owned plots. Natural forest management would require, in this case, polycyclic felling where exploited areas would be subjected to selective logging every 20–35 years (Gorchov 1994). Thus, in most areas of the subsistence zone, timber extraction should cease for the next 20–35 years. Similar to palm extraction, the problem of property rights and open access has partially been responsible for the overexploitation of timber in the RCTT. Thus planting important timber species in privately owned plots would guarantee families' rights to this timber production.

Similar to palms, the relationship between timber extraction and wildlife is important. However, this relationship differs considerably from that of palms. Timber extraction is directly linked to hunting pressure, not indirectly through effects of carrying capacity, as with palms. In fact, the monthly harvests of large game species from the RCTT between July 1985 and June 1986 was in large part conducted by lumbermen. The lumber operations in the Qb. Blanco of the upper Tahuayo accounted for 51 percent of the peccary, deer, and tapir har-

Table 9.9. Number of Ungulates Harvested in the Upper Tahuayo River Basin, July 1985 to June 1986

Species	Subsistence hunters	Lumbermen and commercial hunters
Collared peccary	29	78
White-lipped peccary	29	12
Red brocket deer	23	53
Gray brocket deer	9	14
Lowland tapir	9	5

Source: Bodmer 1989.

vests, while illegal commercial hunters accounted for 11 percent and subsistence hunters for 38 percent (Table 9.9) (Bodmer et al. 1988).

These timber operations, which were small-scale and had considerable financial constraints, decreased costs by supplying workers with shotguns and cartridges in lieu of basic foods. Lumbermen would also sell considerable quantities of game meat in the markets of Iquitos to augment meager salaries. Lumbermen require large quantities of game meat because they spend long periods in the forest extracting timber, unlike people extracting palm products who spend short periods of time in the forest and do not consume large quantities of game meat.

Restricting access to lumbermen was the first wildlife management regulation set up by the communities in conjunction with government officials. In Tahuayo, lumbermen were the most significant hunting class and exerted the major hunting pressure before 1988. The communities expressed their concern to government officials, who promoted the area as a "reserve in study." This land classification made it possible to end timber concessions of noncommunity residents in 1988 and so decrease hunting pressure of game by nonresidents.

Agricultural Lands

Agriculture has been an important economic activity of residents of the Tahuayo basin since its settlement in the late 1800s. Since agriculture is of great financial importance to people living in the permanent settlement zone of the RCTT, its success or failure will directly affect the intensity of extraction of natural resources from the reserve. In addition, the amount of land cleared for agriculture will directly affect populations of wild species.

To understand the current use of agricultural lands, it is important to understand the history of agriculture in the region of the RCTT. The first land title in Tahuayo was granted in 1865 with the rubber boom, which began around 1880, provoking additional agricultural estates throughout the Tahuayo basin. At this time, the population of the basin was around 950 nontribal individuals (Coomes 1995). Much of the agriculture production in the Tahuayo basin during the rubber boom came from the rubber estates. Rubber tappers developed agricultural systems that relied heavily on the production of fruiting species in agroforestry plots. After the collapse of the rubber boom, the population of the Tahuayo basin declined to 600 nontribal individuals by 1940 (Coomes 1995). During the period 1940–60, the Tahuayo basin recovered economically because of increased demands for agricultural and natural resource products. With the Agrarian Land Reform of 1969, the estates were abolished and the land was converted to government property. This enabled many of the tenants to acquire their own private lands, which resulted in an expansion of a peasant agricultural system (Coomes 1995). New crops were introduced into the region, and property owners consolidated more into village communities. Agricultural production was usually in the form of combining short-term crops with fruiting trees, with a family using around 5 ha of land at any one time, in various stages of production. Fruiting trees were usually abandoned after 20–40 years, and fallows were cut for new agricultural production after 60–80 years of abandonment (Denevan and Padoch 1988).

The 1980s saw the beginning of the credit programs, which shifted residents away from the more traditional agroforestry systems to monoculture production (Chibnik 1994). These credit programs stimulated, in part, a population growth of 74 percent between 1981 and 1989, to 3,100 inhabitants in the Tahuayo basin (Coomes 1992a).

Agriculture in upland terra firme habitats has always differed from agriculture in the lowland várzea habitats (Hiraoka 1986). In the várzea of the Tahuayo, many people practice agriculture on seasonally flooded or occasionally flooded (every three to five years) lowlands. Some of these agricultural plots are directly along the riverbank, while other plots are as far as 0.5 km inland. Much of this farming is done on old levee fragments called *restingas* that rarely flood.

Because of unpredictable flooding, lowland and floodplain agriculture is more risky than upland agriculture (Chibnik 1994). However, lowland soil is less acidic and more fertile, and as a result can produce crops that do not grow well in other habitats. Short-term crops (seven months or less), such as cassava, corn, watermelon, beans, and peppers, are produced in the várzea of the upper Tahuayo. Ribereños often depend on income earned from várzea agriculture, especially corn and watermelon. While the returns are high, risks are also high, and farmers often have their economic ambitions ruined by floods. Farmers often abandon these lowlands after losing their crops to floods and return only when they have regained confidence (Lima 1991). However, if plots are left fallow for as little as two years, farmers can lose these lands to ambitious land seekers.

Gardens do not flood in terra firme habitats of the upper Tahuayo, and thus their production is more stable. However, transport to market is often more difficult from upland areas. As a consequence, farmers plant fewer fruit trees in the less accessible areas, as fruit is often difficult to transport.

Crop diversity is necessary in uplands to reduce the risks from disease and pests that can ruin annual crops such as cassava and plantains, as well as fruit trees. For example, a dependence on uvillas (*Pourouma cecropiifolia*) in the upland gardens of the Qb. Blanco left many people without income after a fungal plague destroyed fruit production in 1992. Interestingly, the largest producer had highly diverse gardens and was able to harvest a very successful crop of nescafe (*Canavalia ensiformis*) and overcome the economic hardships that affected other farmers (Penn 1992).

Diversifying production using a wide variety of short- and long-term crops and trees has been a strategy adopted by many farmers in the upland regions of the permanent settlement zone of the RCTT. This has occurred in part because a market exists in Iquitos for more than 190 fruits and scores of other products (Vasquez and Gentry 1989). Thus many farmers believe that diversity is the key to economic stability for upland agriculture.

Traditional agriculture for uplands provides a diverse set of options for farmers (Padoch 1987). For example, farmers begin with basic staples of cassava, bananas, pineapples, rice, and corn, mixed with vegetable crops and papayas. Once these annual crops are harvested, farmers manage forestry systems and fruit production (Hiraoka 1989).

Upland areas are also more suitable for animal production (Casa Campesina 1985). Hogs and cattle are more important in upland areas because these regions lack both fisheries and daily transport, thus mandating nonperishable income sources (Moya and Rimachi 1988; Penn 1993; Puertas 1993). For example, in the Tamshiyacu River basin there are more cattle than people (Moya and Rimachi 1988). In these areas of cattle production, farmers recognize the need to plant diverse species of fruit trees for subsistence to improve diets (Penn 1993; Puertas 1993; R. Leon, pers. comm. 1992), but complain that livestock destroy tree seedlings and compact soil, which hampers cultivation.

While a great diversity of products is harvested from the RCTT, urban influences from Iquitos (market and credit) have in the past lowered the diversity of products of traditional household economies. In the Tahuayo basin for example, half of the households receive the majority of their income from only one product (Coomes 1992a,b). During the late 1980s, the upland agriculture practiced by local residents was increasingly oriented to credit and monoculture practices (Coomes 1991, 1992a; Penn 1992; Rengifo 1994). Upland agriculture on the Tahuayo was dominated from the late 1980s until 1991 by credit programs that promoted monocropping of rice, corn, cassava, and plantains, which were designed to turn fields into pasture for water buffalo. Loan programs greatly distracted local people, who sometimes abandoned agroforestry systems to earn credit. After several years, many lands around villages were degraded by the monocropping programs. When the loan program ended, many local agriculturists were again interested in producing a diverse array of plants in more traditional agroforestry plots.

To enhance sustainable use and enrich productivity in the settlement zone, communities are planting a diversity of species used for subsistence and cash needs on private and community lands (Table 9.10). For example, popular nitrogen-fixing trees and other nutritious fruits that grow in poor soils are planted to improve soils and prevent erosion of hilly lands. This improves soils and future agriculture production. Several durable, transport-resilient fruits are also marketed or traded in upland areas. In lowlands, farmers are retaining possession of plots and diversifying production by planting flood-tolerant trees.

Table 9.10. Fruits Commonly and Occasionally Consumed by Inhabitants of the Upper Tahuayo River Basin

Local name	Scientific name	Source[a]		
		Cultivated	Extracted	Marketed
Aguaje	*Mauritia flexuosa*	O	C	C
Camu camu	*Myrciaria dubia*		C	C
Charichuelo	*Rheedia* spp.	O	C	O
Cocona	*Solanum sessiliflorum*	C		C
Guava	*Inga* spp.	C		C
Mamé	*Syzygium jambos*	C		O
Palta	*Persea americana*	C		C
Papaya	*Carica papaya*	C		O
Pijuayo	*Bactris gasipaes*	C		C
Piña	*Ananas comosus*	C		C
Plátano	*Musa* spp.	C		C
Sandia	*Citrillus lanata*	C		C
Shimbillo	*Inga* spp.	C	C	O
Toronja	*Citrus paradisi*	C		O
Umari	*Poraqueiba sericea*	C		O
Ungurahui	*Oenocarpus bataua*		C	C
Uvilla	*Pourouma cecropifolia*	C		C

[a]C, commonly; O, occasionally.

Once established, trees guarantee land possession, income, and a potential food bank during flooded periods.

The current trend toward greater agroforestry systems in the permanent settlement zone of the RCTT is more sustainable than monocropping systems. These agroforestry systems help maintain nutrients by implanting stable root systems and decreasing the frequency of clear cutting. Most of the areas cleared for agriculture and agroforestry in the permanent settlement zone of the RCTT are abandoned fallows, and there is currently no interest by agriculturists to clear land in the subsistence zone of the RCTT. While the present practice of agriculture is more sustainable than monoculture programs, local farmers still face many problems, including (1) low productivity of land, especially in the upland areas, (2) degradation of upland soil, (3) high pest and weed infestation in floodplain agriculture, (4) low returns for traditional crops, and (5) few viable instruments for savings other than livestock (O. T. Coomes, pers. comm. 1994).

Toward More Sustainable Resource Use in the RCTT

Based on our examination of the use and management of natural resources in the upper Tahuayo region of the RCTT, we conclude that

1. More sustainable use of wildlife would require cessation of hunting of primates, tapirs, marsupials, edentates, and carnivores. Deer, peccary, and large rodent harvests appear to be sustainable at current levels, and harvests should be set at or somewhat below these levels.
2. The current use of fish appears to be sustainable because of a strong community-based approach to managing the fisheries.
3. The current use of palm resources is not sustainable because of the excessive felling of palm trees to collect fruit and palm heart. A more sustainable use of palms would require (a) stopping the cutting of forest plants, (b) stopping the collection of forest fruits, so as not to damage the carrying capacity for game species, and (c) planting palms that are currently harvested in the wild in privately owned agroforestry plots close to villages.
4. The current use of timber is not sustainable because economically important species are being cut at greatly reduced sizes and are not being replanted. A more sustainable use of timber would require (a) decreasing the current level of extraction, (b) regenerating timber species in privately owned agroforestry plots close to villages, and (c) introducing programs of natural forest management.
5. The current use of agricultural lands appears to be more sustainable than previous monoculture systems because inhabitants are planting agroforestry systems using a diversity of species. This will help decrease both nutrient loss and the frequency of clear cutting.

Economics

We examined the economics of establishing more sustainable resource uses in the RCTT by using a financial cost/benefit analysis over short-term and long-term time frames. This analysis considered the following resources: timber, nontimber plants, fish, game, and agricultural and agroforestry products. Ecosystem services (e.g., watershed protection, climate regulation) and other nonmarket benefits of biodiversity (e.g., esthetic and option values) are largely unquantified and therefore have not been included in the analyses. Nor does the analysis include value-added processing, consumer benefits (welfare benefits), or opportunity costs, particularly for labor. Subsistence value is not included because we assume that subsistence uses remain constant between the short and long term as subsistence uses are assumed to be sustainable. We compared the use of these resources over the short and long term using two alternative scenarios, one being current resource use with no management, and the other a system that converts the current resource use to more sustainable uses.

If resource uses are not converted to the more sustainable scenario, then the short-term economic value of natural resources will be similar to the current system. However, in the long term, some of the natural resources, such as timber, nontimber plants, and game animals, would have greatly reduced values because of local depletion.

The scenario is different if more sustainable resource management is established. In the short term, the value of game, timber, and nontimber plants would be less than the current system. However, in the long term, the value of timber and nontimber plants would be similar to the current system because of agroforestry and natural forest management, and the value of game would be similar to the current system because of wildlife management. In all the above cases, the value of fisheries and agriculture would not change as these activities are assumed to remain constant both within and between scenarios because they are apparently sustainable in the current system.

Results of the financial cost/benefit analysis indicate that over the short term (zero to five years) there are economic costs for the local inhabitants if more sustainable resource use is established. These costs are estimated at around 21 percent of the annual financial income that would be earned if the current unsustainable system is maintained (Table 9.11). These costs are incurred mainly from lost hunting revenue, lost timber revenue, and lost revenue from harvesting forest palm resources. Lost hunting revenue would result from a decrease in the harvests of game species that are not sustainably hunted, which include primates, tapirs, edentates, marsupials, and carnivores. Harvests of deer, peccaries, and large rodents would be maintained at current levels. Harvesting of timber and palm resources would be stopped.

Over the long term (6–30 years), results suggest that there would be financial benefits for local people if a more sustainable system were established. These benefits are estimated at around 25 percent above the annual income that would be earned if current practices were continued. These benefits would come from increased revenue earned from

Table 9.11. Cost/Benefit Analysis of Resource Management Systems for Inhabitants of the Settlement Zone of the Reserva Comunal Tamshiyacu-Tahuayo

	Annual revenue (US$)	
Resource management system	Short term (Years 0–5)	Long term (Years 6–30)
Current unsustainable system		
Hunting	18,000	4,500
Timber from natural forest	18,000	0
Nontimber plant extraction from natural forest	12,000	0
Fisheries	43,000	43,000
Agriculture	71,000	71,000
Fruits from agroforestry	0	0
Timber from agroforestry	0	0
Total	162,000	118,500
More sustainable system		
Hunting	13,500	13,500
Timber from natural forest	0	0
Nontimber plant extraction from natural forest	0	0
Fisheries	43,000	43,000
Agriculture	71,000	71,000
Fruits from agroforestry	0	12,000
Timber from agroforestry	0	18,000
Total	127,500	157,500

hunting, fruit production of agroforestry plots, and timber production in agroforestry plots and in naturally managed forests. Hunting would have greater yields than the base case because species would not become locally extinct, and the carrying capacity of large-bodied game would be increased because of the greater abundance of forest palm fruits. Palm fruit and timber production would yield considerable income, whereas these resources will become locally extinct if the current resource uses are maintained. Over the long term, more sustainable use of natural resources would have annual revenues similar to those currently earned.

In contrast, maintaining the current system (the base case) over the long term would generate less annual revenue because there would

Table 9.12. Present Value of Benefits for Two Resource Use Systems for Various Discount Rates in the Reserva Comunal Tamshiyacu-Tahuayo

Discount rate (%)	Present value of benefits (US$ over 30-year period)	
	Unsustainable	More sustainable
5	2,241,364	2,467,260
10	1,444,410	1,499,136
12	1,273,316	1,287,600
15	1,085,815	1,061,060
20	881,374	821,007

be an estimated 75 percent decrease in the revenue earned from hunting (because of effects of habitat destruction and overhunting) and no income derived from either palm resources or from timber, due to overexploitation (Table 9.11). Obviously, the main financial advantage of converting the current resource uses to more sustainable resource uses is that annual revenues from the more sustainable system could be maintained for much longer periods of time and would greatly exceed the costs incurred over the short term.

People of the RCTT should accept the more sustainable resource use system if their discount rates (the degree to which people are willing to postpone economic benefits) are 12 percent or lower, according to the Net Present Value analysis. However, if their discount rates are above 12 percent, then the unsustainable resource use system is more financially profitable (Table 9.12).

The current unsustainable practices of the people of the RCTT might reflect a high discount rate (exceeding 12%), which results from their poverty. It does not appear that the high discount rate is based on any luxury demands. Indeed, the subsistence use of resources by people of the RCTT does not cover all of their basic needs, and products must be sold by the people to acquire goods such as soap, rice, sugar, cooking oil, and kerosene, among other necessities.

However, a high discount rate may not be the only reason why people of the RCTT are currently using an unsustainable system. People of the RCTT might be using certain resources unsustainably because of open access. The resources that are more easily protected by the people, such as agricultural plots and the lakes close to their villages, are not overexploited. Resources that are more distant and for which control of access is more difficult, such as game species, wild palms, and timber, are overexploited. In addition, a lack of information concerning the level of overuse and the potential for converting the unsustainable practices to more sustainable practices may also be a reason for the overexploitation.

Alternatives for Overcoming Short-Term Costs

Local inhabitants can establish more sustainable resource uses only if short-term costs are overcome by decreasing the discount rates of the people. Further lowering revenues would not be acceptable to many families of the RCTT because of their poverty and because it would only increase the discount rate of the people. Yet, if a more sustainable system is not initiated, poverty will eventually increase significantly as resources become depleted. Thus alternative strategies should be developed that allow inhabitants to overcome these short-term costs and realize long-term benefits. There are three possible strategies for overcoming short-term costs: (1) partitioning and staggering management programs for more sustainable resource uses, (2) developing economic alternatives to replace lost revenues, and (3) obtaining economic aid programs.

Staggering management programs would spread the economic costs over a longer period. This would enable local people to absorb economic costs more easily, as opposed to bearing all of the costs at once. Strategies could also be developed that would set up a management program for a given resource only when the economic benefits of a previous management program are realized. For example, the reduction in cutting of forest palm trees might begin only when the production of agroforestry fruits compensates for lost revenue. This might entail a six- to eight-year delay between the planting of agroforestry trees and the cessation of cutting wild trees.

In fact, people of the RCTT have already taken this strategy. For example, fisheries management in the upper Tahuayo began in 1984, timber extraction was reduced in 1989, and people initiated game management in 1994. Thus short-term costs of these management programs have been spread over a decade. The increased fisheries yield that has resulted from the fisheries management program over the past several years has enabled local people to consider other management

Table 9.13. Estimated Annual Household Budget for a Typical Six-Member Family of Upper Tahuayo River Basin, 1989

Activity	Estimated expenditures	
	US$	Percent of total
Expendable supplies	130	36.4
Transportation	66	18.5
Clothing	50	14.0
Social events	40	11.2
Schooling	36	10.1
Health care	20	5.6
Tools	15	4.2

Source: Coomes 1992a.

plans. Likewise the increase in game after the reduction of hunting by lumbermen has enabled people to consider additional game management programs. Once palm fruits become abundant in agroforestry plots, people will be able to reduce the cutting of wild fruit trees.

Another strategy for overcoming short-term costs is to develop economic alternatives. For example, chicken farming developed by the women's clubs might substitute income and protein lost from establishing more sustainable hunting practices. However, many of these economic alternatives require capital investment for their initiation. Credit for such investments has often not been available to people of the RCTT, which in turn has hindered the implementation of economic alternatives for local people. In addition, research must be conducted to assess the sustainability of alternatives and their impact on the ecosystem.

The last alternative would be to obtain an aid program that could offset the losses incurred over the short term. This could be in the form of either cash credit or in-kind aid. Past experiences with cash credit in the RCTT have failed, in part because of corrupt credit programs, increased social disequilibrium of credit holders, and the inability to pay interest requirements (Coomes 1994). Thus cash credit would probably not function as a viable alternative for short-term costs.

Aid in the form of in-kind assistance is a more feasible alternative to cover short-term costs. For example, such aid could cover health, education, or transportation costs. In the Tahuayo River basin, around 34 percent of the household budget is for health, education, and transportation (Table 9.13) (Coomes 1992a). Subsidizing these expenses would cover the costs of establishing more sustainable resource uses. Considerable extension work would be required to link the resource management programs with health, education, and transportation subsidies. To date, in-kind subsidies have not been realized in the RCTT.

Large aid projects should recognize that initiating more sustainable resource use systems and increasing the short-term income of local people are often incompatible. In contrast, setting up more sustainable resource use systems increases the long-term income of rural inhabitants. Thus large aid projects should be concerned with assisting local people in overcoming short-term costs in order for these people to realize long-term economic benefits.

Conclusions

The RCTT is attempting to bridge the socioeconomic needs of local people and biodiversity conservation through sustainable use of natural resources. In turn, sustainable use of natural resources is being approached through community-based resource management. The three zones of the RCTT form a continuum of land uses, from intensive use to full protection, that both conforms with the needs of the local people and helps conserve biodiversity. Ecological sustainability in the Amazon cannot be achieved without establishing different land-use zones, including fully protected areas. Conversely, economic development cannot be realized without intensive land use and natural resource extraction.

This case demonstrates that biodiversity conservation in the RCTT can be compatible with the livelihoods of local communities. However, to accomplish this the following conditions must be met:

1. The majority of people living in the permanent settlement zone must maintain an adequate livelihood from land uses outside the reserve. These people will not need to use the reserve for natural resources if they use land both intensively and sustainably in the permanent settlement zone. This proposition recognizes that the Amazon is a delicate environment and can support only a limited amount of natural resource use, if this use is to be sustainable. Thus the RCTT is similar to other areas where sustainable use of natural resources can support only low human population densities (Salafsky 1994).

2. People currently using the subsistence zone of the reserve should harvest resources sustainably. This will require more attention to nat-

ural resource management by local communities, especially for wildlife hunting, palm harvests, and timber extraction. Using resources sustainably in the subsistence zone will maintain the value of the forest for the people who use this area. These people will maintain an interest in conserving the subsistence zone of the RCTT. In addition, the diversity of products which the people currently harvest from the subsistence zone must be maintained. This should also help conserve biodiversity, as intact habitat will be needed to maintain the diversity of resources used.

3. Ways must be found to overcome the short-term financial costs needed to establish more sustainable resource uses that will lead to long-term financial and economic benefits.

Overall, this case has shown that more sustainable resource use is a real possibility in the Peruvian Amazon, and that more sustainable use will help conserve biodiversity. However, economic and social considerations must be carefully incorporated into the management programs.

Acknowledgments

We wish to thank the people of the RCTT, and Carlos Renifo, Doris Diaz, I. Vilchez, and F. Encarnación for their kind assistance. Logistical and financial support was provided by the World Wildlife Fund, Chicago Zoological Society, the Tropical Conservation and Development Program of the University of Florida, Wildlife Conservation Society, Proyecto Peruano de Primatología "Manuel Moro Sommo," INRENA–Ministerio de Agricultura, the Rainforest Conservation fund of Chicago, the Amazon Conservation Fund, and the Biodiversity Support Program. Oliver Coomes generously allowed us to use material from his dissertation. John Robinson, Curtis H. Freese, Clyde Kiker, and Oliver Coomes provided useful comments. We are particularly grateful to Gretchen Greene, who provided much insight on the economic analysis.

References

Anderson, A. B. 1990. *Alternatives to deforestation: Steps toward sustainable use of the Amazon rain forest.* New York: Columbia University Press.

Ayres, J. M. 1993. *As matas de várzea do Mamirauá.* Brasilia, Brazil: Conselho Nacional de Desenvolvimiento Científico e Tecnologico-CNPq.

Ayres, J. M., and T. H. Clutton-Brock. 1992. River boundaries and species range size in Amazonian primates. *American Naturalist* 140:531–537.

Bartecki, U., E. Heymann, R. E. Bodmer, L. Moya I., and T. Fang. 1986. *Diagnóstico situacional de la Zona de Estudios Quebrada Blanco en el Rio Tahuayo y propuesta para el establecimiento de una Reserva Comunal.* Reporte al Ministerio de Agricultura, Región de Loreto, Iquitos, Peru.

Bayley, P. B., and M. Petrere, Jr. 1989. Amazon fisheries: Assessment methods, current status and management options. In *Proceedings of the International Large River Symposium,* ed. D. P. Dodge, 385–398. Canadian Special Publications in Fisheries and Aquatic Sciences, Vol. 106.

Bendayán A., N. Y. 1990. Influencia socioeconómica de la fauna silvestre en Iquitos, Loreto. Thesis, Universidad Nacional de la Amazonia Peruana.

Bodmer, R. E. 1989. Frugivory in Amazonian Artiodactyla: Evidence for the evolution of the ruminant stomach. *Journal of Zoology* 219:457–467.

Bodmer, R. E. 1990a. Fruit patch size and frugivory in the lowland tapir. *Journal of Zoology* 222:121–128.

Bodmer, R. E. 1990b. Responses of ungulates to seasonal inundations in the Amazon floodplain. *Journal of Tropical Ecology* 6:191–201.

Bodmer, R. E. 1994. Managing wildlife with local communities: The case of the Reserva Comunal Tamshiyacu-Tahuayo. In *Natural connections: Perspectives on community based management,* ed. D. Western, M. Wright, and S. Strum, 113–134. Washington, D.C.: Island Press.

Bodmer, R. E. 1995. Managing Amazonian wildlife: Biological correlates of game choice by detribalized hunters. *Ecological Applications* 5:872–877.

Bodmer, R. E., T. G. Fang, and L. Moya I. 1988. Ungulate management and conservation in the Peruvian Amazon. *Biological Conservation* 45:303–310.

Bodmer, R. E., T. G. Fang, and L. Moya I. 1990a. Fruits of the forest. *Nature* 343:109.

Bodmer, R. E., J. Penn, T. G. Fang, and L. Moya I. 1990b. Management programmes and protected areas: The case of the Reserva Comunal Tamshiyacu-Tahuayo, Peru. *Parks* 1:21–25.

Bodmer, R. E., N. Y. Bendayán A., L. Moya I., and T. G. Fang. 1990c. Manejo de ungulados en la Amazonía Peruana: Análisis de la caza de subsistencia y la comercialización local, nacional e internacional. *Boletín de Lima* 70:49–56.

Bodmer, R. E., T. G. Fang, L. Moya I., and R. Gill. 1994a. Managing wildlife to conserve Amazonian forests: Population biology and economic considerations of game hunting. *Biological Conservation* 67:29–35.

Bodmer, R. E., P. Puertas, L. Moya I., and T. G. Fang. 1994b. Estado de las poblaciones del tapir en la Amazonía Peruana: En el camino de la extinción. *Boletín de Lima* 88:33–42.

Brack, A. 1992. Nontimber forest products of the Peruvian Amazon. In *Sustainable harvest and marketing of rain forest products,* ed. M. Plotkin and L. Famolare, 90–98. Covelo, California: Island Press.

Browder, J. O. 1992. The limits of extractivism: Tropical forest strategies beyond extractive reserves. *BioScience* 42:174–182.

Casa Campesina: COPAPMA, FECADEMA, COPAPLO, COPAPRACAS. 1985. *Analisis y posición sobre el programa regional de bufalos.* Iquitos, Peru: Centro de Estudios Tecnológicos Amazónias Press.

Castro, N. R. 1991. Behavioral ecology of two coexisting tamarin species (*Saguinus fuscicollis nigrifrons* and *Saguinus mystax myatax,* Callitrichidae, Primates) in Amazonian Peru. Doctoral dissertation, Washington University, Saint Louis, Missouri.

CETA (Centro de Estudios Tecnológicos Amazónias). 1986. *Manual para promotores en salud.* Iquitos, Peru: CETA Press.

Chibnik, M. 1994. *Risky rivers: The economics and politics of floodplain farming in Amazonia.* Tucson: University of Arizona Press.

Clutton-Brock, T. H. 1988. Reproductive success: Studies of individual variation in contrasting breeding systems. Chicago: University of Chicago Press.

Coomes, O. T. 1991. Rainforest extraction, agro-forestry, and biodiversity history from the northeastern Peruvian Amazon. Paper presented to the 16th International Congress of the Latin American Studies Association. Washington, D.C.

Coomes, O. T. 1992a. Making a living in the Amazon rain forest: Peasants, land, and economy in the Tahuayo River basin of northeastern Peru. Doctoral dissertation, University of Wisconsin, Madison.

Coomes, O. T. 1992b. Blackwater rivers, adaptation, and environmental heterogeneity in Amazonia. *American Anthropologist* 94:698–701.

Coomes, O. T. 1994. Helping the poor? Agrarian populism and the peasantry in the Peruvian Amazon: Lessons for rural development policy from the APRA experience. Paper presented at the conference Biodiversidad y Desarrollo Sostenible de la Amazonia, en una Economía de Mercado, Pucallpa, Peru.

Coomes, O. T. 1995. A century of rain forest use in western Amazonia: Lessons for extraction-based conservation of tropical forests. *Forest Conservation and History* 39:108–120.

Denevan, W. M., and C. Padoch. 1988. Swidden-fallow agroforestry in the Peruvian Amazon. *Advances in Economic Botany* 5:107.

Dourojeanni, M. J. 1990. *Amazonia? Que hacer?* Iquitos, Peru: Centro de Estudios Tecnológicos Amazónias.

Dourojeanni, M. J., and C. F. Ponce. 1978. *Los parques nacionales del Perú.* Madrid: INCAFO.

Egoavil, E. O. 1992. *Perfil demográfico de la región Loreto.* Iquitos, Peru: Instituto de Investigaciones de la Amazonia Peruana.

Ehrlich, P. R., and G. C. Daily. 1993. Science and the management of natural resources. *Ecological Applications* 3:558–560.

Furch, K. 1984. Water chemistry of the Amazon basin: The distribution of chemical elements among freshwaters. In *The Amazon: Limnology and landscape ecology of a mighty tropical river and its basin,* ed. H. Siolo, 167–199. Dordrecht: Dr. W. Junk Publishing.

Gentry, A. H. 1988. Tree species richness of upper Amazonian forests. *Proceedings of the National Academy of Sciences* 85:156–159.

Gill, R. 1990. *Monitoring the status of European and North American cervids.* Nairobi, Kenya: United Nations Environment Programme.

Ginsberg, J. R., and E. J. Milner-Gulland. 1994. Sex-biased harvesting and population dynamics in ungulates: Implications for conservation and sustainable use. *Conservation Biology* 8:157–166.

Gorchov, D. L. 1994. Natural forest management of tropical rain forests: What will be the "nature" of the managed forest? In *Beyond preservation: Restoring and inventing landscapes,* ed. A. D. Baldwin, Jr., J. de Luce, and C. Pletsch, 136–153. Minneapolis: University of Minnesota Press.

Goulding, M. 1980. *Ecologia da pesca do rio Madeira.* Manaus, Brazil: Instituto Nacional de Pesquisas da Amazonia.

Hiraoka, M. 1986. Zonation of mestizo farming systems in Northeast Peru. *National Geographic Research* 2:354–371.

Hiraoka, M. 1989. Riverine subsistence patterns along a blackwater river in the northeast Peruvian Amazon. *Journal of Cultural Geography* 9:103–119.

Holling, C. S. 1993. Investing in research for sustainability. *Ecological Applications* 3:552–555.

Junk, W. J., and K. Furch. 1985. The physical and chemical properties of Amazonian waters and their relationships with the biota. In *Key environments: Amazonia,* ed. G. T. Prance and T. E. Lovejoy, 3–17. Oxford: Pergamon Press.

Kahn, F. 1988. Ecology of economically important palms in Peruvian Amazonia. *Economic Botany* 6:42–49.

Kahn, F., K. Mejia, and A. de Castro. 1988. Species richness and density of palms in *Terre Firme* forests of Amazonia. *Biotropica* 20:266–269.

Kiltie, R. A., and J. Terborgh. 1983. Observations on the behavior of rain forest peccaries in Peru: Why do white-lipped peccaries form herds? *Zeitschrift für Tierpsychologie* 62:241–255.

Levin, S. A. 1993. Science and sustainability. *Ecological Applications* 3:545–546.

Lima, D. 1991. Kin saints and the forest: The study of Amazonian caboclos in the middle Solimões region. Unpublished doctoral dissertation, University of Cambridge.

Lopes A., J. P., H. Albuquerque M., Y. Silva R., and R. Shrimpton. 1980. Aspectos nutritivos de alguns frutos da Amazônia. *Acta Amazonica* 10: 755–758.

McCullough, D. R. 1987. The theory and management of Odocoileus populations. In *Biology and management of the Cervidae,* ed. C. Wemmer, 415–429. Washington, D.C.: Smithsonian Institution Press.

McDaniel, J. 1994. Fisheries management in the Peruvian Amazon. Unpublished manuscript, University of Florida, Gainsville, Florida.

Moya I., L., F. C. Encarnación, H. A. Velásquez, J. A. Moro, J. W. Penn, and R. E. Bodmer. 1991. *Propuesta técnica para la Reserva Comunal Tamshiyacu-Tahuayo,* Gobierno Regional de Loreto. December 1991.

Moya I., L., and J. Rimachi. 1988. *Diagnostico socio-económico de los centros poblados aledanos a la propuesta de la Reserva Comunal Quebrada Blanco del Río Tahuayo, Loreto, Perú.* Informe Técnico, Ministerio de Agricultura, Loreto, Peru.

Padoch, C. 1987. The economic importance and marketing of forest and fallow products in the Iquitos region. *Advances in Economic Botany* 5:74–89.

Padoch, C. 1988. People of the floodplain and forest. In *People of the tropical forest,* ed. J. S. Denslow and C. Padoch, 127–141. Berkeley: University of California Press.

Penn, J. W. 1992. Agroforestería en la zona de asentamiento de la Reserva Comunal Tamshiyacu-Tahuayo: La reforestación y agroforesteria en el Río Tahuayo. Paper prepared for the First Workshop on Development and Conservation in the Reserva Comunal Tamshiyacu-Tahuayo, Proyecto Peruano de Primatología–IVITA, Iquitos, Peru.

Penn, J. W. 1993. *Plan de Trabajo 1993–1994. Amazon Conservation Fund: El grupo de apoyo para la Reserva Comunal Tamshiyacu-Tahuayo.* Extension work plan submitted to the Dirección Regional de Recursos Naturales y de Medio Ambiente, Región Loreto, Peru.

Penn, J. W. 1994. Agroforestería orientada a la fauna silvestre y necesidades humanas: Desafíos y realidades de la Reserva Comunal Tamshiyacu-Tahuayo. *KANATARI* 490:6–7 and 493:5, 11.

Penn, J. W., and J. A. Alvarez. 1990. Comunidad campesina protege sus cochas. *KANATARI* 258:3,10.

Penn, J. W., C. Rengifo, G. Bertiz, D. Diaz, and G. Saenz. 1993. *Educación ambiental: Planificación para el futuro.* Iquitos, Peru: Amazon Conservation Fund.

Peters, C. M., A. H. Gentry, and R. O. Mendelsohn. 1989. Valuation of an Amazonian rainforest. *Nature* 339:655–656.

Phillips, O. 1993. The potential for harvesting fruits in tropical rainforests: New data from Amazonian Peru. *Biodiversity and Conservation* 2:18–38.

Pires, J. M., and G. T. Prance. 1985. The vegetation types of the Brazilian Amazon. In *Key environments: Amazonia,* ed. G. T. Prance and T. E. Lovejoy, 109–145. Oxford: Pergamon Press.

Puertas, P. 1993. *Aportes para el sistema de vigilancia y apoyo comunal en la RCTT área de influencia—Río Yavari-Miri.* Informe Técnico. Iquitos, Peru: Amazon Conservation Fund.

Puertas, P., and R. E. Bodmer. 1993. Conservation of a high diversity primate assemblage. *Biodiversity and Conservation* 2:586–593.

Redford, K. H., and J. G. Robinson. 1987. The game of choice: Patterns of Indian and colonist hunting in the neotropics. *American Anthropologist* 89:650–667.

Rengifo, C. U. 1994. *Field reports 1993.* Iquitos, Peru: Amazon Conservation Fund.

Robinson, J. G., and K. H. Redford. 1994. Measuring the sustainability of hunting in tropical forests. *Oryx* 28:249–256.

Salafsky, N. 1994. Ecological limits and opportunities for community-based conservation. In *Natural connections: Perspectives in community-based conservation,* ed. D. Western, R. M. Wright, and S. C. Strum, 448–471. Washington, D.C.: Island Press.

Uhl, N. W., and J. Dransfield. 1987. Genera Palmarum: A classification of palms based on the work of Harold E. Moore, Jr. Lawrence, Kansas: Allen Press.

Vasquez, R., and A. H. Gentry. 1989. Use and misuse of forest-harvested fruits in the Iquitos area. *Conservation Biology* 3:350–361.

TEN

Sustainable Utilization of Game at Rooipoort Estate, Northern Cape Province, South Africa

Timothy M. Crowe, Bradley S. Smith, Robin M. Little, and S. Hugh High

Sub-Saharan Africa is endowed with a fabulously rich diversity of wildlife. For example, East Africa alone possesses a greater variety of bird species, with more endemics, than the vast forests of Canada, Scandinavia, and Russia combined (Myers 1979). However, Africa south of the Sahara is, to all intents and purposes, crippled socioeconomically (Africa 1994). Nineteen of the region's 40 countries are placed in the most impoverished ("low income") category by "Caring for the Earth" (IUCN, UNEP, and WWF 1991), and most are encumbered by large foreign debts, averaging 58 percent of their gross national products (IUCN 1990). Since the traditional forms of commodity-based economics which have underpinned Africa's economies over the last century have failed to deal with these financial liabilities, increasing pressure is being placed on the utilization of Africa's natural resources, especially for their potential as sources of foreign exchange.

Such utilization is complicated by a range of factors. First and foremost are the dangers of overutilization driven by the desire for short-term profit, or for the poor, short-term survival. Furthermore, in contrast to the relative predictability of variation in temperate ecosystems of the northern hemisphere, climatic fluctuations in Africa are much more variable and less predictable (Tyson 1986; Onesta and Verhoeff 1976), and tend to influence plant and animal populations in an "event-driven" manner. Thus, unlike many ecosystems in the northern hemisphere, and in certain tropical rainforest biotopes, density-independent factors may be more important than density-dependent factors in controlling resource production and availability. For example, factors such as fire, and frequency and intensity of rainfall, play major roles in the dynamics of animal populations (Crowe et al. 1981), and particularly in the seasonal breeding activities and population fluctuations of passerine birds, for example, the quelea (*Quelea quelea*) (Lack 1966) and gamebirds (Crowe 1978; Berry and Crowe 1985; Little and Crowe 1993). Superimposed on this econo-ecological challenge to African conservation biology is the major problem of satisfying the sociocultural requirements of a mosaic of humanity that partitions Africa more finely than does the present system of political boundaries (Crowe et al. 1994).

This case study investigates the biological suitability and financial potential of a range of antelope species on Rooipoort (28°45′S 24°05′E), a privately owned estate in the Northern Cape Province of South Africa. Rooipoort is unique among African protected areas in that its wildlife has been both utilized and monitored on a regular basis since the early 1900s, and its owners (more recently) have kept records of the financial yields from that utilization. Two major limitations of this study are that (1) inherently the available data sets analyzed herein are limited by the absence of an a priori sampling strategy, and (2) time constraints placed on this study meant that no additional data could be collected.

Overview of the Hunting Market

In the United States, which supplies Africa with most of its foreign hunters, some 16 million people buy hunting licenses annually, representing about US$12.3 billion spent directly on hunting paraphernalia (Fair 1993). In Africa, sport/safari hunting is largely the hunting of game animals for trophies such as horns, tusks, and skins. Clients are usually wealthy foreigners who are willing to pay a high price in foreign exchange for quality trophies and an "African experience." Many of these hunters view themselves as conservationists, contributing through their capital input to the particular system within which they hunt.

Although this may not be their primary motivation for hunting, it certainly implies that, under the current pervading ethos of sustainable utilization held by most international conservation agencies, the demand for hunting is unlikely to decrease internationally in the immediate future.

Within South Africa, the average overseas hunter spends seven times as much per day as a "normal" tourist and requires less "high-tech" facilities (Meiring 1994). Hunting opportunities in the United States are diminishing as rural land is increasingly developed or privatized (Fair 1993), and many of these hunters are looking to Africa for continued hunting experiences. Southern Africans are also becoming interested in game hunting, and local demand is on the increase (various professional hunters, pers. comm.). The South African private sector has reacted to this perceived demand, and many operations are now offering cheaper "local" (southern African) tariffs to hunt quality animals that fall just short of international trophy standards. Furthermore, in some areas, farmers in South Africa are looking to form hunting corporations, where several farms sharing common boundaries are amalgamated to form one large hunting area. This practice is particularly common in drier areas where cattle ranching has always been marginal (Merten 1993), and is probably influenced by the changing South African political climate, in which the best possible utilization of land is a priority (to avoid the threat of expropriation by a future government).

In South Africa, turnover from live game sales for the 1993 season was 10 percent higher than the previous year (Merten 1993), indicating a high demand for live game to restock existing hunting operations. Commercial venison production has a relatively short history in South Africa, where the majority of meat consumers utilize beef, mutton, and pork. The 1992 figures for meat production in South Africa show a total of approximately 770,000 metric tons (MT) for these three categories combined (South African Meat Board 1993). Venison production is not as well regulated as livestock production in South Africa. Consequently, estimates for total production are fairly crude, as informal slaughtering and sales do occur. Nevertheless, for the same period, venison production across the country was an estimated 1,000 MT (Walter Luthley, managing director, SA Venison, pers. comm. 1994), or about a tenth of 1 percent of total meat production. Of this 1,000 MT, only about 180 MT were exported to the European marketplace due to strict European Commission regulations. About 150 MT comprised deboned springbuck (*Antidorcas marsupialis*) meat bound for the German market, where it rates as top-quality venison. South Africa thus supplies a very small percentage of the international venison demand, and the potential for increasing market share is high (Walter Luthley, pers. comm. 1994). Local market demand is also increasing, and the potential for venison production in South Africa is highly underdeveloped, largely due to the monopoly held by beef and mutton producers (Bertie Ackhurst, Public Relations, Meat Hygiene, pers. comm. 1994).

The Northern Cape Community

The Northern Cape consists of approximately 282,000 square kilometers (km^2), making it the second largest region in South Africa. In contrast, it is populated by only 1.1 million people (3.9 persons per km^2), which is the least populated of any of the nine development regions in South Africa and, since 1980, has experienced the lowest rate of population growth (1.31%) of any of the regions. Nearly half of the population lives in urban areas.

Kimberley is the headquarters of De Beers Consolidated Mines Limited and a major center of some of its diamond mining activities (Fig. 10.1). The passage of time has brought a diminution of mining in Kimberley. However, the city continues to serve as an important seat of worldwide operations for De Beers. In addition, Kimberley will serve as the new provincial capital of the Northern Cape Region and, as such, is a major focal point for the agricultural sector which dominates much of the commercial activity of the region.

Real gross geographic product (GGP; comparable to gross domestic product) of the region in 1988, the latest year for which data are available, was US$945 million, US$873 per capita. The region fell in the middle range of all regions in terms of both regional and per capita GGP. It should be noted, however, that the rate of growth in real GGP, between 1980 and 1988, was the next to lowest of any of the nine development regions in the country. Thus, although regional real per capita GGP was in the middle ranges, there was clearly a decline in the region's prosperity relative to that of other regions of South Africa. This is reflected by a labor force participation rate (percent of total labor force over age 16 employed) of 52.5, which is also in the middle

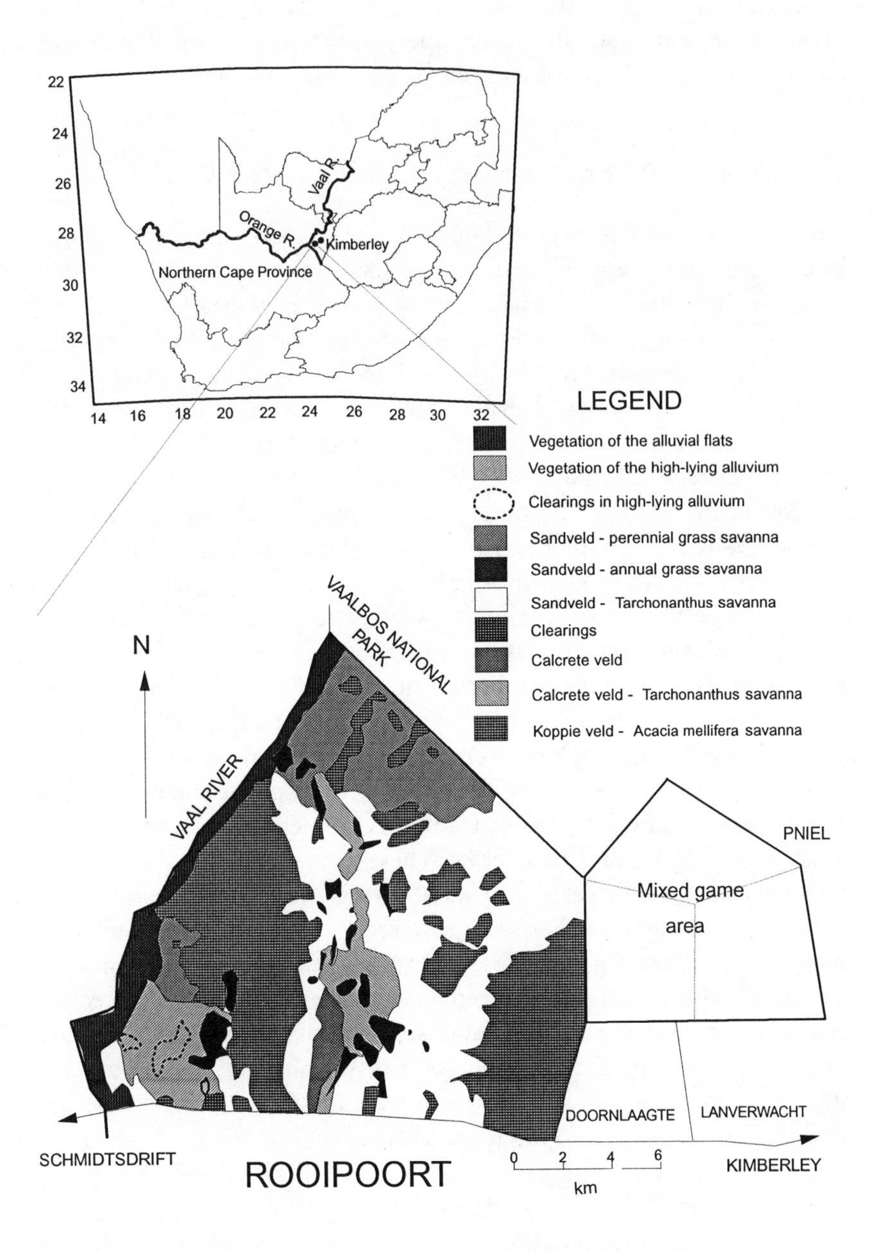

Figure 10.1. Map showing the position of the Northern Cape and Kimberley, and the vegetation types (from Crowe et al. 1981), location, and boundaries of Rooipoort Estate.

ranges of all regions. In summary, the Northern Cape could benefit greatly from commercially viable operations which complement agriculture and keep rural populations on the land.

Rooipoort Estate: History, Habitat, and Wildlife

Although the involvement of South Africa's largest diamond mining house in wildlife management and conservation may seem somewhat anomalous, a series of historical events placed De Beers Consolidated Mines, and its associated companies, in a unique position to contribute toward conservation of biodiversity in South Africa. A brief examination of these historical events facilitates further insight into the methods and management practices currently employed by De Beers in their Farms Department.

In the late 1800s, diamonds were discovered near Kimberley, and the old De Beers Mining Company secured most of the mining rights in the area. De Beers has since been a dominant landowner and employer in the Kimberley region. In 1891, the company formulated a land acquisition policy of systematically buying potentially diamondiferous farms around the Kimberley mine. In 1893, eight properties 65 kilometers (km) west of Kimberley were bought at a public auction from Sheasby Estate. These were to form the nucleus of Rooipoort Estate (Fig. 10.1). A few years later, partly in response to criticism for not using its land productively, De Beers commenced commercial livestock ranching around Kimberley. These activities were often not financially successful, several years showing thousands of pounds in losses. They were nevertheless large in scale and aided in the establishment of stock farming in the district. This ranching continued until the 1930s, when the depression, together with a severe drought, forced De Beers to curtail its farming operations. All farms except Rooipoort and another estate were set under lease to private farmers, and De Beers itself ceased to farm. Some 40 years later this decision was reversed and De Beers returned to intensive cattle ranching.

By the time De Beers had acquired its Kimberley farms, the growing population of the diamond fields was making heavy demands on the surrounding countryside. Most of the trees and shrubs for more than 100 km around Kimberley had been felled for fuel or mining timber, and heavy subsistence shooting had eradicated much of the larger game species. From the outset, the De Beers directors took an active interest in the game on their farms. Many were keen hunters and advocated sustainable hunting as a means of population management. De Beers instituted a conservation policy on its farms to control the public hunting taking place, not an easy task with a population that had grown accustomed to moving freely over the land, hunting what it fancied. In 1909, the company announced its intention to preserve game on its farms, and hunting without company permission became strictly prohibited. From this point until the present, there has been a resident gamekeeper on Rooipoort.

The turn of the century saw the start of a campaign by De Beers to restock its farms with previously eradicated as well as new antelope species. Springbuck, red hartebeest (*Alcelaphus buselaphus*), and kudu (*Tragelaphus strepsiceros*) numbers were initially allowed to build up naturally. From its small-scale inception around 1897, this campaign culminated in black wildebeest (*Connochaetes gnou*), eland (*Taurotragus oryx*), gemsbok (*Oryx gazella*), impala (*Aepyceros melampus*), blue wildebeest (*Connochaetes taurinus*), and giraffe (*Giraffa camelopardalis*) being released onto the De Beers farms by 1965. By the mid-1940s, the De Beers farms around Kimberley had become game-rich islands in a sea of properties that had retained little or no wildlife. At this stage, Rooipoort comprised 42,000 hectares (ha), which is the present area enfenced as a contiguous game reserve.

Following World War II, there was a shift toward a broader utilization policy at Rooipoort, as well as more emphasis on scientific management of utilization activities. This broader utilization today includes culling for venison markets, sport hunting for trophies, social hunting, and live capture of animals for resale.

Large-scale culling began in 1930 when the De Beers board gave permission for 1,500 springbuck to be culled annually on the farms. For several years thereafter, springbuck and small numbers of other antelope were shot and marketed annually. The advent of helicopter cropping in the 1970s led to intensive venison production, mainly of springbuck. Currently 1,500 to 2,000 springbuck are taken annually by appointed commercial contractors on a predetermined sex/age basis, previously for the international venison market, but for the past three years only for the local market.

Social hunting on Rooipoort has always been offered to De Beers directors and distinguished visitors as a noncompensation company privilege. With the rise of hunting associated with international tourism, hunting on the company's properties became managed more scientifically, with prized animals, particularly the larger antelope, being offered to international trophy hunters on a fee basis. De Beers, aware of the large unsatisfied demand for hunting facilities among the South African public, has also opened certain areas for local hunting parties at cheaper rates.

During the 1950s, a general awakening interest in restocking farms with game led to requests from farmers for live animals. De Beers, having well-stocked lands by this stage, began live game capture operations to meet this demand. During the following decades, techniques were refined, and the capture of game became the most important form of utilization. De Beers is now the biggest single private supplier of many antelope species in South Africa, to both state-owned nature reserves and the private game industry.

For the greater part of this century, the De Beers farms were managed centrally by the company Head Office. Management emphasis during this period was on a caretaker role to keep the farms going, thereby enabling social hunts. In the late 1970s, there was a subtle policy shift, and the idea emerged that De Beers had a responsibility to its shareholders to maximize profits across all its operations. Since its farms were a potential profit drain, the appropriately trained people were employed to ensure that the farms became self-sufficient financially, and a semiautonomous Farms Department was established to manage the De Beers farming operations.

The De Beers management has restructured the company into subdivisions or departments, with a high degree of autonomy being granted to the management of each of these "profit centers," subject to the requirement that each department make a profit for the company. However, despite this decentralization, a number of company decisions regarding the farms are made centrally. From our perspective, this is important since, *inter alia,* decisions regarding depreciation and some capital expenditures are made centrally (see the discussion about data and its limitation, below). Thus, by combining livestock production with the preservation of rare species and the utilization of more common ones, De Beers has established itself over the decades as a pioneer in multipurpose land use in South Africa. Furthermore, the continuous block of 42,000 ha, fenced on three sides and bounded on the fourth side by the Vaal River (Fig. 10.1), today is the largest privately owned representative reserve of the Kalahari savanna biome in the Northern Cape (Bigalke 1985; G. Maine, managing director, De Beers Farms Dept., pers. comm. 1994).

From a socioeconomic perspective, Rooipoort supports 28 rural black families (totaling 102 people) on a permanent basis (average salary about US$300 per month) and about 1,600 labor-days for temporary staff (average daily pay US$5). Ten of the 28 permanent employees are involved directly with game operations.

De Beers has recently initiated a pilot program involving tourism and education, whereby limited access is granted to inhabitants of the local area. This program includes guided tours and educational talks on wildlife conservation. While this program is in its infancy, and is currently running at a loss, De Beers management believes that the project will eventually be profitable, while conferring considerable educational and social benefits on the rural community in the region.

Methods

Big Game Data

The species richness of medium-to-large game species currently at Rooipoort was compared with the evidence of historical incidence of mammals (Skead 1989) and with recently described ranges of these species (Smithers 1990) in the Northern Cape. This comparison reflects the level of contribution toward conserving biodiversity afforded by the management of game species at Rooipoort. Hereafter, sustainable utilization of these game species has been evaluated according to available data on annual population counts and annual removal levels.

Springbuck hunting for venison has featured strongly at Rooipoort since the late 1930s, whereas commercial trophy hunting for big game species including springbuck has been recorded only since the 1980s. Indeed, some of the more spectacular species, such as sable (*Hippotragus niger*), have been introduced to the estate only in the last 10 years. Hunting of the species listed in Table 10.1 is controlled strictly under the supervision of a resident certified professional hunter. Bag limits are determined from annual aerial counts made during March and April, and offtake is limited to 20 percent of animals counted.

Annual population estimates are derived as an index of abundance from the number of individuals of a species counted from a helicopter while flying along fixed transects at fixed height and at fixed speed.

Table 10.1. Indices of Abundance of Indigenous Medium and Large Mammals Found Historically and Currently in the Northern Cape and Currently at Rooipoort Estate[a]

Mammal	Historical Northern Cape	Current Northern Cape	Current Rooipoort
Order Primates			
Family Cercopithecidae			
Chacma baboon (*Papio ursinus*)	3	3	3
Order Carnivora			
Family Canidae			
Cape Fox (*Vulpes chama*)	2	1	1
Bat-eared fox (*Otocyon megalotis*)	3	3	3
Black-backed jackal (*Canis mesomelas*)	3	3	3
Families Mustelidae and Vivirridae			
Various species	3	3	3
Family Hyaenidae			
Brown hyena (*Hyaena brunnea*)	1	1	1
Spotted hyena (*Crocuta crocuta*)	2	1	0
Aardwolf (*Proteles cristatus*)	3	3	2
Family Felidae			
Black-footed cat (*Felis nigripes*)	3	3	3
Caracal (*F. caracal*)	2	1	1
Leopard (*Panthera pardus*)	2	1	?
Lion (*P. leo*)	2	0	0
Cheetah (*Acinonyx jubatus*)	2	0	0
Order Tubulidentata			
Family Orycteropodidae			
Antbear (*Orycteropus afer*)	2	2	2
Order Proboscidae			
Family Elephantidae			
African elephant (*Loxodonta africana*)	2	0	0
Order Perissodactyla			
Family Rhinocerotidae			
Black rhinoceros (*Diceros bicornis*)	3	1	0
White rhinoceros (*Ceratotherium simum*)	2	1	0
Order Perissodactyla			
Family Equidae			
Quagga (*Equus quagga*)	3	Extinct	—
Burchell's zebra (*E. burchelli*)	?	2	2
Order Artiodactyla			
Family Suidae			
Warthog (*Phacochoerus aethiopicus*)	1	1	3

Table 10.1. (*continued*)

Mammal	Historical Northern Cape	Current Northern Cape	Current Rooipoort
Family Hippopotamidae			
Hippopotamus (*Hippopotamus amphibius*)	3	0	0
Family Giraffidae			
Giraffe (*Giraffa camelopardalis*)	2	?	1
Family Bovidae			
Gray duiker (*Sylvicapra grimmia*)	1	1	1
Steenbok (*Raphicerus campestris*)	2	1	2
Moutain reedbuck (*Redunca fulvorufula*)	2	1	2
Impala (*Aepyceros melampus*)	2	2	1
Springbuck (*Antidorcas marsupialis*)	1	2	3
Gemsbok (*Oryx gazella*)	2	2	3
Roan antelope (*Hippotragus equinus*)	2	1	0
Sable (*H. niger*)	?	?	1
Blue antelope (*H. leucophaeus*)	?	Extinct	0
Tsessebe (*Damaliscus lunatus*)	?	?	1
Blesbok (*Damaliscus dorcas phillipsi*)	2	2	2
Red hartebeest (*Alcelaphus buselaphus*)	2	2	3
Black wildebeest (*Connochaetus gnou*)	3	2	2
Blue wildebeest (*C. taurinus*)	2	2	3
Kudu (*Tragelaphus strepsiceros*)	3	3	3
Eland (*Taurotragus oryx*)	3	3	3
Cape buffalo (*Syncerus caffer*)	2	1	0

Sources: Historical Northern Cape: Skead 1989; Current Northern Cape: Smithers 1990; Current Rooipoort: De Beers Farms Department, Management Records.
[a]?, Questionable occurrence; 0, absent; 1, uncommon; 2, common; 3, abundant.

Animals are counted within an "effective" strip width determined by the species conspicuousness, assuming that detection decreases with increasing distance from the helicopter. Caution is taken during counts to avoid counting any individual, or group of individuals, twice along any transect.

Analysis of trends in the population indices of 20 species for the period 1982–93, and of trends in the removal of 17 species for trophies (1982–93), and 9 species for live capture (1981–93) were conducted using the Spearman's rank correlation coefficient ($r_s = 1 - 6\Sigma d^2/n^3 - n$; Zar 1984).

Financial Analysis

Extensive financial data were gathered from the De Beers Farms Department Head Offices in Kimberley. Details of assets, income, and expenditure were recorded during confidential access to the Farms Department general ledger. These data were compared with biological data for the game and cattle ranching operations. A direct profit/loss analysis was used to investigate the profitability of each operation. Comparisons were made between the two types of game operations, and between game operations as a whole and cattle operations to determine their relative profitability. Game ranching operations include hunting (trophy and recreational/sport hunting) and game sales (live sales and venison). The rate of return on operating expenses was calculated for each operation. These calculations required that several assumptions about the data be made. These assumptions, along with our methodology, are outlined in the appendix. Personal interviews were conducted with the general manager of the Farms Department, the administration officer, the chief ecologist of De Beers, and farm managers to glean other relevant social information.

Data and Limitations

Through the courtesy of the De Beers farms, we have been given extensive financial data on game ranching operations for 1987–93. Similarly, we have data on sales, and purchases and direct expenses, as well as expenses for capital improvements, for the cattle operation for 1988–93. Thus direct comparisons of the relative profitability of game and cattle operations are possible for a six-year period. Additionally, highly detailed data on both the company's cattle and game operations are available for 1993.

Before proceeding to an analysis of the data, several observations or caveats, made necessary by the limitations of the data, are in order. First, the observations and conclusions are *static,* that is, there is no attempt herein to include time as a factor. We make no apologies for this since this is a case study and, as such, we are simply interested in determining whether game operations such as those at Rooipoort are economically and financially viable. To introduce dynamic analyses would be to invoke a variety of assumptions, many of which may be both contentious and subject to debate.

Second, while there is a robust literature in economics about joint products, and about the multiplant, multiproduct firm, which would, in principle, permit further inquiry into the real cost of game and cattle operations, such sophisticated methods cannot be readily employed with conventional and limited financial and accounting data as are available to us.

Third, we have, additionally, explicitly ignored taxes other than indirect taxation. This is partially defensible because the De Beers Corporation employs a unified, or consolidated, system of accounting. Given this, it would be specious for us to make assumptions regarding tax rates for the company's farming activities.

Fourth, De Beers has chosen, as a matter of company policy, not to depreciate its capital expenditures, but to "write off" capital expenditures in the year in which the expenditures are made. That is, capital expenditures are treated as a direct expense in the year in which they are incurred. This is most unconventional; typically, companies making capital expenditures depreciate them over time, and the depreciation attributable to a particular year is treated as an expense for that year. As capital expenditures are made only intermittently, this means that reported profits or losses for any given year may not reflect underlying profitability of that operation, but may reflect the presence of a large capital expenditure in the given year. Accordingly, in assessing the profitability of an operation, we have measured profits or losses against direct expenses. That is, we have calculated the rate of return on operating expenses (RROE). This is not unconventional, especially where there is a high degree of variability in capital expenditures or depreciation. Moreover, this measure makes sense inasmuch as it provides a measure of the returns to be had from a Rand's expenditure on the activity. While the widely employed current convention is to measure economic activity, for comparative purposes, in U.S. dollars, we suggest that such a measure here would be improper as the monetary unit used by De Beers, and its customers, is the Rand (currently worth about US$0.30), and the activities involved are purely local and, by their very nature, do not enter into international trade.

Finally, we present here two estimates of profitability: one inclusive of imputed rents and one without. Our imputations of land rental values were based on the following observations or assumptions: (1) The selling price of land of the quality employed by De Beers for its cattle and game operations in the general area of the Rooipoort farms ranges between R 200 and R 350 per ha (Department of Agricultural Development, Kimberley; Noordkaap Plase estate agents, Kimberley). (2) The market rental value, per annum, of land selling for R 200 to R 350 per ha was approximately 8–10 percent of these figures (Department of Agricultural Development, Kimberley; Noordkaap Plase estate agents, Kimberley). We have assumed the best-case scenario for someone wishing to rent land in the area and have used a rental value for 1993 equal to R 16 per ha (8% of R 200 per ha). This would be the cost to start a similar operation of the same scale if begun *de novo*. (3) We have further assumed that the underlying value of the land grew by 3 percent per annum since 1988. Thus we assume rental values, by year, per ha, for the land on which there are cattle operations begin, in 1988 at R 13.80 and thereafter grow by 3 percent per annum. Similarly, we assume that the imputed rental value of the land for game was R 13.40 per ha in 1987 and thereafter grew at 3 percent per annum until 1993.

While we have presented for comparative purposes our conclusions as to the profitability of the game and cattle operations with and without rental imputations, we would argue that the results without rental imputations are the better estimates of the profitability of the operations. While recognizing the importance of opportunity cost, we would suggest that, from the company's perspective, the appropriate measure of profitability is the rate of return on operating expenses because (1) the decision to acquire the land, in the first instance, was for "precautionary" or "speculative" reasons, that is, in hopes that alluvial diamonds might eventually be found thereon; (2) the decision to continue to hold the land reflects the fact that the land, once acquired, represents largely a "sunk cost" to the company, rather than a continuing and ongoing one; and finally, (3) from conversations with De Beers management, there appear to be important noneconomic motives for holding the land, dominant among which is a desire to implement the company's environmental and conservation policy. The value of such noneconomic motives is difficult to measure.

Results

The Sustainable Utilization of Big Game

Annual census, trophy hunting, and live capture data for various big game species for Rooipoort during at least the past decade are presented in Tables 10.1–10.3. Long-term venison hunting data for spring-

Table 10.2. Annual Counts of Big Game at Rooipoort Estate, Northern Cape Province, South Africa, 1982–1993

	Year											Trend significance		
Species	1982	1984	1985	1986	1987	1988	1989	1990	1991	1992	1993	r	p	
Springbuck	2,551	1,478	1,272	1,208	1,438	1,498	1,645	2,338	2,024	1,720	3,197	0.49	0.12	
Hartebeest	1,363	724	1,006	721	782	864	846	998	893	787	904	0.02	0.95	
Eland	479	565	487	312	547	456	649	620	867	728	642	0.71	0.03	*
Gemsbok	366	639	476	363	566	742	644	1,000	944	711	1,030	0.82	0.009	**
Blue wildebeest	334	713	340	368	517	425	455	547	681	648	558	0.53	0.10	
Black wildebeest	263	237	188	174	228	205	260	215	256	279	347	0.45	0.15	
Kudu	200	165	192	193	171	226	208	394	252	319	287	0.81	0.01	*
Blesbok	194	188	153	135	120	171	228	254	316	296	328	0.73	0.02	*
Giraffe	21	29	37	34	36	45	65	61	55	71	78	0.94	0.003	**
Zebra	22	45	59	56	67	188	190	195	349	292	386	0.98	0.002	**
Warthog	15	38	58	83	66	98	65	89	85	104	116	0.88	0.005	**
Steenbok	4	3	4	5	7	16	4	22	11	6	11	0.66	0.04	*
Duiker	19	8	29	17	6	19	9	30	35	27	24	0.46	0.14	
Mountain reedbuck	3	3	0	1	0	9	4	9	0	1	1	–0.04	0.89	
Impala	14	9	6	0	1	9	2	0	7	6	3	–0.33	0.29	
Sable	—	—	—	—	—	10	13	15	10	16	17	0.75	0.09	
Tsessebe	—	—	—	—	—	—	—	—	5	9	12	1.00	0.000	***
Jackal	9	22	10	15	12	16	11	13	10	9	18	0.01	0.97	
Baboon	6	5	5	5	2	2	1	5	6	1	7	–0.05	0.87	
Ostrich	241	235	266	209	253	218	246	396	301	320	366	0.69	0.03	*
Total	6,104	5,106	4,588	3,899	4,819	5,217	5,545	7,201	7,107	6,350	8,332	0.71	0.03	*

Source: De Beers Farms Department, Management Records.
Trend significance: $^{*}p < 0.05$, $^{**}p < 0.01$, $^{***}p < 0.001$; Spearman's rank correlation coefficient (Zar 1984).

Table 10.3. Annual Number of Big Game Removed from Rooipoort Estate for Trophies, 1982–1993

Species	Year												Trend significance		
	1982	1983	1984	1985	1986	1987	1988	1989	1990	1991	1992	1993	r	p	
Springbuck	33	5	15	3	16	15	13	5	4	12	5	21	–0.15	0.61	
Hartebeest	19	4	14	6	14	6	7	6	2	12	5	10	–0.28	0.35	
Eland	9	2	5	7	16	5	4	7	5	16	4	9	0.11	0.71	
Gemsbok	15	7	14	9	17	12	12	12	8	16	4	19	0.06	0.83	
Blue wildebeest	11	4	13	7	11	7	5	6	3	11	4	11	–0.22	0.47	
Black wildebeest	13	5	11	9	17	12	4	7	4	7	2	8	–0.53	0.08	
Kudu	18	5	4	10	7	10	10	6	12	10	4	18	0.14	0.65	
Blesbok	17	7	12	6	13	6	7	3	3	9	3	12	–0.41	0.17	
Zebra	0	2	1	3	6	5	12	7	5	10	2	7	0.61	0.04	*
Warthog	0	3	3	0	11	10	9	9	8	14	5	21	0.68	0.02	*
Steenbok	5	0	2	2	4	1	1	0	0	2	1	5	–0.10	0.75	
Duiker	4	0	0	4	3	0	0	0	1	0	0	0	–0.44	0.14	
Mountain reedbuck	3	0	0	1	4	1	1	0	1	1	0	2	–0.03	0.92	
Impala	2	0	0	0	2	2	1	3	3	2	0	2	0.34	0.26	
Jackal	20	2	3	0	4	3	1	1	0	2	0	5	–0.30	0.31	
Baboon	4	0	1	0	0	0	0	1	1	0	0	3	0.06	0.83	
Ostrich	5	2	1	1	2	5	3	0	3	5	1	0	–0.26	0.39	
Total	178	48	99	68	147	100	90	73	63	129	40	153	–0.10	0.75	

Source: De Beers Farms Department, Management Records.
Trend significance: $^{*}p < 0.05$; Spearman's rank correlation coefficient (Zar 1984).

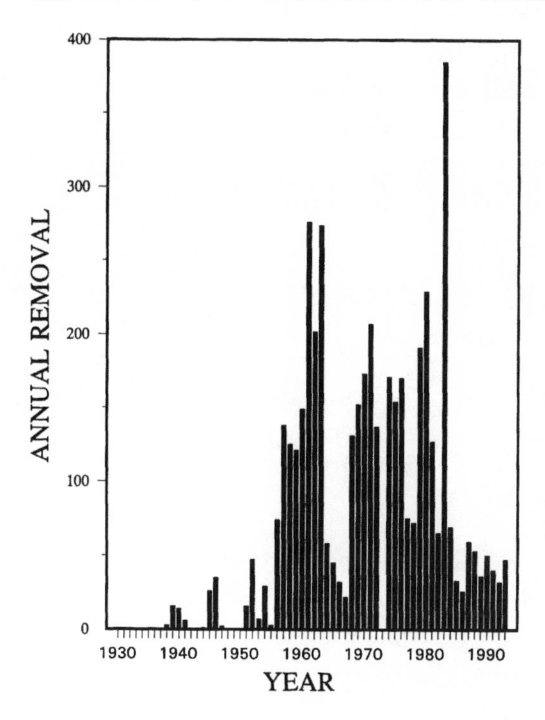

Figure 10.2. Year-to-year variation in hunting bags for springbuck (*Antidorcas marsupialis*) at Rooipoort Estate. (*Source:* DeBeers Farms Department, Management Records.)

buck are presented in Figure 10.2. Seventeen of the 20 species regularly utilized at Rooipoort showed positive trends in annual population numbers during the past decade (Table 10.2). Although two species showed significantly positive trends in annual removal of trophy animals, 10 species showed declining rates of removal during the same period (Table 10.3). The data for annual live capture reflect relatively high numbers of individuals removed for various species, but no significant trends in the temporal pattern of removal, except for eland which reflect a significant positive trend (Table 10.4). Long-term data for springbuck venison hunting show highest rates of removal during the second half of the past century (Fig. 10.2).

At Rooipoort, self-operated trophy hunting, live animal sales, and venison production are the most prolific utilization activities. The annual removal rates for each species for each of the operations are presented in Tables 10.3 and 10.4, and Figure 10.2. For the remainder of

Table 10.4. Annual Number of Animals Caught for Sale (and Donation) at Rooipoort Estate, 1981–1993

Species	Year											Trend significance		
	1981	1983	1984	1985	1987	1988	1989	1990	1991	1992	1993	*r*	*p*	
Hartebeest	401	347	129	95	132	38	66	133	180	29	43	–0.58	0.07	
Eland	—	148	102	100	158	—	167	—	216	137	217	0.83	0.03	*
Kudu	15	74	69	34	—	63	1	15	—	1	20	–0.50	0.15	
Gemsbok	—	—	141	115	30	147	—	132	290	7	72	–0.24	0.53	
Blue wildebeest	—	—	130	80	70	29	58	—	35	267	30	–0.33	0.38	
Black wildebeest	40	32	65	54	61	—	44	—	18	29	76	–0.03	0.93	
Blesbok	—	45	—	30	—	—	—	—	—	101	105	0.80	0.17	
Giraffe	—	—	—	6	—	4	9	9	6	6	8	0.24	0.55	
Zebra	—	—	—		—	—	—	—	—	—	26	—	—	
Total	456	646	636	514	451	281	345	289	745	577	597	0.00	1.00	

Source: De Beers Farms Department, Management Records.
Note: There were no capture operations during 1982 and 1986.
Trend significance: *$p < 0.05$; Spearman's rank correlation coefficient (Zar 1984).

this case study we will evaluate the financial viability and projected potentials of these activities on Rooipoort, and compare these with the profitability of the accompanying livestock program.

As suggested above, we shall not dwell here on methodological or "doctrinal" questions regarding the meaning of "sustainable development." Rather, we pose two relatively more prosaic, but perhaps more fruitful, questions regarding the utilization of the game found at Rooipoort. These questions are: (1) is Rooipoort, as a revenue-generating entity which has services that can be sold for recreational and trophy hunting use, profitable? and, (2) is the estate more profitably employed for game ranching/hunting or conventional livestock (and particularly cattle) ranching? The reasons for posing these questions are obvious. If game ranching, broadly conceived, is not profitable, then it may be taken to be nonsustainable since, after all, no profit-oriented person or company will long continue to subsidize such an undertaking. Furthermore, if Rooipoort is profitable as a game ranch, is that the use which will generate the greatest profit? In short, is the use of the wildlife profitable, and is it more or less profitable than cattle ranching?

Profitability Exclusive of Imputed Land Values

Game Ranching

The most important observation to be made about the game ranching operations of Rooipoort is that game ranching, broadly conceived, is exceptionally profitable as measured by the rate of return on operating expenses. This is observed in Table 10.5, where we have calculated the profitability of game ranching, both absolutely and relative to the direct expenditures on the game operations (RROE).

It will be noted that the rate of return on operating expenses ranges from a high of 222 percent, indicating that the expenditure of one Rand on game farming returned R 2.22 to the company, to a low of 42 percent in 1993, averaging 106 percent during the period. Such returns, while admittedly not inclusive of expenditures on capital and while subject to a high degree of variability by year, indicate a highly successful operation.

It should be noted, however, that the high rates of return on operating expenses owe, in no small part, to the particularly high rates of return on game sales (Table 10.6). There is immense variability by year, in both revenue and expenditures, for both game sales and hunt-

Table 10.5. Returns (Rands) on Operating Expenditures for All Game Operations[a] at Rooipoort Estate, 1987–1993

	Excluding imputed land rent				Including imputed land rent			
Year	Total revenue (1)	Operating expense (2)	Profit or loss (3)	Return on operating expense (4 = 3/2)	Imputed land rent value (5)	Operating expenditures (6 = 2 + 5)	Profit or loss (7)	Return on expenditures (8 = 7/6)
1993	779,924	549,878	230,046	0.418	672,000	1,221,878	–441,954	–0.362
1992	548,532	378,275	170,257	0.450	652,427	1,030,702	–482,170	–0.468
1991	1,039,530	587,417	452,113	0.770	633,424	1,220,841	–181,311	–0.149
1990	427,632	218,077	186,513	0.855	614,975	833,052	–428,462	–0.514
1989	584,625	181,395	403,230	2,223	597,063	778,458	–193,833	–0.249
1988	287,476	156,328	131,148	0.839	579,673	736,001	–448,525	–0.609
1987	321,587	112,705	208,882	1.853	562,789	675,494	–353,907	–0.524
Average				1.058				–0.411

Source: De Beers Farms Department, Management Records.

[a]Hunting, live game sales, and meat sales.

Table 10.6. Returns (Rands) on Operating Expenditures for Game Sales and Hunting Operations at Rooipoort Estate, Excluding Imputed Land Rent, 1987–1993

	Game sales				Hunting operations			
Year	Total revenue (1)	Operating expense (2)	Profit or loss (3)	Return on operating expense (4 = 3/2)	Total revenue (5)	Operating expense (6)	Profit (or loss) (7)	Return on operating expense (8 = 7/6)
1993	485,625	288,019	334,629	0.686	302,330	261,859	40,471	0.155
1992	467,144	292,370	310,796	0.598	81,388	85,905	–4,517	–0.053
1991	768,781	402,532	553,523	0.910	270,749	184,887	85,862	0.464
1990	296,450	138,004	222,651	1.148	108,140	80,073	28,067	0.351
1989	453,440	108,905	395,202	3.164	131,185	72,490	58,695	0.810
1988	179,504	84,209	134,472	1.132	107,972	72,119	35,835	0.497
1987	236,459	70,912	198,538	2.335	85,128	41,793	43,335	1.037
Average				1.425				0.466

Source: De Beers Farms Department, Management Records.

Table 10.7. Revenue, Operating Expenses, and Profit of Game Sales and Hunting Operations as a Percentage of Total Game Operations Revenue and Expenses for Rooipoort Estate, 1987–1993

	Game sales			Hunting operations		
Year	Total revenue (1)	Operating expense (2)	Profit (or loss) (3)	Total revenue (4)	Operating expense (5)	Profit (or loss) (6)
1993	61.63	52.38	83.00	38.37	47.62	17.00
1992	85.16	77.29	100.00	14.84	22.71	00.00
1991	73.95	68.53	81.01	26.05	31.47	18.99
1990	73.27	63.28	84.95	26.73	36.72	15.05
1989	77.56	60.04	85.44	22.44	39.96	14.56
1988	62.44	53.87	72.66	37.56	46.13	27.34
1987	73.53	62.92	79.25	26.47	37.08	20.75
Average	72.51	62.61	83.76	27.49	37.39	16.24

Source: De Beers Farms Department, Management Records.

ing operations. This is particularly the case for game sales. Similarly, revenues derived from the hunting operations also exhibit a high degree of variability, especially for the most recent years, 1991–93. Nonetheless, a line-by-line inspection of the rate of return on operating expenses for game sales and hunting operations (cols. 4 and 8, Table 10.6) suggests that, in general, the game sales yield an exceptionally high return on expenditures on these operations and are more profitable than the hunting operations.

On average, game sales returned R 1.43 for every Rand expended, while the hunting operation, on average, returned only R 0.47 (cols. 4 and 8, Table 10.6). Such returns, while admittedly measured against operating expenses only, compare favorably to those of other industries. The relative profitability of game sales above that of hunting operations is further heightened by inspection of Table 10.7. Focusing on columns 3 and 6, it can be seen that, on average, game sales contribute 73 percent to the profits of the De Beers game ranching operations while accounting for only 63 percent of the expenses of these operations. Specifically, we observe from comparisons of columns 1 and 2 in Table 10.7 that, in each of the years covered by this case study, the contribution to total revenue by game sales exceeded the portion of expenses which these operations were of total expenses.

Similarly, comparisons of columns 2 and 3 in Table 10.7 indicate that the contribution which the game sales make to total game ranching operations well exceeds the pro rata share which these operations are of total expenses. Put differently, and simply, game sales are particularly profitable.

In contrast, while hunting operations generate positive profits, they are less profitable than game sales, both absolutely and relatively. Specifically, and as is observed in column 4 of Table 10.7, on average, hunting operations contribute only 27 percent of all the revenues derived from the game farms. Thus hunting revenues are, on average, only 38 percent of those of game sales. The hunting operations constitute, on average, 37 percent of all expenses. Moreover, as is observed from columns 5 and 6 in Table 10.7, the contribution to profits of hunting is less than its share of expenses. That is, hunting contributes only 16 percent to total profits while accounting for 37 percent of total expenses. This is in pronounced contrast to game sales. Inspection of the underlying accounting data clearly reveals that the relative profitability of these two activities reflects the fact that the major expenditures for game sales are incurred for game capture, while those for hunting include accommodations and similar, more labor-intensive expenses. Accordingly we now turn our attention to hunting operations alone.

The conventional industry measure of hunting profitability is that of revenue, cost, and profit per hunter-day. Rooipoort, like most similar operations, caters to both hunters and nonhunters. Nonhunters are those people accompanying the hunter as observers, for example, family or friends who join the safari but do not hunt. Similarly, the hunting operations cater to both local and international guests. Moreover, the fee structure is such that guests pay separately for accommodations and for game killed by type, that is, hunting fees are "unbundled." Accordingly, while the use of hunter-days introduces the usual problems of averaging across groups, we employ this measure here in deference to industry convention. Inspection of hunting records and accounting data for 1993 shows that, on average, one nonhunter spends approximately 20 percent of that spent by one hunter per day. Accordingly, we have converted nonhunter-days to "full-time hunter-day equivalents" in this proportion.

Table 10.8 shows hunter-days, by year, since 1986. Combining these data with those of Table 10.6 generates Table 10.9, which shows

Table 10.8. Number of Hunter- and Nonhunter-Days at Rooipoort Estate, 1987–1993

	Number of days		
Year	Hunter	Nonhunter	Weighted total
1993	112	83	128.6
1992	19	8	20.6
1991	128	61	140.2
1990	53	29	58.8
1989	80	35	87.0
1988	90	12	92.4
1987	58	16	61.2

Source: De Beers Farms Department, Management Records.
Note: "Nonhunters" include paying guests accompanying the hunt.

Table 10.9. Revenue, Expenses, and Profit (Rands) per Hunter-Day at Rooipoort Estate, Northern Cape Province, South Africa, 1987–1993

Year	Revenue per hunter-day (1)	Expenses per hunter-day (2)	Profit per hunter-day (3)	Rate of return on operating expense (4 = 3/2)
1993	2,350.9	2,036.2	314.7	0.15
1992	3,950.0	4,170.2	–219.3	–0.05
1991	1,931.2	1,318.7	612.4	0.46
1990	1,839.2	1,361.8	477.3	0.35
1989	1,507.9	833.2	674.7	0.81
1988	1,168.5	780.5	388.0	0.50
1987	1,390.0	682.9	708.1	1.04
Average	2,019.9	1,597.6	422.3	0.47

Source: De Beers Farms Department, Management Records.

revenue, expenses, and profitability per hunter-day. Table 10.9 shows that the De Beers hunting operation yielded, on average, R 422.3 in profits per hunter-day, ranging from a low of negative R 219.3 to a high of R 708.1. Just as profits per hunter-day varied widely, so did profits per hunter-day per Rand expended, ranging from negative R 0.05 per Rand expended to R 1.04, while averaging a respectable R 0.47 per Rand expended. Clearly then, the hunting operations are quite profitable.

Our results are similar to those of Berry (1986), who, in an economic analysis of data from Rooipoort and three other De Beers farms, demonstrated venison production to be the most profitable, followed by live sales, nontrophy recreational hunting, and trophy hunting.

Cattle Ranching

As a benchmark, we have also calculated similar measures of profitability for the De Beers cattle operations for the years 1988 to 1993. From column 3 of Table 10.10, we observe that in all years except 1990, which was a year characterized by a severe drought (which increased costs immensely while reducing cattle prices), the cattle operations have been profitable. Further, from column 4 we observe that a one-Rand expenditure on the cattle operations has yielded returns ranging from a negative R 0.43 to a positive R 0.71. For the seven years here considered, while all but that for 1990 have yielded positive profits, no years have yielded Rates of Return on Operating Expenses exceeding 100 percent; that is, in no years has a Rand expenditure yielded in excess of a Rand in profits and, on average, the expenditure of one Rand has yielded only R 0.32 of profits (Table 10.10).

Relative Profitability

As noted above, both the game and the cattle operations are highly profitable. However, it is clear, particularly from inspection of Tables 10.6 and 10.10, that, as measured by the rate of return on operating expenses, game sales are decidedly the most profitable of the three operations, yielding an average of R 1.43 for every Rand expended on these operations, in contrast to R 0.47 for the hunting operations, and a decidedly lower amount, R 0.32, for the cattle operations.

We believe that, with only modest marketing expenditures, sales of hunting services could be increased dramatically. Historically, De Beers has not attempted to market its hunting operations actively. Given that a large measure of the hunting expenditures are of a fixed, or quasi-fixed, nature, for example, accommodation facilities, the returns on increased marketing efforts are likely to be high.

The cattle operations, while profitable, are the least profitable of the three general activities here considered. This reflects two factors:

Table 10.10. Returns (Rands) on Operating Expenditures for Cattle Farming at Rooipoort Estate, 1988–1993

	Excluding imputed land rent				Including imputed land rent			
Year	Total revenue (1)	Operating expense (2)	Profit or loss (3)	Return on operating expense (4 = 3/2)	Imputed land rent value (5)	Operating expenditures (6 = 2 + 5)	Profit or loss (7)	Return on expenditures (8 = 7/6)
1993	1,260,267	853,882	406,404	0.476	645,600	1,499,482	–239,196	–0.160
1992	1,208,132	985,200	222,932	0.226	626,796	1,611,996	–403,364	–0.251
1991	742,328	564,709	177,619	0.315	608,540	1,173,249	–430,921	–0.367
1990	924,939	1,609,251	–684,312	–0.425	590,815	2,200,066	–1,275,127	–0.580
1989	723,393	452,652	270,741	0.598	573,607	1,026,259	–302,866	–0.295
1988	693,339	404,780	288,559	0.713	556,900	961,680	–268,341	–0.279
Average				0.317				–0.322

Source: De Beers Farms Department, Management Records.

(1) the relatively competitive nature of the industry, and (2) the relatively higher variable costs of raising cattle.

Profitability Inclusive of Imputed Land Values

Including imputed land rental values, under the assumed rental values noted above, markedly changes the conclusions regarding the profitability of both the game and cattle operations.

Game Operations

As indicated by Table 10.5, inclusion of imputed rental values for land employed in the game operations markedly alters profitability, by year, and *in toto*. Whereas the overall rates of return on operating expenses for game operations averaged 106 percent and were at all times positive, when we include the imputed rental value of the land used, the average rate of return on expenses falls to –35.3 percent, at all times remaining negative. Partly, this reflects the large variability in revenues from game operations, and the high degree of variability in revenues likely reflects the fact that De Beers has not been able to market its hunting operations consistently and vigorously. It is not too much to speculate that, with an aggressive marketing effort, total revenues from hunting could be increased decidedly with the result that the operation would become a profitable one.

Cattle Operations

Inclusion of imputed rental values leads to results which suggest that cattle ranching, like hunting, is unprofitable. As observed in Table 10.10, inclusion of imputed land values turns the ostensibly positive profits observed in column 3 into consistent losses, with the average rate of return on expenditures falling to a negative 32 percent per year.

Summary of Profitability

Based on these analyses, we conclude that if a similar operation were to be launched *de novo*, requiring land rental or acquisition by borrowing so as to purchase the land, neither game nor cattle ranching would be profitable activities. Initiating such an operation from scratch then is not feasible, particularly if we consider that our calculations explicitly ignore other required capital expenditures. These rather negative conclusions must be tempered by the fact that inclusion of imputed land values is highly questionable. If imputations were required to determine the profitability of a number of economic activities, many business firms currently in operation might not meet such a stringent profitability test or requirement. Additionally, our analyses suggest that the De Beers farms are not, perhaps, being utilized as intensely as they might. That is, the scale of both the game and cattle operations may be suboptimal given the "scale of plant," namely, the size of the farms. With respect to the game operations, De Beers is currently stocking Rooipoort with close to optimum numbers of game. However, further attention to marketing of hunting operations may well enable De Beers to turn the unprofitable hunting operation into a profitable one. If the estate were used to its fullest potential for hunting, ecotourism, and nature education, the overall value of the game operation would be much higher.

Biodiversity

Rooipoort Estate contributes significantly toward the conservation of medium-to-large game species in the Northern Cape Province of South Africa. Nineteen species that were historically common to abundant in the Northern Cape (Skead 1989) are at least common at Rooipoort today (Table 10.1). Three species—springbuck, Burchell's zebra, and warthog—are probably at higher population densities now on Rooipoort than historically in the Northern Cape. Of the ten species that have been lost locally since historic times, two species—blue antelope (*Hippotragus leucophaeus*) and quagga (*Equus quagga*)—are extinct, and seven species have not been fostered or reintroduced because they require large territorial ranges and would possibly escape across the Vaal River, potentially creating problems with neighbors. The tenth species—roan antelope (*Hippotragus equinus*)—has not been reintroduced only because no suitable source of the species from other reserves has been found to date.

The broader questions as to whether biodiversity is secure in conservation areas, and whether this biodiversity can be conserved adequately on well-managed cattle ranches, have not been addressed quantitatively in the past. However, extensive discussions with various biologists in the Northern Cape (M. D. Anderson, H. Erasmus, and J. Koen: Nature Conservation; A. van Rooyen: Animal Rehabilitation Center, Kimberley; T. Anderson and R. Liversidge: McGregor Museum;

A. Anthony: De Beers Cons. Mines) generated anecdotal information on the implications of cattle ranching versus game ranching (nature reserves) on the conservation of biodiversity in the Northern Cape.

These biologists felt that, most importantly, what is conserved on a well-managed game ranch, in comparison to a well-managed cattle ranch, are ecological processes. Furthermore, they suggested that insect diversity is the major difference between these two alternative land-use practices. More specifically, they mentioned a variety of differences caused by the management and behavior of herbivores on livestock versus game ranches. Because cattle are predominantly grazers, grass cover is reduced, which often leads to reduced competition with woody elements (e.g., *Acacia mellifera*); bush encroachment is usually the result. As a consequence, a diversity of grasses (e.g., *Themeda triandra*), forbs (e.g., *Helichrysum* spp.), and succulents (e.g., *Mestoklema, Crassula,* and *Stapelia* spp.) may be lost. However, bare patches caused by overgrazing by cattle favors termite colony establishment, particularly *Hodotermes mossambicus* and *Trinervitermes trinervoides,* and, as a result, a "badly managed" farm would have more termites and consequently favor mammalian myrmecophages such as the aardwolf (*Proteles cristatus*) and the aardvark (*Orycteropus afer*). Rodents are also affected by grazing pressure; for example, heavy grazing pressure would favor some species such as the short-tailed gerbil (*Desmodillus auricularis*), but not benefit others which prefer dense cover such as the striped mouse (*Rhabdomys pumilio*). Cattle farms also usually have a greater number of waterpoints, which with an accompanied higher diversity of seed-bearing, pioneer forbs and grasses may sustain increased densities of certain seed-eating birds, such as sandgrouse (Pteroclidae), waxbills (Estrildidae), and canaries and buntings (Fringillidae). However, trampling by cattle at water holes and wetlands can destroy nests of ground-nesting waterbirds, such as the blacksmith (*Venellus armatus*) and the three-banded *Charadrius tricollaris* and chestnut-banded *C. pallidus* plovers.

More direct effects of managing cattle ranches might include the influences of insecticide dips and predator control. Dipping of cattle with acaricides negatively affects dung beetles, and if organophosphate dips are used, cattle egrets (*Bubulcus ibis*), wattled starlings (*Creatophora cinerea*), and red-billed oxpeckers (*Buphagus erythrorhynchus*), to name but a few, could be affected. Animal control methods for black-backed jackal (*Canis mesomelas*), brown hyena (*Hyaena brunnea*), and caracal (*Felis caracal*) are usually highly effective, thus reducing the population sizes of these species. However, the more nonselective techniques (e.g., poisons, hunting with dogs, and snares) harm nontarget species, such as bat-eared foxes. Finally, because domestic stocks are well managed with few individuals dying on the lands, and carcasses rarely left lying, scavenging birds, such as vultures and crows, have a reduced food supply.

Financial Potential of Gamebird Hunting at Rooipoort

A brief reference to the potential for gamebird hunting at Rooipoort is merited because of its potential to contribute to wildlife-based revenues. The limited amount of gamebird hunting conducted at Rooipoort has traditionally been restricted to senior staff and directors of De Beers and their guests, and has not been charged for at commercial rates. However, hunting of African gamebirds has a marketable mystique about it (Lynn-Allen 1951), and we believe such hunting at Rooipoort could yield a significant unrealized potential.

Gamebirds at Rooipoort occur at medium to low densities relative to other areas in South Africa (Little et al. 1993). At present levels of removal, two species—the resident helmeted guineafowl (*Numida meleagris*) and migratory Namaqua sandgrouse (*Pterocles namaqua*)—are utilizable on a sustainable basis.

Although gamebird hunting at Rooipoort is currently not marketed commercially, comparisons with other commercial operations nearby (Smith 1994), and elsewhere in the United States and the United Kingdom (Johnson and Wannenburgh 1987), and consideration of Rooipoort's magnificent accommodations suggest significant revenue potential. We calculate that an operating profit of US$20,000–$40,000 could be earned from a strictly gamebird-oriented hunting operation conducted at Rooipoort for five hunting sessions (two to three days long) per annum. This is roughly twice the actual profit generated by trophy hunting. If the estate were used to its fullest potential for ecotourism (including broad-spectrum hunting and nature tours), that income could be much higher.

Conclusions

Although these results suggest that De Beers would be well advised to concentrate its efforts on sales of game, this conclusion must be tempered with the observation that game sales are largely sales of live animals to other game farms and, accordingly, there are supply constraints and limitations (which appear not to have been reached) and, particularly, demand constraints. Specifically, the demand by other game farms for the De Beers live animals is a derived demand which could decrease if markets become saturated or depleted for whatever reason. However, the company seems to recognize this: in the last three years it has built its own butchery and plans to move, increasingly, into the sales of meat for the local market. This is reflected in the underlying data to which we have been given access, for meat sales have increased decidedly in the last three years.

Finally, the hunting operation is very profitable (ignoring imputed land values), despite its low-key nature. Thus, with only a modest increase in marketing efforts, the demand for hunting services could be increased significantly, yielding markedly increased profits for the company.

In sum, we conclude that, as measured by rate of return on operating expenses, utilization of the wildlife resources of the De Beers farms is a highly profitable activity. Indeed, it is not too much to suggest that, given its profitability, it is an activity which De Beers has, if anything, not exploited to its fullest potential.

A final caveat is in order: our conclusions are based on profitability as measured by the rate of return on operating expenses. This is not unconventional. However, it must be borne in mind that, were this operation to be started *de novo,* land of a comparable quality would have to be acquired, and this would represent a significant capital outlay. Such an operation would probably be unprofitable unless there were a more aggressive marketing of its hunting operations and, if ecologically sustainable, an increased stocking rate of game or cattle.

Appendix: Methods and Assumptions Employed to Calculate Financial Profitability of Game and Cattle Operations

Total Operating Expenses (TOE)

While De Beers made available to us data extending back to 1987 for game operations and to 1988 for cattle operations, we were only given access to comprehensive general ledger data for 1993. In order for us to make full use of the data for years 1987 to 1992, we make some assumptions regarding distribution of operating costs. These are set out in full below.

The reason for these assumptions is that De Beers does not attribute fully all operating costs to the various game activities on their farms. Rather, two groups of costs, supervision/general expenses and farm operating expenses, are maintained as separate entities and are deducted from total game ranching income at the year's end. The total for these two expense accounts for 1993 was R 536,512. Since there were four game activities recorded in the books for 1993 (game farming, safari/trophy hunting, recreational hunting, tourism/education), we have made the assumption that these expenses were equally attributed to all four operations. Thus for 1993, each operation acquired an extra expense of R 536,512 ÷ 4 = R 134,128.

Now, for each of the years 1987 to 1992 where no data for supervision/general expenses and farm operating expenses were available, we assume that such expenses occur in the same proportions to direct expenses (for which we have data) as they do in 1993 (which for game farming is 0.87, and for safari/trophy hunting is 1.05). These added expenses are then summed with direct expenses to give the TOEs in Tables 10.5, 10.6, and 10.10.

The supervision/general expenses and farm operating expenses for cattle ranching are also kept as a separate entity, both from cattle direct expenses and game ranching supervision/general expenses and farm operating expenses. These expenses totaled R 262,866 in 1993, which is 46 percent of cattle direct expenses for 1993. The same procedure as for game was used to calculate the full TOEs for cattle for 1988 to 1992.

Land Rental Calculations

To estimate whether the operation in question could feasibly be started *de novo*, we needed to impute land rentals for the cattle and game operations independently. The game operations were assumed to occur only on Rooipoort (42,000 ha). While some unaccounted for venison sales do occur on other De Beers farms, we believe that this was offset by the unaccounted for productivity (and hence income) of cattle run on Rooipoort. The total area under cattle utilization is 40,350 ha, most of it off Rooipoort. We assume that this "other" land is of equivalent productivity to Rooipoort to enable a comparison between cattle and game. This assumption is enforced by the official carrying capacity of the land under cattle (1:12 to 1:14 on all cattle land versus 1:12 on Rooipoort; De Beers internal document).

Acknowledgments

We thank De Beers Consolidated Mines Limited for hospitality and logistical support over the years, and access to hunting, stock/game farming information and financial records for Rooipoort. This study was funded by the World Wide Fund for Nature–International, through the Southern African Nature Foundation. We are also grateful to the African Gamebird Research, Education and Development Trust, the Foundation for Research Development (Core and Special Programmes), the FitzPatrick Institute and the University of Cape Town's Research and Equipment Committees for their support of our gamebird research over the years. Mark Berry (De Beers Consolidated Mines Limited) supplied, and helped to interpret, information on the hunting on, and past and present management of, Rooipoort Estate. Graham Main (De Beers Consolidated Mines Limited) provided and helped to explain information relating to the De Beers resource economics activities at Rooipoort. Curtis H. Freese, L. C. Weaver, and Jon Barnes gave constructive criticism of an early draft of the manuscript.

References

Africa: A flicker of light. 1994. *The Economist* March 5, 1994:23–28.

Bautista, L. M., J. C. Alonso, and J. A. Alonso. 1992. A 20-year study of wintering common crane fluctuations using time series analysis. *Journal of Wildlife Management* 56:553–572.

Berry, M.P.S. 1986. A comparison of different wildlife production enterprises in the northern Cape Province, South Africa. *South African Journal of Wildlife Research* 16:124–128.

Berry, M.P.S., and T. M. Crowe. 1985. Effects of monthly and annual rainfall on game bird populations in the northern Cape Province, South Africa. *South African Journal of Wildlife Research* 15:69–76.

Bigalke, R. C. 1985. The De Beers example. *Optima* 33:38–48.

Crowe, T. M. 1978. Limitations of populations in the Helmeted Guineafowl. *South African Journal of Wildlife Research* 8: 117–126.

Crowe, T. M., M. Isahakia, and E. Knox. 1994. Research and training priorities in biological conservation: African solutions to African problems. *South African Journal of Science* 90:517–518.

Crowe, T. M., J. C. Schijf, and A. A. Gubb. 1981. Effects of rainfall variation, fire, vegetation and habitat physiognomy on a northern Cape animal community. *South African Journal of Wildlife Research* 11:87–104.

Fair game only. 1993. *The Economist* August 21, 1993:39.

IUCN (World Conservation Union). 1990. *Biodiversity in sub-Saharan Africa and its islands: Conservation, management and sustainable use.* Species Survival Commission. Gland, Switzerland: IUCN.

IUCN, UNEP, and WWF. 1991. *Caring for the Earth: A strategy for sustainable living.* Gland, Switzerland: IUCN.

Johnson, P. G., and A. Wannenburgh. 1987. *The world of shooting.* Lausanne, Switzerland: Photographex.

Lack, D. 1966. *Population studies of birds.* Oxford: Oxford University Press.

Little, R. M., and T. M. Crowe. 1993. The breeding biology of the greywing Francolin *Francolinus africanus* and its implications for hunting and management. *South African Journal of Zoology* 28:6–12.

Little, R. M., G. Malan, and T. M. Crowe. 1993. The use of counts of Namaqua sandgrouse at watering sites for population estimates. *South African Journal of Wildlife Research* 23:26–28.

Lynn-Allen, B. G. 1951. *Shot-gun and sunlight: The game birds of east Africa.* London: Batchworth Press.

Meiring, J. 1994. Hunting a "conservation tool." *The Argus* March 17:10.

Merten, M. 1993. Cattle ranchers combine farms for game reserve. *Business Day* November 30:3.

Myers, N. 1979. *The sinking ark: A new look at the problem of disappearing species.* Oxford: Pergamon Press.

Onesta, P. A., and P. Verhoeff. 1976. Annual rainfall frequency distribution for 80 rainfall districts in South Africa. *South African Journal of Science* 72:120–122.

Skead, C. J. 1989. *Historical mammal incidence in the Cape Province: The western and northern Cape,* vol. 1. King William's Town: Kaffrarian Museum.

Smith, B. S. 1994. Economically viable utilisation of the Greywing Francolin *Francolinus africanus* in the eastern Cape highlands. Unpublished Masters thesis, University of Cape Town.

Smithers, R.H.N. 1990. *The mammals of the southern African subregion.* Pretoria: University of Pretoria.

South African Meat Board. 1993. South African Meat Board Quarterly Report, Third Quarter 1993. Internal working document.

Tyson, P. D. 1986. *Climatic change and variability in southern Africa.* Cape Town: Oxford University Press.

Zar, J. H. 1984. *Biostatistical analysis.* Englewood Cliffs, New Jersey: Prentice Hall.

ELEVEN

Trophy Hunting as a Conservation Tool for Caprinae in Pakistan

Kurt A. Johnson

Wildlife conservation in developing countries is often hampered by lack of adequate funding and failure to consider the needs of indigenous people when conservation policies and programs are formulated and implemented (Lewis et al. 1990). In recent years, sport hunting of wild birds and mammals has been promoted by some individuals and organizations as a tool for conserving wildlife in developing countries that can also provide an economic return to indigenous people. One subset of sport hunting, trophy hunting, has been especially touted because it is perceived as a wildlife use which generates great profits per unit of investment (i.e., per animal harvested) (Eltringham 1994) and which, if managed properly, has little impact on harvested populations. But it is not without controversy. Antihunting campaigns have been mounted, and sport hunting has been banned in many countries (Eltringham 1994).

Pakistan is a country with formerly rich and diverse wildlife resources (Ferguson 1978) and a long history of hunting for both sport and subsistence (Burrard 1925; Stockley 1936). Sport hunting first became a threat to wildlife during the period of British rule (1858–1947), when the influx of modern weaponry, the breakdown of old kingdoms and princely states, the advent of local militias (see A. A. Khan 1975 for a description of harvest by the Chitral Scouts), and the voracious hunting appetites of British military and civilian rulers (Mallon 1983; Allen 1984) led to substantial increases in the number of animals shot. The trend increased after the British departed and the subcontinent was partitioned into India and Pakistan (Mallon 1983). It has continued to the present day with military officers, wealthy businessmen, and Middle Eastern dignitaries filling the void left by the British (Weaver 1992). Despite the passage of provincial wildlife conservation laws during the mid-1970s, Pakistani authorities have had little success in slowing the decline of important wildlife species such as the markhor (*Capra falconeri*), urial (*Ovis vignei* or *O. orientalis*), and ibex (*Capra ibex*) through law enforcement and traditional management actions (Roberts 1977).

In the mid-1980s, two innovative wildlife conservation programs were begun in Pakistan: the Torghar Conservation Project (TCP) in Baluchistan Province and the Chitral Conservation Hunting Program (CCHP) in the North West Frontier Province (NWFP). A third program, the Bar Valley Project (BVP), was implemented in the Northern Areas in the early 1990s. The TCP was started as a private, "grass-roots" conservation program for the Suleiman markhor (*C. f. jerdoni* of Ellerman and Morrison-Scott [1966] or straight-horned markhor, *C. f. megaceros,* of Schaller [1977]) and Afghan urial (*O. v. cycloceros* or *O. o. cycloceros*) (Roberts 1977) on Pathan tribal lands in the Torghar Hills north of Quetta, Baluchistan. The CCHP, a program of the NWFP government's Wildlife Department, was established to conserve the Pir Panjal markhor (*C. f. cashmirensis* of Ellerman and Morrison-Scott [1966] or flare-horned markhor, *C. f. falconeri* of Schaller [1977]) within and around the Chitral Gol National Park (N.P.) and nearby game reserves (G.R.) in Chitral District. The BVP, initiated by World Wide Fund for Nature–Pakistan (WWF–Pakistan), was intended to be a community-based, integrated conservation and development project designed to conserve Himalayan (or Siberian) ibex (*C. i. sibirica*) and its habitat in the Bar Valley of the Northern Areas.

These so-called conservation hunting programs were all based on the notion that wildlife conservation in developing countries will succeed only if local people are actively involved in and benefit from conservation efforts (Lewis et al. 1990; Murphree 1991; Bond 1994). All three were designed to use funds generated through a limited trophy hunt of Caprinae both to conserve wildlife and to benefit the local people.

Despite the worthiness of these programs' stated goals, their actual conservation value has been questioned by many individuals and organizations. Although some have opposed these programs simply because they involve hunting, others have questioned their value because that value has not been demonstrated through objective and scientific evaluation. To address this deficiency, I undertook a critical review of all three programs in mid-1993. The objectives of my review were (1) to determine which of the three conservation hunting programs have positive conservation value and why; (2) to determine how those programs with marginal or no positive conservation value could be improved; and (3) to recommend conservation alternatives for those programs that could not be improved.

Methods

I address five major questions in this chapter:

1. Are the species of Caprinae and their habitat biologically suitable for sustaining a significant harvest, and is our knowledge adequate for designing and monitoring such a harvest program?
2. What is the ecological impact of harvesting and associated management practices on biodiversity in the area in question?
3. How do the social, economic, and institutional conditions affect incentives for good management in the hunting programs?
4. Are there wild species or wildland uses besides trophy hunting that could benefit Caprinae populations and their habitats?
5. What are the markets for trophy hunting in the three programs, and how do these markets influence management?

I conducted an in-depth review of existing reports and literature pertaining to the three conservation hunting programs, and interviewed the principal persons responsible for implementing each of the three programs. I traveled to Pakistan for two weeks in fall 1993 (November–December) and two weeks in spring 1994 (April–May) to gather existing documentation and conduct interviews. A third trip was made in November 1994 to conduct a field survey of markhor and urial populations in the Torghar Hills.

Each of the three conservation hunting programs is examined and evaluated in regard to several factors. This information is presented concurrently for each factor to lend a comparative perspective to the review. The analysis covers the status of each project through 1994.

Background

The Geographic Setting

Torghar Conservation Project. The TCP is located in the Torghar Hills (*Tor Ghar* means "Black Mountain" in Pushto, the language of the indigenous Pathan people), a chain of ruggedly upturned sedimentary mountains approximately 100 kilometers (km) long and 16–32 km wide in the NE–SW trending Toba Kakar Mountains of the Qila Saifullah District, Zhob Division, Baluchistan. Torghar is characterized by an arid climate, and a shrub-steppe vegetation with scattered junipers (*Juniperus macropoda*) and wild pistachios (*Pistacia khanjak*) (District Gazetteers of Baluchistan 1906).

Chitral Conservation Hunting Program. The CCHP encompasses lands within and around Chitral Gol N.P. and nearby game reserves in Chitral District north of Peshawar, NWFP. The climate of Chitral District is characterized by extremes of temperature—both daily and seasonally—and winds (M. A. Khan 1975). Chitral is north of the "monsoon belt," so there is no fixed rainy season (A. A. Khan 1975). Much of the precipitation is received in the form of snow. A. A. Khan (1975) described five basic vegetation types in the Chitral region: (1) dry temperate coniferous forests; (2) oak forests; (3) open scrub of *Fraxinus xanthoxyloides;* (4) subalpine scrub; and (5) snowfields and adjoining pastures.

Bar Valley Project. The BVP is located in the Nagar Subdivision of the Gilgit District north of Gilgit, Northern Areas. The project area consists of two deep river valleys—the Daintar River valley and Garamsai River valley—that originate in the Karakoram Mountains just west of Batura Mountain and join with each other a few kilometers below the village of Bar. These two river valleys are collectively known as Bar Valley.

Because the Northern Areas north of Gilgit are very dry, vegetation is generally sparse. Lower elevations in valley bottoms are virtually devoid of vegetation. Immediately above this desertlike zone is a sparsely vegetated shrub-steppe zone dominated by *Artemisia maritima* shrubs, with *Juniperus macropoda, Rosa webbiana,* and *Polygonum* sp. on dry slopes. Scrub forests of birch (*Betula utilis*), willow (*Salix* sp.), and poplar (*Populus nepalensis*) occur in the riparian zone along streams and rivers. Above the shrub-steppe are subalpine and alpine meadow zones. These zones have moist pastures consisting of herbaceous flowering plants (e.g., *Primula* sp., *Potentilla* sp.), and dry plateau pastures with several forb, grass, and shrub species (Ahmad 1993). Highest is the nival zone, an unvegetated zone dominated by seasonal and permanent snowfields.

The Species

Torghar Conservation Project. The two species of concern to the TCP are the Suleiman markhor and Afghan urial, both of which are listed on the Third Schedule of the Baluchistan Wildlife Protection Act of 1974. These are "protected animals, i.e., animals which shall not be hunted, killed or captured" except in specific circumstances (Baluchistan Government, Agriculture Department 1977). The Suleiman markhor is listed as "endangered" under the U.S. Endangered Species Act (ESA) (U.S. Fish and Wildlife Service 1993) and is included in Appendix I of the Convention on International Trade in Endangered Species of Wild Fauna and Flora (CITES) (U.S. Fish and Wildlife Service 1992). The Afghan urial is not listed on either of the above.

Chitral Conservation Hunting Program. The species of interest to the CCHP are the Pir Panjal markhor and, to a lesser extent, the Afghan urial and Himalayan ibex. The Pir Panjal markhor is listed in Part II of the First Schedule (Animals Which May Be Hunted on a Big Game Shooting License) of the North West Frontier Province Wild-Life (Protection, Preservation, Conservation and Management) Act, 1975. It is also listed in Appendix I of CITES. It is not listed as endangered or threatened under the ESA. The Himalayan ibex is not on the list of protected species for Pakistan (A. Ahmad, pers. comm. 1994), in any appendix of CITES, nor in the ESA.

Bar Valley Project. The Himalayan ibex is the species of focus for the Bar Valley Project.

Markets

Torghar Conservation Project. The Suleiman markhor and Afghan urial are in high demand as trophy animals among wealthy international hunters from Europe and North America (including Mexico), and opportunities to hunt these species are very limited. Suleiman markhor is not available anywhere else in the world, while Afghan urial has limited availability. High demand and low supply are reflected in the trophy fees that such species command—approximately US$25,000 per markhor and US$11,000 per urial.

Torghar hunters come primarily from Europe. The TCP has an inadequate marketing mechanism which relies on word of mouth from successful hunters and advertising by a few hunting agents connected to other hunters. As a result, the TCP has sometimes had difficulty lining up hunters well enough in advance to plan income (and therefore, expenditures) for a given calendar period, and last-minute cancellations have affected hunting success. In addition, one major market, the United States, has been closed because the ESA prevents American hunters from importing trophies of Suleiman markhor into the United States without a permit that is virtually impossible to obtain because it requires the U.S. secretary of the interior to make an exception to the law. The American hunting community is thus a large but untappable pool of potential supporters of the TCP.

Chitral Conservation Hunting Program. The CCHP also focused on a species—the Pir Panjal markhor—in short supply and high demand among hunters. At the time hunting was suspended in 1991, the CCHP was receiving US$15,000 per trophy. Thus a significant economic gain could be realized from a small harvest, and a small harvest could ensure ongoing demand for hunting. The CCHP also benefited from an exclusive relationship with one organization, the Shikar Safari Club (a small club of wealthy hunters from the United States and Europe). Such a relationship eliminated the costs and uncertainty associated with marketing the hunts.

Bar Valley Project. To date, no Himalayan ibex have been harvested and no income has been generated by hunting. Though the Himalayan ibex is in demand by hunters, hunting opportunities are plentiful elsewhere, including Kazakhstan and Kyrgyzstan (from the Hunting Consortium, Ltd., 1994 hunting brochure), Mongolia (from Ron the Guide

1990 hunting brochure), Nepal, and New Zealand (B. Levitz, pers. comm. 1994). Thus the economic gain realized per animal harvested is not expected to be very high (US$2,000–$3,000). There is legitimate reason to be concerned that international demand for trophy hunting of Himalayan ibex may *not* be sufficient to provide a reliable, long-term source of income for Bar Valley.

The Conservation Hunting Programs

Torghar Conservation Project

The TCP was initiated through the efforts of the local Pathan tribal chieftain, the late Nawab Taimur Shah Jogezai, and one of his relatives, Sardar Naseer A. Tareen. In the early 1980s, these two men became alarmed at what they perceived to be a drastic decline in the Suleiman markhor and Afghan urial populations of the Torghar Hills, primarily, they thought, as a result of the dramatic influx of high-powered weapons and cheap ammunition into the area during the Afghan war (Tareen 1990, n.d.). Jogezai petitioned the Baluchistan Forest Department (BFD) to send wildlife officers to stop the killing because the Baluchistan Wildlife Protection Act of 1974 gives the provincial government responsibility for protecting wildlife. But the BFD did not respond to the request, perhaps because it pertained to tribal lands, and tribal lands in Baluchistan remain largely outside government authority and are still ruled by tribal law in many matters, including access and permission to hunt (N. Tareen, pers. comm. 1994).

In consultation with wildlife biologists from the United States, Tareen and members of the Jogezai family (the local ruling family of Pathans) conceived and developed the TCP. The concept of the TCP was simple: local Pathan tribesmen would be asked to give up hunting in exchange for being hired as game guards to prevent poaching. The TCP would be financed by revenues derived from a limited trophy hunt of markhor and urial. The TCP was instituted in 1986 and developed steadily through fall 1994. The TCP had no formal structure until April 1994, when a nongovernmental organization (NGO)—the Society for Torghar Environmental Protection (STEP)—was established to manage the TCP as a government-recognized, nonprofit conservation effort.

As of November 1994, the TCP was employing 33 local Pathan game guards who were protecting an area of approximately 1,000 square kilometers (km^2). The TCP has virtually halted all poaching of markhor and urial by both locals and outsiders. As a result, the markhor and urial populations of Torghar, which were essentially extirpated by 1983–84, have grown steadily since 1985–86 (Johnson, n.d.). The TCP has been largely self-sufficient since its inception, having relied on an income of approximately US$460,000 generated through the trophy harvest of only 14 markhor and 20 urial over its 10-year history.

Chitral Conservation Hunting Program

The CCHP was initiated in 1983 by the Wildlife Wing of the North West Frontier Province Forest Department (the Wildlife Wing is now the Wildlife Department). In the early 1980s, despite protective laws and the presence of wildlife sanctuaries and game reserves, wildlife of the NWFP was being poached, predators were killed regularly, and ongoing habitat degradation elicited little local concern (M. Malik, pers. comm. 1994).

In 1982, the Wildlife Wing's conservator of wildlife, M. Mumtaz Malik, and representatives of the Shikar Safari Club first discussed the idea of a conservation hunting program for Pir Panjal markhor. Malik felt that such a program would show people that wildlife has great value, and would stimulate government interest and concern for wildlife. While he was concerned about initiating a hunting program in a national park (Chitral Gol), he also knew that markhor hunting permits had been issued in the past for Chitral Gol, without deriving any benefit at all from those permits. He also wanted to focus on the importance of Chitral Gol for the conservation of Pir Panjal markhor. Starting in 1983, two markhor hunting permits per year were issued to members of the Shikar Safari Club. The CCHP was curtailed in 1991, however, when the cabinet of Pakistan recommended that big game hunting and trophy export be banned throughout the country, except as specifically authorized by the prime minister.

During its eight years, the CCHP harvested 16 markhor in and around Chitral Gol National Park and generated approximately US$250,000 in revenues. To date, however, none of the funds generated by the hunts have been expended on either markhor conservation or the local community (M. Malik, pers. comm. 1994). Instead those funds have been deposited into a special government account awaiting a decision on their disposition by the provincial government (M. Malik, pers. comm. 1994). The CCHP has indirectly contributed to increased staffing for the Wildlife Department by increasing the government's awareness and interest in NWFP's protected areas (M. Malik, pers. comm. 1994), and has brought a very limited return to the local community through occasional hunting guide fees.

The NWFP Wildlife Department claims to have achieved a substantial reduction of wildlife poaching in Chitral District, although there are no quantitative data to support this claim. The Pir Panjal markhor population of Chitral District does not appear to have increased in size as a result of any reduction in poaching; the estimated total population size has remained about the same since the late 1970s, according to population figures provided by the NWFP Wildlife Department. (The department conducts periodic wildlife surveys in the province. The last survey was conducted in 1993.)

Bar Valley Project

The BVP was conceived and developed by World Wide Fund for Nature–Pakistan (WWF-Pakistan) (Ahmad, n.d.). By the late 1980s, human population growth, expansion of agriculture and livestock raising, habitat degradation, and hunting by locals and outsiders had combined to put the Himalayan ibex population of Bar Valley in danger of extirpation. The Forest Department of the Northern Areas was too ill-equipped and undermanned to control the hunting, although it had the legal responsibility to do so under the Northern Areas Wildlife Preservation Act of 1975 (A. Ahmad, pers. comm. 1994).

Utilizing the village organization earlier created by the Aga Khan Rural Support Program (AKRSP) as a forum for conducting dialogue with local communities, WWF-Pakistan convinced villagers that they were destroying their wildlife heritage through indiscriminate hunting. A plan to conserve the ibex was prepared under the leadership of a local community leader, with the guidance of WWF-Pakistan. According to the plan, the people of Bar Valley would stop hunting and would protect the ibex until such time as the population recovered sufficiently to allow a limited trophy hunt. To compensate the villagers for the meat they lost as a result of their agreement to stop hunting, WWF-Pakistan would provide a one-time subsistence allowance in the form of a long-term loan. That loan would remain outstanding until revenues began to be generated by trophy hunting. The initial proposal called for half of the permit fee to go to the Northern Areas government for developing other conservation programs, and half to Bar Valley. A portion of the Bar Valley revenues would be used to repay the loan, and the remainder would go to the villagers to compensate their loss of a meat resource and to help with other development needs.

The people of Bar Valley agreed to this plan in fall 1989 and affirmed under formal oath that they would not hunt ibex. The program began in June 1991 when the subsistence loan of Rs. 240,000 (US$8,000) was made to the chief administrator of the Northern Areas government, who distributed it equally among the 240 households of Bar Valley. Local villagers were selected for the "village hunting committee," which was responsible for ibex watch and ward tasks for one year. Members of the committee were to be paid a monthly stipend by WWF-Pakistan.

The BVP appears to have achieved a substantial reduction, if not complete cessation, of unauthorized hunting of Himalayan ibex by both locals and outsiders in Bar Valley, according to A. Ahmad (pers. comm. 1994). The ibex population appears to have stabilized or increased slightly in number as a result. To date, no Himalayan ibex have been harvested in Bar Valley by international hunters, and therefore no revenues have been derived from hunting.

Critical Evaluation

Biological Suitability

The key issue here is whether the caprinid populations in the three conservation hunting projects can sustain a "significant" harvest into perpetuity. A significant harvest is one that generates enough economic return to produce significant benefits for conservation by (1) generating conservation funds or (2) positively influencing conservation awareness, attitudes, and behavior.

Torghar Conservation Project

Detailed ecological studies of Suleiman markhor and Afghan urial have not been conducted in the Torghar Hills. However, the life history characteristics of these two species are known well enough (e.g., Schaller and Khan 1975; Roberts 1977; Schaller 1977) to allow conservative management of the species in Torghar in accordance with the "precautionary principle" (Freese 1994).

Numerous first-person accounts from game guards (Sagzai, Khoshalay, Noordad, pers. comm. 1994), managers of the TCP (N. Tareen, M. Jogezai and pers. comm. 1994), and international biologists (R. Mitchell, pers. comm. 1989) strongly suggest that the Suleiman markhor and Afghan urial populations of the Torghar Hills were virtually extirpated in the early 1980s. Yet, despite an annual trophy harvest of a few animals, the populations of these two species have grown steadily in the past 10 years. In November 1994, STEP conducted a formal, systematic field survey of the markhor and urial populations of Torghar. During the nine-day survey, 135 markhor and 189 urial were counted in five survey areas in the TCP's "core protected area" of 300 km^2 (Johnson, n.d.). Extrapolation of these results to the entire core protected area, an admittedly nonrobust process, yields total population estimates of 390 Suleiman markhor and 670 Afghan urial. Further extrapolation to the entire TCP area yields total population estimates of 695 markhor and 1,175 urial (Johnson, n.d.).

The reason for this population growth is simple: a large source of mortality (uncontrolled hunting) has been replaced by a much smaller source of mortality (controlled, limited trophy hunting). Other than controlling hunting, the TCP has not instituted any management practices (e.g., creation of water holes, elimination of domestic livestock competition) that could contribute to population growth.

Today the Suleiman markhor and Afghan urial populations of Torghar appear to fit the parameters—sufficient total population size, growing population, normal age and sex ratios—of a "viable" caprinid population for both population and genetic processes (e.g., Soule 1987; Hebert 1991). Such a population should be able to sustain an annual "trophy harvest of males, in numbers equivalent to 1–2 percent of the total population size," without negative consequences to the population (assuming no significant mortality from poaching), according to Harris's (1993) review of the literature on similar species. This is true, in part, because many caprinids have a polygynous mating system, and the population's overall reproductive rates would be little affected by the loss of a small number of males (Caughley 1977; Schaller 1977). Assuming a total markhor population of four hundred and a total urial population of seven hundred in the core protected area, sustainable harvest should be 4 to 8 markhor and 7 to 14 urial

(1–2%) per year. Since no more than three adult male markhor and four adult male urial have been harvested in any given year at Torghar, it is clear that the harvest has been conservative.

Chitral Conservation Hunting Program

Studies of the population dynamics and habitat utilization of Pir Panjal markhor in and around Chitral Gol N.P. and Tooshi G.R. were conducted during the 1970s and early 1980s (Schaller and Mirza 1971; Aleem 1977, 1978, 1979; Hess 1986, n.d.). The biology of the Pir Panjal markhor is known well enough (e.g., Roberts 1977) to allow conservative management of the population and its habitat in Chitral District.

Population data presented in the aforementioned publications suggest that the Pir Panjal markhor population of Chitral Gol N.P. increased from a low of around 85 individuals in 1970 (Schaller and Mirza 1971) to around 520 animals in 1979 (Aleem 1979), and then declined again to the most recent estimate of 176 (Wildlife Department, NWFP 1994). The possible causes of these population changes have not been well documented, although predation by snow leopards and dispersal from the area due to an overabundant population have been suggested (A. Ahmad, pers. comm. 1994; M. Malik, pers. comm. 1994). Habitat degradation and human disturbance may also be contributing factors.

The total population of Pir Panjal markhor in Chitral District appears to have remained relatively constant from 1983 through 1993 (unpublished population figures provided by M. Malik, pers. comm. 1994), although individual populations appear to be small and fragmented. According to the figures provided by Malik, only a few populations are of marginally adequate size to sustain a trophy hunt of 1–2 percent of the population per year without negative impact, assuming there is no significant poaching. Those populations are Chitral Gol N.P. (population count of 176 would allow a harvest of two to four animals per year), Tooshi G.R. (population count of 157; allowable harvest of one to two animals), and Gehrait G.R. (population count of 77; allowable harvest of one animal every other year). As a caveat, it has been reported that livestock overgrazing and collection of fuelwood continue to degrade the habitat, both within and near protected areas in Chitral District (Malik 1985, 1987). Such degradation will continue to lower the carrying capacity for markhor, thereby affecting the number that can be hunted. The extent to which hunting by locals (poaching) has been reduced is also open to question.

Thus it appears likely that Pir Panjal markhor populations in Chitral District only marginally satisfy the biological requisites for sustainable hunting. Hunting should be allowed only if several conditions are met: (1) other markhor mortality factors, both natural (e.g., predation) and manmade (e.g., poaching), are minimized; (2) populations are surveyed regularly to monitor their status; (3) harvest is rotated among the three or four largest subpopulations on an annual basis; and (4) habitat degradation is stopped and habitat quality is improved in degraded areas.

Bar Valley Project

Although detailed ecological studies of Himalayan ibex have not been undertaken in Bar Valley, the species' biology is understood adequately (e.g., Roberts 1977; Rasool 1982) to allow conservative management of the ibex population and its habitat in Bar Valley.

No quantitative data on the Himalayan ibex population in Bar Valley are available from before the BVP. However, information obtained from interviews with a number of local hunters strongly indicates that Bar Valley's ibex population was in decline prior to the project (A. Ahmad, pers. comm. 1994; WWF-Pakistan, n.d.). At that time, hunter kill was very high and included females and young. It was likely that this mortality rate exceeded the natality rate of the population. In 1992, the BVP initiated a monitoring program for the Himalayan ibex population of Bar Valley in the form of semiannual surveys. The current population is estimated at eight hundred to one thousand animals, including a substantial number of trophy males (Ahmad 1993; WWF-Pakistan, n.d.). The population decline appears to have been arrested.

The Himalayan ibex population of Bar Valley is now large enough and healthy enough to sustain a trophy hunt of 1–2 percent per year without negative impact to the population, assuming no significant poaching. Given that the current population is estimated at eight hundred to one thousand individuals and there is little poaching, this translates conservatively to a harvest of eight to twenty animals annually.

Biodiversity Impacts

Torghar Conservation Project. The TCP has contributed to maintaining the biodiversity of the Torghar Hills because it has probably prevented the extirpation of Suleiman markhor and Afghan urial populations by eliminating uncontrolled hunting of these species. It is also likely that the TCP has produced a general reduction in hunting pressure in the Torghar Hills, with benefits accruing to those species formerly hunted heavily (especially the Indian wolf [*Canis lupus pallipes*], chukar partridge [*Alectoris graeca*], and seesee partridge [*Ammoperdix griseogularis*]). TCP managers have convinced some tribesmen to stop the practice of spreading poison baits to control porcupines (*Hystrix indica*) which have become a nuisance through their gnawing on economically valuable pistachio trees. As a substitute control measure, TCP managers have raised the idea of reintroducing leopards (*Panthera pardus*) to Torghar. Many tribesmen have reacted favorably to this idea, whereas formerly they would have been universally opposed to such a thought.

The TCP should continue to maintain or even improve the biodiversity of Torghar over the next several years if its plans to reduce grazing pressure from domestic livestock and to plant native trees in cutover habitats are implemented.

Chitral Conservation Hunting Program. Information of actual or potential biodiversity impacts of the CCHP is anecdotal at best. Any conclusions, even preliminary, would be premature.

Bar Valley Project. The Bar Valley Project has contributed to maintaining the biodiversity of Bar Valley by preventing the extirpation of Himalayan ibex. If the project continues for the next several years, it should continue to maintain and may even improve the biodiversity of Bar Valley by reducing grazing pressure from domestic livestock, thereby protecting many plant species from local extirpation (Mustafa 1993), and by planting native trees in cutover ibex habitats, which may help reestablish extirpated species.

Cultural Factors

Torghar Conservation Project. The Torghar Hills are predominantly inhabited by the Jalalzais, a branch of the Sanzar Khail Kakar tribe of Pathans (Tareen 1990). The Pathans are the world's largest tribal society, with more than 15 million members in Afghanistan and Pakistan (Caroe 1958; Spain 1972; Quddus 1987). Pathans are divided along familial lineages into tribes, which are further divided into subtribes, clans, greater families, and so on, down to the level of individual families (N. Tareen, pers. comm. 1994). The Kakars are one of the largest Pathan tribes, and they are divided into two principal subtribes, the Sanzar Khail and Santya. The Sanzar Khail are further subdivided into the Jalalzai, Mardanzai, Abdullazai, and so on, which are themselves further subdivided (N. Tareen, pers. comm. 1994). The Jogezai family exercises authority over the Jalalzais of Torghar.

The head of the ruling family is called the Nawab. In matters of tribal business among the Kakars, the word of the Nawab is virtually equivalent to law. The local tribesmen of Torghar agreed to stop hunting and become game guards largely at the behest of the local Nawab, the late Taimur Shah Jogezai. The tribesmen did not want to go against the Nawab's wishes, out of respect for their leader and tribal custom, and concern about the potential consequences to them and their families if they did not go along. Thus the cultural makeup of Torghar's resident population has been a major contributing factor to the success of the TCP.

Chitral Conservation Hunting Program. The human population of Chitral is comprised of a number of ethnic groups, termed by Biddulph (1971) "a curious and intricate ethnological puzzle." The largest section of the population is Chitrali or Kho (Indo-Aryan), but other groups include the Kalash (Black Kafir), Bashgali (Red Kafir), Tajik, Afghan, Narsati, and Dir Kohistani (Biddulph 1971; M. A. Khan 1975). There are four social classes among the Chitrali people—Adamzadas of the first class, Adamzadas of the second class, Arbadzadas (or Yuft), and Fakir Miskin—each of which is further subdivided (M. A. Khan 1975).

There are no social, religious, or cultural barriers to hunting in Chitral. In fact, Chitralis have a very strong tradition of hunting that has to be overcome if a controlled hunting program is to be successful. Chitralis also have a strong cultural history of traditional use rights on the land (for grazing and fuelwood collection), and are not particularly receptive to restrictions being put on that use, especially by the government. Thus government programs have a difficult time gaining acceptance among the people.

Bar Valley Project. The residents of Bar Valley originated from three tribes: the Rajway, Kachtey, and Babrey (WWF-Pakistan, n.d.). They are Shia Muslims and their language is Shina (Mustafa 1993). About 80 percent of the people in Bar Valley are illiterate, 10 percent are barely literate, and 10 percent are students or graduates.

There are no social, religious, or cultural barriers to hunting in Bar Valley. In fact, Bar Valley has a very strong tradition of hunting that had to be overcome to protect the valley's ibex population. Like the Chitralis, the people of Bar Valley are not particularly cowed by government authority or attempts to limit their traditional use of resources. They are, however, influenced by local leaders and people of respect.

Land Tenure

Torghar Conservation Project. The Jalalzais are divided into a number of greater families, seven of which are represented at Torghar: the Jogezai, Khudzai, Mirozai, Shabozai, Shahizai, Hakimzai, and Mehmanzai (N. Tareen, pers. comm. 1994). Each of these greater families owns a specific area of the mountain, hills, and plain, and individual families own their own land within the greater family area (Tareen 1990). Pathans vigorously defend their land against unwelcomed outsiders (Caroe 1958; Spain 1972; Quddus 1987). This strong system of land tenure has allowed the TCP to succeed. Where previously poachers were of little concern to local residents, they now are not allowed into the mountains.

Chitral Conservation Hunting Program. When Chitral was a princely state, the ruler (the Mehtar) maintained Chitral Gol as a private hunting reserve, but the local people had concessions for grazing and fuelwood collection. The incorporation of Chitral District into the general administration of Pakistan in 1969 broke down this traditional system of land tenure and opened Chitral Gol to the general public (Malik 1985). Seasonal livestock grazing went uncontrolled, and up to 150 cattle and 4,000 sheep and goats foraged in the alpine pastures during summer (Malik 1985). The livestock had a heavy impact on native flora and fauna. Fuelwood collecting also went uncontrolled, and people soon began to fell green trees as well (Malik 1985). Oak trees, a favored fuelwood, suffered heavy losses. Finally, poaching is said to have increased and wildlife populations dwindled alarmingly after the public opening of Chitral Gol (Malik 1985). By 1970, it was estimated that only 100–125 markhor remained (Schaller 1977).

In 1975, Chitral Gol became property of the government with the exception of 8–10 hectares (ha) of cultivated land and several houses which remained property of the ex-Mehtar (Malik 1985). Five families lived on this property year round, along with forty cattle and five hundred goats. Timber and fuelwood for these families also came from the park (Malik 1985). When Chitral Gol N.P. was established, the NWFP government tried to halt the grazing and fuelwood collection and remove the people from the park. In 1984, the NWFP government decided to acquire the private lands and houses in the park (Malik 1985); all tenant families living in the park and their livestock were supposed to be removed in December of that year (Malik 1985). All other livestock grazing in the park was also prohibited in 1984, as was fuelwood cutting and collection (Malik 1987). The NWFP government withdrew all fuelwood concessions from the villages around Chitral Gol, and told residents to get their fuelwood from other areas (Malik 1985). As mitigation for these management actions, the NWFP government proposed a program of compensation for the local people (Malik 1985, 1987).

Although the NWFP government allocated funds to purchase the ex-Mehtar's private lands within the park, the purchase was blocked by the courts. The landowner was allowed to maintain his livestock within the park. In addition, the courts allowed cattle grazing by others to continue within the park. Although other sheep and goats were prohibited from grazing, trespass grazing continues to the present. Finally, collection of dry, downed fuelwood was allowed to continue within the park. Local villagers continue to abuse this privilege by lopping off tree branches and allowing them to dry before collecting them (M. Malik, pers. comm. 1994). In conclusion, disruption of the traditional land tenure system has had rather dire consequences for wildlife conservation around Chitral Gol.

Bar Valley Project. In Bar Valley, individuals own the ground where their residence and small agricultural plots are located. The government owns the forest and high pasture lands, but villages and individuals have traditional use rights (A. Ahmad, pers. comm. 1994). This strong land tenure system has contributed to the success of the BVP.

Local Stakeholders

Torghar Conservation Project. The TCP has succeeded in providing social and economic incentives for local Pathan tribesmen to conserve wildlife. Thirty-three local Pathan tribesmen are employed as game guards, with salaries and benefits coming directly from the TCP. They also receive other benefits such as medical care and assistance with their orchards and irrigation channels (N. Tareen, pers. comm. 1994). Thus these local stakeholders have a major interest in ensuring the success of the TCP. Furthermore, there is a direct linkage between trophy hunting, economic incentives to the locals, and conservation because trophy hunting is allowed only if there are sufficient trophy-sized animals in the mountains. All revenues for the TCP are derived from trophy hunting, and game guards are paid from these revenues. Thus game guards clearly understand that their jobs depend on protecting the markhor and urial. The five TCP managers (three members of the ruling Jogezai family, N. Tareen, and one other relative) also have a direct economic stake in the success of the TCP, because they derive salary and other benefits from the project (for example, the project sponsored and paid all expenses toward a bachelor of science degree in natural resources from a university in the United States for TCP's manager of natural resources, a member of the Jogezai family).

Chitral Conservation Hunting Program. The CCHP has provided only minimal social and economic incentives for local people to conserve wildlife. It may have indirectly contributed to the employment of a few local people as game watchers in Chitral Gol N.P. and other game reserves, and may have contributed to reduced poaching by creating the expectation among locals that they will receive economic benefits from the hunting program (M. Malik, pers. comm. 1994). However, since no funds derived from hunting have yet been returned to the local community, no direct linkages between trophy hunting, economic incentives, and conservation have been established. Thus local people do not have a major economic stake in the success of the CCHP, nor do they have a strong social stake. This is reflected, in part, by ongoing problems associated with Chitral Gol N.P., including trespass grazing and fuelwood harvest.

Bar Valley Project. The BVP, through the WWF-Pakistan loan, has succeeded in providing social and economic incentives for local people to conserve wildlife. It has provided direct employment for 10–14 people per year for three years, as well as a small number of other benefits to local people. It has improved the organization and decision-making capability of local people (A. Ahmad, pers. comm. 1994). The local people have a stake in the success or failure of the project. However, since no trophy hunting has yet taken place in Bar Valley, no direct linkage has been established between trophy hunting, economic incentives to the local people, and conservation. This linkage must be established if trophy hunting is to become a legitimate conservation tool in Bar Valley. If a trophy hunt does not take place soon, the BVP may lose the support of the local people.

Provincial, National, and International Stakeholders

Torghar Conservation Project. The TCP's major shortcoming has been lack of support from provincial (government of Baluchistan), national (government of Pakistan), and international stakeholders. Various influential individuals in Baluchistan have tried to discredit the TCP because it threatens their interests (including their own trophy hunting). In the past, the Baluchistan Forest Department was reticent and slow to issue the necessary hunting permits for the TCP, forcing the project to operate outside the strict boundaries of provincial law in order to survive. Likewise, export permits for trophy animals have, in the past, been consistently denied because of national government conflicts with the province over interpretation of provincial law and because of the national "ban" on big game hunting. Good progress has recently been made toward resolving these problems and complying with all formal requirements. Provincial and national governments now support and endorse the TCP.

Among international stakeholders, the government of the United States has had the most influence on the TCP because, as noted earlier, U.S. law makes it difficult for American hunters to import markhor trophies.

Chitral Conservation Hunting Program. Because it was created and run by the Wildlife Department of NWFP, the CCHP enjoyed the support of the government of NWFP. Within the Wildlife Department, however, it appears that the CCHP was regarded as important, but not essential, for markhor conservation in Chitral District because no additional field or administrative staff were hired nor major financial resources committed to implementing it. Thus it appears that the Wildlife Department had less stake in ensuring the continuation and success of the CCHP than it might have had the program been considered essential to markhor conservation. When the CCHP was curtailed in 1991 as a result of the national government's ban on big game hunting, the Wildlife Department made only a modest effort to revive it.

The CCHP's downfall was lack of support at the national level (government of Pakistan). Both the minister of food, agriculture, and cooperatives and the president opposed the program because they believed—incorrectly—that it violated CITES. This situation has yet to be resolved.

Bar Valley Project. The BVP was the first of the three conservation hunting programs to garner both provincial and national government support. Eight hunting permits have been approved by the Northern Areas administration and the government of Pakistan. The BVP's success should make it easier for the other conservation hunting programs in Pakistan to obtain permits.

Controls on Use

Torghar Conservation Project. As noted above, the Pathan game guards now have a strong economic incentive to limit poaching both by outsiders and locals. Furthermore, the game guards have become zealous protectors of "their" animals, and will not allow any to be killed if they can prevent it. There are also strong sociocultural pressures that limit poaching. Tribesmen do not want to be ostracized from their community by killing a markhor or urial without permission.

Given that the TCP is a private initiative on tribal lands, that the government of Baluchistan has little control over what happens in tribal areas, and that "bakshish" (payoff) is common throughout modern Pakistan, TCP managers have faced few real controls on their use of the wildlife resource. They could have exploited the TCP for personal gain by harvesting more animals or returning a lower percentage of income to the local tribesmen.

But TCP managers started the project because of their personal concern for the survival of markhor and urial, not to make a profit. They did not want to lose these significant symbols of their natural and cultural heritage. Their control was entirely self-imposed: they chose not to exploit the resource but to harvest only enough animals to cover immediate TCP costs. Over the nine-year history of the TCP, fewer than 35 animals have been harvested. Management salaries have averaged 23 percent of income. All remaining funds have gone to game guard salaries and benefits, educational costs, costs for conducting hunts, and physical improvements to Torghar (e.g., roads). TCP managers have a vested interest in seeing the project succeed because they are committed to saving the wildlife resources of Torghar, and because they have invested considerable time, energy, and personal prestige in the project.

Chitral Conservation Hunting Program. As an official program of the NWFP government, the CCHP was under the full control of the conservator of wildlife, who had the full authority to determine the number of animals harvested by the CCHP. That authority appears to have been exercised conservatively with an annual harvest of two animals.

More problematic in Chitral has been the control of poaching by locals. Local people were promised some economic return from the CCHP if they agreed to limit their hunting, but to date they have received nothing from the provincial government. As they currently have little stake (economic or otherwise) in the CCHP, they have little incentive to limit poaching.

Bar Valley Project. Bar Valley is a confined, isolated area with a relatively small human population. Most residents know each other; many are related by birth or marriage. The people have been organized for years, initially through the "village organization" created by AKRSP and subsequently through organizational activities of the BVP (A. Ahmad, pers. comm. 1994). Thus there are strong social pressures associated with participation in the project, especially in regard to violations of the no-hunting agreement. These social pressures have been important to the success of the BVP during its early years (A. Ahmad, pers. comm. 1994).

Effects of the Programs on Other Resource Uses

Torghar Conservation Project. There are no other viable consumptive uses of markhor and urial that provide the same economic return per animal harvested. Formerly, the markhor and urial of Torghar were valued only as challenging targets for local hunters and as a supplemental meat source for locals and nomadic herders passing through the area twice a year. A trophy fee of US$25,000 for markhor and US$11,000 for urial thus maximizes the "per individual" value.

The only potentially viable nonconsumptive use of markhor and urial is tourist viewing/photography. Unfortunately, the Torghar Hills are remote and access is difficult, so tourism opportunities are limited. Even if tourism were to become popular, it would not be incompatible with a small, controlled harvest of markhor and urial because harvest would not reduce the number of animals available for viewing, nor interfere with the esthetics of viewing as harvest takes place at a time of year (fall) when tourists are not likely to be present because of poor weather.

The TCP provides enough benefits to local people to preclude their converting the land to other uses, such as increased grazing or agriculture, that could decrease the biological diversity of Torghar.

Chitral Conservation Hunting Program. The only possible viable nonconsumptive use of markhor in Chitral District is tourist viewing/photography. Tourism in Chitral has been increasing steadily over the past few years, and an appropriately targeted promotional campaign could increase tourism visitation to the park and bring needed revenues to the area. In any event, a controlled harvest of one to two markhor per year would not significantly affect opportunities to use the markhor population of Chitral Gol or the other game reserves as tourism objectives. Harvest would not reduce the number of animals available for viewing, nor interfere with the esthetics of viewing as harvest takes place at a time of year when tourists are not likely to be present.

Since conservation programs such as the CCHP have provided few, if any, economic benefits to the local people of Chitral, land uses that compete with good conservation, such as ongoing heavy livestock grazing and fuelwood collection, and poaching by local people, continue. As a consequence, wildlife habitat and biodiversity continue to decline in Chitral District.

Bar Valley Project. As in the other two programs, the only potentially viable nonconsumptive use of ibex is tourism and tourist viewing/photography. Tourism in the Northern Areas of Pakistan has been increasing steadily over the past few years, and WWF-Pakistan has undertaken a promotional campaign to help increase tourist visitation to Bar Valley. In any event, a controlled harvest of ibex would not significantly affect opportunities to use the ibex population as a tourism objective. Harvest would not reduce the number of animals available for viewing, nor interfere with the esthetics of viewing, as harvest takes place at a time of year (fall) when tourists are not likely to be present because of poor weather.

WWF-Pakistan has temporarily helped prevent alternative land uses and resource degradation in Bar Valley through its "loan" and underwriting the cost of management. This situation cannot continue indefinitely, and the transition to self-sufficiency through sport hunting and, potentially, tourism will be critical to ensure that local villages do not turn to alternative land uses that degrade ibex habitat and biodiversity.

Conclusions

Of the three conservation hunting programs evaluated in this case study, only the Torghar Conservation Project can be considered a success because there is clear evidence that it has achieved its objectives—conserving the markhor and urial populations of the Torghar Hills and improving the well-being of the local Pathan tribesmen—through a limited sport hunt of trophy animals. The Chitral Conservation Hunting Program cannot be considered a success because there is no clear evidence that it has helped conserve markhor in Chitral District, nor has the program provided any substantial benefits to the local people. The jury is still out on the Bar Valley Project. Although it has helped conserve the ibex of Bar Valley, and has provided economic benefits to local people, it has not accomplished these tasks with revenues derived from sport hunting (because no revenues have yet been generated through this means). The BVP is now in a critical period in its development—the transition from being completely subsidized by a large conservation NGO to being self-sufficient with funds generated by trophy hunting. How this transition is accomplished may ultimately determine the long-term success of the BVP. Only when the program generates enough revenue through sport hunting to become self-sufficient will it be a success. Without a long-term source of

Table 11.1. Factors Determining the Success or Failure of Conservation Hunting Programs in Pakistan[a]

Factor	TCP	CCHP	BVP
1. Population size	+	–	+
2. Management	+	+/–	+
3. Demand	+	+	–
4. Marketing	–	+	–
5. Revenue	+	–	+
6. Linkage	+	–	–
7. Controls on use	+	+/–	+
8. Type of control	+	–	+
9. Barriers	–	–	–
10. Other stakeholders	+/–	–	+

[a]TCP, Torghar Conservation Project; CCHP, Chitral Conservation Hunting Program; BVP, Bar Valley Project.

demand/income for ibex hunting, the BVP will have to explore alternative sources of income generation if the project is to survive. The next two to three years will be critical for Bar Valley.

Which factors are important in determining the success or failure of a conservation hunting program in Pakistan? Based on this study, the following factors appear to be most significant. Table 11.1 compares how each of the three conservation hunting programs rates (positive, negative, or uncertain) in relation to these factors.

1. The target species' population is of adequate size and condition to sustain a significant harvest in perpetuity.
2. Management activities—especially population monitoring and habitat management—are adequate to ensure that the target species' population is being maintained.
3. The target species is in high enough demand that a biologically sustainable harvest can generate adequate funding to cover the management costs of the program and provide strong socioeconomic incentives for good resource management by local stakeholders.
4. The hunting program has an adequate marketing mechanism to ensure long-term demand for the resource.
5. Revenue distribution is such that local people are deriving enough benefits from the program to prevent poaching and alternative land uses which are destructive to biodiversity.

6. There is a direct linkage between economic gain to the local people, hunting as a source of that economic gain, and conservation of wildlife as a basis for hunting, so that people understand they must maintain populations in order to benefit.

7. Actual controls on use are adequate to minimize poaching. Controls can be local (i.e., by the local people) or external (i.e., by the government), but local control appears to be more effective.

8. Strong systems of land tenure (e.g., private property or strong traditional use rights) and strong cultural bonds among local people (e.g., tribal groups) are important factors in achieving local controls on use.

9. There are no social/cultural barriers among the local people toward instituting a conservation hunting program.

10. Opposition from provincial, national, or international stakeholders does not substantially impact the program.

Recommendations

Torghar Conservation Project

STEP could take several actions to improve the TCP. First, STEP must convince the government of Baluchistan to issue hunting permits for both markhor (two per year) and urial (four per year) for the next five years. This can be legally accomplished under the Baluchistan Wildlife Protection Act of 1974.

Second, STEP should work with the government of Baluchistan to resolve the impasse with the government of Pakistan regarding issuance of hunting and export permits for markhor and urial. The national government must be lobbied to develop and implement an objective and binding policy for the approval of conservation hunting programs in Pakistan. That policy must unambiguously set out the biological and harvest control requirements a conservation hunting program must meet in order to be approved, the minimum permit fees, and the minimum economic return that must go to local people.

STEP must locate and cultivate new international markets for Suleiman markhor hunting. Opening the U.S. market will be critical to long-term survival of the TCP. However, since Suleiman markhor is listed as "endangered" under the ESA, it will be necessary either to downlist the Torghar population to "threatened" or to obtain an exemption from the secretary of the interior under section 10 of the act to allow import of trophies into the United States.

STEP should explore opportunities for marketing ecotourism in the Torghar Hills. Wildlife viewing and photography could attract a small number of potentially affluent international tourists.

STEP needs to begin implementing its five-year plan. One priority is implementation of annual population surveys for markhor and urial. Annual surveys are needed to determine population trends and allowable harvests accurately.

Chitral Conservation Hunting Program

A first priority for the government of NWFP is to decide how to distribute the funds earned from trophy hunting from 1983 to 1991. According to M. Malik (pers. comm. 1994), half of these funds are currently designated to go to local communities, but local expectations will turn to disappointment if no decision is made in the near future.

What remains to be determined is which communities receive the funds, the manner by which funds will be distributed, and what activities will be supported. It is logical that a significant portion (perhaps 40%) of the funds go to the seven communities that lost some of their traditional use rights when Chitral Gol N.P. was established. A portion (perhaps 15%) could then go to communities near Tooshi G.R. and to Gehrait Gol G.R. (another 15%) to solidify their support of the reserves. The final portion (perhaps 30%) could go to other communities in important habitat areas as incentives for habitat conservation. Participatory rural appraisals should be conducted in each village to determine how the funds are to be used. All villagers should have a chance to voice an opinion, and the final decision should be taken by the village organization.

The government of NWFP should amend its existing rules to allow the establishment of "private game reserves" on government lands in NWFP (M. Malik, pers. comm. 1994). The Wildlife Department should then encourage establishment of such reserves in important habitat areas. A limited harvest of wildlife would be allowed on these reserves, and the funds generated would be divided among local villages and the NWFP government. Thus the reserves would serve a dual function of providing income for local people while at the same time protecting wildlife populations and habitat.

The government of NWFP should resolve its impasse with the national government regarding hunting/export permits for markhor. The national government must be convinced that export of sport-hunted trophies of Appendix I species is not prohibited by CITES. Furthermore, NWFP should join with Baluchistan in lobbying the national government to develop and implement an objective and binding policy regarding approval of conservation hunting programs.

The international trophy value of Pir Panjal markhor has undoubtedly appreciated since the last permits were issued to the Shikar Safari Club for US$15,000 each. When the Chitral Conservation Hunting Program is reinitiated, the charge for markhor hunting permits should be fixed by the NWFP Wildlife Department at a minimum of US$22,500.

The NWFP Wildlife Department needs to begin monitoring markhor populations and habitat conditions on an annual or biennial basis.

Bar Valley Project

What can be done to improve the BVP? First, the Northern Areas administration must move aggressively to locate international hunters for the Himalayan ibex in Bar Valley. Because the ibex can be hunted legally in a number of other countries, the BVP must do something to increase hunter interest in coming to Bar Valley. Two possibilities are (1) charging a trophy fee substantially less than the going rate, or (2) coupling ibex hunting with hunting for a more desirable species such as markhor.

The Northern Areas administration should locate a foreign hunting agent willing to take a reduced fee for the sake of conservation, and should also consider reducing the permit fee to US$2,000–$2,250 to stimulate hunter interest.

WWF-Pakistan might consider developing a project to conserve Astor markhor (*C. f. falconeri*) and linking ibex hunting to markhor hunting. Because markhor are in much greater demand among international hunters than ibex, permit fees are much higher (US$22,000–$25,000), and only one or two animals would have to be harvested to provide a substantial income. The markhor hunting could be linked with ibex hunting in Bar Valley (i.e., markhor hunters would be required to take an ibex permit, which they could use or not as desired). In this way, markhor hunting would provide a long-term source of income for both programs. This same linkage could also be established with WWF-Pakistan's ongoing Suleiman markhor conservation project in the Suleiman Range.

Finally, the marketing of ecotourism opportunities in Bar Valley should be further pursued as a way to increase and broaden wildlife-based revenues in the region.

Acknowledgments

I wish to thank WWF-International and Curtis H. Freese for providing the primary financial support needed to prepare this case study. Additional thanks are due to the Office of International Affairs, United States Fish and Wildlife Service (David Ferguson), and the Society for Torghar Environmental Protection (Naseer Tareen) for supplemental financial and logistical support.

For providing infomation related to the Torghar Conservation Project, thanks are due to STEP/TCP managers (Naseer Tareen, Maboob Jogezai, Mirwais Jogezai, Paind Khan) and game guards (especially Khoshalay, Abdullah, Sagzai, Noordad, Safferkhan, Khodaidad, Baqidad, Mohammad Afzal, Abdul Raziq, Janan, Piao, Mohammad Khan), without whose tireless assistance the fieldwork would not have succeeded. For information on the Chitral Conservation Hunting Program and hospitality in Peshawar, I thank M. Mumtaz Malik and his staff in the NWFP Wildlife Department. For information related to the Bar Valley Project, thanks go to Ashiq Ahmad, Iftikhar Ahmad, and Dawwod Ghaznavi of World Wide Fund for Nature–Pakistan.

Drafts of this manuscript benefited from a critical review by Curtis H. Freese.

References

Ahmad, A. n.d. Environmental evaluation of a conservation project in the Bar Valley. Handwritten report. World Wide Fund for Nature–Pakistan, Peshawar.

Ahmad, I. 1993. Report of the Bar Valley 3rd visit. Unpublished progress report. World Wide Fund for Nature–Pakistan, Peshawar.

Aleem, A. 1977. Population dynamics of markhor in Chitral Gol. *Pakistan Journal of Forestry* 27:86–92.

Aleem, A. 1978. Markhor, population dynamics, and food availability in Chitral Gol game sanctuary. *Pakistan Journal of Forestry* 28:159–165.

Aleem, A. 1979. Markhor, population dynamics and food availability, in Chitral Gol wildlife sanctuary. *Pakistan Journal of Forestry* 29:166–181.

Allen, C. 1984. *Lives of the Indian princes.* London: Century Publishing Co.

Baluchistan Government, Agriculture Department. 1977. *The Baluchistan Wildlife Protection Act 1974.* And rules notified thereunder (with Urdu translation, as amended up to December 13, 1977).

Biddulph, J. 1971. *Tribes of the Hindoo Koosh.* Reprint edition. Karachi, Pakistan: Indus Publications.

Bond, I. 1994. The importance of sport-hunted African elephant to CAMPFIRE in Zimbabwe. *TRAFFIC Bulletin* 14(3):117–119.

Burrard, G. 1925. *Big game hunting in the Himalayas and Tibet.* London: Herbert Jenkins.

Caroe, O. 1958. *The Pathans 550 B.C.–A.D. 1957.* London: Macmillan.

Caughley, G. 1977. *Analysis of vertebrate populations.* New York: John Wiley & Sons.

District Gazetteers of Baluchistan. 1906. *The gazetteer of Baluchistan (Zhob).* Quetta, Baluchistan: Government of Baluchistan.

Ellerman, J., and T. Morrison-Scott. 1966. *Checklist of Palaearctic and Indian mammals 1758 to 1946.* 2d edition. London: British Museum.

Eltringham, S. K. 1994. Can wildlife pay its way? *Oryx* 28(3):163–168.

Ferguson, D., ed. 1978. *Protection, conservation, and management of threatened and endangered species in Pakistan.* Report of the U.S. Fish and Wildlife Service–National Park Service study mission to Pakistan, February 5–16, 1978. Washington, D.C.: International Affairs Staff, U.S. Fish and Wildlife Service.

Freese, C., ed. 1994 (October). *The commercial, consumptive use of wild species: Implications for biodiversity conservation.* WWF International Interim Report. Gland, Switzerland.

Harris, R. B. 1993. Wildlife conservation in Yeniugou, Qinghai, China: Executive summary. Doctoral dissertation, University of Montana, Missoula.

Hebert, D. 1991. Appendix 4: Theoretical considerations in determining proper ESA listing for argali. In *Status review, critical evaluation, and recommendations on proposed threatened status for argali (Ovis ammon)*, ed. M. Lillywhite, 120–121. Prepared by Domestic Technology International, Evergreen, Colorado, for Safari Club International.

Hess, R. 1986. *Number of flare-horned markhor (Capra falconeri falconeri) and Siberian ibexes (Capra ibex sibirica) in the areas of the Chitral Gol, the Shinghai Nallah, and the Barpu Glacier in summer 1985 and winter*

1985/86. Report to the National Council for Conservation of Wildlife in Pakistan, Islamabad.

Hess, R. n.d. Wildlife in Pakistan: Extinction or recovery? In *Himalayan crucible: North Pakistan in transition,* ed. N.J.R. Allan. New York: St. Martin's Press.

Johnson, K. A. n.d. *Status of Suleiman markhor and Afghan urial populations in the Torghar Hills, Baluchistan Province, Pakistan.*

Khan, A. A. 1975. *Report on the wildlife of Chitral.* Peshawar, Pakistan: NWFP Forest Department.

Khan, M. A. 1975. *Chitral and Kafiristan: A personal study.* Self-published. Reprinted by Printing Corporation of Frontier Ltd., Pakistan Museum of Natural History, Islamabad.

Lewis, D., G. B. Kaweche, and A. Mwenya. 1990. Wildlife conservation outside protected areas: Lessons from an experiment in Zambia. *Conservation Biology* 4:171–180.

Malik, M. M. 1985. Management of Chitral Gol National Park, Pakistan. In *People and protected areas in the Hindu Kush, Himalaya,* ed. J. A. McNeely, J. W. Thorsell, and S. R. Chalise, 103–106. Kathmandu, Nepal: King Mahendra Trust for Nature Conservation and International Centre for Integrated Mountain Development.

Malik, M. M. 1987. Management plan for wild artiodactyls in North West Frontier Province, Pakistan. Unpublished master of science thesis, University of Montana, Missoula.

Mallon, D. 1983. The status of Ladakh urial (*Ovis orientalis vignei*) in Ladakh, India. *Biological Conservation* 27:373–381.

Murphree, M. W. 1991. *Communities as institutions for resource management.* Center for Applied Social Sciences, University of Zimbabwe: Occasional Paper Series. Harare, Zimbabwe.

Mustafa, G. 1993. *Comparative study of the vegetation, floristic composition and economic use of plants of Nasir Abad Valley, Hunza District Gilgit in relation to animal and human use.* Department of Botany, University of Peshawar, Pakistan.

Quddus, S. A. 1987. *The Pathans.* Lahore, Pakistan: Ferozsons.

Rasool, G. 1982. The Himalayan ibex. *Pakistan Journal of Forestry* 32:46–51.

Roberts, T. J. 1977. *The mammals of Pakistan.* London: Ernest Benn.

Schaller, G. 1977. *Mountain monarchs: Wild sheep and goats of the Himalaya.* Wildlife Behavior and Ecology Series. Chicago: University of Chicago Press.

Schaller, G., and S. Khan. 1975. The status and distribution of markhor (*Capra falconeri*). *Biological Conservation* 7:185–198.

Schaller, G., and Z. Mirza. 1971. On the behaviour of Kashmir markhor (*Capra falconeri cashmiriensis*). *Mammalia* 35:548–566.

Soule, M., ed. 1987. *Viable populations for conservation.* Cambridge: Cambridge University Press.

Spain, J. W. 1972. *The way of the Pathans.* 2d edition. Karachi, Pakistan: Oxford University Press.

Stockley, C. 1936. *Stalking in the Himalayas and Northern India.* London: Herbert Jenkins.

Tareen, N. A. 1990. Torghar Conservation Project. Unpublished manuscript.

Tareen, N. A. n.d. Sworn declaration regarding the Torghar Conservation Project.

U.S. Fish and Wildlife Service. 1992. *Appendices I, II, and III to the Convention on International Trade in Endangered Species of Wild Fauna and Flora, September 30, 1992.* Washington, D.C.: U.S. Department of the Interior.

U.S. Fish and Wildlife Service. 1993. *Endangered and threatened wildlife and plants.* 50 Code pf Federal Regulations 17.11 and 17.12, August 23, 1993. Wshington, D.C.: U.S. Department of the Interior.

Weaver, M. A. 1992. Hunting with the sheikhs: How Arab royals spend millions in the pursuit of a rare bird. *The New Yorker* December 14:51–52, 54, 56, 58–64.

Wildlife Department, NWFP. 1994. *Distribution and status of wildlife in N.W.F.P.* Peshawar: Wildlife Department, North West Frontier Province.

WWF-Pakistan. n.d. Untitled report dealing with the Bar Valley Project, including "Report of the Bar Valley second visit." Unpublished progress report, World Wide Fund for Nature–Pakistan, Peshawar.

TWELVE

Management of Ungulates and the Conservation of Biodiversity

James G. Teer

This chapter discusses effects of management or lack of management of large mammals (big game) on biodiversity. Biodiversity is defined for this purpose as ecological processes, species composition, richness, and genetic makeup in the community. Most attention is given to the ecological role of ungulates as herbivores, and especially to their impacts on their habitats and that of companion species.

> Many managers now appreciate the fundamental importance of maintaining the integrity of whole systems. An important consequence of this change in focus (from single products) is the realization that the components of ecosystems that were once viewed as outputs or products must now be seen as entry points or opportunities for management action. Ungulates offer such opportunities because they play an important role in many ecosystem processes and because managers can influence their abundance and distribution by harvest. This opportunity allows managers of natural resources to alter the state and trajectory of ecosytems. (Hobbs 1996)

Ungulates comprise most big game species and are hunted, legally or illegally, throughout most of the world. They belong to the orders Artiodactyla and Perissodactyla; however, other herbivorous large mammals, including members of the order Probiscidia (elephants), are referenced in the following discussion.

Management Strategies for Ungulates

Except for ungulates residing in parks and other protected areas, management of ungulates has typically been directed at single species. In some of the great assemblages of plains herbivores, such as in Africa, ungulates have been managed as guilds of species, usually by providing protected areas and by regulating their harvests. Management, especially on public lands, has increased from local populations to ecosystem and landscape scales. Current management strategies for producing and protecting ungulates are trending toward holistic or ecosystem approaches. Subsistence for indigenous people and recreational hunting were and remain the primary purposes of management of ungulates. Recreational hunting is by far the most important use of ungulates by humans throughout the world. Literally millions of deer, antelope, pigs, sheep, goats, and other hoofed animals are taken each year by sport hunters. In subsistence cultures, ungulates are exploited for meat, hides, and animal parts which are used for food, clothing, medicines, tools, and totems. Protection of biodiversity is now an added goal of management.

In the following text, control of ungulate numbers is presented as desirable, if not essential, to protect habitats and biodiversity. Hunting and cropping of surplus animals are the primary tools used to control ungulate numbers. Overexploitation of ungulates through hunting often has serious effects not only on their own numbers but also on ecosystem structure and function. Uncontrolled numbers of ungulates likewise may disrupt normal processes and functions, perhaps with even greater and more profound effects on biodiversity.

Land-Use Practices and Loss of Biodiversity

It is important to put into perspective disruptions of ecosystems and losses of life in them caused by human development activities and those losses and disruptions that result from efforts to manage single species.

Interventions by humans into ecosystems worldwide have so modified wild ungulate habitat that many species of ungulates and biodiversity have been enormously reduced or eliminated entirely. The primary reasons are the settlement and use of land for row crop and animal agriculture and for lumbering.

Agriculture was the precursor or cause of invasions of woody vegetation into the great savannas of East Africa and the Great Plains grasslands of North America. At least half of the former natural habitat of wildlife in the 48 contiguous United States has declined to the point of endangerment (Noss et al. 1995). Europe experienced these changes long before they occurred in the New World. Grazing, farming, burning, tree felling, coppicing, and turf cutting have changed landscapes throughout Europe to seminatural ecosystems (Spellerberg et al. 1991). Nature conservation in western Europe faces the continuous loss and fragmentation of habitats for wild species, and current conservation measures have not been successful in countering threats to biological diversity (Wallis de Vries 1994).

Great areas of tropical forests have been destroyed, putting at risk the richest assemblages of life in any ecosystem in the world. Out of 15 million square kilometers (km^2) of tropical forest, only 9 million km^2 remain (Myers 1988). Between 76,000 and 92,000 km^2 are eliminated each year with another 100,000 km^2 grossly disrupted (Myers 1988). Clearly, human development activities have had, and continue to have, a far greater impact on biodiversity than human efforts to manage single species.

Management Practices

Wild ungulate populations are managed by the following practices which may affect biodiversity:

1. management of habitat,
2. recreational hunting of ungulates,
3. control of predators,
4. introduction of non-native species,
5. game farming and ranching, and
6. parks and protected areas.

All have the potential to impact community structure and function and the life that inhabits ecosystems. Of all management practices, however, lack of control of populations of wild ungulates poses the most serious threat to other species and to their habitats.

Unfortunately, few studies have been conducted to document the effects of management of ungulates on other life. I have examined the question by reviewing case studies of management practices and their effects on biodiversity.

Habitat Management

Garrott et al. (1993) note that while conservation biologists give considerable attention to the invasion of exotic organisms, little attention is given to the "closely related problem of controlling locally overabundant or expanding native species that negatively affect other native species." Herbivores, especially big game, are potentially devastating to other life if allowed to go unchecked in numbers and distribution. For this reason, control is necessary, especially when natural checks such as predation and dispersal of young are absent from community processes.

McNaughton (1979) and Danell et al. (1994) describe responses of plants to grazing and browsing in grasslands and woodlands. They demonstrate that different levels or intensities of browsing result in different responses of woody plants: (1) decreased productivity with heavy browsing, (2) maintenance of normal productivity with moderate levels of browsing, and (3) low to moderate increases in productivity with low to moderate levels of browsing.

Ungulates must be considered dominant or keystone animals in either single-species populations or as members of communities or guilds such as occur in the great savannas of East Africa. Despite ecological and behavioral processes and phenomena of plants and animals that prevent overgrazing, uncontrolled populations of herbivores in any complex of species can, and often do, change vegetation composition and structure and reduce biodiversity (Lack 1954; McShea and Rappole 1994). The result is often a decrease in primary productivity and habitat quality and, ultimately, decreases and changes in biodiversity.

White-Tailed Deer in Texas

Of all species of big game animals in the world, none is more abundant and none has been more intensively managed than the white-tailed deer (*Odocoileus virginianus*). An ecological generalist, 38 subspecies are widespread in many ecosystems and habitats in North, Central, and South America (Baker 1984). The species responds readily to population and habitat management and, given time, is forgiving and resilient to overexploitation. Its history provides clear evidence of effects of management or lack of management on other ecosystem life and processes.

In the 1920s, the total number of white-tailed deer in the United States was about 500,000 (Seton 1909). Trefethen (1975) suggested its numbers may have been as low as 350,000 from overexploitation. Presently, white-tailed deer numbers are estimated at more than 26 million in the United States, which supports an annual offtake of nearly 5,600,000 (Jacobson and Kroll 1994).

Some of the densest populations, up to one deer per 2 hectares (ha), occur on ranches in Texas (Teer et al. 1965; Cook 1984; Teer 1984). Trends in numbers and harvests have been inexorably upward since the 1940s. Harvests of white-tailed deer in the state between 1981–82 and 1990–91 have fluctuated between 300,052 and 504,953 of both sexes (Texas Parks and Wildlife Department 1991). Changes in pristine habitats are clearly responsible.

Numbers of white-tailed deer increased with changes in their habitats wrought by humans. Primarily a forest-dwelling species, white-tailed deer occurred in small numbers in riparian habitats threading through grasslands and oak savannas. The carrying capacity of its habitat was increased by logging and, in some instances, by uses of the land for animal and even row-crop agriculture. As a midlayer of shrubs and groundcover became established in cutover forests, browse, mast, and herbaceous forage were brought within reach of deer. More cover and food were provided by the encroachment of woody vegetation onto prairies and plains due to grazing of domestic livestock. White-tailed deer increased to all-time highs.

Pioneers that came into Texas in the early 1800s discovered a cattleman's paradise, but a hundred years later would not have recognized the same rolling grasslands, especially those in south and west Texas. They are now almost completely covered with mesquite (*Prosopis glandulosa*), cactus (*Opuntia* spp.), several species of *Acacia,* and another dozen or so of other woody plants (Bray 1906; Cook 1908; Tharp 1939; Parker 1959; Johnston 1963).

Changes in grassland habitats were caused by overgrazing, seed scattering and trampling by livestock, cessation or control of wild and purposefully set fires (Bray 1901, 1904, 1906; Cook 1908; Tharp 1926, 1939; Johnston 1963; Brown and Archer 1989, 1990), and climatic changes (Archer 1989, 1990). While deer thrived under these changes, other species were lost. Bison (*Bison bison*), pronghorn antelope (*Antelocapra americana*), prairie chickens (*Tympanuchus cupido*), and many other vertebrates of prairies and savannas were replaced by brush-dwelling species, namely, white-tailed deer, Rio Grande turkeys (*Meleagris gallapavo*), and collared peccaries (*Tayassu tajacu*).

Clearly, Texas ranges are overstocked with white-tailed deer despite extremely liberal hunting regulations. Only 10–15 percent of the annual increment is taken each year from the state's population (Teer et al. 1965; Cook 1984), though an average offtake of 20–25 percent is possible. In Wisconsin, for example, an overwinter population of 575,000 whitetails is capable of providing an autumn population in excess of 800,000, a 40 percent increase (Creed et al. 1984). Depending on the severity of winter, this population can support an offtake of 150,000 animals.

Moreover, increases in numbers have been stimulated by the commercial or fee-hunting system in Texas (Teer et al. 1965; Teer and Forrest 1968; Burger and Teer 1983; Teer 1984). The inadequate offtake is exacerbated somewhat by the conservatism of private landowners. Although prices of leases for hunting (access to ranches with deer in Texas) vary greatly with quality and quantity of wildlife, the average lease price in Texas is near US$9.88 to $12.40/ha under a season-lease agreement (Teer et al. 1983). Leases were purchased by 89 percent of deer hunters surveyed by Thomas et al. (1990, table Q11) in the state in 1989.

Fee-hunting is increasing throughout North America as private landowners and government agencies seek to increase personal income and funds for management through sale of hunting privileges (Gartner and Severson 1972; Burger and Teer 1983; Dill et al. 1983; Thomas 1984; Loomis and Fitzhugh 1989). Commercialization of hunting recreation has been the norm in western Europe, Africa, and in many other parts of the world for many years.

Although fee-hunting is increasing in North America, not all conservationists endorse commercialization of wildlife resources. Geist (1993) argues that commercialization is dangerous to the successful system of wildlife management developed in North America. Debate centers around the efficacy of placing an economic value on a species, which puts it at risk, and whether North Americans should be expected to pay for something that is generally perceived as a part of their heritage (Teer and Forrest 1968).

Despite the popularity of white-tailed deer hunting and liberal regulations, it is often difficult to match offtake with annual recruitment of deer. Conservatism in harvest rates is promoted by economic values (Teer and Forrest 1968). Enormous losses of deer through poor nutrition on food-short ranges are frequent events in rangeland and forest habitats (Leopold et al. 1947; Krausman et al. 1992; Jacobson and Kroll 1994; Teer 1996).

Conversion of prairies to brush-covered woodlands, which increased habitats for various brush species of wildlife, cannot be attributed to management of wild ungulates. Overgrazing by domestic livestock is a major cause of woody-plant invasions in Texas and other arid and semiarid systems in North American (Archer 1989). Landowners are unlikely to return woodlands to prairies and savannas due to the commercial value of woodland wildlife, especially white-tailed deer (Teer 1996).

High costs of vegetation management also prohibit major changes in brush-covered habitats (Scifres 1980; Workman 1986; Valentine 1989; Krausman 1996). Landowners with commercial hunting operations are careful to integrate management of wildlife and livestock lest they destroy habitat of commercially valuable wildlife (Burger and Teer 1983; Bryant 1989; Teer 1993). Combinations of livestock and wildlife produce higher yields than do one or other class of herbivore in Zimbabwe (Kreuter and Workman 1996). Harvest of wood products and management of forests can be tailored to meet the needs of large mammals and other wildlife (Hoover and Wills 1984; Cooperrider et al. 1986).

The biodiversity of the major ecosystems of North America has clearly changed from that of pristine, pre-Columbian times. In some areas such as the Great Plains and prairies of the central United States, there is now no resemblance to the vegetation and attendant life known before European settlement. This poses substantial difficulty in any attempt to assess the effects of ungulate management on native biodiversity.

Elk in the Northern Range of Yellowstone National Park

Effects of overgrazing and overbrowsing of habitat by ungulates are clearly demonstrated by the history of elk (*Cervus elaphus*) and other ungulate populations in the Greater Yellowstone Ecosystem (GYE) of the United States.

Management of elk in the GYE, of which Yellowstone National Park (YNP), Grand Teton National Park, and several national forests are parts, has been a constant and heavily debated problem since YNP was gazetted in 1872 (Chase 1986; Ralston 1989; Yellowstone National Park 1989; Kay 1990, 1993a,b, 1994; Wagner et al. 1995).

Elk numbers and other ungulates were held at low densities by aboriginal hunting and predation by large carnivores in pre-Columbian times (Kay 1990). They have since fluctuated widely in response to policies and management by park personnel.

Wolves (*Canis lupus*), mountain lions (*Felis concolor*), and coyotes (*Canis latrans*) were controlled early in this century by park personnel to protect the elk herd. Hay was fed to wintering elk, bison (*Bison bison*), bighorn sheep (*Ovis canadensis*), pronghorn antelope (*Antilocapra americana*), white-tailed deer (*Odocoileus virginianus*), and mule deer (*Odocoileus hemionus*) (Houston 1982). Elk numbers erupted to about 35,000 by the 1930s, and the population overbrowsed the vegetation, especially in riparian habitats. More than 11,500 elk were removed by trapping and transplanting to other areas of the park from 1934 to 1968, and another 13,500 were shot by hunters and park personnel from 1939 to 1968 (Erickson 1981). Sport hunters killed another 43,700 between 1934 and 1968 outside the park.

Caughley's (1970, 1976) natural regulation paradigm was adopted as policy by YNP (Houston 1982). Interactions between elk and their habitat, primarily food, were projected to produce an accommodation or balance resulting in stability of the herd within the range's ability to carry it. Numbers of elk reached near 25,000 animals in the early 1990s, a population level much beyond projections of numbers of elk on the Northern Range. Overuse of the range by elk continued.

Repeated and heavy browsing by elk, moose (*Alces alces*), and beaver (*Castor canadensis*) have reduced tall willows (*Salix* spp.), aspen (*Populus tremuloides*), and cottonwood (*Populus deltoides*) communities approximately 95 percent since the 1800s (Kay 1993b, 1994; Chadde and Kay 1988, 1991). Aspen, a favored food of elk, moose, and beaver, has been cropped to low heights and has not regenerated vegetatively. It has sexually reproduced following the great YNP wildfires of 1988 (Kay 1993a). Photographs taken by Yellowstone National Park Service personnel at intervals of several years from 1932 through 1986 graphically depict the difference in growth of aspen in and outside of exclosures, the difference alleged to be due to elk browsing and barking (Kay 1990). Coughenour and Singer (1991) and Singer et al. (1994) concluded that a more xeric climate and locally reduced water tables influenced tall willow declines on the Northern Range, a position opposed by Kay (1990) and Wagner et al. (1995). Photographs of vegetation along major drainages such as the Lamar River revealed an almost complete loss of willows and other riparian species (Kay 1990). Streambanks have been denuded with bare, eroding banks apparent in many sections.

The elimination of tall, screening vegetation by ungulates along streams has caused decreases in the distribution and numbers of native trout (Kay 1993b). Hydrological regimes were changed by elimination of the food supply of beaver. These changes caused erosion, produced deeper cutbanks, and either eliminated or changed the character and composition of vegetation along streams. Tall forbs and shrubs in riparian habitats were reduced by grazing and trampling by ungulates. These were replaced by higher proportions of grasses and sedges (Kay 1993b). Loss of shading vegetation at water's edge contributed to raised water temperatures which produced unsuitable habitat for cold-water species of fish.

Whatever the cause of the elk herd's increase, clearly several species of browse have been seriously damaged, which has affected other animals and ecological processes in YNP. Kay (1990) avers that under current policy and management, aspen and willow—two highly favored foods of elk, moose, and beaver—will be completely lost from the Northern Range in time.

Beaver, once plentiful in the park, were all but extirpated from the Northern Range by overbrowsing of willows and aspen by elk and moose (Kay 1994). Keating (1982) reported that elk competed with sympatric populations of bighorn sheep for food and were partially responsible for regulating their numbers. Chase (1986) blamed elk numbers on declining populations of mule deer. White-tailed deer, once fairly common, have been virtually extirpated from YNP (Singer 1989) primarily due to absence of tall willow communities and other tall deciduous shrub habitats (Kay 1990). Avian densities and species diversity in the Northern Range were much lower in severely browsed riparian vegetation than in unbrowsed or lightly browsed sites (Jackson 1992, 1993).

Wagner et al. (1995) concluded from their examination of the history of elk and park policy in YNP that "the balance of evidence indicates that a large wintering herd of elk during park history has profoundly altered the northern-range ecosystem, reducing species, habitat and landscape diversity."

African Elephant

Animals of the size and bulk-food requirements of elephants (*Loxodonta africana*) or huge densities of ungulates such as the more than 1,500,000 blue wildebeest in the Greater Serengeti Ecosystem have the potential to and sometimes do devastate vegetation communities. Elephants are keystone or apex animals in African savannas and forests (Owen-Smith 1988) and in forests of India. Changes in vegetation communities caused by elephants are well documented (Laws 1970; Cumming 1982; McNaughton and Sabuni 1988; Owen-Smith 1988; Joubert 1991).

Elephants and other large mammals are increasingly crowded into smaller areas through pressures of human development and activities (Buechner and Dawkins 1961; Buss 1961; Laws and Parker 1968; Field 1971; Douglas-Hamilton 1987). Some six thousand elephants died from starvation in Tsavo National Park in Kenya (Corfield 1973). Crowded into the park, they destroyed their habitat and that of other animals with it. Many species were reduced or completely extirpated. African elephants that congregate in Kenya's Amboseli National Park during the dry season are putting at risk vegetative communities and other wildlife in the park.

When some animal populations reached peak numbers in Kruger National Park in South Africa in the 1960s, and their effects on other species were considered undesirable, culling of zebra (*Equus burchelli*), impala (*Aepyceros melampus*), and blue wildebeest (*Connochaetes taurinus*) was instituted to protect their habitat (Joubert 1991). Culling of these species was stopped because their numbers were judged to be subject to short-term climatic cycles. On the other hand, elephants and African buffalo (*Syncerus caffer*), by virtue of their size, competitive nature, and adaptability, are routinely cropped in Kruger to protect habitats from overuse.

Numbers of elephants in several other southern African nations are also reduced by culling for the same reasons reported for Kruger.

Zimbabwe has a surplus of elephants, and the ban on trade in ivory, which in the past helped fund wildlife management programs, has put at risk conservation efforts in the country (Cumming 1981; Bond 1993).

Birds and Small Mammal Communities in North America

Declines of birds and small mammals are largely caused by human development activities; however, management practices and lack of control of eruptive populations of ungulates are also involved. The typical pattern is for a large increase in ungulate numbers to cause changes in the vegetation communities, which is followed by losses of animals that inhabit them. The following examples illustrate this pattern.

Changes in Species Composition of Plant Communities

Changes in species composition of northern coniferous and eastern deciduous forests in the United States have been striking. Changes were mainly due to cutting of mature forests in the eighteenth and nineteenth centuries and, more recently, by overuse of the vegetation by large numbers of ungulates (Halls 1984).

Alverson et al. (1988) reported serious losses of regeneration of Canada yew (*Taxus canadensis*), white cedar (*Thuja canadensis*), and several herbaceous species from browsing by white-tailed deer in northern Wisconsin. White-tailed deer utilization reduced total plant species on study areas in Tennessee (Bratton 1980). The 3-centimeter (cm) diameter and smaller stem classes were favored browse for the deer, which reduced regeneration of several species.

Hemlock (*Tsuga canadensis*) is an important timber species in the northern forests of the Lake States and northeastern United States. It is also a preferred food of white-tailed deer. Before 1900, deer were not numerous enough to suppress reproduction and growth of the species. According to Behrend et al. (1970), when deer increased to a density of 27 deer per square mile, regeneration of hemlock virtually ceased and very few hardwoods grew over three feet tall in the Adirondack Mountains of northern New York. Failure of regeneration of hemlock and change in dominance of hemlock to sugar maple (*Acer saccharum*) in remnant virgin hemlock forests in Michigan were also attributed to browsing of white-tailed deer (Frelich and Lorimer 1985). A major cause of failure of regeneration and changes in composition of Allegheny hardwood forests in Pennsylvania and New York was attributed to browsing by dense populations of white-tailed deer (Hough 1965; Marquis 1974; Marquis and Brenneman 1981; Tilghman 1989).

Use of balsam fir (*Abies balsamea*) by eruptive populations of moose and other large ungulates repressed height growth and regeneration of fir on Isle Royale National Park in Michigan (Risenhoover and Maass 1987; Brandner et al. 1990). Similar effects of moose browsing on balsam fir and white hemlock in Newfoundland were reported by Bergerud and Manuel (1968).

Vegetation dynamics and small rodent and insect populations were surveyed on two adjoining 5.6-ha enclosures in the New Forest, Hampshire, England (Putman et al. 1989). After 22 years of use by fallow deer (*Dama dama*) stocked at a rate of one deer per hectare, composition, diversity, and biomass of the ground and shrub layers in grazed plots were greatly different from those in control plots. Regeneration of trees in the browsed plots was absent, and palatable and browse-sensitive species were absent and replaced by species resistant to browsing. Small mammal and ground invertebrate communities reflected the vegetation under the browsed and ungrazed regimes. Overuse of browse by white-tailed deer may have negative consequences to the successful establishment of snowshoe hares in the northern hardwood forests (Scott and Yahner 1989).

Changes in Structure of Vegetation Communities

Effects of grazing and browsing of woodland habitats by large mammals are well documented. Ungulates, especially in uncontrolled numbers, influence biodiversity by altering habitat of nesting and resident birds and small rodents. Composition, physical structure, layering in vertical zones, and the density of the understory vegetation are often affected. MacArthur and MacArthur (1961), MacArthur (1964), and Recher (1969) demonstrated relationships of foliage height diversity to components of bird species diversity. Overuse of vegetation by wild ungulates may therefore in some cases be expected to affect bird species diversity. Following are case studies of effects of ungulate uses of woodland habitats and how management of ungulate populations may affect biodiversity.

Overbrowsing reduced nesting substrates and food of breeding bird populations in forests in the eastern United States (Hooper et al. 1973;

Balda 1975; Casey and Hein 1983; McShea and Rappole 1992, 1994; and McShea et al. 1995). McShea and Rappole (1992) reported both significant increases in densities of several migratory bird species, and no effect on others, in forests isolated from heavy deer browsing.

White-tailed deer in Pennsylvania are a major problem in altering habitats of neotropical migrants and resident species. Though deer hunting is extremely popular in Pennsylvania, deer numbers there remain quite high (Barber 1984; Mattfield 1984), averaging greater than 11 deer/km^2 in 1992 (Witmer and deCalesta 1992). DeCalesta (1994) found in Pennsylvania that browsing by white-tailed deer at densities between 7.9 and 14.9 deer/km^2 affected habitat and birds that used it.

At Rachelwood, an experimental forest in Pennsylvania, Casey and Hein (1983) compared the vegetation and species richness of the avian community in experimental plots with avian fauna on adjacent forests stocked at much lower rates with wild ungulates. White-tailed deer, elk, and mouflon sheep (*Ovis musinon*) were stocked at approximately one animal per 1.2 ha. Plots outside Rachelwood in adjacent forest land were browsed only by white-tailed deer and at a stocking rate of one deer per 5–10 ha. Understory and groundcover were virtually eliminated, and a browse line was produced on the midlayer of the vegetation in the experimental plots inside Rachelwood. Bird species diversity was higher in the heavily grazed plots, but with different species than in adjacent forest.

Of 10 migratory bird species captured on 12 four-ha sites within deciduous forests in the ridge-and-valley region of Virginia in the eastern United States, several species—American redstart (*Setophaga ruticilla*) and hooded warbler (*Wilsonia citrina*) among them—increased after exclusion of deer (McShea and Rappole 1994). Other species such as the wood thrush (*Hylocicla mustelina*) and ovenbird (*Seiurus aurocapillus*) were not affected. Kentucky warblers were not randomly distributed within the forests of northwestern Virginia. Density of deer, forest type, and streams were significant variables governing warbler distribution (McShea et al. 1995).

Jackson (1992, 1993) reported that browsing by elk and moose reduced assemblages of breeding birds in willow communities of the Northern Range in Yellowstone National Park. Food abundance, willow species composition, hydrology, type and gradient of adjacent communities, and riparian-zone width were affected by overbrowsing. The relationship between total bird densities and the frequency of severe browsing suggested that birds have a threshold of tolerance to browsing beyond which the number of species and their densities decrease.

Healy et al. (1987) in Massachusetts, Brooks and Healy (1989) in West Virginia and Massachusetts, and DeGraaf et al. (1991) in New England reported that forest habitat overbrowsed by white-tailed deer contained fewer young trees and forbs but more grasses than habitat with low densities of deer. Southern red-backed voles (*Clethrionomys gapperi*) and short-tailed shrews (*Blarina brevicauda*) were lower in abundance in areas with high densities of deer; white-footed mice (*Peromyscus leucopus*) were higher in the high-density areas. Greater changes in composition of the communities were observed as a result of overbrowsing by deer than from silvicultural practices. Warren (1991) reported that deer overbrowsing does not affect numbers and biodiversity indices as much as it does species composition.

Through competitive exclusion, high densities of deer limited small mammals in areas of low acorn crops at Front Royal, Virginia (McShea and Schwede 1993), and numbers of small mammals were correlated with mast crops of the previous autumn in Virginia (McShea and Rappole 1992).

Fragmentation of Habitats and Changes in Landscape Patterns

Fragmentation of habitats by manipulating vegetation for forestry and agricultural products can result in changes in biodiversity, especially the faunal elements, of ecosystems. It is generally accepted that effects of fragmentation are especially severe in clear-cut or block-cleared forests and in rangelands which are managed to reduce woody vegetation.

Management of vegetation for ungulates is usually aimed at opening up large, unbroken tracts of woody vegetation or forests to create edge and diversity in vegetation communities. The result can be islands of habitat that tenuously support isolated populations (metapopulations) of plant and animal species. Ecological sinks and metapopulations are created, which result in low productivity and in some cases population extinction. Genetic isolation, speciation, and population viability are evolutionary concepts that are expressed in isolated populations.

Questions remain on the long-term influences on gene pools and ecological health of such populations. Lynch and Whigham (1984) concluded that highly isolated forest habitats adversely affect some bird species, but "structural and floristic components of the forest are more important than patch size and isolation of many species." (See Fahrig et al. 1983, Saunders et al. 1990, and Simberloff et al. 1992 for useful reviews.)

Management of ungulates can fragment habitats, but seldom is such management practical on a large scale. Ordinarily, large unbroken tracts of vegetation are opened (i.e., fragmented) for the production of agricultural and forest products. Thus, though little information is available regarding the biodiversity effects of habitat fragmentation due to ungulate management, it probably plays a very minor role compared to other human-induced factors.

Effects of Selective Hunting on Ungulate Populations

Prolonged and intensive selection of either sex or of trophy animals in recreational hunting is commonly practiced throughout much of the world. Animals with large antler size and mass are preferred objects of management of big game populations, and harvests of large mammals for meat and other products usually are directed at the larger individuals or of one or the other sex. It is now becoming clear that selective removal of particular cohorts from a population can skew sex and age ratios which, in turn, can result in reduction of genetic variability or heterozygosity in gene pools.

Decreases in antler size and mass of Old World deer are alleged to be due to selection of large stags and bulls over long periods. Moose in Prussia were subjected to trophy hunting for centuries, and by World War I, when "eugenic culling" began to be practiced, most had lost their palms (Kramer 1963). By selectively removing moose with atypical antlers and those without palms, moose recovered palmated, large antlers. This practice is sometimes termed "management with the rifle." Fervert's (1977) report of spectacular increase in body and antler sizes of red deer (*Cervus elaphus*) in Rominten, Germany, was achieved in part by eugenic culling. However, he also introduced much better feeding conditions and changed the social structure of the animals. Thus his results cannot clearly be charged to eugenic culling as environmental effects were involved.

Selective removal of tusked female elephants has been responsible for increases in tuskless females in many regions of Africa. Between

1969 and 1989, the percentage of adult tuskless females increased from 10.5 to 38.2 percent in the Eastern Province of Zambia as a result of selective hunting (Jachmann et al. 1995).

Some research has linked aberrations in population parameters to genetic changes. Skewed sex and age ratios that differ from those of naturally functioning populations affect size and configuration of antler and body sizes of ungulates. Experiments with captive breeding of white-tailed deer in North America, for example, demonstrate conclusively that body mass and size and configuration of antlers are heritable characters (Harmel 1982; Harmel et al. 1989). Moreover, reduced variability in genetic material can also affect reproductive viability and production of young (Ginsberg and Milner-Gulland 1994).

Separation of genetic from environmental effects on antlers and body size is difficult. Age and nutritional levels on which the individual is maintained also affect them. Depending on the species, cervids usually do not attain full body size and antler development until they are three to eight years old. Feeding experiments have shown that they respond to nutritional intakes on a scale which their physiological or genetic makeup will permit (French et al. 1955, 1956; Ullrey 1982; Wolfe 1982). For a synoptic review of nutritional effects on antler and body size of Old World deer, see Geist (1986). His review clearly identifies nutrition as the major determinant of antler and body mass.

Several field studies and simulation experiments have been conducted to demonstrate the effects of selective removal of particular cohorts of a population. Overall genetic variability was significantly correlated with total number of antler points in 1.5-year-old white-tailed deer on the Savannah River Plant in South Carolina. Other significant interactions occurred between heterozygosity and antler characteristics in 1.5- and 2.5-year-old males (Scribner et al. 1989).

Selective removal of large-horned rams in bighorn sheep populations is reported to be responsible for loss of variability and fitness in the Rocky Mountains of the United States (Fitzsimmons et al. 1995). Horn growth was significantly higher in the more heterozygous rams in 6-, 7-, and 8-year-old members.

Ryman et al. (1981) concluded from simulation studies that hunting can reduce genetic variation in moose and white-tailed deer. Effective population size and generation interval were reduced when extensive male (or female) hunting produced a skewed sex ratio. Inbreeding and genetic drift may occur, which decreases effective population size below that in which population parameters (sex and age ratios and densities) are naturally distributed. Ginsberg and Milner-Gulland (1994) demonstrated from simulation studies that skewed sex ratios induced by selective removal of impala males can lead to reduced numbers of young, which in time will lead to population collapse.

Impacts on ungulates by selective hunting are not always expressed in reduced body size or antler mass. Trophy management of cervids usually involves selective removal or "sanitization" of herds through culling of small-antlered bucks, reducing competition for forage by reducing population densities, and by permitting males to attain ages of physiological and physical maturity.

Predator Control

Predators are integral elements of functional ecosystems in which ungulates are prey species. No practice has been so disruptive in ecosystem and population dynamics or as politically debated as control of predators to protect livestock or produce big game for recreational hunting.

At one time or another in most countries of the world, large carnivores have been relentlessly killed; consequently, carnivores have declined in numbers and distribution. The result has often been eruptions of their prey. The dogma that predators cannot control populations of large ungulates has been questioned, and evidence is mounting to refute it. Connolly (1978) reviewed the debate and concluded that carnivores can and do control populations of *k*-selected species, and numerous studies are now supporting this thesis. The following review of selected studies provides some of this evidence.

The Kaibab mule deer herd in the state of Arizona has been held and frequently cited as *the* example of how human intervention has severely disrupted an ecosystem process, that is, the balance of predators and prey (Allee et al. 1949). Rasmussen (1941) reported that the Kaibab herd increased to 100,000 animals by 1924, after 816 mountain lions, 30 wolves, 7,388 coyotes, and 863 bobcats were removed between 1906 and 1939. The herd crashed to less than 10,000 animals by 1939, the inference being that removal of predators was the cause.

Russo (1964) and others questioned the validity of some of the estimates of deer numbers and stated, "Extensive livestock use of the range over a number of years, principally from 1887 to 1923, undoubtedly contributed to what later became a deer problem." Mule deer carrying capacity was increased due to a change in the grassland vegetation to more palatable and plentiful forbs, annuals, and shrubs.

What caused the eruptions and crash? Lack of predators or increase in forage that was eventually overused? The answer is confounded. Nonetheless, control of predators contributed to the extirpation of one carnivore, the wolf, in Arizona. Certainly predator control and overuse by deer and livestock interrupted an ecological balance that had been operative long before human intervention occurred.

The role of coyotes in controlling deer numbers is also an ongoing debate. Recent studies suggest that coyotes do indeed control large mammal prey. Coyotes have not been controlled and deer have not been recreationally hunted on the 3,545-ha Welder Wildlife Foundation Refuge since the refuge was established in 1954 (Teer et al. 1991). Deer have stabilized and fluctuated around a mean density of about 22/km^2 (Blankenship et al. 1994); coyote numbers have also fluctuated around 2/km^2 (Kie 1977; Kie and White 1985). Although not the sole factor in controlling deer numbers on the Welder Refuge (up to 40 deer per year are taken for research), coyotes and weather, primarily drought, have given some stability to the herd. Recommendations for management of overabundant deer in south Texas call for protection of coyotes for the roles they play in ungulate management (Teer et al. 1991).

Operational policies of the National Parks Board of South Africa permitted predator control to favor large mammals. African hunting dogs (*Lycaon pictus*), lions (*Felis leo*), cheetahs (*Acinonyx jubatus*), black-backed jackals (*Canis mesomelas*), side-striped jackals (*C. adustus*), spotted hyenas (*Crocuta crocuta*), and some of the smaller cats were reduced, the objective being to increase ungulates and other wildlife in the park (Joubert 1991). The lion's chief prey were blue wildebeest, impala, zebra, waterbuck (*Kobus ellipsiprymnus*), greater kudu (*Tragelaphus strepsiceros*), giraffe (*Giraffa camelopardalis*), and buffalo in that order of preference. Decreased predation was a major factor in the large increases in prey species and the overstocked habitats that resulted (Pienaar 1963). Control of predators was stopped in the 1950s, when the Parks Board decided that the "predator community represents an integral part of the ecosystem and fulfills a vital role in sustaining ecological stability and resilience" (Joubert 1991).

The National Parks Board of South Africa now permits culling of species that, confined by fencing and park boundaries, build to population levels that damage the habitats for themselves and other species. Impala and other antelope were formerly cropped; now, however, elephants and buffalo are the major dominant species sustainably cropped. From an elephant population that ranged between 5,590 and 7,486 individuals between 1967 and 1989, 12,787 were culled in Kruger National Park to manage habitats (I. J. Whyte, files of Kruger National Park, pers. comm. 1990).

Reintroduction of wolves into the northern Rocky Mountains and the Greater Yellowstone Ecosystem (Tilt et al. 1987; Peek et al. 1991; Mech 1995) and the control of wolves and bears in Alaska to provide more hunting for moose in predator-limited moose populations (Gasaway et al. 1992; Boertje et al. 1993; van Ballenberghe and Ballard 1994) are recent issues in predator control and reintroductions. Debates continue over the role and propriety of controlling predators to promote welfare and survival of species of big game.

Translocations of Non-Native Ungulates

Translocation of native mammals is an important tool in reestablishment of species that have been extirpated from native ranges, to increase heterogeneity in gene pools of populations, and to augment populations that are threatened by low reproductive viability (IUCN 1987; Teer 1989). At least 93 species of native birds and mammals were translocated between 1973 and 1986 in Australia, Canada, Hawaii, New Zealand, and the United States (Griffith et al. 1989). Ninety percent were game species, but endangered and threatened species were also relocated. Of the average of 700 translocations of native game made annually, 98 percent were made in North America between 1973 and 1986.

Introductions and translocations of non-native ungulates have been a common practice throughout the world (Laycock 1966). Settlers into new continents have brought with them animals and plants which they used or which they simply enjoyed in their surroundings. Most translocations of large mammals are now made for recreational hunting. Translocated animals, alien to their new environments, often lack natural controls that operated in their native habitats. Consequently, some explode in numbers in new environments.

Translocated ungulates exhibit at least one but often three results: (1) explosions in numbers of the exotic species beyond the habitat's ability to support them, (2) reduction of food resources leading to poor pasturage and mortality of native species, and (3) interbreeding between closely related species and subspecies (Teer 1979).

Numerous examples document the results of translocations. Among them are the Himalayan tahr (*Hemitragus jemlahicus*) (Caughley 1970) and red deer (Howard 1965) in New Zealand, the Axis deer (*Axis axis*) (Ables 1977), nilgai antelope (*Boselaphus tragocamelus*) (Sheffield et al. 1983), and the blackbuck antelope (*Antilope cervicapra*) (Mungall 1978) in Texas, and the red deer in Argentina (Veblen et al. 1989).

Impacts of non-native ungulates onto islands and into small, isolated habitats are well known from the New Zealand experience (Howard 1965; Caughley 1983). When non-native species are released into ecosystems that have not co-evolved with grazing and browsing herbivores, as is the case with New Zealand and island habitats, plants have little or no defense to herbivory. Tough epidermal cuticles, thorns and spines, woody protuberances, and prostrate growth forms prevent cropping of photosynthetic parts of plants that have evolved with herbivores. Others, especially woody species, develop aversive odors and tastes from chemicals in plant tissues such as alkaloids, resins, tars, and terpenes (Wagner 1978, 1987, 1989).

Many species of non-native deer and antelope have been released since the early 1920s on ranches in Texas where they are produced for hunting and for sale of brood stock for further translocations (Mungall and Sheffield 1994). As commercial and fee-hunting systems developed in the state, introductions of non-native big game species increased.

The first survey of exotics in Texas, made in 1963 by Jackson (1964), recorded 13 species totaling 13,000 head. By 1988, numbers of non-native ungulates in the state had increased to 123 races of 67 species totaling 164,257 animals on 486 ranches (Traweek 1989). Of the total, 73,857 animals were established animals in free-ranging, unhusbanded populations of three cervids and three bovids: the axis deer, sika deer (*Cervus nippon*), fallow deer, nilgai antelope, blackbuck antelope, and Aoudad or Barbary sheep (*Ammotragus lervia*). In 1994, when the most recent estimate of exotics was made, numbers of exotic ungulates had increased to 71 species totaling 195,483 animals on 637 ranches (Traweek 1995).

These numbers are seemingly insignificant when measured against the total numbers of white-tailed deer in the state, estimated at 5,211,184 animals in 1990–91 (Texas Parks and Wildlife Department 1992). Exotics are nonetheless important competitors with native game and domestic livestock on many ranches. Axis, fallow, and sika deer are competitors with white-tailed deer in the Edwards Plateau of Texas (Butts 1979; Butts et al. 1982; Baccus et al. 1985). They have become more numerous than white-tailed deer in some regions. When forage becomes short as a result of overgrazing by any species of game or class of livestock, axis deer outcompete native deer because they take a larger diversity of forage in their diets (Butts et al. 1982).

Nilgai antelope, a large antelope translocated from the Indian subcontinent, have reached more than 37,000 individuals in south Texas, all from a release of less than 30 animals on King Ranch in the early 1920s and 1930s (Sheffield et al. 1983). On a 2,146-ha ranch adjoining King Ranch in south Texas, nilgai were about two times as abundant in 1995 as they were in 1978 when the first aerial census was made (Teer, unpublished data). Control of nilgai on several large cattle ranches in south Texas is now being practiced by shooting from helicopters. The carcasses are collected, butchered, and chilled in portable abattoirs, inspected by government veterinarians, and sold in specialty restaurants as venison.

Game Farming and Ranching

Confined and translocated populations of ungulates, especially of exotic or non-native species, have serious implications for other ungulates and ecosystem processes. Game farming and ranching industries have been responsible for moving big game intercontinentally, regionally, and between ecosystems. Ungulates are produced in closely husbanded conditions for production of meat, leather, medicines (antler velvet and bone, musk glands, and so on), curios, and stock for recreational hunting (Caughley 1983; Bothma 1989; Renecker and Hudson 1991; Teer et al. 1993).

The development of effective game fences has played a major role in the growth of the game farming and ranching industries. From very meager beginnings, the industry has grown almost exponentially in the past 20 years. In New Zealand, where the industry began and by which the impetus for worldwide interest was stimulated, more than

700,000 deer were husbanded on 4,000 registered deer farms in New Zealand in 1988 (Muir 1988). About 3,000 local farms were enclosed by game fences in South Africa in 1992 (Grobler and van der Bank 1992). Data on the industry in North America is patchy; however, the industry is growing and more and more ungulates are being penned and husbanded under fence. Estimates of the total number of hectares fenced for white-tailed deer and non-native ungulates now exceed 1,620,000 ha in Texas (E. Young and J. G. Teer, unpublished data).

Few fenced game ranges are secure, and escapes of penned animals usually occur. Some establish themselves in free-ranging populations, while others may succumb to inadequacies of wild environments. When contact of non-native with closely related, native species occurs, interbreeding may be the unwanted result. Members of the family Cervidae, especially of the genus *Cervus,* are major species on game farms. When confined, red deer, elk, sika, and sambar (*C. unicolor*) interbreed (Dratch 1993). Biochemical tests showed that 30 percent of elk and red deer in New Zealand game farms had genetic markers of both species. Red deer and sika deer have been hybridized in Scotland to the extent where no true individuals of either species may exist (Wheaton et al. 1993). Intergeneric crosses of wild sheep are also known to occur, and even crosses of domestic and wild species of these bovids occur.

Translocations of diseases and parasites with non-native ungulates are also potentially serious problems. Brucellosis, tuberculosis, anthrax, and hoof and mouth disease affect most hoofed mammals. Bovine tuberculosis was diagnosed in two fallow deer in 1989, which led to the slaughter of 350 animals in British Columbia. Discoveries of tuberculosis in several herds of elk in Alberta resulted in condemnation of these captive herds (Ireland and Lewis 1992). Meningeal worms (*Parelaphostrongylus tenuis* and *Elaphostrongylus cervi*) have caused serious problems in cervids and bovids in Canada and the United States (Teer et al. 1993). Interjurisdictional movements of ungulates has been and is a topic of keen interest to state, provincial, and federal governments (Teer 1991).

The genetic implications of confining metapopulations of ungulates, native or non-native, on fenced game ranches are not well known. Captive or confined populations can be expected ultimately to lose genetic variability because gene flow is prevented or through genetic drift. Small populations such as those in zoos and confined pastures lose genetic diversity mainly through genetic drift (Lacy 1987).

Meat, hides, and antlers produced on game farms and ranches have also put wild stocks of ungulates at risk because wild and farmed products cannot easily be differentiated. To protect native ungulates and domestic animals, states of the United States and provinces of Canada are examining their regulations and redefining import and management procedures for farming and ranching of especially non-native wildlife (Teer 1991).

Protected Areas

Of all strategies for protecting wildlife, none has been practiced longer or more universally than the uses of parks and equivalent reserves. The global network of protected areas larger than 1,000 km^2 grew from none in 1900 to 9,832 comprising 926,349,646 km^2 or 3 percent of the earth's surface in 1990 (World Conservation Monitoring Unit, IUCN Commission on National Parks and Protected Areas 1994). Many were established to protect large game animals, with little interest in other species or biodiversity.

Are protected areas fulfilling their purpose? Unfortunately, many protected areas, especially in the developing nations, are little more than paper parks. Funds and personnel are often not available to operate and protect them from outside incursions of needy people. Further, (gap) analyses of protected area systems show that species do not now occur, or perhaps never occurred, in viable numbers in some of them (Scott et al. 1991, 1993). More than 80 percent of the ungulates in Kenya's extensive national park system move outside in seasonal migrations where they depredate on crops and cause hardships to local subsistence farmers (Lusigi 1981). Two hundred thousand elephants range over 22 percent of southern Africa; only 40 percent occur in the protected area network (Taylor and Cumming 1993). One and a half million blue wildebeest make seasonal movements out of the Greater Serengeti Ecosystem in response to rainfall patterns and food availability (Sinclair and Norton-Griffiths 1979), where they are exposed to poaching by local people (Kurji 1979).

Wallis de Vries (1994) suggested that nature conservation in Europe has failed to stop the continuous loss and fragmentation of habitat and loss of species. He argues that nature reserves should be enlarged to accommodate large mammals because they can serve as "umbrellas" to other creatures in the ecosystem. Simonetti (1995) cautions that conservation interests must protect large mammals beyond parks, "that is, to rely on both protected areas and the unprotected lands surrounding parks and reserves."

Despite the shortcomings of parks, they have been effective in protecting many species and examples of communities and ecosystems. Nevertheless, it seems clear that the use and management of ungulates and other large mammals beyond park boundaries has a crucial role to play in biodiversity conservation in many regions of the world.

Discussion and Conclusions

The building blocks of the biosphere are species. Species, their gene pools, the processes to which they contribute and in which they participate, and the functions they serve define communities and ecosystems. That species are valuable in their own right and for the benefit of man, and that biodiversity is important in stability of ecosystems, are tenets almost universally accepted by ecologists (Wilson 1988, 1992). Leopold (1949) emphasized the importance of all life in natural systems: "the first principle of intelligent tinkering is to save all the parts."

Conservation of living resources and environmental quality are now major interests throughout the world. Losses of animals through management of others are to be avoided, and much more attention is being given to ecosystem and landscape scales of management. However, Baskin (1994a) reports that little is known about the function of and requirement for a species in a community, and the effects of the loss of a species on ecosystems are unpredictable. She examined the decisions of ecologists in a workshop sponsored by the Scientific Committee on Problems of the Environment (SCOPE), and reported that "function is as much an anthropocentric concept as aesthetics because creatures often fill multiple roles in different settings and their only biological function is to persist." She further states that ecosystems are likely to be compromised and disrupted more by anthropic influences such as swamp draining and forest clearing than by losses of species (Baskin 1994b).

Not all species are equal in the roles they play in natural systems. However, the "rivet species" concept (Erlich and Erlich 1981) is an appropriate analogy for the ultimate effects of loss of biodiversity on ecosystem processes. When large numbers of species are lost, as often occurs from man's interventions and machinations, the natural system of which they were a part may collapse as Baskin (1994b) suggests. Further, when an apex or keystone species is lost or allowed to dominate a community, disruption of ecological processes is often severe.

Carrying capacity is food-determined for most ungulates (Lack 1954; Sinclair and Norton-Griffiths 1979; Moen 1980–82). Vegetation is the source of primary productivity on which subsequent trophic levels depend; it governs composition and numbers of animals and community processes. Ungulates also serve as food for carnivores, and predator-prey interactions are important if not essential to functional ecosystems in which ungulates are keystone or dominant animals. When set free from predation and other sources of natural mortality, ungulates can increase to levels beyond the capacity of their habitats to support them. Conversely, without large prey, most large carnivores will be unable to continue as functional parts of ecosystems.

A conservative position, the one adopted by most ecologists, is that all species have roles to play, some undoubtedly more important than others. Losses of species as a result of management of large mammals, which may be dominant or keystone in their ecological roles, can be serious. However, few studies have set out to document direct conflicts of management of ungulates with biodiversity and ecosystem function.

Many species are yet unidentified. How to conserve ecosystem processes in ecosystem scales of management is yet to be resolved. Walker (1995) applauds ecological redundancy wherein processes or functions in an ecosystem are carried out by more than one species. He observes that by maintaining ecosystem structure, "stability (the probability of all species persisting) is enhanced if each important functional group of organisms (important for maintaining function and structure) comprises several ecologically equivalent species, each with different responses to environmental factors."

Yahner (1989), in discussing forest management for wildlife, indicts the wildlife profession's frequent assumption that habitat management for a game species has beneficial effects on coexisting species: "More studies are needed to assess the impact of management for a featured species on abundance and distribution of other species and, in particular, on the relationships between characteristics of managed stands (size, shape, and distribution) and their effects on wildlife."

When domestic livestock and translocated non-native ungulates are overlays of grazing pressure on ungulate ranges, as is typically the case, vegetation and all life dependent on it are placed at increased risk. Destruction of essential habitat and disruption of traditional migration routes and corridors by agriculture and other human industries result in overstocked habitats. Protected areas are useful in maintaining populations of ungulates; however, few are self-contained systems on which ungulates remain throughout their lives. Consequently, as events described above take place, changes in the habitat occur and biodiversity is changed or compromised.

The obvious solution to overabundance of ungulates is control of their numbers through natural mortality. Lacking that, management becomes a necessity. What, then, is the role in management of recreational hunting or cropping of excess animals through shooting? Legal subsistence and recreational hunting has been the answer in the past, and it continues to be favored by most wildlife biologists and managers.

Changes in the habitat by herbivores may result in loss of species, disruption of ecological processes, and loss or deprivation of heterogeneity in gene pools of especially scarce and sensitive species (Sinclair and Norton-Griffiths 1979; Warren 1991; Wallis de Vries 1994). Conversely, by influencing succession, large herbivores are keystone species and can increase quality and quantity of habitat for other animals and plants (Wallis de Vries 1994).

Where habitats are protected and managed for big game and where numbers are kept within their carrying capacities, cervid and bovid populations are maintained and produce recreational hunting for hundreds of thousands of hunters. However, hunting as a tool to control eruptive populations of herbivores is increasingly in conflict with societal values. Recreational hunting usually does not mimic natural mortality in fashioning population composition as occurs from predation, diseases, accidents, and other sources. Further, selective hunting for certain age groups and sexes can distort normal population composition and processes, and perhaps even gene pools. Large mammals have always been hunted by subsistence and recreational hunters. Numerous examples of long-term sustainable utilization of ungulate populations and their habitats are available to demonstrate effective management and utility. However, very few studies exist of management effects on other animal species or of the genetic effects of selective removal from free-ranging populations. Currently, hunting remains the most effective method of controlling ungulate numbers.

Protectionists and animal rightists champion hands-off positions, that is, many are antihunting and antimanagement in philosophical bent. Population control through antifertility drugs and translocations of excess numbers to less-populated ranges are usually offered as options to hunting. Efficacy and expense are not important dimensions to animal rightists. Their usual recommendation is to let nature take care of too many animals and that fluctuations in numbers of herbivores follow natural laws.

Recreational hunters, on the other hand, see this position as waste. Usually overlooked in such debates is the practical certainty of deterioration of habitat and subsequent losses of animals from perhaps less humane sources of mortality.

Finally, human interventions and the absolute need for benefits to accrue to those who control habitat are keys to conservation of ungulates and biodiversity. Competition between agronomic and wildlife interests are at the heart of conservation of terrestrial habitat. In the developing nations of the world, and even in the more affluent western nations, management of the land now more than ever translates to economic values. While all values and all species need not be put into the marketplace, conservation of all living resources on both private and public lands is at the doorstep of the market.

References

Ables, E., ed. 1977. *The axis deer in Texas.* Kleberg Studies in Natural Resources no. KS2. College Station, Texas: Texas A & M University.

Allee, W. C., A. E. Emerson, O. Park, T. Park, and K. P. Schmidt. 1949. *Principles of animal ecology.* Philadelphia and London: W. B. Saunders Co.

Alverson, W. W., D. M. Waller, and S. L. Solheim. 1988. Forests too deer: Edge effects in northern Wisconsin. *Conservation Biology* 2:348–358.

Archer, S. 1989. Have Southern Texas savannas been converted to woodlands in recent history? *American Naturalist* 134:545–561.

Archer, S. 1990. Development and stability of grass/woody mosaics in a subtropical savanna parkland, Texas, USA. *Journal of Biogeography* 17:453–462.

Baccus, J. T., D. E. Harmel, and W. E. Armstrong. 1985. Management of exotic deer in conjunction with white-tailed deer. In *Game harvest man-*

agement, ed. S. L. Beasom and S. F. Roberson, 213–226. Kingsville, Texas: Caesar Kleberg Wildlife Research Institute, Texas A&I University.

Baker, R. H. 1984. Origin, classification and distribution. In *White-tailed deer: Ecology and management,* ed. L. K. Halls, 1–18. Harrisburg, Pennsylvania: Stackpole Books.

Balda, R. P. 1975. Vegetation structure and breeding bird diversity. In *Proceedings of the Symposium on Management of Forest and Range Habitats for Nongame Birds,* ed. D. R. Smith, 59–80. General Technical Report WO-1. Washington, D.C.: United States Department of Agriculture Forest Service.

Barber, H. L. 1984. Eastern mixed forest. In *White-tailed deer: Ecology and management,* ed. L. K. Halls, 345–354. Harrisburg, Pennsylvania: Stackpole Books.

Baskin, Y. 1994a. Ecosystem function of biodiversity. *BioScience* 44:657–660.

Baskin, Y. 1994b. How much does diversity matter? *Renewable Resource Journal* 12:9–11.

Behrend, D. F., G. F. Mattfeld, W. C. Tierson, and J. E. Wiley III. 1970. Deer density control for comprehensive forest management. *Journal of Forestry* 68:695–700.

Bergerud, A. T., and F. Manuel. 1968. Moose damage to balsam fir-white birch forests in central Newfoundland. *Journal of Wildlife Management* 32:729–746.

Blankenship, T. L., D. L. Drawe, and J. G. Teer. 1994. Reproductive ecology of white-tailed deer on the Welder Wildlife Foundation. *Proceedings of the Conference of the Southeastern Association of Fish and Wildlife Agencies* 48:69–77.

Boertje, R. D., D. G. Kelleyhous, and R. D. Hayes. 1993. Methods for reducing natural predation on moose in Alaska and the Yukon: An evaluation. *Proceedings of the 2nd North American Wolf Symposium, Edmonton, Alberta.* Juneau: Federal Aid in Wildlife Restoration, Alaska Department of Fish and Game. Multilithed.

Bond, I. 1993. *The economics of wildlife and land use in Zimbabwe: An examination of current knowledge and issues.* WWF Multispecies Animal Production Systems Project Paper no. 33. Harare, Zimbabwe: World Wild Fund for Nature.

Bothma, J. du P., ed. 1989. *Game ranch management.* Pretoria, Republic of South Africa: J. L. van Schaik, Publisher.

Brandner, T. A., R. O. Peterson, and K. L. Risenhoover. 1990. Balsam fir on Isle Royal: Effects of moose herbivory and population density. *Ecology* 71:155–164.

Bratton, S. P. 1980. Impacts of white-tailed deer on the vegetation of Cades Cove, Great Smoky Mountains National Park. *Proceedings of the Annual Conference of the Southeastern Association of Fish and Wildlife Agencies* 33:305–312.

Bray, W. L. 1901. The ecological relations of the vegetation of western Texas. *Botanical Gazette* 32:262–291.

Bray, W. L. 1904. *Forest resources of Texas.* Bulletin no. 47. Washington, D.C.: United States Bureau of Forests.

Bray, W. L. 1906. *Distribution and adaption of the vegetation of Texas.* Bulletin no. 82. Austin: University of Texas.

Brooks, R. T., and W. M. Healy. 1989. Responses of small mammal communities to silvicultural treatments in eastern hardwood forests of West Virginia and Massachusetts. In *Proceedings of the Symposium on Management of Amphibians, Reptiles, and Small Mammals in North America,* technical coordinators R. C. Szaro, K. E. Severson, and D. R. Patton, 313–318. General Technical Report RM-166. Flagstaff, Arizona: United States Department of Agriculture Forest Service.

Brown, J. R., and S. Archer. 1989. Woody plant invasion of grasslands: Establishment of mesquite (*Prosopis glandulosa*) on sites differing in herbaceous biomass and grazing history. *Oecologia* 80:19–26.

Brown, J. R., and S. Archer. 1990. Water relations of a perennial grass and seedling versus adult woody plants in a subtropical savanna, Texas. *Oikos* 57:366–374.

Bryant, F. C. 1989. Economic implications of wildlife. *Proceedings, Western Section, American Society of Animal Science and Western Branch, Canadian Society of Animal Science* 40:500–502.

Buechner, H. K., and H. C. Dawkins. 1961. Vegetation changes induced by elephants and fire in Muchinson Falls National Park, Uganda. *Ecology* 42(4):752–766.

Burger, G. V., and J. G. Teer. 1983. Economic and socioeconomic issues influencing wildlife management on private land. In *Proceedings, Symposium: Wildlife Management on Private Lands,* ed. R. T. Dumke, G. V. Burger, and J. R. March, 252–278. Madison, Wisconsin: Wisconsin Chapter Wildlife Society.

Buss, I. O. 1961. Some observations on food habits and behavior of the African elephant. *Journal of Wildlife Management* 25:131–148.

Butts, G. L. 1979. The status of exotic big game in Texas. *Rangelands* 1(4): 152–153.

Butts, G. L., M. J. Anderegg, W. E. Armstrong, D. E. Harmel, C. W. Ramsey, and S. H. Sorola. 1982. *Food habits of five exotic ungulates on Kerr*

Wildlife Management Area, Texas. Technical Series 30. Austin: Texas Parks and Wildlife Department.

Casey, D., and D. Hein. 1983. Effects of heavy browsing on a bird community in deciduous forest. *Journal of Wildlife Management* 47:829–836.

Caughley, G. 1970. Eruptions of ungulate populations, with emphasis on Himalayan thar in New Zealand. *Ecology* 51:53–72.

Caughley, G. 1976. Wildlife management and the dynamics of ungulate populations. *Applied Biology* 1:183–246.

Caughley, G. 1983. *The deer wars: The story of deer in New Zealand.* Auckland, New Zealand: Heinemann Publishers.

Chadde, S., and C. Kay. 1988. *Willows and moose: A study of grazing pressure, Slough Creek Exclosure, Montana, 1961–1986.* Research note no. 24, Montana Forest and Conservation Experiment Station, University of Montana, Missoula.

Chadde, S., and C. Kay. 1991. Tall willow communities on Yellowstone's northern range: A test of the "natural regulation" paradigm. In *The Greater Yellowstone Ecosystem: Redefining America's wilderness heritage,* ed. R. B. Keiter and M. S. Boyece, 231–262. New Haven, Connecticut: Yale University Press.

Chase, A. 1986. *Playing God in Yellowstone: The destruction of America's first national park.* Boston: Atlantic Monthly Press.

Connolly, G. E. 1978. Predators and predator control. In *Big game of North America: Ecology and management,* ed. J. L. Schmidt and D. L. Gilberts, 369–394. Harrisburg, Pennsylvania: Stackpole Books.

Cook, O. F. 1908. *Change of vegetation of the south Texas prairies.* Bureau of Plant Industry, Circular 14. Washington, D.C.: United States Department of Agriculture.

Cook, R. L. 1984. Texas. In *The white-tailed deer: Ecology and management,* ed. L. K. Halls, 457–474. Harrisburg, Pennsylvania: Stackpole Books.

Cooperrider, A. Y., R. J. Boyd, and H. R. Stuart. 1986. *Inventory and monitoring of wildlife habitat.* Denver, Colordado: United States Department of the Interior, Bureau of Land Management, Service Center.

Corfield, T. F. 1973. Elephant mortality in Tsavo National Park, Kenya. *East Africa Wildlife Journal* 10:91–115.

Coughenour, M. B., and F. J. Singer. 1991. The concept of overgrazing and its application to Yellowstone's northern range. In *The Greater Yellowstone Ecosystem: Redefining America's wilderness heritage,* ed. R. B. Keiter and M. S. Boyece, 209–230. New Haven, Connecticut: Yale University Press.

Creed, W. A., F. Haberland, B. E. Kohn, and K. R. McCaffery. 1984. Harvest management: The Wisconsin experience. In *White-tailed deer: Ecology and management*, ed. L K. Halls, 243–260. Harrisburg, Pennsylvania: Stackpole Books.

Cumming, D.H.M. 1981. The management of elephant and other large mammals in Zimbabwe. In *Problems in management of locally abundant wild mammals*, ed. P. A. Jewell, S. Holt, and D. Hart, 91–118. New York: Academic Press.

Cumming, D.H.M. 1982. The influence of large herbivores on savanna structure in Africa. In *Ecology of tropical savannas*, ed. B. J. Huntley and B. H. Walker, 217–245. New York: Springer-Verlag.

Danell, K., R. Bergstrom, and L. Edenius. 1994. Effects of large mammalian browsers on architecture, biomass, and nutrients of woody plants. *Journal of Mammalogy* 75:833–844.

deCalesta, D. S. 1994. Effect of white-tailed deer on songbirds within managed forests in Pennsylvania. *Journal of Wildlife Management* 58:711–718.

DeGraaf, R. M., W. M. Healy, and R. T. Brooks. 1991. Effects of thinning and deer browsing on breeding birds in New England oak woodlands. *Forest Ecology and Management* 41:179–191.

Dill, T. O., J. Menghini, S. S. Waller, and R. Case. 1983. Fee hunting for Nebraska big game a possibility. *Rangelands* 5:24–27.

Douglas-Hamilton, I. 1987. African elephants: Populations trends and causes. *Oryx* 21:11–24.

Dratch, P. A. 1993. Genetic tests and game ranching: No simple solutions. *Transactions of the North American Wildlife and Natural Resources Conference* 48:479–486.

Erickson, G. L. 1981. The northern Yellowstone elk herd: A conflict of policies. *Western Association of Fish and Game Agencies* 61:92–108.

Erlich, P. R., and A. H. Erlich. 1981. *Extinction: The causes and consequences of the disappearance of species*. New York: Random House.

Fahrig, L., L. Lefkovitch, and G. Merriam. 1983. Population stability in a patchy environment. In *Analysis of ecological systems: State-of-the-art in ecological modelling*, ed. W. K. Lauenroth, G. V. Skogerboe, and M. Flug, 61–67. New York: Elsevier.

Fervert, W. 1977. *Rominten*. 6th ed. Munich: BLV Verlagsgesellschaft.

Field, C. R. 1971. Elephant ecology in the Queen Elizabeth National Park, Uganda. *East Africa Wildlife Journal* 9:99–123.

Fitzsimmons, N. N., S. W. Buskirk, and M. H. Smith. 1995. Population history, genetic variability, and horn growth in bighorn sheep. *Conservation Biology* 9:314–323.

Frelich, L. E., and C. G. Lorimer. 1985. Current and predicted long-term effects of deer browsing in hemlock forests in Michigan, USA. *Biological Conservation* 34:99–120.

French, C. E., L. C. McEwen, N. D. Magruder, R. H. Ingram, and R. W. Swift. 1955. *Nutritional requirements of white-tailed deer for growth and antler development.* Pennsylvania State University, Agricultural Experiment Station Bulletin no. 600. University Park, Pennsylvania: Penn State University.

French, C. E., L. C. McEwen, N. D. Magruder, R. H. Ingram, and R. W. Swift. 1956. Nutrient requirements for growth and antler development in the white-tailed deer. *Journal of Wildlife Management* 20:221–232.

Garrott, R. A., P. J. White, and C. A. Vanderbilt White. 1993. Overabundance: An issue for conservation biologists. *Conservation Biology* 7:946–949.

Gartner, F. R., and K. E. Severson. 1972. Fee hunting in western South Dakota. *Journal of Range Management* 25:234–237.

Gasaway, W. C., R. D. Boertje, D. V. Grangaard, D. G. Kellyhouse, R. O. Stephenson, and D. G. Larsen. 1992. The role of predation in limiting moose at low densities in Alaska and Yukon and implications for conservation. *Wildlife Monographs* 120:1–59.

Geist, V. 1986. Super antlers and pre-World War II European research. *Wildlife Society Bulletin* 14:91–94.

Geist, V. 1993. Great achievements, great expectations. Successes of North American wildlife management. In *Commercialization and wildlife management: Dancing with the devil,* ed. A.W.L. Hawley, 47–72. Malabar, Florida: Krieger Publishing.

Ginsberg, J. R., and E. J. Milner-Gulland. 1994. Sex-based harvesting and population dynamics in ungulates: Implications for conservation and sustainable use. *Conservation Biology* 8:156–157.

Griffith, B., J. M. Scott, J. W. Carpenter, and C. Reed. 1989. Translocation as a species conservation tool: Status and strategy. *Science* 245:477–480.

Grobler, J. P., and F. H. van der Bank. 1992. Do game fences affect the genetic diversity in commercially utilized game populations? In *Proceedings, Wildlife Ranching: A Celebration of Diversity,* ed. W. van Hoven, H. Ebedes, and A. Conroy, 238–240. Pretoria, Republic of South Africa: University of Pretoria.

Halls, L. K., ed. 1984. *White-tailed deer: Ecology and management.* Harrisburg, Pennsylvania: Stackpole Books.

Harmel, D. E. 1982. Effects of genetics on antler quality and body size in white-tailed deer. In *Antler development in cervidae,* ed. R. D. Brown, 339–348. Kingsville, Texas: Caesar Kleberg Wildlife Research Institute, Texas A & I University.

Harmel, D. E., J. D. Williams, and W. E. Armstrong. 1989. *Effects of genetics and nutrition on antler development and body size of white-tailed deer.* Austin, Texas: Texas Parks and Wildlife Department, Wildlife Division.

Healy, W. M., R. T. Brooks, and P. J. Lyons. 1987. Deer and forests on Boston's municipal watershed after 50 years as a wildlife sanctuary. In *Proceedings of the Symposium on Deer, Forestry, and Agriculture: Interactions and Strategies for Management,* ed. D. A. Marquis, 3–21. Warren, Pennsylvania: Plateau and Northern Hardwood Chapters, Allegheny Society of American Foresters.

Hobbs, N. T. 1996. Modification of ecosystems by ungulates. *Journal of Wildlife Management* 60:695–713.

Hooper, R. G., H. S. Crawford, and R. F. Harlow. 1973. Bird density and diversity as related to vegetation in forest recreation areas. *Journal of Forestry* 71:766–769.

Hoover, R. L., and D. L. Wills, eds. 1984. *Managing forested lands for wildlife.* Denver, Colorado: Colorado Division of Wildlife in cooperation with United States Department of Agriculture Forest Service, Rocky Mountain Region.

Hough, A. F. 1965. A twenty-year record of understory vegetational change in a virgin Pennsylvania forest. *Ecology* 46:370–373.

Houston, D. B. 1982. *The northern Yellowstone elk: Ecology and management.* New York: Macmillan.

Howard, W. E. 1965. *Control of introduced mammals in New Zealand.* New Zealand Department of Scientific and Industrial Research Information Service, no. 45.

Ireland, D. B., and R. J. Lewis. 1992. Game farming in British Columbia, Canada. In *Wildlife ranching: A celebration of diversity,* ed. W. van Hoven, H. Ebedes, and A. Conroy, 63–65. Pretoria, Republic of South Africa: Center for Wildlife Management, University of Pretoria.

IUCN (International Union for the Conservation of Nature). 1987. *The IUCN position statement on translocation of living organisms.* Gland, Switzerland: IUCN Council.

Jachmann, H., P.S.M. Berry, and H. Imae. 1995. Tusklessness in African elephants: A future trend. *African Journal of Ecology* 33:230–235.

Jackson, A. 1964. Texotics. *Texas Game and Fish* 22:7–11.

Jackson, S. G. 1992. Relationships among birds, willows, and native ungulates in and around northern Yellowstone National Park. Master of science thesis, Utah State University, Logan.

Jackson, S. G. 1993. The effects of browsing on bird communities. *Utah Birds* 9:53–62.

Jacobson, H. A., and J. C. Kroll. 1994. The white-tailed deer: The most managed and mismanaged species. *Proceedings of the Third International Congress on the Biology of Deer,* ed. H. W. Reid. Edinburgh, Scotland: Moredun Research Institute.

Johnston, M. C. 1963. Past and present grasslands of southern Texas and northeastern Mexico. *Ecology* 44:456–465.

Joubert, S.C.J. 1991. Management of the Kruger National Park: Principles, policies and strategies. *Transactions of the North American Wildlife and Natural Resources Conference* 56:27–39.

Kay, C. E. 1990. Yellowstone's northern elk herd: A critical evaluation of the "natural regulation paradigm." Doctoral dissertation, Utah State University, Logan.

Kay, C. E. 1993a. Aspen seedlings in recently burned areas of Grand Teton and Yellowstone National Parks. *Northwest Science* 67(2):94–104.

Kay, C. E. 1993b. Browsing by native ungulates on shrub and seed production in the Greater Yellowstone Ecosystem. In *Proceedings: Wildlife Shrub and Arid Land Restoration Symposium.* Compilers, B. A. Roundy, E. D. McArthur, J. S. Haley, and D. K. Mann, 310–320. Ogden, Utah: United States Department of Agriculture Forest Service Intermountain Research Station.

Kay, C. E. 1994. The impact of native ungulates and beaver on riparian communities in the Intermountain West. *Natural Resources and Environmental Issues* 1:23–44.

Keating, K. A. 1982. Population ecology of Rocky Mountain bighorn sheep in the upper Yellowstone River drainage, Montana/Wyoming. Master of science thesis, Montana State University, Bozeman.

Kie, J. G. 1977. Effects of predation on population dynamics of white-tailed deer in south Texas. Doctoral dissertation, University of California, Berkeley.

Kie, J. G., and M. White. 1985. Population dynamics of white-tailed deer (*Odocoileus virginianus*) on the Welder Wildlife Refuge, Texas. *Southwestern Naturalist* 30:105–118.

Kramer, H. 1963. *Elchwald.* Munich: BLV Verlagsgesellschaft.

Krausman, P. R., ed. 1996. *Rangeland wildlife.* Denver, Colorado: Society for Range Management.

Krausman, P. R., L. K. Sowls, and B. D. Leopold. 1992. Revisiting overpopulated deer ranges in the United States. *California Fish and Game* 78:1–10.

Kreuter, U. P., and J. P. Workman. 1996. Cattle and wildlife ranching in Zimbabwe. *Rangelands* 18:44–47.

Kurji, F. 1979. *Conservation areas and their demographic settings in Tanzania.* Research Report no. 18. Dar es Salaam, Tanzania: Bureau of Land-Use Practice, University of Dar es Salaam.

Lack, D. 1954. *The natural regulation of animal numbers.* London: Oxford University Press.

Lacy, R. C. 1987. Loss of genetic diversity from managed populations: Interacting effects of drift, mutation, immigration, selection, and population subdivision. *Conservation Biology* 1:143–158.

Laws, R. M. 1970. Elephants as agents of landscape change in East Africa. *Oikos* 21:1–15.

Laws, R. M., and I.S.C. Parker. 1968. Recent studies on elephant populations in East Africa. *Symposium of the Zoological Society of London* 21:319–359.

Laycock, G. 1966. *The alien animals: The story of imported wildlife.* New York: Ballantine.

Leopold, A. 1949. *A Sand County almanac.* New York: Oxford University Press.

Leopold, A., L. K. Sowls, and D. L. Spencer. 1947. A survey of over-populated deer ranges in the United States. *Journal of Wildlife Management* 11:162–177.

Loomis, J. B., and L. Fitzhugh. 1989. Financial returns to California landowners for providing hunting access: Analysis and determinants of returns and implications to wildlife management. *Transactions of the North American Wildlife and Natural Resources Conference* 54:196–201.

Lusigi, W. J. 1981. New approaches to wildlife conservation in Kenya. *Ambio* 10:87–92.

Lynch, J. F., and D. F. Whigham. 1984. Effects of forest fragmentation on breeding bird communities in Maryland, USA. *Biological Conservation* 28:287–324.

MacArthur, R. H. 1964. Environmental factors affecting bird species diversity. *American Naturalist* 98:387–397.

MacArthur, R. H., and J. W. MacArthur. 1961. On bird species diversity. *Ecology* 42:594–598.

McNaughton, S. J. 1979. Grazing as an optimization process: Grass-ungulate relationships in Serengeti. *American Naturalist* 113:691–703.

McNaughton, S. J., and G. A. Sabuni. 1988. Large African mammals as regulators of vegetation structure. In *Plant form and vegetation structure,* ed. M.J.A. Werger, P.J.M. Van der Aart, H. J. During, and J.T.A. Verhoeven, 339–354. The Hague: Academic Publishing.

McShea, W. J., and J. H. Rappole. 1992. White-tailed deer as keystone species within forest habitats of Virginia. *Virginia Journal of Science* 43(1B): 177–186.

McShea, W. J., and J. H. Rappole. 1994. Forest understory, white-tailed deer and understory dependent species. In *Proceedings, The Science of Overabundance: The Ecology of Unmanaged Deer Populations,* ed. W. J. McShea, 15. Front Royal, Virginia: Conservation and Research Center, National Zoological Park, Smithsonian Institution.

McShea, W. J., and G. Schwede. 1993. Variable acorn crops: Responses of white-tailed deer and other mast consumers. *Journal of Mammalogy* 74(4):999–1006.

McShea, W. J., M. V. McDonald, E. S. Morton, R. Meier, and J. H. Rappole. 1995. Long-term trends in habitat selection by Kentucky Warblers. *Auk* 112:375–381.

Marquis, D. A. 1974. *The impact of deer browsing on Allegheny hardwood regeneration.* United States Department of Agriculture Forest Service Research Paper NE-308. Washington, D.C.: United States Government Printing Office.

Marquis, D. A., and R. Brenneman. 1981. *The impact of deer on forest vegetation in Pennsylvania.* United States Department of Agriculture Forest Service General Technical Report NE-65. Washington, D.C.: United States Government Printing Office.

Mattfield, G. F. 1984. Northeastern hardwood and spruce/fir forests. In *White-tailed deer: Ecology and management,* ed. L. K. Halls, 305–330. Harrisburg, Pennsylvania: Stackpole Books.

Mech, L. D. 1995. The challenge and opportunity of recovering wolf populations. *Conservation Biology* 9:270–278.

Moen, A. N. 1980–82. *The biology and management of wild ruminants,* vols. 1–7. Lansing, New York: ConerBook Press.

Muir, P. D. 1988. Overview of wildlife ranching in New Zealand. In *Proceedings of the First International Wildlife Ranching Symposium,* ed. Raul Valdez, 24–29. Las Cruces, New Mexico: New Mexico State University Press.

Mungall, E. C. 1978. *The Indian blackbuck antelope: A Texas view.* Kleberg Studies in Natural Resources no. KS3. College Station: Texas A&M University Press.

Mungall, E. C., and W. J. Sheffield. 1994. *Exotics on the range: The Texas example.* College Station, Texas: Texas A & M University Press.

Myers, N. 1988. Tropical forests and their species: Going going . . . ? In *Biodiversity,* ed. E. O. Wilson, 28–35. Washington, D.C.: National Academy Press.

Noss, R. F., E. T. LaRoe III, and J. M. Scott. 1995. *Endangered ecosystems of the United States: A preliminary assessment of loss and degradation.* Biological Report no. 28. Washington, D.C.: United States Department of the Interior, National Biological Survey.

Owen-Smith, R. N. 1988. *Megaherbivores: The influence of very large body size on ecology.* Cambridge, England: Cambridge University Press.

Parker, R. L. 1959. Brush control and range rehabilitation in Texas. *Texas Business Review* 33(3):1–2.

Peek, J. M., D. E. Brown, S. R. Kellert, L. D. Mech, J. H. Shaw, and V. van Ballenberghe. 1991. *Restoration of wolves in North America.* Technical Review 91–1. Bethesda, Maryland: Wildlife Society.

Pienaar, U. de V. 1963. The large mammals of the Kruger National Park: Their distribution and present-day status. *Koedoe* 6:1–37.

Putman, R. J., P. J. Edwards, J.C.E. Mann, R. C. How, and S. D. Hill. 1989. Vegetational and faunal changes in an area of heavily grazed woodland following relief of grazing. *Biological Conservation* 47:13–32.

Ralston, H. 1989. Biology and philosophy in Yellowstone. *Biology and Philosophy* 4:1–18.

Rasmussen, D. I. 1941. Biotic communities of Kaibab Plateau, Arizona. *Ecological Monographs* 11:229–275.

Recher, H. F. 1969. Bird species diversity and habitat diversity in Australia and North America. *American Naturalist* 103:75–80.

Renecker, L. A., and R. J. Hudson, eds. 1991. *Wildlife production: Conservation and sustainable development.* Agricultural and Forestry Experiment Station Miscellaneous Publication 9–16. Fairbanks, Alaska: University of Alaska.

Risenhoover, K. L., and S. A. Maass. 1987. The influence of moose on the structure and composition of Isle Royale forests. *Canadian Journal of Forest Research* 17:357–364.

Russo, J. P. 1964. *The Kaibab North deer herd: Its history, problems and management.* Wildlife Bulletin no. 7. Phoenix, Arizona: Arizona Game and Fish Department.

Ryman, N., R. Baccus, C. Reuterwall, and M. H. Smith. 1981. Effective population size, generation interval, and potential loss of genetic variability in game species under different hunting regimes. *Oikos* 36:257–266.

Saunders, D. A., R. J. Hobbs, and C. R. Margules. 1990. Biological consequences of ecosystem fragmentation: A review. *Conservation Biology* 5:18–32.

Scifres, C. J. 1980. *Brush management: Principles and practices for Texas and the Southwest.* College Station, Texas: Texas A & M University Press.

Scott, D. P., and R. H. Yahner. 1989. Winter habitat and browse use by snowshoe hares, *Lepus americanus,* in a marginal habitat in Pennsylvania. *Canadian Field Naturalist* 103:560–563.

Scott, J. M., B. Csuti, and S. Caicco. 1991. Gap analysis: Assessing protection needs. In *Landscape linkages and biodiversity,* ed. W. E. Hudson, 15–26. Washington, D.C.: Defenders of Wildlife.

Scott, J. M., F. Davis, B. Csuti, R. Noss, B. Butterfield, C. Groves, J. Anderson, S. Caicco, F. D'Erchia, T. C. Edwards, J. Ulliman, and R. G. Wright. 1993. Gap analysis: A geographical approach to protection of biological diversity. *Wildlife Monographs* 123:1–41.

Scribner, K. T., M. H. Smith, and P. E. Johns. 1989. Environmental and genetic components of antler growth in white-tailed deer. *Journal of Mammalogy* 70:284–291.

Seton, E. T. 1909. *Life histories of northern mammals,* vol. 1. New York: Charles Scribner's Sons.

Sheffield, W. J., B. A. Fall, and B. A. Brown. 1983. *The nilgai antelope in Texas.* Caesar Kleberg Research Program in Wildlife Ecology no. KS5. College Station, Texas: Texas A & M University.

Simberloff, D., J. A. Farr, J. Cox, and D. W. Mehlman. 1992. Movement corridors: Conservation bargains or poor investments. *Conservation Biology* 6:493–504.

Simonetti, J. A. 1995. Wildlife conservation outside parks is a disease-mediated task. *Conservation Biology* 9:454–465.

Sinclair, A.R.E., and M. Norton-Griffiths, eds. 1979. *Serengeti: Dynamics of an ecosystem.* Chicago: University of Chicago Press.

Singer, F. J. 1989. Yellowstone's northern range revisited. *Park Science: A Resource Management Bulletin* 9(5):18–19. National Parks Service, United States Department of the Interior.

Singer, F. J., L. C. Mark, and R. C. Cates. 1994. Ungulate herbivory of willows on Yellowstone's northern winter range. *Journal of Range Management* 47:435–443.

Spellerberg, I. F., F. B. Goldsmith, and M. G. Morris. 1991. *The scientific management of temperate communities for conservation.* Oxford, England: Blackwell Scientific Publications.

Taylor, R. D., and D.H.M. Cumming. 1993. *Elephant management in southern Africa.* WWF Multispecies Animal Production Systems Project Paper no. 40. Harare, Zimbabwe: World Wide Fund for Nature.

Teer, J. G. 1979. Introduction of exotic animals. In *Wildlife conservation: Principles and practices,* ed. R. D. Teague and E. Decker, 172–177. Washington, D.C.: The Wildlife Society.

Teer, J. G. 1984. Lessons from the Llano Basin. In *White-tailed deer: Ecology and management,* ed. L. K. Halls, 261–290. Harrisburg, Pennsylvania: Stackpole Books.

Teer, J. G. 1989. Exotic animals: Conservation implications. In *Proceedings of the 44th Annual Conference, International Union of Directors of Zoological Gardens, San Antonio, Texas,* ed. Steve Kinswood, 31–36. International Union of Directors of Zoological Gardens.

Teer, J. G. 1991. Non-native large ungulates in North America. In *Wildlife production: Conservation and sustainable development,* ed. L. A. Renecker and R. J. Hudson, 55–66. Agricultural and Forestry Experiment Station Miscellaneous Publication 91–6. Fairbanks, Alaska: University of Alaska.

Teer, J. G. 1993. Commercial utilization of wildlife: Has its time come? In *Commercialization of wildlife management: Dancing with the devil,* ed. A.W.L. Hawley, 73–82. Malabar, Florida: Krieger Publishing Co.

Teer, J. G. 1996. The white-tailed deer: Natural history and management. In *Rangeland wildlife,* ed. P. R. Krausman, 193–210. Denver, Colorado: Society for Range Management.

Teer, J. G., and N. K. Forest. 1968. Bionomic and ethical implications of commercial game harvest programs. *Transactions of the North American Wildlife and National Resources Conference* 33:192–204.

Teer, J. G., J. W. Thomas, and E. A. Walker. 1965. Ecology and management of white-tailed deer in the Llano Basin of Texas. *Wildlife Monographs* 15. Bethesda, Maryland: The Wildlife Society.

Teer, J. G., G. V. Burger, and C. Y. Deknatel. 1983. State-supported habitat management and commercial hunting on private lands in the United States. *Transactions of the North American Wildlife Conference* 48:445–456. Washington, D.C.: Wildlife Management Institute.

Teer, J. G., D. L. Drawe, T. L. Blankenship, W. F. Andelt, R. S. Cook, J. G. Kie, F. F. Knowlton, and M. White. 1991. Deer and coyotes: The Welder experiments. *Transactions of the North American Wildlife and Natural Resources Conference* 56:550–560.

Teer, J. G., L. A. Renecker, and R. J. Hudson. 1993. Overview of wildlife farming and ranching in North America. *Transactions of the North American Wildlife and Natural Resources Conference* 58:448–460.

Texas Parks and Wildlife Department. 1991. *Big game harvest survey results: 1981–82 thru 1990–91.* Austin, Texas: Texas Parks and Wildlife Department.

Texas Parks and Wildlife Department. 1992. *Texas white-tailed deer population and harvest trends, 1974–91.* Austin, Texas: Texas Parks and Wildlife Department.

Tharp, B. C. 1926. *Structure of Texas vegetation east of the 98th meridian.* University of Texas Bulletin 2606. Austin: University of Texas Press.

Tharp, B. C. 1939. *The vegetation of Texas.* Series no. 1, Texas Academy of Science Publication, Natural History. Houston, Texas: Anson Jones Press.

Thomas, J. K., C. E. Adams, and J. Thigpen. 1990. *Texas hunting leases: Statewide and regional summary.* Technical Report no. 90-4. College Station, Texas: Department of Rural Sociology, Texas Agricultural Experiment Station, Texas A & M University.

Thomas, J. W. 1984. Fee hunting on the public's land: An appraisal. *Transactions of the North American Wildlife and Natural Resources Conference* 49:455–468.

Tilghman, N. G. 1989. Impacts of white-tailed deer on forest regeneration in northwestern Pennsylvania. *Journal of Wildlife Management* 53:524–532.

Tilt, W., R. Norris, and A. S. Eno. 1987. *Wolf recovery in the northern Rocky Mountains.* Washington, D.C.: National Audubon Society and National Fish and Wildlife Foundation.

Trammell, M. A., and J. L. Butler. 1995. Effects of exotic plants on native ungulate use of habitat. *Journal of Wildlife Management* 59:808–816.

Traweek, M. S. 1989. *Statewide census of exotic big game animals.* Federal Aid Project no. W-109-R-12, Job 21. Austin, Texas: Texas Parks and Wildlife Department.

Traweek, M. S. 1995. *Statewide census of exotic big game animals.* Federal Aid Project no. W-127-R-3, Job 21. Austin, Texas: Texas Parks and Wildlife Department.

Trefethen, J. B. 1975. *An American crusade for wildlife.* New York: Winchester Press and the Boone and Crockett Club.

Ullrey, D. E. 1982. Nutrition and antler development in white-tailed deer. In *Antler development in cervidae,* ed. R. D. Brown, 49–59. Kingsville, Texas: Caesar Kleberg Wildlife Research Institute, Texas A & I University.

Valentine, J. F. 1989. *Range development and improvements.* San Diego, California: Academic Press.

van Ballenberghe, V., and W. B. Ballard. 1994. Limitation and regulation of moose populations: The role of predation. *Canadian Journal of Zoology* 72:2071–2077.

Veblen, T. T., M. Mermoz, C. Martin, and E. Ramilo. 1989. Effects of exotic deer on forest regeneration and composition in northern Patagonia. *Journal of Applied Ecology* 26:711–724.

Wagner, F. H. 1978. Livestock grazing and the livestock industry. In *Wildlife and America: Contributions to an understanding of American wildlife and its conservation,* ed. H. P. Brokaw, 121–145. Washington, D.C.: Council on Environmental Quality.

Wagner, F. H. 1987. North American terrestrial grazing. *Revista Chilena de Historia Natural* 60:245–263.

Wagner, F. H. 1989. Grazers, past and present. In *Grassland structure and function: California annual grassland,* ed. L. F. Huenneke and H. Mooney, 151–162. Dordrecht, Netherlands: Kluwer Academic Publishers.

Wagner, F. H., R. Foresta, R. B. Gill, D. R. McCullough, M. R. Pelton, W. F. Porter, and H. Salwasser. 1995. *Wildlife policies in the U.S. national parks.* Washington D.C.: Island Press.

Walker, B. 1995. Conserving biological diversity through ecosystem resilience. *Conservation Biology* 9:747–752.

Wallis de Vries, M. F. 1994. Large herbivores and the design of large-scale nature reserves in western Europe. *Conservation Biology* 9:(1)25–33.

Warren, R. J. 1991. Ecological justification for controlling deer populations in eastern national parks. *Transactions of the North American Wildlife and Natural Resources Conference* 56:56–66.

Wheaton, C., M. Pybus, and K. Blakely. 1993. Agency perspectives on private ownership of wildlife in the United States and Canada. *Transactions of the North American Wildlife and Natural Resources Conference* 58: 487–494.

Wilson, E. O. 1992. *The diversity of life*. Cambridge, Massachusetts: The Belknap Press of Harvard University Press.

Wilson, E. O., ed. 1988. *Biodiversity*. Washington, D.C.: National Academy Press.

Witmer, G. W., and D. S. deCalesta. 1992. The need and difficulty of bringing the Pennsylvania deer herd under control. *Proceedings of the Eastern Wildlife Damage Control Conference* 5:130–137.

Wolfe, G. J. 1982. The relationship between age and antler development in wapiti. In *Antler development in cervidae,* ed. R. D. Brown, 29–36. Kingsville, Texas: Caesar Kleberg Wildlife Research Institute, Texas A & I University.

Workman, J. P. 1986. *Range economics*. New York: Macmillan.

World Conservation Monitoring Center and the IUCN Commission on National Parks and Protected Areas. 1994. *1993 United Nations list of national parks and protected areas.* Gland, Switzerland: IUCN–The World Conservation Union.

Yahner, R. H. 1989. Forest management and featured species of wildlife: Effects on coexisting species. In *Proceedings of the Symposium on Timber Management and Its Effects on Wildlife,* ed. J. Finley and M. Brittingham, 146–161. University Park, Pennsylvania: School of Forest Resources, Pennsylvania State University.

Yellowstone National Park. 1989. *Grazing influences on Yellowstone's Northern Range. Summaries of research studies.* F. J. Singer, compiler. Washington, D.C.: National Parks Service, United States Department of the Interior.

THIRTEEN

The Commercial Consumptive Use of the American Alligator (*Alligator mississippiensis*) in Louisiana

Its Effects on Conservation

Ted Joanen, Larry McNease, Ruth M. Elsey, and Mark A. Staton

The commercial consumptive use of some wildlife species, when managed on a sustainable basis, is considered by many to be not only compatible with their conservation but also an effective conservation tool with the potential to benefit a species, its habitat, and associated biodiversity (IUCN/UNEP/WWF 1991; Swanson 1992). However, this view, which is widely held by crocodilian biologists and managers, and most IUCN Crocodile Specialist Group members (e.g., Palmisano et al. 1973; Child 1987; Gorzula 1987; Hines and Abercrombie 1987; Hollands 1987; Jenkins 1987; Joanen and McNease 1987a; Webb et al. 1987a,b; Whitaker 1987; Messel 1991; Thorbjarnarson et al. 1992), is not universal and is under attack in various quarters, including the public media (e.g., Robinson 1993). Much of the argument in this high-profile debate results from emotional and political points of view (e.g., *U.S. News and World Report*, Nov. 15, 1993) rather than objective reviews of biological and socioeconomic data relevant to concerned species and the implications of sustainable use management for conservation.

To evaluate the proposition that sustainable use can provide incentives to conserve a wild species and its supporting ecosystem, we develop here a case study of the American alligator (*Alligator mississippiensis*) in Louisiana. This species is exemplary because knowledge of trade in alligator products, and the effects of commercial harvests on alligator populations, have been reported for almost two centuries. During this time, commercial use of the Louisiana alligator population evolved from unregulated overexploitation to present-day sustainable use management involving a wild harvest and an extensive egg ranching/farming program. Relevant historical and scientific information affords the opportunity to evaluate objectively the biological suitability of alligators in a sustainable use program, the effects of such use on wetland biodiversity, the efficacy of incentives and controls in the commercial consumptive use of alligators, and the benefits of current use and any alternative use of the species.

The Uses, Markets, and Harvest of Alligators in Louisiana

This section discusses the main alligator products—skins and meat—and the markets for those products. The response of Louisiana alligator populations to the demands of those markets and the development of strategies to protect alligator populations are also reviewed.

Alligator Skins

The commercial value of alligators is derived primarily from their skins. Leather made from the skins of American alligators, along with that of several crocodile species, is considered the "classic" exotic leather. Placed in the hands of skilled designers and leathersmiths, leather made from alligator skins can be crafted into some of the most valuable leather goods available. Although the strength and durability of the leather contribute to its value, the real appeal of alligator leather, compared with common leathers, is akin to that of gold compared to common metals or diamonds compared to semiprecious stones. The beauty is in the eyes of the beholder. Many people do not see any special appeal in alligator leather goods, while others cannot or do not care to pay the prices these products command. Some prefer not to purchase alligator products because it is counter to their personal (animal-welfare-based) beliefs. Others fear that alligators are still endangered or that alligator products are illegal. Nevertheless, there are many people of sufficient means who see an attraction in alligator products—enough

Table 13.1. Markets (Percent) for Louisiana Salted Alligator Skins, Based on the Countries to Which Skins Were Exported in 1992–1993

Country	Wild harvest (25,864)	Farms (125,511)
France	27	44
Italy	16	15
Japan	19	8
Singapore	28	16
Switzerland	8	—
United States	1	17

Source: Louisiana Department of Wildlife and Fisheries.

to create the market for the exclusive leather made from the skins of alligators which are sustainably and legally produced.

The market for Louisiana alligator skins, leather, and leather goods is international in scope. Relatively few skins find a market or end use within Louisiana, and most enter international trade. Ischii (1990) estimated that 70 percent of all classic crocodilian skins found a market in Japan, either as raw skins, tanned skins, or finished leather goods. Thus the largest single market for alligator leather goods are Japanese consumers buying them domestically or abroad. Other major markets are the centers of wealth in Europe, the emerging economies of Asia and Latin America, and the United States. Alligator leather goods are produced throughout these same areas, but manufacturers in Europe are considered to be the most exclusive.

The market for raw (salted) alligator skins is much more limited. Relatively few major tanneries (several dozen) are capable of tanning/finishing alligator skins into the highly valued leather. These are mainly centered in France, Italy, Japan, and Singapore; others are found in other European and Pacific Rim nations, Latin America, and the United States. Most Louisiana alligator skins are tanned and finished in these countries (Table 13.1).

Although leather made from alligator skins is currently assigned a high value, this was not always the case. The earliest record of the use of alligator leather was around 1800 in the manufacture of relatively rough articles of trade such as boots, shoes, and saddles (Stevenson 1904). In the first quarter of the nineteenth century, "many thousands" of alligators were killed for this trade, with obvious destructive effects on the alligator population (Audubon 1827). Audubon reports that when it was found that the leather was "not sufficiently firm and close grained to prevent water passage or dampness," widespread hunting of alligators stopped.

The following half century saw only sporadic demand for alligator skins, first as a novelty fashion leather around 1855 and later in the Civil War (1860–65) to supply Confederate soldiers with shoe and boot leather. Immediately after the war, alligator skin was again without a use, as trade in regular shoe leather was restored (Stevenson 1904; Joanen and McNease 1991). This proved to be a "lull before the storm" as the modern era of alligator utilization began in the late 1860s, when alligator leather "rose to the top of the fashion scale of all leathers" (Joanen and McNease 1991). Stevenson (1904) wrote that by 1870 the large demand resulted in the slaughter of thousands of alligators per year. Alligator populations were placed under tremendous hunting pressure, and "demand soon exceeded the productive capacity of the U.S. and a large number of skins were imported from Mexico and Central America" (Stevenson 1904).

McIlhenny (1935) estimated that between 3 and 3½ million alligators were taken from Louisiana alone between 1880 and 1933—approximately 60,000/year on average—but with far fewer alligators being taken in the later years of this period. Kellogg (1929), for example, reported 21,885 and 36,041 skins taken in the years 1925 and 1926, respectively. The continued decline of the alligator population is recorded by Louisiana severance tax records (Joanen and McNease 1991), which show that 414,126 skins were sold between 1939 and 1960—down to less than 20,000/year. During this period, the size of alligators taken also decreased, further reflecting excessive pressure on the Louisiana population.

In 1962, Louisiana banned the hunting of alligators, and the state's Department of Wildlife and Fisheries (LDWF) turned its attention to professional management of this valuable wildlife resource, five years before the species was listed in the federal Endangered Species Act in 1967. During a period of total protection in Louisiana (1962–72), extensive research on alligators was conducted, and state and federal laws governing the taking, possession, and transportation of alligators and their products were enacted (Joanen et al. 1983; Joanen and McNease 1991). In 1970, the Louisiana legislature made provisions for a closely regulated, experimental commercial harvest (Palmisano et al.

1973). During this period of total protection, the institutional and political framework for the implementation of a sustainable use program, based on sound scientific information, was developed. The goals of the program were to manage and conserve Louisiana's alligators, as one of many components in the state's vast (more than 2 million hectares [ha]) wetlands ecosystems, to the benefit of the species, its habitat, and associated wildlife (i.e., biodiversity). Inherent in the philosophy of managing the species was to allow and encourage sustainable harvest of alligators to benefit economically the citizens of Louisiana. Since most (approximately 80%) of the state's wetlands are privately owned, this would provide an incentive to landowners, and hunters who lease land, to maintain and enhance the alligator's wetlands habitats (e.g., Joanen and McNease 1973a, 1974, 1975, 1981, 1987a, 1991). Also during this period of total protection, alligator populations rebounded dramatically, to the extent that harvest of some populations was possible. In 1972, a year before the federal Endangered Species Act took effect, the LDWF began modestly to implement its program of sustainable use management. While keeping the conservation of the alligators as the primary objective, the state authorized an experimental hunt, confined to one parish in southwestern Louisiana. A total of 1,350 alligators were taken by 59 hunters. No hunt was allowed in 1974, when declassification of the alligator in Louisiana to the status of "threatened due to similarity of appearance" began. Hunting in three coastal parishes resumed in 1975 and was gradually expanded until 1981, when all Louisiana alligators were reclassified and the season was opened statewide. Presently, about 25,000 wild alligators are currently harvested annually from Louisiana wetlands (Table 13.2).

Provisions in the law were also made for a carefully regulated program of skin production from the farming of alligators originating from captive breeding stock or from eggs harvested from wild populations. The latter source of hatchlings has been the primary source of alligators for the Louisiana "farming" program (known as "egg ranching" in the lexicon of wildlife managers). The egg ranching program began experimentally with eggs harvested from state-owned lands at Rockefeller Refuge, was expanded statewide in 1986, and has grown substantially. In 1972, a total of 35 skins were sold by three of eight licensed farms. By 1992, 85 of 125 licensed farms sold approximately 128,300 skins (Table 13.3).

Table 13.2. Statistics from the September Wild Alligator Harvest in Louisiana, 1973–1993[a]

Year	Commercial hunters	Tags issued[b]	Number taken	Success (%)	Average length (cm)	Average value of skins per 30.5 cm (US$)	Total value of skins (US$)	Area hunted (ha)	Amount meat sold (kg)	Value of meat (US$)
1972	59	1,961	1,350	68.8	211	8.10	75,505	111,254	N/P	N/P
1973	107	3,243	2,921	90.1	213	13.13	268,994	216,513	N/P	N/P
1975	191	4,645	4,420	95.2	229	7.88	258,791	329,144	N/P	N/P
1976	198	4,767	4,389	92.1	216	16.55	512,240	327,038	N/P	N/P
1977	236	5,760	5,474	95.0	224	12.23	488,499	395,645	N/P	N.P
1979	708	17,516	16,300	95.0	211	15.00	1,711,500	1,047,168	45,400[c]	125,000
1980	796	19,134	17,692	92.5	201	13.00	1,609,972	1,313,456	45,400[c]	125,000
1981	913	15,534	14,870	95.7	211	17.50	1,821,575	1,415,394	45,400[c]	125,000
1982	1,184	18,188	17,142	94.2	208	13.50	1,621,633	1,617,408	45,400[c]	125,000
1983	945	17,130	16,154	94.3	211	13.00	1,452,568	1,416,974	45,400[c]	125,000
1984	1,104	18,386	17,389	94.6	213	21.00	2,556,183	1,428,192	45,400[c]	125,000
1985	1,076	17,466	16,691	95.6	216	21.00	2,482,619	1,415,313	68,100[d]	625,000
1986	1,207	23,267	22,429	96.0	211	23.00	3,611,000	1,539,000	140,740[d]	1,395,000
1987	1,370	24,635	23,892	97.0	216	40.00	6,689,760	1,579,500	227,000[d]	2,250,000
1988	1,545	24,111	23,526	98.0	221	48.00	7,905,024	1,741,500	272,400[d]	3,000,000
1989	1,769	25,492	24,846	97.4	221	50.00	9,006,675	1,766,580	339,040[d]	3,000,000
1990	1,921	26,051	25,575	98.2	221	57.00	10,568,869	1,765,800	318,000[d]	3,000,000
1991	1,995	24,532	23,870	97.3	227	32.00	5,686,025	1,766,000	310,310[d]	2,935,000
1992[e]	1,686	25,378	24,000	94.0	221	23.00	4,002,000	1,766,000	312,000[d]	2,951,520
1993	1,690	24,381	23,500	96.4	221	23.00	3,916,410	1,707,000	305,500[d]	2,890,000
Total	20,700	341,577	326,430				66,925,842	24,664,879	2,565,490	22,846,520
Average				95.6	216					

Source: Louisiana Department of Wildlife and Fisheries.
[a]N/P, sale of meat not permitted; Louisiana Health Department regulations first allowed meat sales in 1979.
[b]Does not include Salvador and Marsh Island experimental and farm harvest.
[c]Bone in.
[d]Deboned.
[e]Predicted December 28, 1992.

Table 13.3. Statistics from the Alligator Farming/Ranching Program in Louisiana, 1972–1993

	Number of farms				Value of skins (US$)		Value of meat (deboned)[a]	
Tag year	Licensed	Sold skins	Number of skins sold	Average length (cm)	Average per 30.5 cm	Total	Amount (kg)	Value (US$)
1972	8	3	35	152.4	8.10	1,417	N/P	N/P
1973	8	5	103	193.0	13.13	8,560	N/P	N/P
1975	8	3	83	167.6	7.88	3,597	N/P	N/P
1976	8	3	360	175.3	16.55	34,258	N/P	N/P
1977	8	4	376	160.0	12.23	24,142	N/P	N/P
1980	8	1	191	142.2	13.00	11,595	434	3,342
1981	8	3	360	142.2	17.50	29,421	817	6,300
1982	8	1	113	121.9	13.50	6,102	205	1,582
1983	14	6	1,449	139.7	13.00	86,273	3,290	25,357
1984	12	7	2,836	129.5	21.00	253,113	5,150	39,704
1985	15	12	4,430	129.5	21.00	395,377	8,045	79,740
1986	22	15	5,925	137.2	23.00	613,237	12,105	119,983
1987	30	23	10,670	134.6	24.00	1,131,873	21,800	216,067
1988	47	38	27,749	129.5	36.00	4,245,597	50,392	554,980
1989	83	68	66,737	121.3	32.00	8,499,624	136,477	1,202,362
1990	123	79	88,220	122.9	24.00	8,534,396	180,410	1,786,059
1991[b]	134	93	119,000	126.1	15.00	7,385,400	243,300	2,380,000
1992[c]	125	85	128,300	122.9	12.00	6,206,128	262,310	2,566,000

Source: Louisiana Department of Wildlife and Fisheries.

[a]N/P, sale of meat not permitted; Louisiana Health Department regulations first allowed meat sales in 1979.

[b]Revised October 12, 1992; subject to further revision.

[c]Revised November 15, 1993; subject to further revision; includes the time period September 10, 1992 through September 10, 1993.

In contrast to the supply of alligator skins over the previous century, the renewed availability beginning in the 1970s carried with it the promise of a continued supply of skins resulting from a sustainable use program. Markets reacted with a steady price increase, which continued through most of the 1980s (Tables 13.2 and 13.3) until supply met market demands. Since the late 1980s, the major economies of the world, including that of Japan, have been in recession to one extent or another. Consequently, at the time that the ability of the farming industry to produce skins reached an all-time high (Table 13.3), prices dropped significantly (Tables 13.2 and 13.3). During this time, the industry has been marked by consolidation of and improvements in the production efficiency. Groups such as the Louisiana Fur and Alligator Advisory Council, the Louisiana Department of Agriculture, the Southern U.S. Trade Association, and the American Alligator Council have engaged in efforts to expand and diversify the markets for alligator skins. These efforts, coupled with a reduction of the backlog of farm and tannery stocks, and economic improvements around the world, led to a 25–40 percent rebound in prices in 1994.

Unlike markets of the past, which were not built around sustainable use programs, the present availability of affordable alligator leather in sustainable quantities has set the stage for market expansion and diversification due to competition. This competition will come primarily from within the alligator industry itself, as there are no true substitutes for alligator skin. Leather pressed with the scale pattern of alligator skin is very popular, reflecting alligator leather's general appeal among people who cannot or will not pay the very high prices of the genuine article. However, this leather belongs to a far less exclusive marketplace and cannot replace genuine alligator leather.

Other exotic leathers would appear to have the potential to "substitute" for alligator leather through competition in the marketplace. However, the noncrocodilian exotics (e.g., snake, lizard, and ostrich) do not directly compete with crocodilian skins because of price structure and consumer choice (Durland 1990; Yamanaka 1990). The supply of other classic crocodilian leathers, such as from saltwater crocodile (*Crocodylus porosus*), Nile crocodile (*C. niloticus*), and Papua New Guinea freshwater crocodile (*C. n. novaequineae*), is smaller and less reliable, and competition from these sources is of less consequence than that within the alligator industry itself. In recent years, alligator skin production has grown to represent the preponderance of classic crocodilian skins produced worldwide. For example, excluding alligators, the total number of classic crocodilian skins produced in 1990 was estimated at 102,687 skins (R. Luxmore, pers. comm. of data presented at a workshop on trade at the IUCN Crocodile Specialist Group Meeting, August 1992, Victoria Falls, Zimbabwe). In the same year, alligator production in Louisiana alone, excluding nuisance alligators, was 113,795 skins (Tables 13.2 and 13.3). Since that time, alligator production increases in Louisiana (152,681 in 1992) and other states (Florida, Texas) have greatly exceeded gains in production of crocodile species. This contrasts sharply with the situation only a decade earlier. For example, alligators represented no more than 26 percent of the classic crocodilian skins traded worldwide in the period 1983–85 (Luxmore 1990). Despite this decline in market share of other crocodilian exotic leathers, competition from these sources can only serve to improve efficiency in the industry and stimulate markets for all exotic leathers.

Alligator Meat

The alligator industry is based on the sale of skins, but the meat produced is a valuable by-product. The commercial sale of alligator meat has been allowed in Louisiana since 1979, when relevant health department requirements took effect. "Controlled by the Food and Drug Control Unit, Office of Health Services and Environmental Quality of the Department of Health and Human Resources, alligator skinning facilities were licensed and approved for the sale of both farm and wild alligator meat for human consumption under Louisiana Law, Chapter VI of the State Sanitary Code" (Joanen and McNease 1991).

During the period 1979–93, more than 2,500 MT of alligator meat from the wild harvest and more than 900 MT of meat from the ranching/farming program were produced and sold. Current annual production has grown to about 500 tons of alligator meat from the harvest of both wild and farmed alligators (Tables 13.2 and 13.3). Obviously, only a very small portion of the meat produced from Louisiana's sustainable use program is consumed in Louisiana by those who produce it, and it is not important in their diet as might be the case in an underdeveloped nation. Instead, its value comes from its sale into markets for human consumption and the economic benefits conferred on the hunters, ranchers and farmers, and their employees.

During the period 1979–93, approximately US$32 million was generated from the sale of alligator meat within the state, other states, and overseas, including Taiwan, Canada, and several European countries. As supply increased during the period 1979–93, meat prices steadily declined (Tables 13.2 and 13.3).

The nutritional attractiveness—low in fat and cholesterol (Johnson et al. 1983)—and the ease of cooking alligator meat (Moody et al. 1988) ensure that markets for alligator meat will always exist. However, supply of those markets will likely be limited by the demand for skins in the leather markets because alligator meat is essentially produced as a by-product. In the United States, alligator meat is primarily a "novelty" on restaurant menus. In the Far East, it is consumed as "dragon meat," and many customers eat it because they believe that by doing so their spirit will be imbued with that of the dragon. Because the supply of alligator meat is now significant, steady, and intrinsically limited, and because of alligator meat's unique nature, the stage is set for development of further processed, "value-added" alligator meat products for the retail market. Limited efforts along this line are evidenced by the development of alligator sausages in Louisiana seafood markets and experiments with canning (Moody et al. 1988).

Biological Suitability

This section raises and attempts to answer two important questions: (1) Are alligators in Louisiana, and their wetland habitats, capable of sustaining annual harvests? (2) Is the current Louisiana harvest program well designed and adequately monitored?

It was recognized in the late 1960s and early 1970s that one of the greatest threats to alligators in Louisiana was the potential for unmanaged development of wetlands, with the likelihood of habitat loss to agricultural, commercial, and residential uses. A policy of managing alligator populations as a renewable natural resource on a sustained yield basis to provide an incentive to landowners to "manage wetlands as wetlands," rather than for other commercial purposes, was adopted (Joanen and McNease 1973a, 1974, 1981, 1987a, 1991; Joanen et al. 1983). As a consequence, alligators have been managed under sustainable use programs in Louisiana for more than 20 years (Joanen and McNease 1991). An essential component of this effort was an extensive research program, initiated by the Louisiana Department of Wildlife and Fisheries in 1964, to study the natural history, management, and captive propagation of this species (see reviews in Joanen and McNease 1981, 1987a,b, 1991; Joanen et al. 1983). Over the past 30 years, the state of Louisiana has invested extensive resources into this effort, resulting in steady growth of the alligator population, the development and publication of much scientific information, and —as an unintended benefit—the most highly developed crocodilian industry in the world. Furthermore, the work done by LDWF on the management of alligators, both wild and under farm conditions, has been applied to crocodilian conservation programs throughout the world (e.g., Webb et al. 1987b). Nevertheless, the goal of the Louisiana program has always been to manage the alligator populations, as a component of the state's wildlife resources, to the benefit of alligators, other species, and the people of Louisiana.

Central to this research program are the exhaustive and ongoing studies on the size, structure, dynamics, and distribution of the alligator population in the vast (more than 2 million ha) Louisiana wetlands, which includes an estimated 1.7 million ha of coastal marsh (Chabreck 1970; Joanen and McNease 1981) and more than 400,000 ha of nonmarsh wetlands (Taylor 1980; Joanen and McNease 1991: LDWF data). Alligators are present in 63 of the 64 parishes in Louisiana and inhabit the entire range of wetland habitats throughout the state.

Coastal marsh habitats in Louisiana are categorized geographically: (1) the Chenier Plain, 560,000 ha east of the Sabine River to Vermilion Bay; (2) the active delta marshes, 97,000 ha of very recent origin near the Mississippi River delta; and (3) subdelta marshes, 1.1 million ha west of the Chenier Plain except the active delta region. Throughout these regions, approximately 404,000 ha are salt or highly brackish marshes and are not considered to be alligator habitat. Most of the approximately 404,000 ha of nonmarsh wetlands are alligator habitat, including 10,925 ha of lake habitat and more than 385,000 ha of cypress-tupelo swamps, which are comprised mostly (64%) of the Atchafalaya Basin (Joanen and McNease 1981, 1991; LDWF data). Thus there are approximately 1.7 million ha of alligator habitat in the state.

Another classification of Louisiana coastal wetlands is that based on salinity and resultant vegetative types (e.g., Chabreck et al. 1968; Joanen and McNease 1972a). The intrusion of saline water, both naturally and as a result of human activities, is the greatest abiotic factor affecting alligator populations. Salinity of the environment is known to affect feeding habits (Chabreck 1972; McNease and Joanen 1977; Lauren 1985), movements (Joanen and McNease 1972b), reproduction (Joanen 1969; Joanen and McNease 1972a; Ruckel and Steele 1985), physiology (Lauren 1985; Dunson and Mazzotti 1988), and growth rates (Mazzotti 1982; Lauren 1985). Prolonged exposure of young alligators to salinities of 3.5 parts per thousand (ppt) results in cessation of feeding and to salinities greater than 5–13 ppt can result in death (Joanen and McNease 1972a; Lauren 1985).

Alligator population densities are consequently related to salinity conditions (Joanen and McNease 1970, 1972a,b, 1978; Table 13.4),

Table 13.4. Tag Allotments in Louisiana in Relation to Alligator and Nest Densities, 1993

Habitat type	Acres per: Alligator[a]	Nest[b]	Tag allotment[c]
Marsh			
Fresh	6.9	146	75–250
Intermediate	5.3	70	75–325
Brackish	10.2	105	100–700
Nonmarsh			
Cypress-tupelo lakes	4.1	N/C	Avg. 125
Cypress-tupelo swamps	6.4	N/C	Avg. 200
Atchafalaya Basin	18.0	N/C	Avg. 1,280

[a]1981 Louisiana Department of Wildlife and Fisheries data.
[b]1988–92 Louisiana Department of Wildlife and Fisheries data. N/C, not compiled.
[c]1993 Louisiana Department of Wildlife and Fisheries data.

being greatest in intermediate marsh (0.5–8.3 ppt salinity), and slightly less in fresh marsh (0.1–3.4 ppt salinity) and brackish marsh (1.0–18.4 ppt salinity). The latter is a band of marsh averaging 13.36 kilometers (km) statewide (Joanen and McNease 1972a) between intermediate and salt marshes (8.17–29.4 ppt salinity). Nesting does not occur on the gulf side of the 10 ppt isohaline line (Joanen and McNease 1972a).

Maps of the isohaline lines and associated vegetation of Louisiana are prepared about every 10 years by the LDWF (Chabreck et al. 1968; Chabreck and Linscombe 1978). These maps are used in assessment of populations within each parish by wetland habitat type, including subdivisions by salinity (e.g., Chabreck 1970; Joanen and McNease 1973b; Ensminger and Linscombe 1980; Table 13.4), and resulting management decisions. In addition to salinity, factors such as the amount of open water, cover, water depth, and interspersion, as developed in the alligator habitat suitability index model (Newsom et al. 1987), are considered.

This detailed knowledge of alligator distribution in Louisiana wetlands shows the versatility and adaptability of this species to diverse habitats. Although a notable member of wetlands fauna, the alligator shows no unique association with specific habitats or other species. Alligator populations are densest in intermediate marsh (Table 13.4), where they feed heavily on nutria (*Myocastor coypus*), an introduced species which also prefers this habitat.

The alligator plays the role of a large top predator throughout the diverse wetland habitats it occupies. Numerous studies on the feeding habits of alligators in Louisiana have been conducted (e.g., Kellogg 1929; Giles and Childs 1949; Chabreck 1972; Valentine et al. 1972; McNease and Joanen 1976; Taylor 1986; Wolfe et al. 1987). Prey species for alligators include insects, amphibians, mollusks, crustaceans, fish, birds, reptiles, and mammals. As McIlhenny (1935) pointed out, alligators are opportunistic feeders, consuming most animals smaller than themselves. The importance of prey species in an alligator's diet depends largely on the relative abundance of the prey, compared with other prey species, in habitat selected by individual alligators.

Based on the extensive published record and experience of LDWF personnel, it is considered unlikely that alligators play a dominant role in the regulation of populations of any of these species. For example, nutria and muskrat are the most important food items for large alligators, but these mammals have reproductive potentials that vastly exceed the ability of alligators to control. For example, the Louisiana nutria population, which produced in excess of 13,500,000 pelts from 1970 to 1979 (Ensminger and Linscombe 1980), started from the accidental release of 13 individuals into a marsh of high alligator density in 1937 (Kinler et al. 1987). Similarly, other common prey such as crustaceans and fish are characterized by almost incalculable biomass and reproductive potential, and any regulation of their populations by alligator predation would be marginal and isolated.

Martin and Hight (1977) provide an unusual example of prey population numbers being regulated by alligators. In ponds at Florida's Loxahatchee National Wildlife Refuge, where other prey species apparently were relatively sparse, alligators fed almost exclusively on apple snails, large aquatic snails important in the diet of the Everglade kite. The number of snails in the ponds was inversely proportional to alligator density. When alligators were excluded from ponds, snail population indexes increased.

Perhaps the only species on which alligator predation has a serious impact is alligators themselves. Alligators of virtually all sizes suffer from intraspecific predation (Nichols et al. 1976b; Hines and Abercrombie 1987; Rootes 1989). Rootes (1989) estimated that cannibalism accounted for 50.2 percent of hatchling mortality and 70.1 percent of mortality of alligators 11 months and older. "Total cannibalism losses were an estimated 2.13 prey alligators per predator size alligator in the standing crop per year" (Rootes 1989). For alligators and other crocodilians, cannibalism is believed to be the major population density-related mortality factor (Nichols et al. 1976b; Polis and Myers 1985).

The only suggestion that alligators play an influential ecosystem role is in southern Florida wetlands (Craighead 1968; Rhoads and Pope 1968; Kushlan 1974, 1979), where, when marshes dry, alligator burrows/dens serve as refuges for fish species and help to maintain fish diversity. Alligator activity at these ponds prevented encroachment of successional vegetation. While alligator holes are present in Louisiana (e.g., McIlhenny 1935; Chabreck 1965), such a benefit is now viewed as minimal in most Louisiana marshes because of active water management by landowners, which is designed to prevent stressful drought conditions, as discussed below.

The Louisiana coastal marsh alligator population has been quantified annually by aerial nesting surveys (Chabreck 1966) along transects in representative habitat since 1970 (LDWF annual survey data; Fig. 13.1; Table 13.5), validated by actual counts along selected transects. Consistent annual growth has led to the conservative estimate (Taylor et al. 1991) of approximately 750,000 alligators in Louisiana wetlands (Table 13.5), including a quadrupling of the coastal marsh alligator population. This growth occurred over a 20-year period when 800,000 alligators were harvested as a result of the wild hunt, the nuisance alligator program, and as part of the egg ranching program (Tables 13.2 and 13.3).

One reason alligators show remarkable population growth numbers is a high reproductive rate. It has been conservatively estimated (Taylor et al. 1991) that nesting females represent 5 percent of the nonhatchling population (Chabreck 1966). Over a four-week period in June and early July, these females build a nest and deposit an average of 39 eggs (Joanen 1969; Joanen and McNease 1976, 1978, 1979, 1980). Thus immediately after nesting there are more eggs in nests than there are alligators in the entire population. Given that alligators are long-lived, that as many as 63 percent of adult females nest each year (Joanen and McNease 1980), and that a relatively small percentage of the adult male population is required to mate each year (adult males can mate with three to five females), the intrinsic capacity of population growth, even under enormous pressures, is considerable.

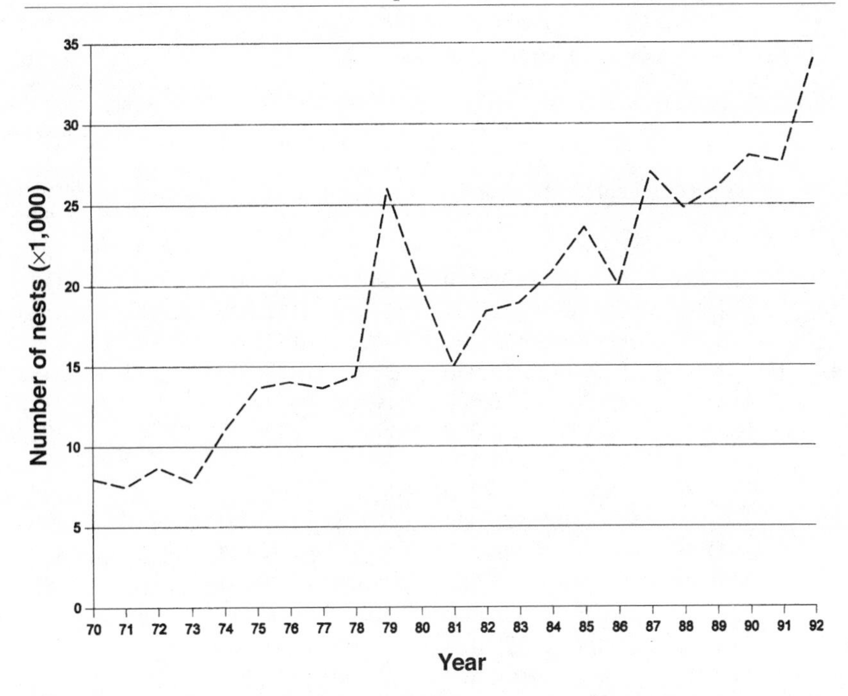

Figure 13.1. Louisiana costal marsh alligator nest projections based on aerial surveys, 1970–92. (*Source:* Louisiana Department of Wildlife and Fisheries data.)

Balancing this reproductive potential is the high natural mortality of alligator eggs and hatchlings (Taylor and Neal 1984). Egg mortality is primarily due to flooding (Joanen et al. 1976) and predation (Fleming et al. 1976; Nichols et al. 1976b). Droughts, desiccation, overheating, and hurricanes (Waldo 1957; Ensminger and Nichols 1958; LDWF data) are influential factors. Predation of hatchlings is maximized under conditions of drought, as suitable habitat is reduced and the ability of predators to find their prey is enhanced (Hines et al. 1968; Fleming et al. 1976). This is particularly damaging because, under drought conditions, many adult females will not nest (Joanen and McNease 1989); as a result, eggs or hatchlings lost to predation are a disproportionately large portion of the population's reproductive output, as compared to the loss of an equal number of eggs or hatchlings in years of normal rainfall and nesting.

Estimates of survivorship, from the time of egg deposition through a hatchling's first years of life, range from 5 to 17 percent (Nichols et

Table 13.5. The Louisiana Alligator Population

Year	Wild[a]					Farm[c]	State total
	Marsh			Non-marsh[b]	Total wild		
	Private	Public	Total				
1970	104,180	67,900	172,080	N/A	172,080	N/C	172,080
1971	77,191	57,218	134,000	N/A	134,000	N/C	134,000
1972	102,460	79,940	182,000	N/A	182,000	N/C	182,000
1973	93,096	60,394	153,000	N/A	153,000	N/C	153,000
1974	121,549	92,445	213,000	N/A	213,000	N/C	213,000
1975	152,773	119,703	272,000	N/A	272,000	N/C	272,000
1976	186,668	95,644	282,000	N/A	282,000	N/C	282,000
1977	182,615	91,733	274,000	N/A	274,000	N/C	274,000
1978	191,520	93,635	285,000	N/A	285,000	N/C	285,000
1979	368,706	150,720	520,000	N/A	520,000	N/C	520,000
1980	295,178	107,760	400,000	N/A	400,000	N/C	400,000
1981	213,940	76,039	289,000	168,000	457,000	N/C	457,000
1982	225,207	136,500	368,000	168,000	368,000	N/C	536,000
1983	256,432	119,560	379,000	168,000	379,000	N/C	547,000
1984	315,712	94,120	410,000	168,000	578,000	N/C	578,000
1985	322,100	146,000	468,000	168,000	636,000	N/C	636,000
1986	299,420	97,500	397,000	168,000	565,000	N/C	565,000
1987	384,078	138,460	523,000	168,000	691,000	46,579	737,579
1988	367,900	115,320	483,000	168,000	651,000	86,446	737,446
1989	372,860	145,200	518,000	168,000	686,000	194,807	880,807
1990	423,500	134,980	558,500	168,000	726,500	325,451	1,051,951
1991	N/C	N/C	550,000	168,000	718,000	318,177	1,036,177
1992	N/C	N/C	690,000	168,000	585,000	291,983	1,149,983
1993	N/C	N/C	566,000	168,000	734,000	258,314	992,314

Source: Louisiana Department of Wildlife and Fisheries.
[a]Does not include hatchlings.
[b]N/A, not available until conservative estimate by Joanen et al. 1981.
[c]N/C, not compiled.

al. 1976a,b; Taylor and Neal 1984; Carbonneau and Chabreck 1990). These represent averages, and in years of either extreme flooding or drought, mortality can reach 100 percent for some populations. For example, in 1985 all nests produced in the Marsh Island population were lost in a hurricane (LDWF data). Alternatively, low mortality for any one cohort of eggs or hatchlings can lead to significant increases in population numbers. Because of this, the reproductive capacity of alligator populations can be considered to be compensatory.

The management implication is that controlling marsh water levels avoids extremes and decreases mortality of eggs and young. This both stabilizes the recruitment of young into the natural population and makes available a steady supply of eggs for the egg ranching program, which further serves to compensate for extremes in natural mortality.

Both components of the Louisiana alligator harvest target the removal of specific components from the population. In the case of the egg harvest, members of a specific age class (eggs and resulting hatchlings) are removed. In the wild hunt, adult males and nonreproductive females are targeted by way of sex- and size-specific habitat specificity based on telemetry studies (Joanen and McNease 1970, 1972b; McNease and Joanen 1974). The goal is to protect the nesting female segment of the population at a time of nest-tending by breeding females. Hunting takes place in canals, bayous, and lakes—habitats that are avoided by reproductive females during the postnesting period (Giles and Childs 1949; Joanen and McNease 1970). This is important to the success of a year's crop of eggs or hatchlings, which often receive protection by the parental female. Harvest results (Joanen and McNease 1987a) show that adult males routinely comprised 60–80 percent of the total take and 70–85 percent of adults hunted. This trend has not changed. In 1993, for example, males constituted 74 percent of the total take (LDWF data).

The Louisiana alligator ranching program is based on the abovementioned combination of high reproductive rate and high natural mortality. Under the program, a quota of eggs can be picked up under permit by LDWF, as long as a number of the resulting alligators are subsequently released to the marsh where the eggs were collected. The number to be released is equal to a percentage that would be expected to survive under natural conditions, based on their size and survivorship data (Taylor and Neal 1984). This percentage ranges from 29.6 percent for 91.44-cm alligators, to 9.8 percent for 152.40-cm animals. Published (Elsey et al. 1992a,b) and unpublished data from ongoing LDWF research show that these released alligators eat and grow (and presumably survive and reproduce) as well as their wild counterparts. The ranching program therefore does nothing to restrict the inherent diversity of the wild population. To the contrary, taking a portion of the most fragile part of the population into captivity serves to "buffer" the Louisiana alligator population during periods of unmanageable habitat stress, for example, hard freezes or hurricanes. The number of eggs harvested annually reached about 275,000 in 1990, and averages about 20 percent of eggs laid. Natural events therefore continue to shape genetic diversity of the vast majority of the wild population.

Alligators are characterized by very low genetic variability; in fact, they are widely recognized as being among the most homogeneous of vertebrate species (Dessauer 1983; Menzies et al. 1979; Adams et al. 1980). There are no recognized subspecies, and systematic studies have shown little intraspecific variation over the range of the species (Ross 1977; Ross and Roberts 1979). None of the periphery of the species' geographical range lies within Louisiana. There are no isolated populations within the state which have differentiated themselves genetically. Because of the widespread distribution of alligators in diverse habitats throughout the state, and the regulation of the harvest according to habitat type, there is no danger that the hunt or egg harvest affects the population by virtue of its being genetically distinct or unique on a geographical basis.

Research by LDWF on the rate at which populations can be harvested is ongoing. Taylor and Neal (1984) suggested that an 8 percent harvest rate was too liberal. At the Marsh Island Refuge, an experimental hunt conducted periodically since 1986, with few restrictions and a target harvest of 20 percent of the adult female population, has significantly changed the age-class structure, in contrast to the relatively constant average size of alligators hunted statewide (Table 13.6).

In practice, the maximum harvest rate is 5 percent for a population. However, the rate for any given population—managed to habitat type within parish boundaries—takes into consideration the population density and trends in the area, and can therefore be lower (Table 13.4). In marsh habitats, quotas (tag allotments) range from one per 30 ha in the fresh marshes of St. Charles and St. John the Baptist parishes to one per 283 ha in the brackish marshes of St. Bernard and Orleans parishes. In nonmarsh habitats, quotas average one per 50 ha in cypress-tupelo lakes and one per 518 ha in the cypress-tupelo swamps of the Atchafalaya Basin.

Currently, the actual harvest rate is about 3 percent of the state's population (approximately 25,000 alligators harvested from a popu-

Table 13.6. Percentage of Adults and Juveniles in the Statewide Alligator Hunt and the Experimental Hunt at Marsh Island Refuge

Harvest	Year	Percentage of adults	Percentage of juveniles	Sample size
Public hunt	1972	64	36	303
	1973	71	29	843
	1974	87	13	782
	1975	73	27	591
	1976	75	25	281
	1972–76	76	24	2,800
	1972–90	80	20	231,713
Marsh Island	1986	83	17	2,930
	1987	43	57	1,261
	1988	32	68	166

Source: Joanen et al. 1987a (statewide hunt); Louisiana Department of Wildlife and Fisheries (experimental hunt).

lation of approximately 750,000; Tables 13.2 and 13.5). This conservative harvest level provides a hedge against unknown and unpredictable factors, ensuring that the population will not be overharvested. The success of this strategy is validated by the continued growth of the population, by the consistent size of alligators taken in the statewide hunt over the past 22 years (Tables 13.2 and 13.3), and by comparisons of sex and size-class structure from hunted and nonhunted populations. The alligators of Marsh Island were totally protected and unhunted, that is, "preserved" for a period of 24 years between 1962 and 1986. The size-class structure of this population (as reflected in proportion of adults in the first hunt in 1986) and that of the statewide population are very similar (Table 13.6). Over the period 1972–90, approximately 80 percent (Joanen and McNease 1991) of the 231,713 alligators hunted in Louisiana were adults (1.83 m or more in total length), virtually identical to the percentage of adults (83%) taken at Marsh Island in 1986. Therefore, populations hunted under LDWF guidelines are not being excessively harvested, as alligators in unhunted populations died at about the same rate as those from hunted populations.

Not only is the harvest rate low compared to the much higher (20–30%) harvest rates prescribed for species such as rabbit, quail, squirrel, and deer, but the offtake of alligators is also regulated to a degree rarely seen in wildlife management—to the level of the individual alligator. This is achieved as a result of the tagging requirements developed by the LDWF to regulate effectively the alligator harvest and comply with CITES (Conference on International Trade in Endangered Species) requirements. A locking tag (specific to Louisiana alligators, with unique serial numbers, and monitored under CITES permitting procedures) must be affixed to any alligator skin before it leaves the site on which the animal was killed, whether as a result of the wild harvest or farming. This tagging system is a part of the licensing and hunting regulations developed by LDWF. Any skin not bearing the Louisiana/CITES tag will be confiscated; fines can be imposed and licenses revoked. Furthermore, to conform with and assist CITES, this tag must remain on the skins, whether salted or processed as leather, or traded internationally.

There can be no doubt that the alligator in Louisiana can be managed sustainably without damaging the viability of the resident population. The continued monitoring of the population, the dedication of LDWF to professionally managing the Louisiana population, and the natural resiliency of the species ensure that any indication of overharvest or damage to the population from other factors, even those unknown and unpredictable, will be responded to in such a way as to protect the Louisiana alligator population and ensure its long-term viability.

Impacts on Biodiversity

This section discusses Louisiana's coastal wetlands and the benefits of managing and harvesting alligator populations for the biodiversity of this region. The flora, fauna, habitats, and ecosystems of Louisiana are amply described in the general scientific literature and are the concern of state and federal wildlife and environmental agencies as well as private conservation groups. The Louisiana Natural Heritage Program, and interagency efforts of the LDWF and the state's Department of Natural Resources (LDNR) have described the natural communities (Craig et al. 1987) and listed the plants and animals "of special concern" in the coastal zone of the state (Lester 1988), which is where most alligators are found. Included in this list are two invertebrate species, six reptiles, nine avian species, and five mammal species classified as either threatened or endangered under the federal Endangered Species Act. An additional ten plant species, four invertebrates, ten fish species, one reptile, and three birds are considered potential candidates for the list (summarized by Louisiana Department of Culture, Recreation, and Tourism 1988). The species on these various lists show no close association with alligators; nor would it appear that they are in any way dependent on alligators or negatively impacted by their management and harvest per se. To the contrary, efforts to maintain or restore Louisiana's historical wetland diversity benefit the native biodiversity.

Louisiana's wetlands are of national and international importance as part of two important migratory bird routes extending from Canada to South America. For example, Louisiana's coastal zone provides the wintering grounds for 66 percent of the geese in the Mississippi flyway and 20–25 percent of the nation's total puddle duck population (Louisiana Department of Culture, Recreation, and Tourism 1988). The United States government has signed treaties with several other countries to ensure the conservation of migratory bird habitat (Louisiana Department of Culture, Recreation, and Tourism 1988). It is not inconsequential then that waterfowl habitat is also alligator habitat, and that much of this land is privately owned. Louisiana's wetlands are dynamic, productive ecosystems which have historically received significant annual inflow of freshwater, sediments, and nutrients from the Mississippi River and, to a lesser degree, from the Red River. This has led to marsh building or accretion. Operating against this inflow of inland freshwater, sediments, and nutrients is soil erosion and saltwater intrusion from the Gulf of Mexico. Over the past century, historical/natural patterns of sedimentation and hydrology in coastal Louisiana have been altered significantly by the levying of the Mississippi River. This has diverted freshwater and sediments, which under natural conditions would have countered the actions of erosion and saltwater intrusion into the Gulf of Mexico. Furthermore, the past 50 years have seen the construction of ship channels and numerous canals which have facilitated saltwater intrusion into coastal marshes. Also, large-scale pumping from freshwater-bearing sediments have further altered subsurface hydrology (Louisiana Geological Survey 1967). Consequently, coastal marshes have been characterized by salinification, vegetational alteration and losses, subsidence, and both inland and coastal erosion (Beek and Meyer-Arendt 1982), causing significant alteration and loss of habitat for alligators and other coastal zone species. For example, an estimated 155 km^2 of coastal Louisiana are eroded into the Gulf annually (Templet and Meyer-Arendt 1986). Thus Louisiana's historical wetland diversity cannot merely be "maintained"; it must be "restored."

Throughout geologic history, and particularly over the past century of manmade alterations to wetlands, alligators adapted to the variety of habitats formed. The current policy of LDWF and other state and federal agencies is to attempt to maintain isohaline lines throughout Louisiana wetlands and prevent further habitat loss. The adaptability of alligators, as well as LDWF policies, ensures the viability of the species. However, continued encroachment of saltwater into fresh, intermediate, or brackish marshes puts into question just how much alligator habitat will exist in future years and what its carrying capacity will be.

Alligator managers in Louisiana (Joanen and McNease 1975) and other states (Schortemeyer 1972; Schemnitz 1974; Hines 1979; Potter 1981) realized long ago that the greatest potential problem in managing alligators was that of possible habitat loss to agricultural, commercial, or residential uses. Louisiana adopted the approach of cooperating with landowners to manage alligators as a renewable natural resource. As most alligators in Louisiana are found in the coastal marshes (Table 13.5), alligator habitat management largely means marsh management. The economic value of wildlife—waterfowl, furbearers, crawfish, shrimp, and gamefish, as well as alligators—makes it in the landowners' interests to manage wetlands in such a way as to maintain their integrity and prevent extremes such as flooding, drought, and salinity. With LDWF and Soil and Conservation Service advice, many landowners have taken a multispecies approach to marsh management (e.g., Joanen et al. 1989), investing significantly in resources in hopes of reaping economic benefits. At the same time, wetland habitats have been stabilized or improved, and the native biodiversity has benefited.

Of prime importance in the multispecies approach to marsh management is the control of water levels and associated habitat, with the use of water control structures such as weirs, levees, and impoundments. It may be argued that some species are dependent on habitat extremes and that their interests are not served by marsh management. However, as discussed above, modifications to Louisiana's wetlands, vis-à-vis historical wetland diversity, are extreme, and there is abundant habitat for such species. Furthermore, these species benefit from periodic "draw downs" practiced as part of good marsh management. Nevertheless, control of flooding, drought, and salinity serve to stabilize marsh habitats to the advantage of the vast majority of wetlands species.

Waterfowl species inhabiting Louisiana marshes provide excellent examples of the benefits of a multispecies approach to marsh management. Duck populations at Rockefeller Refuge grew from 75,000 to 500,000 over a period of several decades of concentrated marsh management (Joanen 1982; LDWF data). Alligator populations also benefited from the water control measures, which created an abundance of nesting sites and food supply for alligators (Chabreck 1960). Waterfowl that use Louisiana wetlands to overwinter before migrating to summer breeding grounds in Canada provide another good example. Their reproductive success in summer is related to their feeding success at winter sites in Louisiana's coastal marsh managed for their benefit as well as for alligators (Williams and Chabreck 1986; Chabreck et al. 1989).

Alligator habitats are not manipulated in an unnatural manner in order to meet harvest objectives. Instead, the harvest of alligators is improved by increasing the quantity and quality of wetlands habitat available, that is, restoring and maintaining historical wetlands diversity. Following the pre-1962 depletion of many alligator populations, stocking was conducted in the 1960s and early 1970s to reestablish and replenish those populations (LDWF data). This augmentation of the population was conducted as a conservation measure, with no consideration for eventual harvests. Although it has been noted that the alligator productivity of marshes could be increased by systematic egg collection and subsequent restocking to minimize high natural mortality of juveniles (Nichols et al. 1976a,b), such measures have not been taken.

Marsh management recommendations allow for sustainable harvests of some species (e.g., otters, raccoons, large predatory fish) that prey upon alligator eggs or young. Direct benefits (e.g., lowered predation rates) to alligator populations are, however, probably minimal. Any such benefits are probably negated by a reduction in food supply to the adult alligator population.

Louisiana is known as the "Sportsman's Paradise," and wetlands throughout the state are regularly traversed by hunters, trappers, and fishermen throughout the year. Alligator hunting and egg gathering are only a small percentage of the human activity in the species' habitat, and cannot be considered intrusive or damaging to it. The harvest of alligators is species-specific, and the incidental taking of other species occurs only rarely. For example, turtles occasionally take baits intended for alligators.

Perry et al. (1993) have shown how a restored, well-managed marsh can reap approximately twice the economic return of a degraded marsh, as well as provide better habitat for wildlife. Thus fauna and flora native to Louisiana's wetlands can benefit from the economic incentive provided by sustainable use of alligators and other wildlife. The maintenance and restoration of the area's wetlands and associated biodiversity can obviously be served by economic forces associated with the wildlife inhabiting them. However, management of Louisiana's expansive marshlands is expensive and could not be undertaken on a large-scale basis without a way to pay for the various professional and capital requirements.

Incentives and Controls

Substantial investments in personnel, equipment, and land management are required to enhance alligator habitat. Rather than being mandated, these investments are made willingly because of the promise of economic return. This incentive, however, must be balanced by careful monitoring in order to prevent overexploitation.

Most large landowners in the Louisiana coastal zone receive royalties from oil and gas production on their land. However, oil and gas are nonrenewable natural resources, and this source of income is both finite and in decline. The reduction in this revenue base adds emphasis to the need to encourage investment in wetlands management schemes that enhance habitat for alligators and other wildlife. Many landowners have responded positively to LDWF efforts and guidance, and, within the above-discussed limitations imposed by considerable modifications to marsh formation and salinity profiles of the state's wetlands, the results are generally well-managed wetlands thriving with populations of alligators and other wildlife.

Although landowners benefit from the sustainable use of alligators and other wetlands species with economic value, the impacts—economic and cultural—are perhaps even greater for the community at large. Approximately fifteen hundred alligator trappers, one hundred alligator farmers, and five thousand furbearer trappers derive a significant portion of their livelihood from what is essentially a subsistence economy. These people and others living in the communities where they live receive most of the economic benefit generated by various components (wild hunt, egg harvest, farming, skin and meat sales, and so on) of the alligator harvest. Direct industry sales, which have grown to as much as US$25 million per year, account for the equivalent of 231 full-time jobs (Brannan et al. 1991; Tables 13.2 and 13.3). Economic modeling shows that an additional US$13 million in goods and services and 172 full-time jobs are indirectly created by the Louisiana alligator industry (Brannan et al. 1991). This economic activity provides tax revenues to underwrite state and local government services.

Though the total value of skins and meat shown in Table 13.2 results in a return of US$4–$6 per ha hunted (not including eggs), it is important to understand that the alligator is only one of several species that are harvested and provide economic value for wetlands in Louisiana. For example, Perry et al. (1993) estimated that waterfowl hunting (guide fees and leasing), muskrat and nutria trapping, shrimp and crab harvests, and alligator hunting and egg collecting in a restored brackish marsh in Louisiana could yield approximately US$48 per ha. Revenues from alligators represented 11 percent of this total.

Beyond economic considerations, alligators and other wildlife native to Louisiana's swamps, marshes, bayous, rivers, and lakes are integral to the centuries-old cultural heritage of Louisianians inhabiting these areas (e.g., Glasgow 1991). Louisiana was pioneered by trappers, resulting in the establishment of the city of New Orleans as a fur trading center (Ensminger and Linscombe 1980). The prominence of alligators in local culture, reflecting the heritage of subsistence living and "closeness to the land," continues today despite the system of land tenureship that has evolved in Louisiana. Although control of the land lies in the hands of relatively few landowners, actual use of the land and associated wildlife remains with the descendants of generations of trappers and hunters, perpetuating the subsistence economy of Louisiana's wetlands. By and large, these hunters willingly adhere to LDWF regulations in order to prevent overharvesting, ensuring their take in the future. Furthermore, they frequently act as agents for the landowners, watching over and protecting the land and the alligators they anticipate hunting. In some cases, they are paid for these efforts, but generally they are acting out of self-interest in maintaining the alligator population and the suitability of its habitat.

The influence of alligators on the culture of Louisiana is formalized at many local fairs and festivals, such as the Alligator Festival in Boutte, the Alligator Harvest Festival in Grand Chenier, and the International Alligator Festival in Franklin (Glasgow 1991). Alligator skinning is commonly displayed at these events, and alligator meat is cooked in a variety of ways. Furthermore, numerous alligator meat recipes and cookbooks have been published (Glasgow 1991; Moody et al. 1988), and alligator meat is a favorite item in restaurants catering to both tourists and the local population.

Thus the association between man and alligators, which began almost seven hundred years ago and led to the depletion of the state's alligator population, continues today in the form of a sustainable harvest. The stewardship of most of the estimated 1.7 million ha of alligator habitat in Louisiana, upon which alligators ultimately depend, lies with landowners. However, the use of alligators is regulated by and managed under state laws, federal laws, and international convention (CITES).

The 1970 Louisiana legislature established LDWF as the state regulatory agency with principal responsibility for managing the wild population, controlling the harvest, and the ranching/farming program. The state regulations are embodied in Title 56 of Louisiana Public Law. The stated policies and plans regarding the management and harvest of alligators, as described throughout this chapter, have been published (e.g., Palmisano et al. 1973; Joanen et al. 1983; Joanen and McNease 1987a, 1991). Louisiana gained federal authority to manage alligators with the approval of Louisiana's Alligator Management Plan. The state's performance under this management plan is reviewed annually by way of the nondetriment statements filed with the U.S. Fish and Wildlife Service (USFWS) acting as the CITES authority in the United States. Shipments of alligator products are permitted and inspected by LDWF officials, who work in tandem with USFWS to enforce CITES inspection and permitting requirements on shipments entering international trade. Other agencies involved with alligators are the Louisiana Department of Agriculture, which is active in marketing (e.g., Louisiana Department of Agriculture 1989) and the Louisiana Department of Human and Health Services, which regulates the production of meat in slaughterhouses.

In addition to actions taken by the landowner with guidance from LDWF, the management of alligator habitat rests with a host of other state and federal agencies, including the Louisiana Department of Natural Resources, the U.S. Environmental Protection Agency, the U.S. Fish and Wildlife Service, the National Marine and Fisheries Service, and the U.S. Corps of Engineers. The controlling law influencing Louisiana wetlands is the U.S. Clean Water Act. Section 404 of this law has made it difficult and expensive for owners of wetlands in Louisiana to commercially develop their land in any way, especially for pursuits other than those based on agricultural or wildlife use. Wetlands management, in general, is receiving much attention statewide and nationally in terms of research, funding, and legislation, and is beyond the scope of this report.

While federal and state agencies address the problems associated with channelization, hydrology, and sedimentation in Louisiana's coastal marsh (e.g., Louisiana Department of Natural Resources 1991, 1992, 1993), many practical actions to restore Louisiana's historical wetland diversity will have to be taken by landowners acting out of self-interest. The revenues derived from the harvest of alligators help landowners in this effort. Discussions with landowners point to a willingness to invest in wetlands in amounts, at a minimum, commensurate with the economic return that can be expected from the use of alligators and other wildlife species. Some landowners would invest more, out of a sense of responsibility toward land stewardship. Landowners who have not taken active measures to maintain and improve their marshes point to the cumbersome, expensive, and time-consuming permit requirements under Section 404 of the Clean Water Act as the primary impediment. Large land companies, with land management staff wholly or partially funded from revenues generated by wildlife use are able to meet and afford these permitting requirements and in some cases procure private and public funds to improve wetlands habitat. Unfortunately, this is beyond the means of small and even medium-sized landowners.

The income derived by landowners from the alligator harvests makes the difference between profitability and financial loss for some land companies. Income and expenses associated with alligators are carefully budgeted. Operating decisions—such as when to schedule construction activities or to permit seismic or mineral exploration activities —are often taken only after consideration of potential impacts on the biology of alligators or other wildlife species and how revenues could be affected.

The financial incentives made available to landowners, hunters, farmers, and others directly associated with the alligator industry have obviously been sufficient to engender a sense of responsibility toward the resource. They recognize that they are essentially cropping a renewable natural resource and want to do so within the ability of the resource to renew itself. The vast majority willingly work within the limits set by LDWF to prevent overexploitation and maintain healthy, dynamic wetlands. This ethic is consistent with the heritage of wetlands utilization predominant in the culture of many rural Louisianians. A major difference in the present consumptive use of alligators, as contrasted with the pre-1962 years, is the guidance available to the public. The knowledge gleaned from the extensive monitoring and research conducted by LDWF is embodied in management policy and state law.

To ensure that alligator use is under professional control and management, all important aspects of the harvest of alligators and alligator eggs are regulated under state law. Harvests are highly organized and restricted in time in order to maximize the ability of LDWF to monitor the various activities involved. The statewide alligator hunt is conducted in September when the impact on the reproductive output of the population is minimal. Eggs are gathered by or for farmers with the permission (at a price ranging from US$2–$10/egg) of the landowner to whom the permit is issued. Eggs may be gathered only during the established egg collection season, between sunrise and sunset. Furthermore, eggs must be transported so as to maximize survival of eggs, and permits are not issued to any alligator farmer who fails to hatch at least 70 percent of viable eggs for two consecutive years.

LDWF regulations govern the taking, possession, selling, raising, and propagation of alligators in the wild and in captivity. Alligators may be hunted only after hunters complete orientation in the procedures to be followed, and obtain required licenses and tags. Tags are allotted to landowners and are then issued to hunters in exchange for an amount that customarily equals 25–40 percent of the value of the alligator. Approved wild harvest methods include hook and line, long bow and barbed arrow, and firearms. Alligators may be taken only during the hours between official sunrise and official sunset. Alligator hunters are required to inspect any hooks and lines they are utilizing on a daily basis. They are not allowed to cut alligators loose from hooks and lines for the purpose of selecting larger (more valuable) animals. In the event that an alligator is hooked and the hunter's quota has been reached, the hunter must release the alligator in the most humane way possible.

The ongoing alligator research and monitoring programs conducted by the LDWF are integrated into the department's alligator management efforts and constitute an important part of the controls needed to ensure sustainability of the harvests. The conservative rate (less than 5%) at which alligators are being harvested is based solely on scientific and monitoring information, not on any harvest goals. Harvest details and the results of the monitoring are reviewed annually by the USFWS, acting as the CITES management authority in the United States, by way of nondetriment submissions by LDWF. Despite the economic and cultural benefits of the alligator industry, political considerations bear no influence on decision making in the management of alligators and the biodiversity of their wetlands habitat. This has been LDWF's position over the last 30 years, beginning with the initial decision to take strong and effective measures to safeguard the depleted alligator population even before federal protection was enacted.

Legislative and regulatory safeguards established to protect alligators, their habitat, and associated wildlife communities from overexploitation are enforced by more than four hundred LDWF enforcement agents. There are no exceptions made in the strict enforcement of these laws. Aside from the coercive force of laws and enforcement agencies, the policy of sustainable, multispecies land use promoted by LDWF generates even more effective protection of alligators and wetland habitats by landowners and hunters acting in their own self-interest. Without such cooperation, it is doubtful the state could afford all the law enforcement capabilities required to prevent poaching and unnecessary killing of alligators.

Given that the alligator population is still growing, a cessation of all harvesting for several years would increase their occurrence around dwellings, and the nuisance alligator program would have new and growing demands placed on it. The law enforcement capabilities of the LDWF would be severely stressed. Funds required for the LDWF to manage alligators—which today are generated almost entirely by a US$4 tag fee on each alligator skin harvested in the state—would be lost and have to be found elsewhere in a state budget that is already running at a deficit. The 30-year investment by LDWF in its alligator programs, which have cost a minimum of US$10 million in operating expenses, would be jeopardized. Personnel and infrastructure important in the sustainable use of alligators, as it exists today, would be lost and difficult to replace.

Loss of an economic incentive to landowners to manage marshes would be highly detrimental to habitat and to alligators and other wildlife. Pressure for other uses of the land, for example, agricultural or residential use, would increase. Large-scale agricultural use of marshes would require their drainage and would result in significant loss of habitat. Management of land as wetlands would decrease. The carrying capacity for most, if not all, species would decrease. Some areas would in all likelihood be developed, but in ways disadvantageous to alligator habitat.

Perhaps the most profound influence would be the loss in income to the many Louisiana residents partially or totally dependent on the alligator industry. There would be less sentiment on the part of local people and public officials to protect alligators or their habitats. Poaching would increase. Protection of alligators by hunters who adhere to LDWF regulations would be lost. Such problems would be greatly magnified under a more prolonged (five years or more) cessation of the alligator harvest. The fallback conservation position is one of pure regulation and law enforcement. Without the cooperation of landowners and the public at large, this would be extremely expensive and much less effective. In the courts, cases involving alligators or their habitat would be less likely to be ruled in favor of the interest of alligator populations. Alligators would lose fear of humans, and attacks on humans could well increase. In Florida, for example, where dense human and alligator populations coexist, a number of alligator attacks on people, including fatalities, have occurred in recent years. Irate citizens have threatened to kill alligators, and on numerous occasions have done so (T. C. Hines, pers. comm. 1994). Under similar circumstances in Louisiana, political pressure against alligators would increase, and courts may not act to protect them. In Florida, more than ten thousand nuisance alligator responses per year have been recorded (Hines 1990), and wildlife officials have spent as much as 50 percent of their time answering nuisance alligator calls (T. C. Hines, pers.

comm.). Similar problems could be expected in Louisiana if alligators were not hunted.

In summary, the growth of the Louisiana alligator population during the last few decades—at a time when alligator skin production from all sources is comparable to the maximum number ever taken from the state—makes it obvious that the commercial consumptive use of alligators in Louisiana as a renewable natural resource is both highly successful and, what is more important, sustainable. The program is made possible by the (regulated) economic incentives available to landowners, hunters, farmers, traders, and so on, balanced by extensive monitoring and enforcement efforts of the LDWF. The results of the annual surveys, along with the knowledge resulting from the ongoing 30-year research program, ensure that overexploitation will not occur.

Opportunity Costs for Biodiversity Conservation

Here we discuss the benefits of consumptive use of alligator populations in terms of conservation options. Are there alternative uses and management options for maintenance of alligator populations and habitat that are preempted by developing and implementing sustainable use management programs?

As presented above, the direct incentives for consumptive use of alligators are largely economic, and the beneficiaries are multiple: landowners, hunters, farmers, employees, traders, dealers, manufacturers, tanners, the communities in and around the sites of economic activity, and both local and state governments. Furthermore, sustainable use of alligators is not conducted at the expense of biodiversity conservation. On the contrary, as structured in Louisiana, consumptive use of alligators and other wildlife supports biodiversity conservation by providing landowners with incentives and revenues required to "manage wetlands as wetlands" to the benefit of native wildlife and flora.

Alternative management options have been evaluated in Louisiana. A lower rate of harvest, or no harvest, would do nothing to enhance alligator populations and in fact would ultimately damage wetland biodiversity, as landowners would be deprived of the economic means of maintaining them. Although a higher rate of harvest might increase economic returns initially, LDWF's conservative harvest strategy inherently protects against unknown and unpredictable factors and ensures against "overharvesting"—which is within Louisiana's experience with the alligator.

The success of Louisiana's sustainable use program at a time of, and in conjunction with, a remarkable recovery in the state's alligator populations has paved the way for a nonconsumptive use of alligators, that is, tourism. Louisiana has a small but growing tourist market, one that is largely dependent on the state's unique heritage and its reputation as the "Sportsman's Paradise." The hunting of alligators and other wildlife is prominent in Louisiana's wetlands heritage, which is imbued with a healthy appreciation for the state's wetlands and the harvest of numerous species they harbor. A number of tourist attractions (alligator farms, marsh tours, festivals) in the state are built around both the success of the alligator industry and the special mystique of the alligator in its natural environment. Tour guides tell tall tales of giant alligators while at the same time pointing out facts about the biology of alligators and other species, and the economic benefits that harvest of these species bring. Tourists enjoy seeing well-managed wetlands and the wildlife associated with it. This nonconsumptive use of alligators and their habitat is built around the success of the sustainable consumptive use program, and in no way are they mutually exclusive.

Section 404 of the Clean Water Act limits the commercial use of the vast majority (an estimated 80%) of Louisiana wetlands to sustainable use of alligators and other wildlife. No alternative use of personnel or financial resources could result in greater conservation benefits. The only other alternative is to do nothing, that is, no utilization of the resource. This would carry no conservation benefits and would be counterproductive in many ways discussed in the previous section.

The alligator harvests are part of a multiuse approach to wildlife conservation, management, and utilization which is dependent on restoring and maintaining Louisiana's historical wetland habitats and associated biodiversity. The sustainable use of alligators in Louisiana can only be construed to have a positive impact on opportunities to conserve the state's native biodiversity. This has resulted from decades of research and development of elaborate methods of control. Nevertheless, the conservation strategy is a simple and time-tested one—to maintain the native habitat and its biodiversity, and to harvest nature's "generous surplus," which if not harvested will die of other causes (e.g., Clepper 1966). It is truly remarkable that the sustainable harvest of those alligator eggs and young which would normally die, to satisfy an international market resulting from the appreciation of alligator leather goods by relatively few people, is sufficient to do so much for the conservation of this and other species in the state's wetlands.

The bottom line is that conservation must be paid for. The price paid in the form of overexploitation of alligators followed by a period of total protection is unacceptably high. Similarly, the price that would be paid by a cessation of sustainable use of Louisiana's alligator population would also be very high in terms of economic losses to the citizens of the state, additional costs of law enforcement to the state government, cultural degeneration, and detrimental effects to alligator habitats and, therefore, alligator populations themselves. The commercial consumptive use of alligators as structured in Louisiana, on the other hand, provides a way to pay the costs of conserving alligators, their wetland habitats, and the biodiversity associated with these while at the same time conferring economic and cultural benefits to the citizens of the state.

Acknowledgments

We greatly appreciate the interest, assistance, and input of Kermit Coulon, John Donahue, Judge Edwards, Frank Ellender, David Richard, Ned Simmons Jr., and Ned Simmons Sr. Leisa Theriot provided capable technical support in the preparation of this manuscript. We thank Lee Caubarreaux and James Manning of the Louisiana Department of Wildlife and Fisheries for administrative support with this project.

References

Adams, S. E., M. H. Smith, and R. Baccus. 1980. Biochemical variation in the American alligator. *Herpetologica* 36:289–296.

Audubon, J. J. 1827. Observations on the natural history of the alligator (in a letter to Sir William Jardine, Baronet and Prideaux John Selby, Esq.). *The Edinburgh New Philosophical Journal* (n.s.) 2:270–280.

Beek, J. L., and K. J. Meyer-Arendt. 1982. *Louisiana's eroding coastline: Recommendations for protection.* Baton Rouge, Louisiana: Louisiana Department of Natural Resources, Division of Coastal Management.

Brannan, D., K. Roberts, and W. Keithly. 1991. *Louisiana alligator farming: 1990 economic impact.* Baton Rouge, Louisiana: Louisiana Sea Grant College Program and Louisiana Department of Wildlife and Fisheries.

Carbonneau, D. A., and R. H. Chabreck. 1990. Population size, composition, and recruitment of American alligators in freshwater marsh. In *Crocodiles. Proceedings of the 10th Working Meeting of the Crocodile Specialist Group Meeting,* vol. 1, 32–40. Gland, Switzerland: World Conservation Union.

Chabreck, R. H. 1960. Coastal marsh impoundments for ducks in Louisiana. *Proceedings of the Annual Conference of the Southeastern Association of Game and Fish Commissioners* 14:24–29.

Chabreck, R. H. 1965. The movement of alligators in Louisiana. *Proceedings of the Annual Conference of the Southeastern Association of Game and Fish Commissioners* 19:102–110.

Chabreck, R. H. 1966. Methods of determining the size and composition of alligator populations in Louisiana. *Proceedings of the Annual Conference of the Southeastern Association of Game and Fish Commissioners* 20:105–112.

Chabreck, R. H. 1970. Marsh zones and vegetative types in the Louisiana coastal marshes. Doctoral dissertation, Louisiana State University, Baton Rouge.

Chabreck, R. H. 1972. The foods and feeding habits of alligators from fresh and saline environments in Louisiana. *Proceedings of the Annual Conference of the Southeastern Association of Game and Fish Commissioners* 25:117–124.

Chabreck, R. H., T. Joanen, and A. W. Palmisano. 1968. *Vegetative type map of the Louisiana coastal marshes.* New Orleans: Louisiana Wildlife and Fisheries Commission.

Chabreck, R. H., and R. G. Linscombe. 1978. *Vegetative type map of the Louisiana coastal marshes.* New Orleans, Louisiana: Louisiana Department of Wildlife and Fisheries.

Chabreck, R. H., T. Joanen, and S. L. Paulus. 1989. Southern coastal marshes and lakes. In *Habitat management for migrating and wintering waterfowl in North America,* ed. L. M. Smith, R. L. Pederson, and R. M. Kaminski, 249–277. Lubbock, Texas: Texas Technical University Press.

Child, G. 1987. The management of crocodiles in Zimbabwe. In *Wildlife management: Crocodiles and alligators,* ed. G.J.W. Webb, S. C. Manolis, and P. J. Whitehead, 49–62. Chipping Norton, New South Wales, Australia: Surrey, Beatty and Sons.

Clepper, H. 1966. *Origins of American conservation.* New York: Ronald Press.

Craig, N. J., L. M. Smith, N. M. Gilmore, G. D. Lester, and A. Williams. 1987. *The natural communities of coastal Louisiana: Classification and description*. Baton Rouge, Louisiana: Louisiana Natural Heritage Program, Louisiana Department of Wildlife and Fisheries.

Craighead, F. C., Sr. 1968. The role of the alligator in shaping plant communities and maintaining wildlife in the southern Everglades. *Florida Naturalist* 41:2–7, 69–74, 94.

Dessauer, H. C., and L. D. Densmore. 1983. Biochemical genetics. In *Alligator metabolism: Studies of chemical reactions in vivo*, ed. R. A. Coulson and T. Hernandez, 6–13. New York: Pergamon Press.

Dunson, W. A., and F. J. Mazzotti. 1988. Some aspects of water and sodium exchange of freshwater crocodilians in fresh water and sea-water: Role of the integument. *Comparative Biochemistry and Physiology* 90A(3): 391–396.

Durland, D. 1990. The trade in other exotics: Their effect on the crocodilian market, ostrich and elephant trade in the USA. In *Crocodiles, Proceedings of the 10th Working Meeting of the IUCN Crocodile Specialist Group Meeting*, vol. 2, 334–336. Gland, Switzerland: World Conservation Union.

Elsey, R. M., T. Joanen, and L. McNease. 1992a. Growth rates and body condition factors of alligators in coastal Louisiana wetlands: A comparison of wild and farm-released juveniles. *Comparative Biochemistry and Physiology* 103A(4):667–672.

Elsey, R. M., T. Joanen, L. McNease, and N. Kinler. 1992b. Feeding habits of juvenile alligators on Marsh Island Wildlife Refuge: A comparison of wild and farm-released alligators. In *Crocodiles, Proceedings of the 11th Working Meeting of the Crocodile Specialist Group of the Species Survival Commission of the IUCN*, vol. 1, 96. Gland, Switzerland: World Conservation Union.

Ensminger, A. B., and L. G. Nichols. 1958. Hurricane damage to Rockefeller Refuge. *Proceedings of the Annual Conference of the Southeastern Association of Game and Fish Commissioners* 11:52–56.

Ensminger, A. B., and G. Linscombe. 1980. *The fur animals, the alligator, and the fur industry in Louisiana.* Wildlife Education Bulletin No. 109. Baton Rouge, Louisiana: Louisiana Wildlife and Fisheries Commission.

Fleming, D. M., A. W. Palmisano, and T. Joanen. 1976. Food habits of coastal marsh raccoons with observations of alligator nest predation. *Proceedings of the Annual Conference of the Southeastern Association of Game and Fish Commissioners* 30:348–357.

Giles, L. W., and V. L. Childs. 1949. Alligator management of the Sabine National Wildlife Refuge. *Journal of Wildlife Management* 13:16–28.

Glasgow, V. L. 1991. *A social history of the American alligator.* New York: St. Martin's Press.

Gorzula, S. 1987. The management of crocodilians in Venezuela. In *Wildlife management: Crocodiles and alligators,* ed. G.J.W. Webb, S. C. Manolis, and P. J. Whitehead, 91–101. Chipping Norton, New South Wales, Australia: Surrey, Beatty and Sons.

Hines, T. C. 1979. The past and present status of the alligator in Florida. *Proceedings of the Annual Conference of the Southeastern Association of Fish and Wildlife Agencies* 33:224–232.

Hines, T. C. 1990. An updated report on alligator management and value-added conservation in Florida. In *Crocodiles. Proceedings of the 10th Working Meeting of the Crocodile Specialist Group,* vol. 1, 186–199. Gland, Switzerland: World Conservation Union.

Hines, T. C., and C. L. Abercrombie III. 1987. The management of alligators in Florida, USA. In *Wildlife management: Crocodiles and alligators,* ed. G.J.W. Webb, S. C. Manolis, and P. J. Whitehead, 43–47. Chipping Norton, New South Wales, Australia: Surrey, Beatty and Sons.

Hines, T. C., M. J. Fogarty, and L. C. Chappell. 1968. Alligator research in Florida: A progress report. *Proceedings of the Annual Conference of the Southeastern Association of Game and Fish Commissioners* 22:166–180.

Hollands, M. 1987. The management of crocodiles in Papua New Guinea. In *Wildlife management: Crocodiles and alligators,* ed. G.J.W. Webb, S. C. Manolis, and P. J. Whitehead, 73–89. Chipping Norton, New South Wales, Australia: Surrey, Beatty and Sons.

IUCN/UNEP/WWF. 1991. *Caring for the earth: A strategy of sustainable living.* Gland, Switzerland: IUCN/UNEP/WWF.

Ischii, N. 1990. Global market perspectives: Japanese market implications of species and size class changes in the crocodilian trade. In *Crocodiles. Proceedings of the 10th Working Meeting of the Crocodile Specialist Group,* vol. 2, 322–326. Gland, Switzerland: World Conservation Union.

Jenkins, R.W.G. 1987. The world conservation strategy and CITES: Principles for the management of crocodilians. In *Wildlife management: Crocodiles and alligators,* ed. G.J.W. Webb, S. C. Manolis, and P. J. Whitehead, 27–31. Chipping Norton, New South Wales, Australia: Surrey, Beatty and Sons.

Joanen, T. 1969. Nesting ecology of alligators in Louisiana. *Proceedings of the Annual Conference of the Southeastern Association of Fish and Wildlife Agencies* 23:141–151.

Joanen, T. 1982. *Rockefeller Refuge: Haven for wildlife.* Wildlife Education Bulletin No. 105. Baton Rouge, Louisiana: Louisiana Department of Wildlife and Fisheries.

Joanen, T., and L. McNease. 1970. A telemetric study of nesting female alligators on Rockefeller Refuge, Louisiana. *Proceedings of the Annual*

Conference of the Southeastern Association of Fish and Wildlife Agencies 24:175–193.

Joanen, T., and L. McNease. 1972a. Population distribution of alligators with special reference to the Louisiana coastal marsh zones. *Proceedings of the Symposium of the American Alligator Council,* Lake Charles, Louisiana. Mimeograph.

Joanen, T., and L. McNease. 1972b. A telemetric study of adult male alligators on Rockefeller Refuge, Louisiana. *Proceedings of the Annual Conference of the Southeastern Association of Fish and Wildlife Agencies* 26:175–193.

Joanen, T., and L. McNease. 1973a. Developments in alligator research in Louisiana since 1968. *Proceedings of the Symposium of the American Alligator Council,* Winter Park, Florida. Mimeograph.

Joanen, T., and L. McNease. 1973b. *Population estimates and recommendations for experimental harvest program on alligators for Cameron and Vermilion parishes, Louisiana.* Report to the Louisiana Department of Wildlife and Fisheries, New Orleans.

Joanen, T., and L. McNease. 1974. *A summary of population surveys of the American alligator in the Louisiana coastal marshes, 1970–1974.* Report to the Louisiana Department of Wildlife and Fisheries, Grand Chenier.

Joanen, T., and L. McNease. 1975. Louisiana's alligator research program. *Proceedings of the 68th National Audubon Convention,* New Orleans, Louisiana. Mimeograph.

Joanen, T., and L. McNease. 1976. Notes on the reproductive biology and captive biology of the American alligator. *Proceedings of the Annual Conference of the Southeastern Association of Game and Fish Commissioners* 29:407–415.

Joanen, T., and L. McNease. 1978. Time of nesting for the American alligator. *Proceedings of the IUCN Crocodile Specialist Group Meeting,* Madras, India. Mimeograph.

Joanen, T., and L. McNease. 1979. Time of egg deposition for the American alligator. *Proceedings of the Annual Conference of the Southeastern Association of Fish and Wildlife Agencies* 33:15–19.

Joanen, T., and L. McNease. 1980. Reproductive biology of the American alligator in southwest Louisiana. In *Reproductive biology and diseases of captive reptiles,* ed. J. B. Murphy and J. R. Collins, SSAR Contribution to Herpetology (1) 153–159. Lawrence, Kansas: Society for the Study of Amphibians and Reptiles.

Joanen, T., and L. McNease. 1981. Management of the alligator as a renewable natural resource in Louisiana. Georgia Department of Natural Resources Technical Bulletin 5:62–72.

Joanen, T., and L. McNease. 1987a. The management of alligators in Louisiana, USA. In *Wildlife management: Crocodiles and alligators,* ed. G.J.W. Webb, S. C. Manolis, and P. J. Whitehead, 33–42. Chipping Norton, New South Wales, Australia: Surrey, Beatty and Sons.

Joanen, T., and L. McNease. 1987b. Alligator farming research in Louisiana, USA. In *Wildlife management: Crocodiles and alligators,* ed. G.J.W. Webb, S. C. Manolis, and P. J. Whitehead, 329–340. Chipping Norton, New South Wales, Australia: Surrey, Beatty and Sons.

Joanen, T., and L. McNease. 1989. Ecology and physiology of nesting and early development of the American alligator. *American Zoologist* 29: 987–998.

Joanen, T., and L. McNease. 1991. The development of the American alligator industry. In *Proceedings Intensive Tropical Animal Production Seminar,* Townsville, Australia, 208–215.

Joanen, T., L. McNease, and G. Perry. 1976. Effects of simulated flooding on alligator eggs. *Proceedings of the Annual Conference of the Southeastern Association of Fish and Wildlife Agencies* 31:33–35.

Joanen, T., L. McNease, and D. Taylor. 1981. Alligator management plan—State of Louisiana. Unpublished report to the Louisiana Department of Wildlife and Fisheries, New Orleans.

Joanen, T., L. McNease, G. Perry, D. Richard, and D. Taylor. 1983. Louisiana's alligator management program. *Proceedings of the Annual Conference of the Southeastern Association of Fish and Wildlife Agencies* 38: 201–211.

Joanen, T., L. McNease, G. Perry, D. Richard, and T. Hess. 1989. Marsh management policy. Unpublished report to the Louisiana Department of Wildlife and Fisheries, New Orleans.

Johnson, C. C., S. L. Biede, J. C. Close, M. W. Moody, and P. D. Coreil. 1983. Nutrient composition of selected portions of Louisiana alligator. Presented at the 43rd meeting of the Institute of Food Technologists, New Orleans, Louisiana.

Kellogg, R. 1929. *The habits and economic importance of alligators.* Technical Bulletin 147. Washington, D.C.: United States Department of Agriculture.

Kinler, N. W., G. Linscombe, and P. R. Ramsey. 1987. Nutria. In *Wild furbearer management and conservation in North America,* ed. M. Novak, J. A. Baker, M. E. Obbard, and B. Mallach, 326–343. Ontario, Canada: Ministry of Natural Resources.

Kushlan, J. A. 1974. Observations on the role of the American alligator (*Alligator mississippiensis*) in the southern Florida wetlands. *Copeia* 1974: 993–996.

Kushlan, J. A. 1979. Temperature and oxygen in an Everglades alligator pond. *Hydrobiologia* 67:267—271.

Lauren, D. J. 1985. The effect of chronic saline exposure on the electrolyte balance, nitrogen metabolism, and corticosterone titer in the American alligator (*Alligator mississippiensis*). *Comparative Biochemistry and Physiology* 81A:217–223.

LDNR (Louisiana Department of Natural Resources). 1991. *Preliminary assessment of the Louisiana coastal management program.* Baton Rouge, Louisiana: Department of Natural Resources, Division of Coastal Management.

LDNR. 1992. *Coastal wetlands conservation and restoration plan.* Prepared for House and Senate Subcommittees. Baton Rouge, Louisiana: Louisiana Department of Natural Resources, Division of Coastal Restoration.

LDNR. 1993. *Status report for coastal wetlands conservation and restoration program.* Prepared for House and Senate Subcommittees. Baton Rouge, Louisiana: Louisiana Department of Public Works, Division of Coastal Restoration.

Lester, G., ed. 1988. *Plants and animals of special concern in the Louisiana coastal zone.* Louisiana Natural Heritage Program Special Publication no. 2. Baton Rouge, Louisiana: Louisiana Department of Wildlife and Fisheries.

Louisiana Department of Agriculture. 1989. *Louisiana: The source. A world leader in production.* Baton Rouge, Louisiana: Louisiana Department of Agriculture and Forestry, Office of Marketing.

Louisiana Department of Culture, Recreation, and Tourism. 1988. *Louisiana wetlands priority conservation plan.* Baton Rouge, Louisiana: Louisiana Department of Culture, Recreation, and Tourism, Division of Outdoor Recreation.

Louisiana Geographical Survey. 1967. *Effects of ground-water withdrawals on water levels and salt-water encroachment in southwestern Louisiana.* Baton Rouge, Louisiana: Department of Conservation and Department of Public Works.

Luxmore, R. 1990. International crocodilian skin trade: Trends and sources. In *Crocodiles. Proceedings of the 10th Working Meeting of the Crocodile Specialist Group Meeting,* 302–305. Gland Switzerland: The World Conservation Union.

McIlhenny, E. A. 1935. *The alligator's life history.* Boston: Christopher Publishing House.

McNease, L., and T. Joanen. 1974. A telemetric study of immature alligators on Rockefeller Refuge, Louisiana. *Proceedings of the Annual Conference of the Southeastern Association of Game and Fish Commissioners* 28:482–500.

McNease, L., and T. Joanen. 1976. Alligator diets in relation to marsh salinity. *Proceedings of the Annual Conference of the Southeastern Association of Fish and Wildlife Agencies* 31:36–40.

McNease, L., and T. Joanen. 1977. Distribution and relative abundance of the alligator in Louisiana coastal marshes. *Proceedings of the Annual Conference of the Southeastern Association of Fish and Wildlife Agencies* 32:182–186.

Martin, T. W., and A. R. Hight. 1977. Alligator food habits and their relationship to Everglade kite habitat productivity. Unpublished report to the Department of the Interior, United States Fish and Wildlife Service.

Mazzotti, F. J. 1982. Effects of temperature and salinity on growth rates of hatchling crocodiles and alligators: Implications for their distributions in southern Florida. *Bulletin of the Ecological Society of America* 63:83 (Abstract).

Menzies, R. A., J. Kushlan, and H. C. Dessauer. 1979. Low degree of genetic variability in the American alligator (*Alligator mississippiensis*). *Isozyme Bulletin* 12:61.

Messel, H. 1991. Why sustainable use? (A letter to the editor of *Wildlife Conservation Magazine.*) Reprinted in *Proceedings of the Symposium on Conservation and Utilization of Wildlife, Part II, World Environment and Sustainable Use,* Tokyo, Japan, May 31, 1991, 22–30.

Moody, M. W., P. D. Coreil, and T. Joanen. 1988. *Alligators: Harvesting and processing*. Baton Rouge, Louisiana: Louisiana Cooperative Extension Service.

Newsom, J. D., T. Joanen, and R. J. Howard. 1987. *Habitat suitability index models: American alligator.* Biology Report 82(10.136). Washington, D.C.: United States Department of the Interior, Fish and Wildlife Service.

Nichols, J. D., R. H. Chabreck, and W. Conley. 1976a. The use of restocking quotas in crocodilian harvest management. *Transactions of the North American Wildlife and Natural Resources Conference* 41:385–395.

Nichols, J. D., L. Viehman, R. H. Chabreck, and B. Fenderson. 1976b. *Simulation of a commercially harvested alligator population in Louisiana.* Bulletin 691. Baton Rouge, Louisiana: Louisiana State University.

Palmisano, A. W., T. Joanen, and L. McNease. 1973. An analysis of Louisiana's 1972 experimental alligator harvest program. *Proceedings of the Annual Conference of the Southeastern Association of Game and Fish Commissioners* 27:184–206.

Perry, G., T. Joanen, and L. McNease. 1993. Values of wildlife and selected commercial activities associated with a coastal restoration project in southwest Louisiana. *Proceedings of the Louisiana Academy of Sciences* 56:18–25.

Polis, G. A., and C. A Myers. 1985. A survey of intraspecific predation among reptiles and amphibians. *Journal of Herpetology* 19(1):99–107.

Potter, F. E., Jr. 1981. *Status of the American alligator in Texas.* Special report. Austin, Texas: Texas Parks and Wildlife Department.

Rhoads, P. B., and R. Pope. 1968. A comparison of phytoplankton of open Everglades and an alligator hole. *Association of Southeastern Biologists Bulletin* 15:51.

Robinson, J. G. 1993. The limits to caring: Sustainable living and the loss of biodiversity. *Conservation Biology* 7:20–28.

Rootes, W. L. 1989. Behavior of the American alligator in a Louisiana freshwater marsh. Doctoral dissertation, Louisiana State University, Baton Rouge.

Ross, C. A. 1977. Scale variation in the American alligator. *American Society of Ichthyology and Herpetology Program,* 1977 Annual Meeting (Abstract).

Ross, C. A., and C. D. Roberts. 1979. *Scalation of the American alligator.* Special Science Report (Wildlife) 225. Washington, D.C.: United States Department of the Interior, Fish and Wildlife Service.

Ruckel, S. W., and G. W. Steele. 1985. Alligator nesting ecology in two habitats in southern Georgia. *Proceedings of the Annual Conference of the Southeastern Association of Fish and Wildlife Agencies* 38: 212–221.

Schemnitz, S. D. 1974. Populations of bear, panther, alligator, and deer in the Florida Everglades. *Florida Scientist* 37:157–167.

Schortemeyer, J. L. 1972. Destruction of alligator habitat in Florida. *Proceedings of the Symposium on the Status of the American Alligator, 1972.*

Stevenson, O. H. 1904. Utilization of the skins of aquatic animals. *United States Commissioner of Fish and Fisheries Report* 1902:281–352.

Swanson, T. M. 1992. Wildlife and wildlands, diversity and development. In *Economics for the wilds: Wildlife, wildlands, diversity, and development,* ed. T. M. Swanson and E. B. Barbier, 1–14. London: Earthscan.

Taylor, D. 1980. An alligator population model and associated minimum population estimate for nonmarsh alligator habitat. Unpublished report to the Louisiana Department of Wildlife and Fisheries, Baton Rouge, Louisiana.

Taylor, D. 1986. Fall foods of adult alligators from Cypress Lake habitat, Louisiana. *Proceedings of the Annual Conference of the Southeastern Association of Fish and Wildlife Agencies* 40:338–441.

Taylor, D., and W. Neal. 1984. Management implications of the size class frequency distributions in Louisiana alligator populations. *Wildlife Society Bulletin* 12:312–319.

Taylor, D., N. Kinler, and G. Linscombe. 1991. Female alligator reproduction and associated population estimates. *Journal of Wildlife Management* 55:682–688.

Templet, P., and K. Meyer-Arendt. 1986. Louisiana wetland loss and sea level rise: A regional management approach to the problem. Paper presented at the National Wetland Symposium: Mitigation and Impacts of Losses, New Orleans, October.

Thorbjarnarson, J., H. Messel, F. W. King, and J. P. Ross. 1992. *Crocodiles: An action plan for their conservation.* Gland, Switzerland: World Conservation Union.

U.S. News and World Report. 1993. Wildlife's last chance. Should endangered big game be killed in order to save the species? November 15 (19):68–72.

Valentine, J. M., Jr., J. R. Walther, K. M. McCartney, and L. M. Ivy. 1972. Alligator diets on the Sabine National Wildlife Refuge, Louisiana. *Journal of Wildlife Management* 36:809–815.

Waldo, E. 1957. Hurricane damages refuges. *Louisiana Conservationist* 9: 16–17.

Webb, G.J.W., P. J. Whitehead, and S. C. Manolis. 1987a. Crocodile management in the Northern Territory of Australia. In *Wildlife management: Crocodiles and alligators,* ed. G.J.W. Webb, S. C. Manolis, and P. J. Whitehead, 107–124. Chipping Norton, New South Wales, Australia: Surrey, Beatty and Sons.

Webb, G.J.W., S. C. Manolis, and P. J. Whitehead, eds. 1987b. *Wildlife management: Crocodiles and alligators.* Chipping Norton, New South Wales, Australia: Surrey, Beatty and Sons.

Whitaker, R. 1987. The management of crocodilians in India. In *Wildlife management: Crocodiles and alligators,* ed. G.J.W. Webb, S. C. Manolis, and P. J. Whitehead, 63–72. Chipping Norton, New South Wales, Australia: Surrey, Beatty and Sons.

Williams, S. O., III, and R. H. Chabreck. 1986. *Quantity and quality of waterfowl habitat in Louisiana.* Research Report 8. Baton Rouge, Louisiana: Louisiana Agricultural Experiment Station.

Wolfe, J. L., D. K. Bradshaw, and R. H. Chabreck. 1987. Alligator feeding habits: New data and a review. *Northeast Gulf Science* 9:1–8.

Yamanaka, T. 1990. The trade in other exotics: Their effect on the crocodilian market. Lizard, snake, shark. In *Crocodiles. Proceedings of the 10th Working Meeting of the Crocodile Specialist Group Meeting,* vol. 2, 330–331. Gland, Switzerland: World Conservation Union.

FOURTEEN

The Effects of Recreational Waterfowl Hunting on Biodiversity

Implications for Sustainability

Des A. Callaghan, Jeff S. Kirby, and Baz Hughes

The loss of global biodiversity is a major problem with profound repercussions for present and future human generations (Clark 1993). In an attempt to stem the loss of biodiversity outside protected areas, sustainability has been hailed the new paradigm of conservation (Salwasser 1990). This is consistent with the developing concept of "wise use of wetlands" (Holland 1988; Pierce 1988; Ramsar 1990). In response to this, resource management is in the midst of rapid change in an attempt to integrate the goals of biodiversity conservation and resource utilization (Pletscher and Hutto 1991).

Recreational waterfowl hunting is practiced by more than 6 million people around the world, and well in excess of 23 million waterfowl are harvested annually (Table 14.1). This high level of use is often accompanied by intensive management of wetlands to support waterfowl production, and by considerable monitoring and research to improve waterfowl management. Waterfowl hunting and management should therefore provide fertile ground for examining ways in which the long-term consumptive use of wild species may affect biodiversity.

Table 14.1. Estimates of Numbers and Density of Waterfowl Hunters, Trends in Their Numbers, and Annual Harvest of Ducks and Geese within Countries for Which Data Are Available

Country	Number of hunters	Hunters per km^2	Trend in hunter numbers	Annual harvest
Denmark	88,500	2.05	Stable	817,000
Luxembourg	2,000	0.77	?	?
Greece	100,000	0.76	Stable	?
Netherlands	29,700	0.72	?	365,000
France	375,000	0.69	Declining	2,242,000
Ireland	45,960	0.67	Increasing	?
United Kingdom	160,000	0.66	Stable	1,085,000
Finland	190,000	0.56	Increasing	481,500
Liechtenstein	80	0.50	?	?
Italy	148,000	0.49	Declining	?
Germany	151,600	0.42	?	406,000
Czech and Slovak Republics	52,500	0.41	Declining	141,900
Bulgaria	40,000	0.36	Stable	56,000
Sweden	150,000	0.33	Increasing	168,000
Japan	100,000	0.27	Increasing	695,000
Hungary	17,000	0.18	Stable	94,500
Poland	55,000	0.18	Increasing	660,000
New Zealand	39,000	0.15	Declining	1,063,869
United States	1,300,000	0.14	Declining	9,600,000
Switzerland	5,500	0.13	Stable	20,000
Former USSR	2,800,000	0.13	?	3,293,000[a]
Austria	10,000	0.12	Increasing	75,000
Norway	35,000	0.11	?	115,000
Belgium	3,000	0.10	?	46,000
Spain	30,000	0.06	Increasing	229,000
Rumania	10,000	0.04	Declining	80,000
Canada	300,000	0.03	Declining	2,300,000
Portugal	2,100	0.02	?	?
Iceland	600	0.01	?	?
Australia	34,000	>0.01	Declining	?
Senegal	600	>0.01	?	?
Tunisia	500	>0.01	Increasing	?
Svalbard	100	>0.01	?	?
Algeria	0	0.00	—	0
Total	6,275,740			>23,800,000

Sources: Lampio 1983; Harradine 1985; Tamisier 1985; Ueno 1989; Bell and Owen 1990; Landry 1990; Pain 1992b; Buchanan and Barker 1993; Lévesque et al. 1993; Martin and Padding 1993; F. I. Norman, pers. comm. 1994; B. Parkes, pers. comm. 1994; Fawcett 1996; Dobrowolski and Dmowski in press.

[a]Figure relates only to European former Russia.

Two questions are central to this study:

1. What effects are the sustainable offtake, and associated hunting and management practices, having on biodiversity?
2. If biodiversity is suffering any negative impacts from this activity, what measures can be taken to mitigate these effects?

We synthesize current knowledge regarding waterfowl management and biodiversity in an attempt to formulate answers to these questions.

Synthesis

Identifying and Measuring Effects

Maximum biodiversity, or ecological integrity, occurs under pristine ecological conditions, and its loss should be assessed by comparing the status of biotic elements and processes against this benchmark condition (Angermeier 1994; Grumbine 1994). This forms a workable concept for assessing effects on biodiversity not only in natural and seminatural wetlands, but also in artificial wetlands where the goal of biodiversity conservation should be to emulate pristine conditions typical of that region. Given the problems of measuring ecological integrity, the measurement and interpretation of these effects often incur a heavy dose of subjectivity. Before examining these effects, we summarize the concept of sustainable yield as generally practiced in waterfowl management.

Recreational Waterfowl Hunting and Sustainable Yields

The sustainable yield of waterfowl is reliant on natural density-dependent factors, which, in effect, largely compensate for the harvest. These compensatory factors are related to mortality or reproduction (although emigration/immigration may be important in certain situations), but within any single population they can only compensate for hunting mortality up to a threshold level. Above this level, hunting mortality becomes entirely additive and nonsustainable, resulting in a rapid population decline if remedial action is not taken (Anderson and Burnham 1976; Nichols and Hines 1983; Burnham and Anderson 1984; Burnham et al. 1984; Nichols et al. 1984; Caughley 1985; Conroy and Krementz 1990; Trost et al. 1993). Threshold levels, or maximum sustainable yields, vary considerably according to species and time, owing to variations in the size, sex, and age structure of a population, and the quantity and quality of habitat available (Anderson and Burnham 1976; Conroy and Krementz 1990). Because of these complexities, it has not been possible to identify accurately the threshold level for any particular waterfowl population. Nonetheless, it has often been shown that certain mallard (*Anas platyrhynchos*) populations can sustain yields of more than 20 percent (i.e., more than 20 percent of a population alive at the start of year *i* can be harvested in year *i*), while on the other hand populations of the canvasback (*Aythya valisineria*) are often unable to sustain even very low yields (Anderson 1975; Buffington and MacLauchlan 1993).

Populations of waterfowl are among the best monitored of all wild organisms, particularly in areas with the highest densities of recreational waterfowl hunters (e.g., North America and western Europe). Yields of waterfowl are frequently adjusted according to the dynamics of the target populations, particularly in North America, and so overhunting rarely occurs for any significant length of time (cf., Boyd 1983; Trouvilliez et al. in press). Harvests are regulated by, for example, imposing limits on (1) the species that can be hunted; (2) those days that hunting may occur within each year (hunting seasons); (3) the number of birds of a species, sex, or group that may be taken each day by one hunter (bag limits); (4) those times during the day/night when birds may be hunted (hunting hours); and (5) restrictions on weapons (after Lampio 1983; USFWS 1988; Lecocq 1993).

Waterfowl Introduction and Stocking

Population Introduction: Hybridization and Competition

Species of waterfowl are often introduced to areas outside their natural range for recreational hunting purposes, but the Canada goose (*Branta canadensis*) and the mallard are the only ones that have been successfully established in significant numbers (Table 14.2). A proportion of Canada geese introduced and now breeding in Scandinavia winter in southern Sweden, Denmark, Germany, and the Netherlands (Owen and Salmon 1988a). These birds will probably soon spread and establish breeding populations in countries such as Denmark, the Commonwealth of Independent States, the Baltic Republics, and Poland (Madsen and Andersson 1990). Canada geese probably have the potential to hybridize introgressively with most native, congeneric wild

Table 14.2. Established Breeding Populations of the Canada Goose (*Branta canadensis*) and Mallard (*Anas platyrhynchos*) in Countries outside Their Native Range Where They Were Introduced Primarily for Hunting Purposes

Species population	Country	First releases	Last releases	Current population size	Trend
Canada goose	Russia	1987	Ongoing	>300	Increasing
	Britain	1665[a]	1960s	64,000	Increasing
	Norway	1963[b]	Ongoing	10,000	Increasing
	Sweden	1933	1950s	30–50,000	Increasing
	Finland	1960	Ongoing	2–3,000	Increasing
	New Zealand	1905[c]	1923	35–40,000	Increasing
Mallard	Australia	1862	1970s	?	Increasing
	New Zealand	1867	1950s	>5,000,000	Increasing

Sources: Weller 1969; Imber 1971; Fabricius 1983; Vikberg and Moilanen 1985; Heggberget 1987; Gabuzov 1990; Kear 1990; Madsen and Andersson 1990; Marchant and Higgins 1990; Nichols et al. 1990; Caithness et al. 1991; Delany 1993; A. D. Fox, pers. comm. 1994; M. Williams, pers. comm. 1994; Allan et al. 1996.

[a]First releases were for ornamental purposes, and the birds were restricted to a few areas until they were translocated to many new sites for hunting during the 1950s and 1960s.

[b]Unsuccessful releases occurred in the late 1930s.

[c]Unsuccessful releases occurred in the 1870s.

geese (Kear 1990). However, the breeding grounds of introduced Canada geese and native *Branta* geese (confined to the northern hemisphere) are separated by considerable distances, and native birds only spend the winter period within the range of feral Canada geese. There is evidence, however, for interspecific aggression between feral Canada geese and other native waterfowl during the breeding season. For example, in Britain there is circumstantial evidence that the Canada goose competes for food with the mute swan (*Cygnus olor*) during the nonbreeding season, but that there is no competition during the breeding season (seemingly due to the territorial and aggressive nature of the mute swan during this period) (Owen et al. in press). Fabricius et al. (1974) documented considerable interspecific aggression between Canada geese and native greylag geese (*Anser anser*) when nesting together on islands off the Swedish coast, but found no evidence of negative effects on the numbers of breeding pairs of either species. However, the study was performed when both species were still colonizing the study area, and it is possible that the breeding numbers of either species will be negatively influenced by the presence of the other once the carrying capacity of the area is approached. Master and Oplinger (1984) suggested that mallard productivity may be negatively affected by increasing nesting densities of Canada geese, while Giroux (1981) proposed that the nesting success of mallard, gadwall (*Anas strepera*), and lesser scaup (*Aythya affinis*) may increase because of the protection from predators afforded by neighboring Canada geese. In New Zealand, Canada geese may be competitively excluding paradise shelduck (*Tadorna variegata*) from some feeding and breeding areas, but this remains to be investigated scientifically (M. Williams, pers. comm. 1995). At least in some areas then, it would appear that introduced Canada geese are causing alterations to the structure of waterbird communities, which of course may have broader implications for ecosystem structure, but this remains undocumented.

Extensive release programs for mallards in New Zealand, coordinated by hunting organizations, have resulted in the species now being widespread and abundant on artificial and natural wetlands (Fig. 14.1). Widespread introgressive hybridization has occurred between this species and the Pacific black duck (*Anas superciliosa*), although the magnitude of introgression is confused, and perhaps underestimated, because hybrids are often difficult to recognize (Williams and Roderick 1973; Haddon 1984; Gillespie 1985; Williams 1994). Nonetheless, it has been estimated that about 15–20 percent of the current total mallard/Pacific black duck population in New Zealand consists of pure Pacific black duck genotypes (M. Williams, pers. comm. 1995), compared to an estimate of 95 percent in 1960 (Green 1992). The mallard is now the dominant waterbird in the wetlands of the agricultural environment of mainland New Zealand, a niche that would have probably been exploited by the Pacific black duck in the absence of competition from mallards (Gillespie 1985; M. Williams, pers. comm. 1994). In addition, it seems that dispersal of mallards from New Zealand has led to their establishment on the islands of Campbell, Auckland, Chatham, Lord Howe, Norfolk, and Macquarie, where hybridization with Pacific black ducks has, or is likely to have, occurred (Norman 1987, 1990; Marchant and Higgins 1990; Williams 1994). It seems inevitable that introgressive hybridization will continue between these species in New Zealand until Pacific black ducks are either extinct or confined to largely unmodified, isolated, wooded catchments or islands.

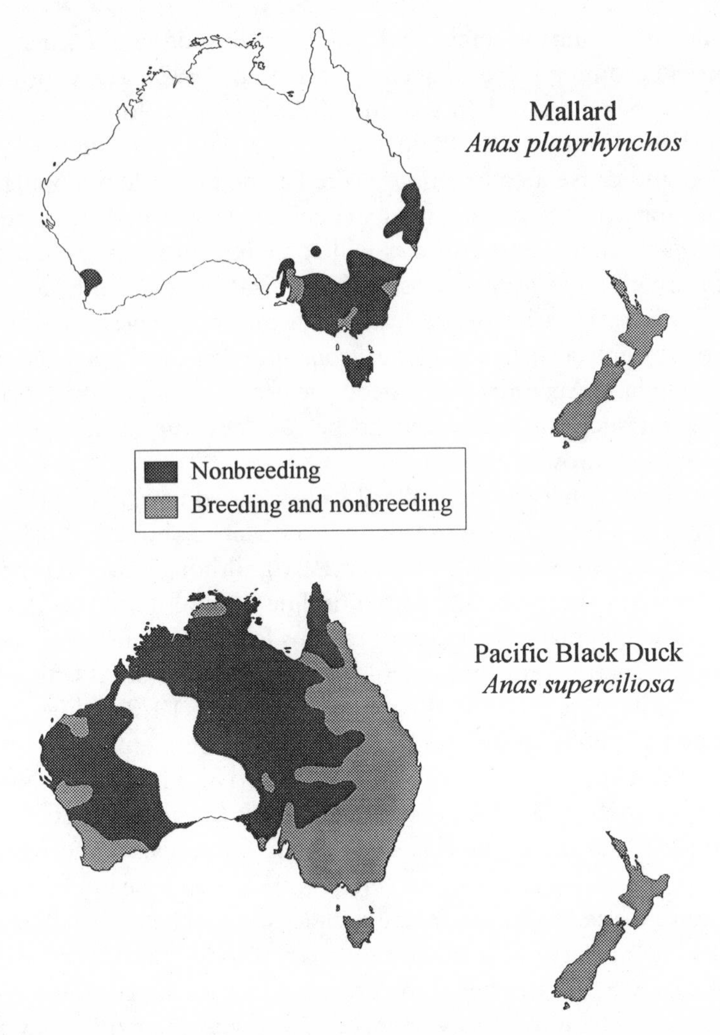

Figure 14.1. Distribution of mallard (*Anas platyrhynchos*) and Pacific black duck (*Anas superciliosa*) in Australia and New Zealand. (*Source:* Marchant and Higgins 1990.)

In Australia, the release of mallards for hunting has not occurred to the same magnitude as it has in New Zealand, and, partly as a consequence, the species has a limited distribution. It is almost totally confined to wetlands in urban or intensely farmed districts, being only occasionally recorded on natural wetlands, mainly in the southeast (Braithwaite and Miller 1975; Marchant and Higgins 1990; F. I. Norman, pers. comm. 1994) (Fig. 14.1). Nonetheless, the species is slowly expanding its range in some areas, and hybridization with Pacific black ducks is becoming more frequent (Paton et al. 1992). The continuing modification of Australian wetlands (through agricultural expansion and urbanization) will facilitate further expansion of the mallard population and cause a contraction of the Pacific black duck population (as the former is better adapted to such environments). It therefore seems likely that the two species will come into increasing contact in the future, thus elevating the risk of hybridization (cf. Braithwaite and Miller 1975). Hybrids between mallards and both chestnut teal and Australian grey teal (*Anas gibberifrons gracilis*) have been reported from mainland Australia and Macquarie Island, respectively (Norman 1987), but such instances seem to be rare and the fertility of their progeny is unknown.

In North America, the mallard has extended its range noticeably east and, to a lesser extent, north and south since about 1900 (Johnsgard and Di Silvestro 1976; Palmer 1976), although over recent decades this has corresponded with a decline in total numbers (USFWS 1994). The two main (interrelated) reasons for this expansion are large-scale introductions for hunting and the species' very successful adaption to an increase in artificial habitats (Studholme 1972; Palmer 1976; Dennis et al. 1984; Rogers and Patterson 1984; Heusmann 1974, 1988). This range expansion has brought the mallard into increasing contact with the closely related American black duck (*Anas rubripes*), Mexican duck (*Anas diazi*), Florida duck (*Anas f. fulvigula*), and mottled duck (*Anas fulvigula maculosa*) (Fig. 14.2). The American black duck has been experiencing a substantial decline, and numbers have decreased by about 60 percent over the last 40 years (Brodsky and Weatherhead 1984; USFWS 1994). One of the principal factors responsible for this is thought to be introgressive hybridization with the expanding mallard population (Johnsgard 1967; Heusmann 1974; Grandy 1983; Brodsky and Weatherhead 1984; Ankney et al. 1987, 1989; Rusch et al. 1989; Seymour 1990; Merendino et al. 1993). No doubt this problem has been compounded by the release of hybrid stock in at least two regions of the United States (see Palmer 1976). Also, competitive exclusion from nutrient-rich wetlands by mallards is thought to have contributed to the decline of the American black duck (Merendino and Ankney 1994).

The southward spread of the mallard in North America has already caused the extinction of the pure Mexican duck genotype in the

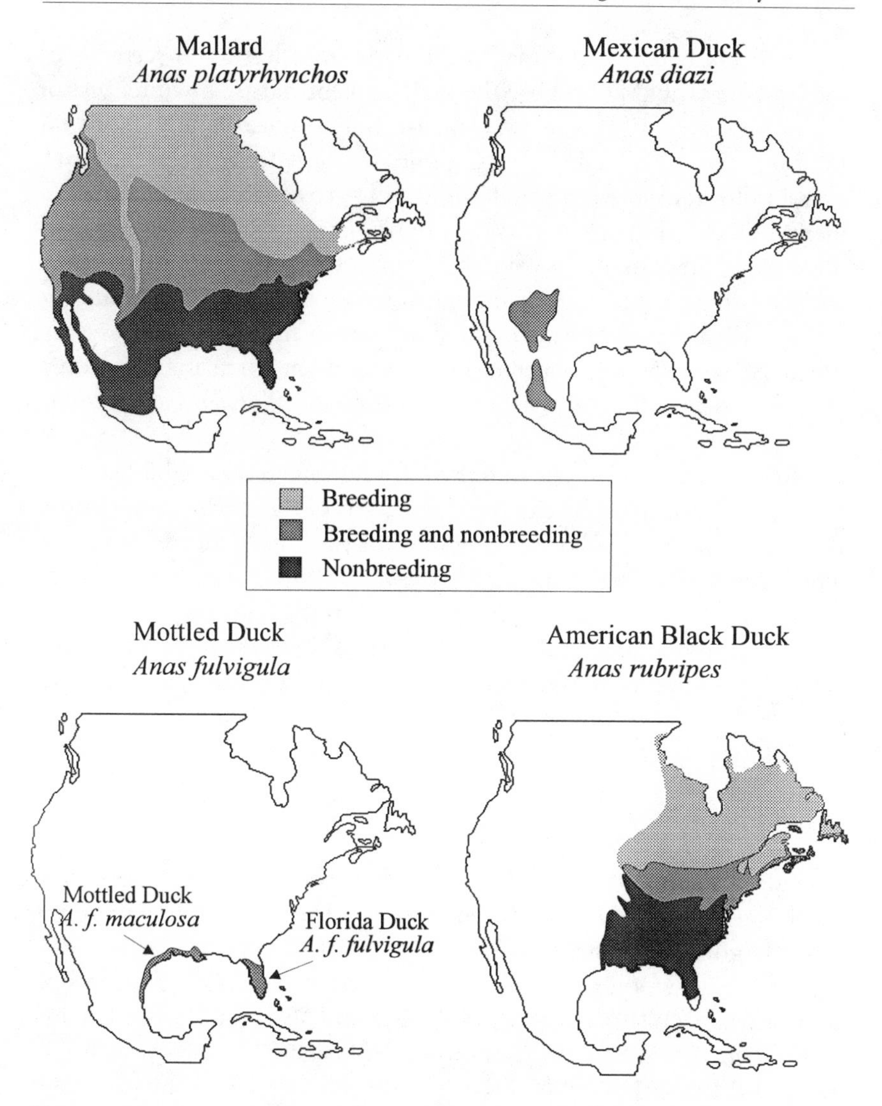

Figure 14.2. Distribution of the mallard-complex in North America. (*Source:* Palmer 1976.)

United States (originally numbering about five thousand individuals) owing to hybridization ("Mexican Duck" 1978; Greig 1980; Callaghan and Green 1993). In Mexico, hybridization has probably occurred in the north (northeast Sonora and north Chihuahua), but apparently pure Mexican ducks are still common in central and south Chihuahua (S. Howell, pers. comm. 1994).

Hybrids between the mallard and the mottled duck seem to be infrequent, probably because the mallard is primarily a winter visitor within the range of this taxon (Fig. 14.2), and releases of mallard for hunting seem to have been small-scale events within this area. However, in Florida there is a small but rapidly growing population of resident mallards that are increasingly hybridizing with the resident Florida ducks. These mallards originate from releases for both ornamental and hunting purposes, even though the former is illegal. Despite the serious threat to the Florida duck, there are several hunting reserves in northern Florida that continue to release game-farm mallards on a regular basis (Hill 1994; Mazourek and Gray 1994; D. R. Eggeman, pers. comm. 1995).

In summary, releases of waterfowl for hunting have established populations of introduced species over large parts of the world, and continue to do so in parts of Europe. These introductions have inevitably caused alterations to the structure of native waterbird communities, but the broader effects that this has had on ecosystem structure remain undocumented. Introductions of the mallard have caused substantial erosion of the genetic integrity of five waterfowl taxa, and the survival of at least three of these seems unlikely if present trends continue.

Population Stocking: Autecology and Genetic Integrity

A wide variety of waterfowl species have been used in substantial stocking programs (i.e., > 500 individuals) for hunting, including the Canada goose, greylag goose, mallard, gadwall, pintail (*Anas acuta*), chestnut teal, wood duck (*Aix sponsa*), redhead (*Aythya americana*), and canvasback (Brakhage 1953; Weller and Ward 1959; Borden and Hochbaum 1966; Fog 1971; Norman 1971; Bishop and Howing 1972; Doty and Kruse 1972; Studholme 1972; Sellers 1973; Johnson 1983; Lee et al. 1984; Harradine 1985; Soutiere 1989; Batt and Nelson 1990; Urbánek, in press). By far the most commonly used species is the mallard, particularly in North America and western Europe. There are no reliable regionwide or nationwide totals for the numbers of mallards stocked per annum in any area, although rough estimates include 100,000 in Maryland (USA) (Soutiere 1989), 200,000 in the Czech Republic (Urbánek, in press), 300,000 in Denmark (Madsen 1995), and 500,000 in the United Kingdom (Harradine 1985). Some concern has arisen because the breeding stock of released mallards are invariably of game-farm origin (i.e., having been under captivity for a number of generations), and differences in morphology, physiology, behavior, and ecology suggest that these birds may be genetically different from those in the wild (Cheng et al. 1978, 1979; Heusmann 1981; Figley and VanDruff 1982; Byers and Cary 1991). As such, the genetic integrity of wild populations may be at risk from hybridization with their farmed congeners, this being a frequent concern among other groups of organisms (e.g., Blanco et al. 1992; Tulgat and Schaller 1992; Balharry et al. 1994; Cowx 1994). Banks (1972) argued that introgressive hybridization between the two mallard types has occurred to such an extent in North America that the conservation of the genetic integrity of wild mallards is largely rhetoric, although Cheng et al. (1978, 1979) showed that mate preference is a potential barrier to their panmixia (i.e., mallards of a particular strain preferentially mated with birds of the same strain). There is some indication that a hybrid genotype has evolved in the eastern United States, which may be particularly successful in the urban environment (Shoffner 1972), and it could be that the resident, "urban" mallard populations of North America and Europe are largely derived from released birds. Indeed, many game-farm birds released in more natural habitats soon move to urban environments (Heusmann 1988).

In North America, the use of "pure" wild-strain mallards (less than two generations removed from the wild) in release programs has been promoted and practiced in an attempt to increase wild breeding populations. However, relatively few birds survive for more than two weeks following release, and numbers soon return to prerelease levels (Brakhage 1953; Fog 1971; Schladweiler and Tester 1972; Soutiere 1989). It is now generally considered that wild populations are limited by habitat availability/quality rather than the availability of breeding birds (Batt and Nelson 1990).

There seems to be little doubt that the stocking of game-farm mallards for hunting has had a profound effect on the autecology of this species, although this has been augmented by accidental escapes and releases for esthetic purposes. In particular, many resident populations have been established, which may be particularly damaging to biodiversity in areas where the species was formerly only a seasonal visitor (see the following two sections).

Disease

Infectious diseases can have a profound effect on the distribution and abundance of their host populations, and generally account for 20–40 percent of all deaths in birds (Dobson and May 1991; May 1994). Outbreaks of duck virus enteritis (DVE), commonly known as duck plague, in wild birds have almost exclusively followed contact with captive-reared or feral waterfowl in North America and Eurasia (Brand 1987; Brand and Docherty 1988). In Great Britain, the sporadic outbreaks in wild waterfowl have always followed contact with captive or released waterfowl (Gough 1984; Gough and Alexander 1990; Mitchell et al. 1993). Birds that are infected with or have survived exposure to infectious diseases such as tuberculosis, salmonellosis, and DVE can act as vectors of such diseases for a number of years (Burgess et al. 1979; Burgess and Yuill 1983; M. J. Brown, pers. comm. 1995), and so it is conceivable that the release of birds for hunting may promote the incidence of such diseases in wild populations. Waterfowl released for hunting are neither screened nor inoculated prior to release, and so it is impossible to determine their role as disease vectors.

The release of waterfowl can greatly increase waterfowl usage of a wetland, at least in the short term and often in the long term. Concentrated waterfowl use can result in the rapid spread of infectious diseases (e.g., avian cholera and botulism), since such diseases are transmitted more efficiently when host populations are relatively dense. This can result in high mortality among susceptible species (Holmes 1982; USFWS 1993), which has not only a serious effect on the infected population but also on other associated species (e.g., predators, mutualists, competitors, and parasites) (Ratcliffe 1979; Osbourne 1982; Dobson and May 1986), and consequently ecosystem structure and function.

Although there is little conclusive evidence of any relationships between waterfowl releases and infectious diseases, this is probably due to a lack of research and the tremendous difficulties associated with observing the incidence and effects of disease in wild populations.

Nutrient Dynamics and Energy Flow

Aquatic birds have a relatively high metabolic rate, and consequently they significantly influence nutrient dynamics and energy flow in wetlands (Gere and Andrikovics 1992, 1994). Data provided by Marion et al. (1994) suggest that waterbirds account for up to 10 percent of all nitrogen (N) and 56 percent of all phosphorus (P) entering Lake Grand-Lieu (France) from external sources when inputs from sewage effluent and agricultural runoff are excluded, while Manny et al. (1994) estimated that waterfowl were responsible for 69 percent of all carbon (C), 27 percent of all N, and 70 percent of all P entering Wintergreen Lake (USA) from external sources. At Lake Esrom (Denmark), Woollhead (1994) showed that waterbirds consumed 20 percent of the annual macrophyte and 4 percent of the annual zoobenthos production, while Kiørboe (1980) calculated that waterfowl consumed about 30 percent of the annual macrophyte production at Tipper Grund (Denmark). In the southern Baltic Sea, Nilsson (1980) estimated that diving ducks consumed about 26 percent of the annual production of their food animals (mainly the mollusca [*Mytilus edulis* and *Macoma baltica*]). These studies show that waterbirds often have a significant influence on nutrient dynamics and energy flow in wetlands, which is fundamentally important to considerations of biodiversity in relation to waterfowl hunting.

Unfortunately, the influence of waterfowl releases with regard to these factors has never been assessed. Nevertheless, the information presented above strongly suggests that excessive releases of waterfowl, and in particular the establishment of feral and introduced populations, can disrupt these functions and seriously erode biodiversity.

Hunting Pressure

It has been suggested that waterfowl stocking helps to reduce the hunting pressure on wild populations (Swift and Laws 1982). Apart from there being no evidence for this, such a statement is only valid when hunting pressure is excessive. In such situations, the implementation of regulations to establish sustainable yields is obviously the priority action required, rather than the release of captive-reared birds.

Lead Poisoning and Relevant Management Practices

Lead is a highly toxic heavy metal, whose potential toxicity does not diminish over time because it is resistant to oxidation and disintegration (Thomas 1994). Birds are exposed to lead via many sources (e.g., lead in gasoline and fishing weights), but the vast majority of lead

Table 14.3. Frequency (Percent) of Lead Shot in Gizzards of Hunter-Killed Grazing, Dabbling, and Diving Waterfowl in Europe and North America

Waterfowl group	Europe[a]	North America
Grazing species	2.2 (n = 1,939)	3.4 (n = 35,480)
Dabbling species	6.0 (n = 12,564)	7.7 (n = 169,940)
Diving species	23.6 (n = 1,134)	16.7 (n = 22,238)

Sources: Sanderson and Bellrose 1986; Anderson and Havera 1989; Butler 1990; Pain and Handrinos 1990; Pain 1991a; Pain et al. 1992; Schricke in press.
[a]Data from Denmark, Finland, France, Greece, Ireland, the Netherlands, Norway, Sweden, Switzerland, and the United Kingdom.

poisoning cases in avian species arises from birds having ingested spent gunshot pellets (Pain 1992a). About 9,000 metric tons (MT) of lead shot are added to wetlands in western Europe and North America each year by waterfowl hunters, and this frequently results in lead shot concentrations in wetland sediments in excess of one million pellets per hectare (ha) (Amiard-Triquet et al. 1992; Thomas 1994). Lead poisoning in waterfowl has been recognized for more than a century, and has so far been recorded in 27 countries, where in some areas it is the primary cause of nonhunting mortality (Grinnell 1894; Humburg and Babcock 1982; Pain 1992b; Ákoshegyi 1994; Thomas 1994).

The presence of shot in the gizzards of hunter-killed waterfowl has long been used as an index for qualifying the tendency of a species to ingest shot (Pain et al. 1992). It is well documented that diving species have the highest incidence of shot ingestion, followed by dabbling species, with shot ingested far less frequently by grazing species (Table 14.3). Most dabbling species feed on plants and invertebrates in the water column or on the surface of the lake bottom, while most diving species penetrate the substrate by up to 15 centimeters (cm) depth in search of food. Thus dabbling species are generally less likely to ingest shot that has settled into the substrate, and grazing species even less so. However, there is considerable interspecific variation within these groups (Fig. 14.3) owing to the diversity of feeding habits.

Following ingestion, the toxicity of lead is influenced by many variables, although most important is the type of food consumed. A diet high in protein can considerably suppress the toxic symptoms of lead, while calcium and phosphorus can also reduce toxicosis (Jordan and

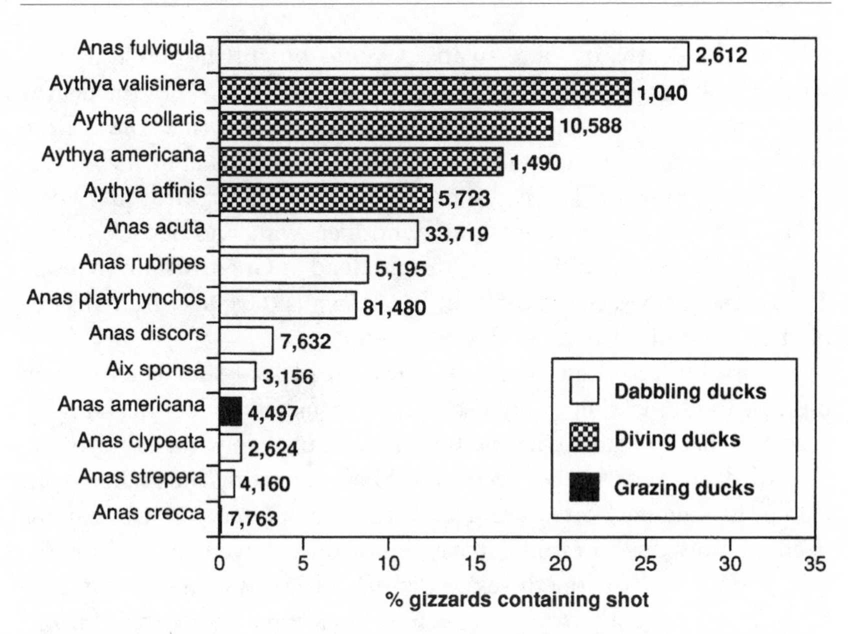

Figure 14.3. Frequency of lead shot in gizzards of hunter-killed game ducks in the United States, 1973–84. Data labels refer to sample sizes. (*Source:* Sanderson and Bellrose 1986.)

Bellrose 1951; Godin 1967; Koranda et al. 1979). Thus, although the mottled duck has the highest rate of shot ingestion of any species in North America (Fig. 14.3), the high level of vegetable protein in its diet probably militates against the catastrophic losses that might otherwise occur (Sanderson and Bellrose 1986). With consideration to variables such as diet, Sanderson and Bellrose (1986) concluded that lead toxicosis in the United States poses the greatest threat to the mallard, followed, in lessening degrees, by the American black duck, mottled duck, pintail (*Anas acuta*), canvasback (*Aythya valisineria*), redhead (*Aythya americana*), and ring-necked duck (*Aythya collaris*), thus emphasizing the fact that shot ingestion cannot be used to qualify lead toxicosis.

Information concerning lead poisoning in swans is rather limited, as samples can rarely be obtained due to their protected status. Nonetheless, results from several studies suggest that lead poisoning can account for a considerable proportion of the total mortality in some swan populations. Necropsies of birds found dead revealed that lead poisoning from spent gunshot ingestion accounted for 22 percent (n =

447) of the deaths of mute swans (*Cygnus olor*) (Clausen and Wolstrup 1979; Spray and Milne 1988), 47 percent (n = 57) of the deaths of whooper swans (*Cygnus cygnus*), and 35 percent (n = 188) of the deaths of trumpeter swans (*Cygnus buccinator*) (Blus et al. 1989; Degernes and Frank 1991). Lead shot was found in the gizzards of 19 percent (n = 57) of swans (mute swan, whooper swan, and Bewick's swan [*Cygnus colombianus bewickii*]) found dead in Great Britain (Mudge 1983), and 12 percent (n = 50) of black swans (*Cygnus atratus*) sampled in Australia (Koh and Harper 1988).

Mortality resulting from the ingestion of lead shot is extremely difficult to estimate, not only because many factors influence a species' susceptibility to lead toxicosis, but also because only a small proportion of dead birds is ever recorded. High waterfowl mortality from lead poisoning may often occur unnoticed by hunters and game managers, mainly due to rapid carcass removal by scavengers and predators (Pain 1991b). Nonetheless, "die-offs" (deaths of more than 100 birds at a time in a particular area) are occasionally reported, for example, in the United States (USFWS 1986a), Australia (Harper and Hindmarsh 1990; Whitehead and Tschirner 1991), and Scotland (Davis 1988). In Missouri (USA), Humburg and Babcock (1982) estimated that 37 percent of snow goose mortality was caused by lead poisoning, while throughout the United States it has been estimated that 2–3 percent of the autumn population of all waterfowl was lost annually as a direct result of lead poisoning (Bellrose 1959). In the former Federal Republic of Germany, about thirty thousand waterfowl are thought to die each year from lead poisoning (Billings 1990).

Many waterfowl shot by hunters are not retrieved. These crippled losses in the United States have been estimated at 19 percent of ducks and 15 percent of geese shot (USFWS 1985). Other studies in the United States have shown that approximately 30 percent of healthy waterfowl carry lead pellets embedded in their flesh (USFWS 1986a). The prevalence of lead shot embedded in the tissues of waterfowl exposes predators and scavengers to lead poisoning. This is a particular concern with avian predators such as raptors (Falconiformes), whose population dynamics are very sensitive to increases in adult mortality because of their small populations and low reproductive rates (Grier 1980). A number of raptor species frequent wetland habitats, and studies indicate that lead toxicosis can be a significant problem. In the United

States, lead poisoning accounted for 5.2–7.6 percent of the total mortality of the bald eagle (*Haliaeetus leucocephalus*) between the mid-1960s and early 1980s (Pattee and Hennes 1983; Reichel et al. 1984), while in British Columbia (Canada) 14 percent of 65 bald eagles found dead had died from lead poisoning and a further 23 percent had been exposed to sublethal doses of lead (Whitehead and Elliot 1993). Of 94 marsh harriers (*Circus aeruginosus*) trapped in the Camargue and Charente-Maritime (France), 14 percent had blood lead concentrations (PbB) indicative of clinical lead poisoning (> 60 micrograms [μg]/deciliter [dl]), while a further 17 percent had elevated PbB concentrations (> 30 μg/dl) (Pain et al. 1993). In parts of their range, golden eagles (*Aquila chrysaetos*) frequently forage over wetlands, and although lead poisoning has often been recorded in this species, data concerning its significance are lacking (Meister and Kösters 1981; USFWS 1986a; Pain and Amiard-Triquet 1993). Lead poisoning has not been recorded in any other raptor species that regularly frequent wetlands, although Pain et al. (1993) considered that the white-tailed eagle (*Haliaeetus albicilla*) may be particularly susceptible. The incidence of lead poisoning among mammalian predators and scavengers of wetlands (e.g., *Canis* spp., *Vulpes* spp., *Mustela* spp., and *Sus scrofa*) has not been documented but undoubtedly occurs.

Lead toxicosis caused by shot ingestion is very poorly understood among waterbirds other than waterfowl and raptors. Information presented in Table 14.4, however, shows that a wide variety of species are susceptible, and suggests that lead toxicosis can be a significant mortality factor for some of them. Large sample sizes are available only for game species, but it is likely that a wide variety of nongame species also frequently ingest lead.

Little is known of the sublethal effects of lead on avian species, although evidence suggests that it can cause a decline in fecundity, an increased susceptibility to hunting and predation, and an increased susceptibility to infectious diseases (Elder 1954; Del Bono et al. 1975; Sanderson et al. 1981; Birkhead 1983; USFWS 1986a).

Similarly, the effect of lead shot on water quality, wetland plants, and invertebrates is poorly understood, although available information suggests that impacts are minor (Behan and Kinraide 1979; USFWS 1986a; Lund et al. 1991). Adverse effects of lead poisoning in fish are well documented (e.g., Johansson-Sjobeck and Larsson 1979; Haux

Table 14.4. Records of Lead Poisoning and Ingested Shot for Waterbirds Other Than Waterfowl (Anatidae) and Raptors (Falconiformes)

Species	Ingestion frequency (%)[a]	Sample size
Black-tailed godwit (*Limosa limosa*)	16.5	91
American coot (*Fulica americana*)	14.9	801
Ruff (*Philomachus pugnax*)	10.6	66
Sora rail (*Porzana carolina*)	10.2	1,497
Dunlin (*Calidris alpina*)	9.3	54
Common snipe (*Gallinago gallinago*)	7.6	263
Common coot (*Fulica atra*)	4.0	720
Common moorhen (*Gallinula chloropus*)	2.5	285
Sandhill crane (*Grus canadensis*)	2.3	704
Clapper rail (*Rallus longirostris*)	1.4	423
King rail (*Rallus elegans*)	1.0	96
Virginia rail (*Rallus limicola*)	0.8	128
Dusky moorhen (*Gallinula tenebrosa*)	0.6	159
Wood sandpiper (*Tringa glareola*)	0.0	34
Greater flamingo (*Phoenicopterus ruber*)	—	ca. 66[b]
American avocet (*Recurvirostra americana*)	—	?
Black-necked stilt (*Himantopus mexicanus*)	—	14
White-faced ibis (*Plegadis chichi*)	—	4
Marbled godwit (*Limosa fedoa*)	—	?
Pectoral sandpiper (*Calidris melanotos*)	—	?
Western sandpiper (*Calidris mauri*)	—	?
Long-billed dowitcher (*Limnodromus scolopaceus*)	—	8
Herring gull (*Larus argentatus*)	—	?
Glaucous-winged gull (*Larus glaucescens*)	—	?

Sources: Jones 1939; Artmann and Martin 1975; Thomas 1975; Norman 1976; Kaiser et al. 1980; Stendell et al. 1980; Mudge 1983; Wallace et al. 1983; Windingstad et al. 1984; Hall and Fisher 1985; Bayle et al. 1986; USFWS 1986a; Taris and Bressac-Vaquer 1987; Pain 1990; Schmitz et al. 1990; Locke and Friend 1992; Pain 1992b; Pain et al. 1992.

[a]Ingestion rates are shown only for large (i.e., in excess of 30), random samples.
[b]Nonrandom sample.

and Larsson 1982; Tewari et al. 1987), although its significance in relation to the ingestion of spent shot or secondary effects on piscivores has not been investigated.

A variety of methods have been suggested or tried in an attempt to alleviate the problem of lead poisoning from spent shot (e.g., manipulation of water levels, provision of grit and suitable foods, and treatment of sick birds), but the only approach to offer a demonstrably effective solution, while maintaining the traditions of waterfowl hunting, is conversion to the use of nontoxic shot (Mudge 1992). For example, Calle et al. (1982) reported a decrease in lead shot ingestion rates in mallards and American black ducks from 11.2 percent to 5.6 percent just three years after lead shot was banned in Pennsylvania (USA).

One of the most controversial issues surrounding the use of nontoxic shot is its ability to kill birds cleanly. Results from a plethora of tests show this to be an insignificant problem, particularly once a hunter is acquainted with the ballistics of the new shot (Feierabend 1983; Brister 1992; Morehouse 1992a). Another issue causing controversy is damage to firearms (barrel scratching, barrel erosion, and choke expansion) caused by the relatively hard steel pellets. However, firing steel from modern, relatively thick-barreled shotguns with full chokes causes no significant expansion, while problems of barrel scratching and erosion, associated with earlier steel shot loads, have been completely eliminated. These have been achieved through the combined improvements of shot-shell design and increased shotgun barrel hardness (Feierabend 1983). Damage may occur to some of the older, thin-walled shotguns with tight chokes (more than ¾ choke) (Thomas 1994), which are still frequently used in Europe. Recent developments in the United Kingdom suggest that a nontoxic shot alloy suitable for use in these guns will be available in the near future.

Within the 27 countries where lead-poisoned waterfowl have been recorded, governmental action on the use of lead shot for waterfowl hunting ranges from nationwide mandatory bans to no action (Table 14.5). In Denmark, Finland, the Netherlands, Norway, and the United States, nationwide mandatory bans have been implemented, while the Ukraine will follow suit in the near future. In eight nations (Australia, Mexico, Switzerland, Sweden, Finland, Canada, France, and the United Kingdom), voluntary phase-outs or mandatory bans in certain areas have been implemented. No action to limit the use of lead shot

Table 14.5. Current Governmental Action on Use of Lead Shot for Waterfowl Hunting in Countries Where Lead Poisoning in Waterfowl Has Been Recorded and Where Hunting Is Still Practiced

Country	Action
Denmark	Nationwide ban
Finland	Nationwide ban
Netherlands	Nationwide ban
Norway	Nationwide ban
United States	Nationwide ban
Australia	Ban in South Australia and two sites in Northern Territory
Canada	Ban in certain areas (nontoxic shot zones)
Mexico	Ban in Yucután
Sweden	Ban in Ramsar sites and voluntary phase-out elsewhere
Switzerland	Ban in Thurgau and voluntary phase-out elsewhere
Germany	Voluntary phase-out and research in progress
United Kingdom	Voluntary phase-out and research in progress
Ukraine	No current action, but planned nationwide ban
France	Research in progress
Japan	Research in progress
Albania	No current action
Belgium	No current action
Czech Republic	No current action
Greece	No current action
Hungary	No current action
Ireland	No current action
Italy	No current action
Latvia	No current action
New Zealand	No current action
Poland	No current action
Russian Federation	No current action
Spain	No current action

Sources: Billings 1990; Pain 1992b; Ákoshegyi 1994; Larsson 1994; F. I. Norman, pers. comm. 1994; B. Parkes, pers. comm. 1994; Thomas 1994; Fawcett 1996; Golovatch in press.

has been taken in the other 14 nations, although in two of these, research into the problem and possible solutions is being conducted. Canada, Sweden, and parts of the Northern Territory (Australia) are unique in that they are the only areas where nontoxic zones are currently established, rather than nationwide or regionwide bans. A recent assessment of the situation in Canada (Kennedy and Nadeau 1993) suggests that these zones need to be applied on a much broader basis than they are at present if the risk of lead poisoning is to be reduced to acceptable levels. In addition, such zones are difficult and expensive to enforce (Thomas 1994), and their use in the United States from 1976/77 to 1990/91 proved to be an inadequate solution to the problem (Bishop and Wagner 1992; Morehouse 1992b). Further experience in the United States has shown that mandatory bans on the use of lead shot are the best policy for ensuring hunter compliance with phase-out programs, since voluntary phase-outs resulted in very low compliance (Havera et al. 1994). Nonetheless, it may be necessary to preempt mandatory bans with voluntary phase-outs so that industry and hunters can adjust fluently.

Shooting Disturbance and Related Management Practices

The effects of shooting disturbance on biodiversity can be analyzed in two ways: (1) the impact that this has on the distribution and abundance of waterbirds; and (2) the effect that any changes in waterbird distribution/abundance have on ecosystem structure and function. The most obvious effects of disturbance are manifest within the former topic, and so studies within this area are by far the most common. Before analyzing these topics, it is important to note that the impact of shooting disturbance is compounded by the shyness it invokes upon birds, to the extent that they become cautious of any human presence. Studies have shown that geese increase their flight distance from humans by 100–350 meters (m) during hunting activities (Owens 1977; Gerdes and Reepmayer 1983; Madsen 1985, 1988; Rudfeld 1990).

Disturbance to birds from shooting has probably received more attention in Denmark than in any other country. Increased shooting disturbance on the main traditional autumn staging site for pink-footed geese (*Anser brachyrhynchus*) in Denmark has resulted in the geese bypassing the country and flying straight to the Dutch and Belgian wintering grounds (Madsen 1985; Madsen and Jepsen 1992). About

20 percent of western Jutland has been declared no-shooting zones, within which Meltofte (1982) found about 90 percent of the region's ducks. One particular area held a much higher proportion of the birds after the end of the open season, while the numbers of birds increased markedly after another area was declared a no-shooting area in 1976 (Jepsen 1983). Meltofte (1989) investigated the effects of shooting disturbance on waders by comparing their distribution outside the shooting season with that during the open season. The results showed that the number of Golden plover (*Pluvialis apricaria*) and curlew (*Numenius arquata*) in intensively hunted areas was significantly reduced during the shooting season (by up to 90% on some sites), but such redistribution was not evident for other waders. Madsen (1993, 1995) showed, via a series of experiments on two large wetlands, that hunting disturbance not only affected the distribution of the birds (Fig. 14.4), but also severely depressed total numbers of quarry (by four to twenty times) and nonquarry species (by two to five times). Furthermore, following the control of hunting, the staging periods of most quarry species were prolonged by several months. Following the control of shooting within the two areas, national autumn staging populations of both wigeon (*Anas penelope*) and shoveler (*Anas clypeata*) have nearly doubled, and the sites are now among the most important for coastal waterfowl in Denmark.

Shooting disturbance has also received a significant amount of attention in Britain, where the protected area system has developed rapidly since the 1960s. Owen (1993) analyzed the trend in the number of wigeon within and outside refuges during 1961–89, the results of which suggested that as refuges have progressively become established, more and more birds have moved to them from hunted areas (Fig. 14.5). In addition, it was found that the birds stay longer in refuge areas (see also Owen and Thomas 1979). At the Ouse Washes, Thomas (1978) showed that 86 percent of waterfowl usage was on refuges during the hunting season, but this dropped to 46 percent after the hunting season. In the same area, Bell and Owen (1990) recorded much lower densities of wigeon in areas that were hunted than on refuges, and similar results were demonstrated on the Exe Estuary for wigeon and Brent geese. Mudge (1989) found decreased feeding activity of wigeon at times of night shooting in the Moray Firth, and although the birds did not leave the region, they did redistribute themselves within the Moray complex. Gomes (1981, 1982) recorded a sharp decline in the number of mallard and teal (*Anas crecca*) using hunted areas on the Dee Estuary during the beginning of the hunting season, and marked increases in the number of wigeon and pintail using hunted areas during bans on hunting introduced as a result of cold weather. Hirons and Thomas (1993) provided data showing two- to threefold increases in numbers of wildfowl in four estuarine sites in England following the creation of refuges (all areas were hunted prior to refuge establishment). Owen and Salmon (1988b) examined changes in numbers of wildfowl on nine sites in Britain before and after the restriction or prohibition of shooting. In five cases there were large increases in numbers of protected as well as quarry species following the change, whereas in the other four there was no discernible immediate effect.

Species that have difficulty in meeting their energy requirements in winter may be most sensitive to disturbance, and all species may be vulnerable at times of greatest energy demand and least feeding opportunity. Mayhew (1988) plotted the number of hours needed for feeding in winter by a number of duck species against the proportion found on refuges in Britain (Fig. 14.6). There was a clear positive relationship, with herbivorous wigeon and gadwall needing the greatest time to feed and occurring in the highest proportion in protected areas. The diving ducks are least vulnerable not only because they can gather their daily energy in a rather short time but also because they invariably feed nocturnally. Though hunting is not the only factor responsible for the relationship in Figure 14.6, it is generally accepted that hunting displaces waterfowl at greater ranges than most other recreational activities and creates large waterfowl exclusion zones (Bell and Owen 1990).

Studies of shooting disturbance in countries other than Britain and Denmark are rather limited. In France, Tamisier and Saint-Gerand (1981) examined the numbers and distribution of waterbirds in 13 coastal departments where night shooting was allowed compared with that in the 12 where it was banned. The number of ducks and coot (*Fulica atra*) in the departments without night shooting was seven to ten times greater, for which the principal factor responsible was thought to be shooting disturbance. In the Bassin d'Arcachon, numbers of waders increased over an order of magnitude (from 20,000 to 220,000) following the protection of a roost site from shooting (Campredon 1979).

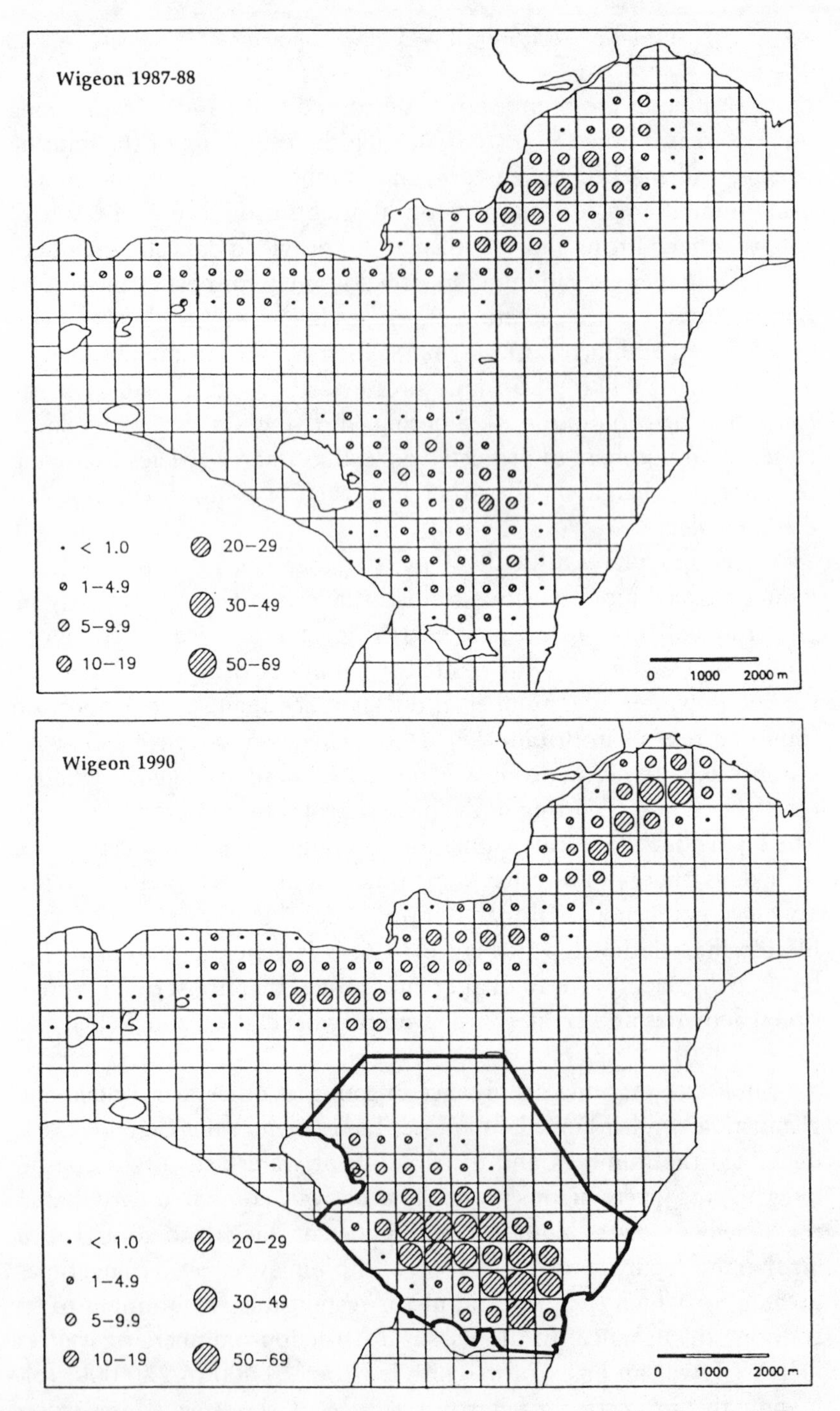

Figure 14.4. Distribution of wigeon-days during autumn at Nibe-Bredning (Denmark), 1987–91. In 1987–88, no refuges existed; in 1989 there was a ban on shooting from mobile punts in the whole area; in 1990, there was a complete

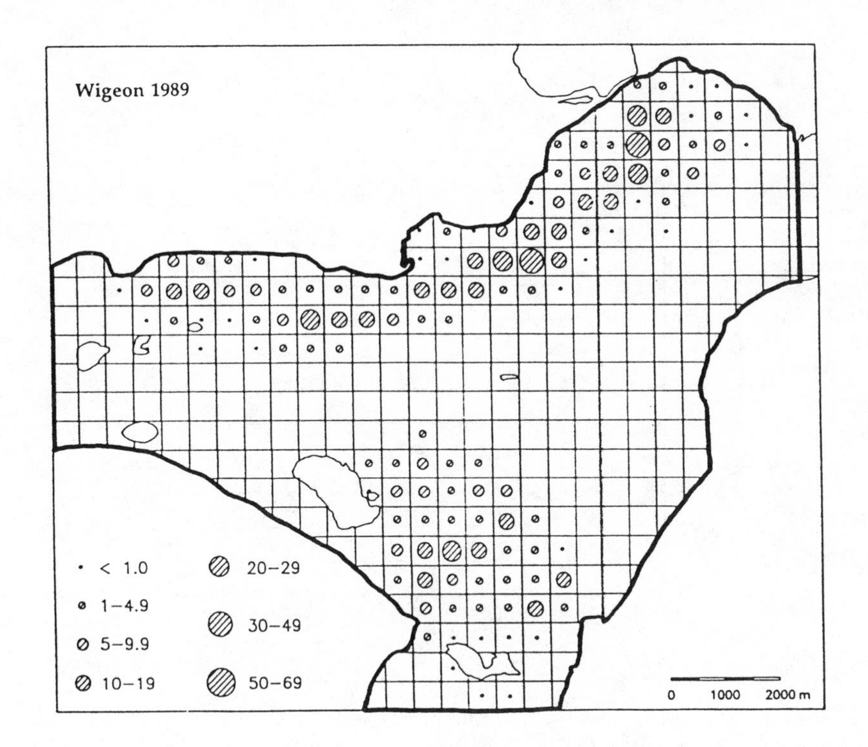

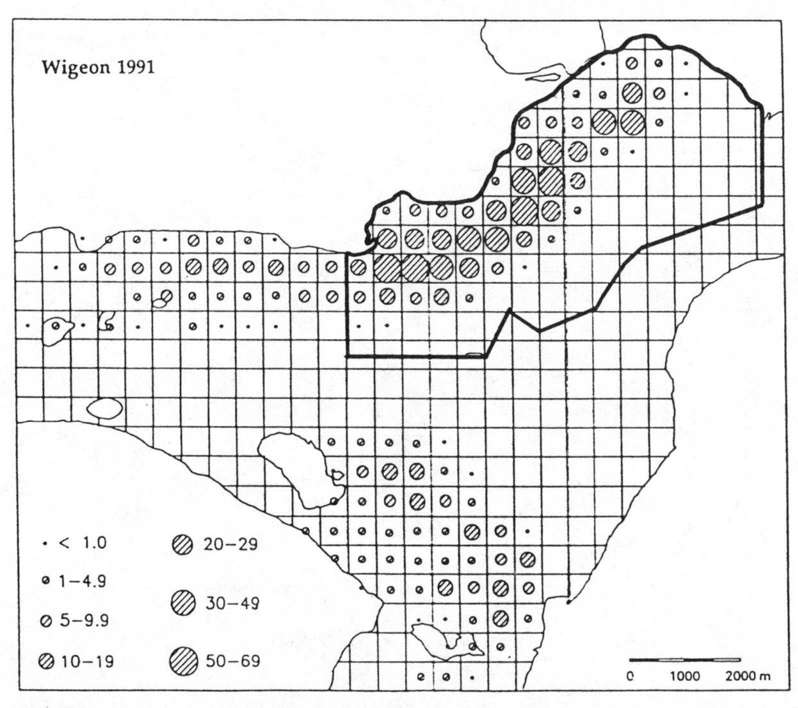

shooting ban in the southern part of the area; in 1991, there was a complete shooting ban in the northern part. Refuge boundaries are shown by bold lines. (*Source:* Madsen 1995.)

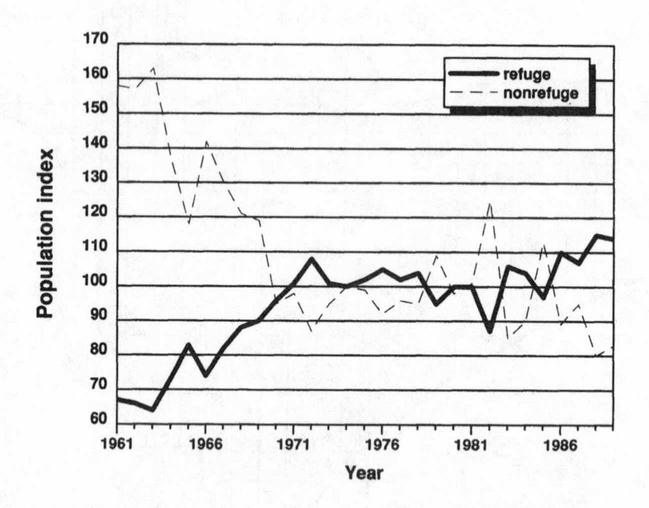

Figure 14.5. January trend in the number of wigeon in Britain on refuge and nonrefuge sites. (*Source:* Owen 1993.)

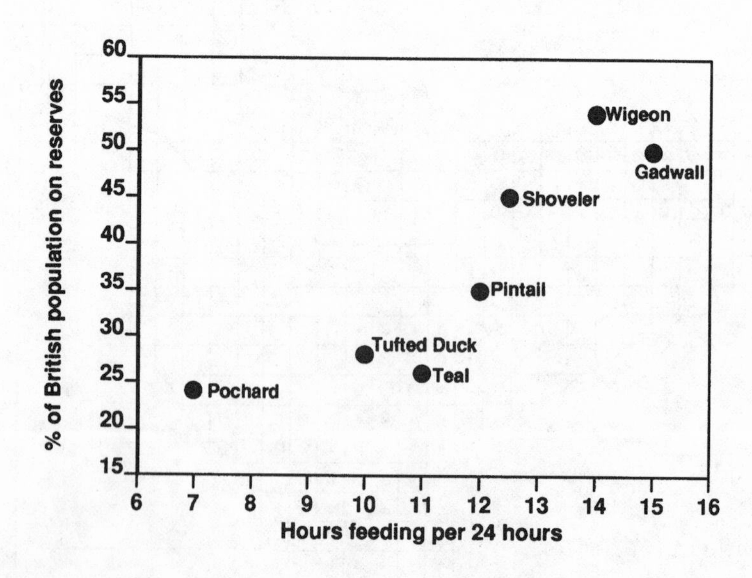

Figure 14.6. Proportion of seven species of duck found on reserves in Britain in relation to their need for feeding. (*Source:* Mayhew 1988.)

Maher (1982) investigated the movements of ducks in response to the opening of the duck shooting season in New South Wales (Australia), and concluded that four out of the five species studied actively sought refuge in protected areas. Mikkola and Lind (1974) found that the behavior and local distribution of ducks altered abruptly at the start of the shooting season at a site in Finland, while Geroudet (1967) documented large increases in the number of waterfowl present on a stretch of the river Rhone (Switzerland) following the creation of a refuge. Reichholf (1973) calculated that disturbance caused a 60 percent reduction in waterfowl numbers at several sites in Switzerland, and at one site where 200 ducks were shot, some twenty thousand abandoned the site as a result of the disturbance. In Germany, an experimental shooting regime was introduced to an area, whereby shooting was allowed in some weeks and not in others (Ziegler and Hanke 1988). Numbers of the two species monitored, mallards and tufted ducks, were significantly higher within the reserve only after more than a week of no shooting, as the birds took more than a week to build up to peak numbers. Counts outside the reserve showed corresponding decreases. In Belgium and along the lower Rhine in Germany, major wintering concentrations of European white-fronted geese have developed partly as a result of bans on shooting (Meire and Kuijken 1991; Mooij 1991), while Faralli (in press) documented a considerable increase in the abundance and diversity of waterfowl utilizing Lake Montepulciano (Italy) following a hunting ban, including the colonization of the site by the globally threatened ferruginous duck (*Aythya nyroca*).

In Ireland, the creation of no-shooting areas led to increases in flock sizes of Greenland white-fronted geese (*Anser albifrons flavirostris*) and has allowed the geese to exploit food resources more efficiently (Norriss and Wilson 1988). Similarly, Schneider-Jacoby et al. (1991) recorded a substantial increase in the number of whooper swans wintering within an area of Lake Constance (Switzerland) and concluded that the increase was mainly due to the termination of shooting disturbance (following a shooting ban), which allowed the swans to exploit previously unavailable food resources. A model produced by Frederick et al. (1987) showed that an increase in hunter activity not only increased the kill of lesser snow geese (*Anser c. caerulescens*) directly, but also caused greater movement of geese, increasing their accessibility to hunters outside the refuge, with consequent further increase in the kill. Disturbance from hunting caused disruption of feeding patterns, which reduced energy gains and hastened emigration. It was concluded that the offtake from the population by hunting was less important in reducing the population size of the geese than the associated disturbance that led to early emigration. Follestad (in press) concluded that increasing hunting disturbance caused the premature emigration of greylag geese from Vega (Norway) by four to five weeks, although the establishment of nonhunting zones and modifications to the timing of hunting seem to have since solved this problem.

Another approach to assessing the effects of hunting disturbance has been to examine the consequences on individual birds in order to determine whether changes in behavior or feeding opportunity have an effect on survival or breeding (Bell and Owen 1990). Geese tend to pair for life, and the cost of separation can be high in terms of future breeding success (Owen et al. 1988). In addition, Owen and Black (1989) found that being in a family conferred survival advantages on young geese. Prevett and MacInnes (1980) showed that shooting disturbance caused the disintegration of family groups of snow geese, which resulted in lower survival of young birds, while Madsen (1995) showed that disturbance of pink-footed geese (by farmers) on a spring staging site in Norway caused a significant decline in subsequent breeding success because disturbed birds could not lay down as much fat as undisturbed birds.

The redistribution of waterbirds as a result of disturbance will inevitably lead to a shift in community structure, although the extent of change will depend on the tightness of coupling of the redistributed birds to others in the food web. Several studies from the marine environment provide useful insights to such interactions. Paine (1974, 1980) has shown that a particular starfish (*Pisaster* sp.) plays a key role in regulating the abundance of several other species in the associated rocky intertidal community. When the starfish was experimentally removed, the community became dominated by a mussel (*Mytilus* sp.) and species richness nearly halved. An equally dramatic shift in community composition has been observed when the sea otter (*Enhydra lutris*) was removed or reintroduced to marine sites (Estes and Palmisano 1974; Simenstad et al. 1978; Duggins 1980). In Switzerland, Reichholf (1973) has addressed this issue in relation to shooting disturbance and waterfowl management. The study measured the percentages of primary production and invertebrate fauna consumed by waterfowl in marshes where shooting was allowed and in marshes where it was prohibited. The results showed that primary production and invertebrate abundance were much higher in hunted areas because of lower waterfowl abundance caused by shooting disturbance. Thus shooting disturbance was directly responsible for a substantial alteration in community structure and primary productivity.

The principal mechanism for alleviating the impacts of disturbance is modification of shooting duration. It is generally accepted that prolonged disturbance from night shooting (i.e., about an hour after sunset until an hour before dawn) can be extremely damaging to bird populations (Tamisier and Saint-Gerand 1981; Mudge 1989; Smit and Visser 1993). One would predict that crippling losses and shooting of protected species (owing to misidentification) are more likely to occur at night, but evidence is lacking and would be extremely difficult to collect. An important distinction needs to be made between true "flight shooting," where birds are intercepted while flying between places in the course of their normal activities, and more intrusive shooting, where night feeding or roosting grounds are disturbed intentionally and birds shot while flying following this disturbance. This latter practice may have significant impacts, yet it is a legal pursuit in seven of the eight European countries where night shooting remains legal (Fig. 14.7); in North America, night shooting has been banned since 1916 (Tamisier 1985).

Regulation of daytime shooting can also be important, especially for terrestrial diurnal feeders such as geese (Raveling et al. 1972). Organized groups of hunters usually exercise restrictions voluntarily, such as limiting the number of shooting days or the number of participants. Even if this increases the success of shooting, it lessens the impact of disturbance on both quarry and nonquarry species (Bell and Owen 1990).

The timing and length of hunting seasons is critical if the effects of disturbance are to be minimized. Hunting seasons should not open before all juveniles have fledged, which in Europe would correspond to September for northeastern countries and October for southwestern countries (Cramp and Simmons 1977; Tamisier 1985). However, hunting of waterbirds (either ducks, geese, rails, or waders) occurs in Poland throughout the year, while in Germany it occurs in every month except May, and in Sweden, the Netherlands, Belgium, and France it begins in mid-July (Fig. 14.8). Hunting seasons should close at the

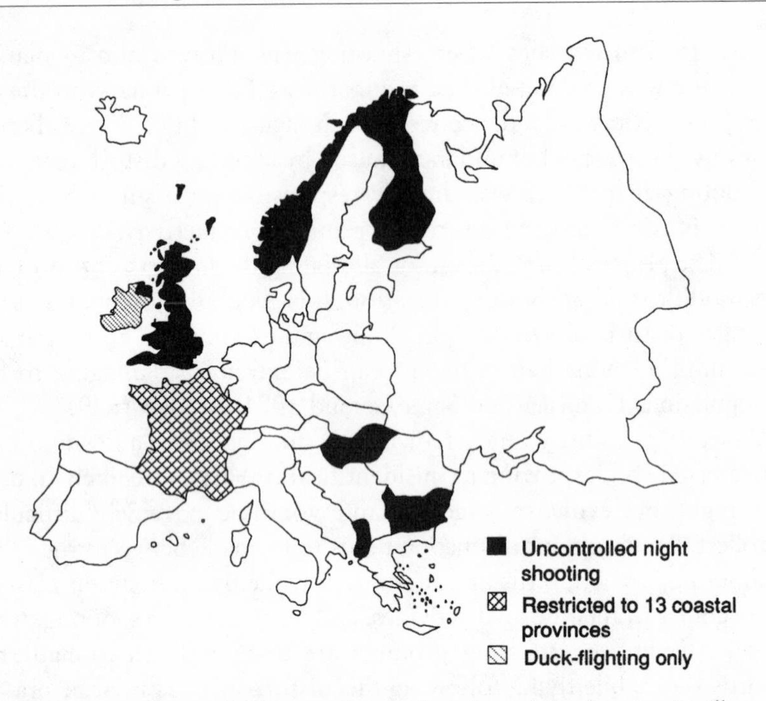

Figure 14.7. Distribution in Europe of countries where night shooting is still permitted. (*Source:* Bell and Owen 1990.)

point when disturbance is likely to have a significant impact on the breeding success of the birds during the following season. Arctic breeding waterfowl generally maintain their weight during early winter, but by January the effect of short days, cold weather, and reduced food availability necessitates some fat reserves be used (Owen and Black 1990). Therefore, it is from January onward that most species are particularly susceptible to disturbance, and even more so from March onward when many species begin to lay down endogenous fat reserves. Indeed, several studies of disturbance during spring staging have shown population effects, including reductions in the numbers of birds using staging areas (e.g., Pfister et al. 1992) and substantial increases in daily energy expenditure that exceeded the compensatory capacity of the birds (Belanger and Bedard 1990). In realization of this, the hunting season, already closed at the latest on January 31 for half a century in all the American states, has been progressively brought forward there to the middle of January, and eventually to the end of December without any compensation in the opening time

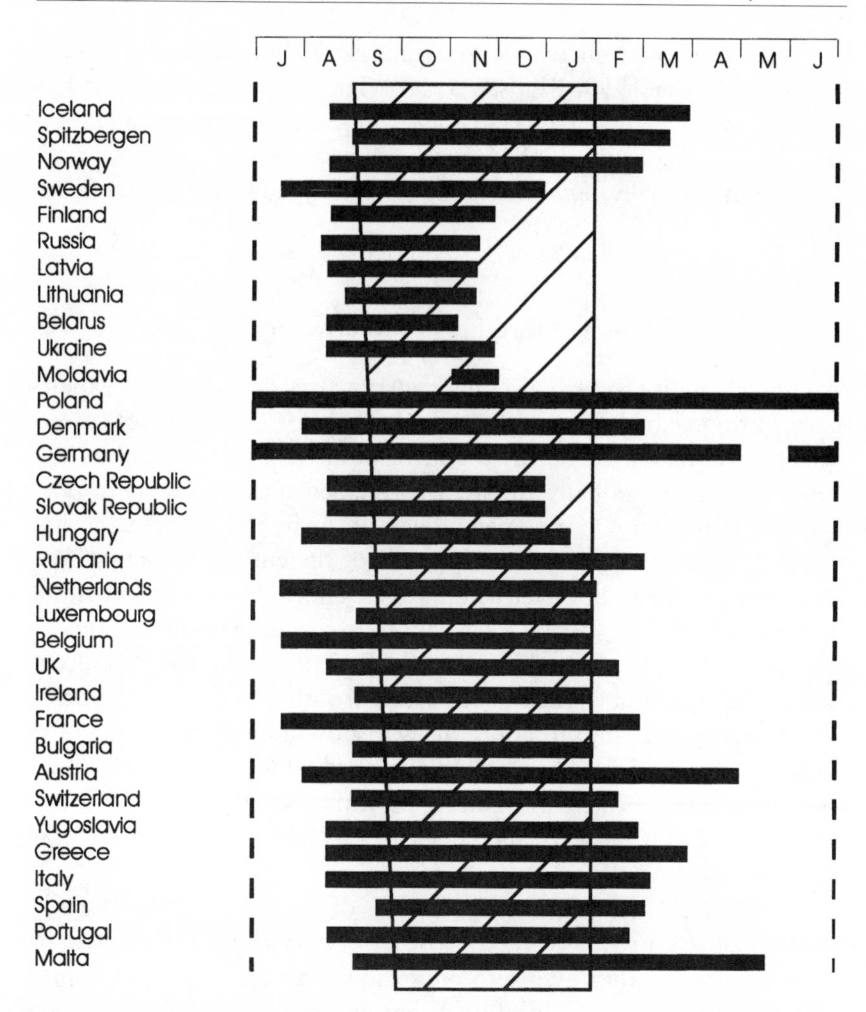

Figure 14.8. Temporal distribution of waterbird hunting (either ducks, geese, rails, or waders) in European countries (bold lines), with suggested distribution indicated (hatched box). (*Sources:* Lampio 1983; Tamisier 1985.)

(Tamisier 1985). Based on the above information, hunting seasons in Europe ought to open during the beginning of September in northern countries and the end of this month in southern countries, while the season should close at the end of January. Many countries, however, employ hunting seasons that occur well outside these dates (Fig. 14.8).

Finally, the provision of refuges within hunting zones is becoming a cornerstone of waterfowl management because the number of birds using an area often increases, allowing better hunting opportunities (Bell and Owen 1990; Madsen 1993). The Danish government has recently taken a convincing lead concerning such policy, and will soon be establishing a network of 55 nonhunting sanctuaries, most of which have been mutually agreed with the hunting fraternity (Meltofte in press).

Hunting Mortality of Protected Species

A hunter's ability to recognize his quarry varies depending on experience, location, and light conditions. Perhaps the best-documented case in which this poses a particular threat to biodiversity concerns hunting mortality of the globally threatened freckled duck (*Stictonetta naevosa*) of Australia. On the opening day of the hunting seasons during 1979–82, an estimated 4.5–8.2 percent of the entire population was shot, and numbers shot over whole seasons may have been three times as high (Martindale 1984; Green 1992). In response to this, counts of waterfowl have been conducted since 1987 prior to the hunting season in Victoria, and wetlands supporting freckled ducks are temporarily closed to shooting. In addition, waterfowl identification tests for hunters (using videos) have been established, along with other educational material. As a result, hunting mortality of this species has been greatly reduced (Callaghan and Green 1993; Holmes 1994; R. T. Kingsford, pers. comm. 1994).

In Russia, one of the principal causes of the decline of the globally threatened lesser white-fronted goose (*Anser erythropus*) is thought to be hunting mortality owing to confusion with the European white-fronted goose (*Anser a. albifrons*), but as yet nothing has been done about this situation (Morozov in press). Similarly, the globally threatened slender-billed curlew (*Numenius tenuirostris*) and ferruginous duck are frequently shot due to confusion with curlew and tufted duck, respectively (Anon., pers. comm. 1995). At a number of localities in Denmark, Meltofte (1978) showed that 7.9 percent (n = 1,006) of shots were fired at protected species. At Slimbridge (England), 34 percent (n = 272) of live Bewick's swans (*Cygnus colombianus bewickii*) examined during 1970–73 had lead shot embedded in their tissues (Evans et al. 1973). Further monitoring during 1989–92 and 1995 showed a similar proportion (30%, n = 304) (Wildfowl and Wetlands Trust, unpublished data). These proportions are similar to those for important quarry species of comparable size in North America, such as the Canada goose (see USFWS 1986a), yet the Bewick's swans visiting Slimbridge are protected throughout their range, and within which there are no quarry species remotely similar in appearance (Evans et al. 1973). Furthermore, Anderson and Sanderson (1979) showed that for every single Canada goose carrying embedded shot, a further three birds were shot dead, which would suggest that the northwest European Bewick's swan population is suffering substantial hunting mortality.

Although the whooper swan (*Cygnus cygnus*), mute swan, and trumpeter swan are protected throughout their ranges in Europe and North America, there is evidence of considerable illegal hunting. Examination of live whooper swans at Caerlaverock (Scotland) during 1988–89 showed that 12 percent (n = 150) of birds had embedded lead shot (Wildfowl & Wetlands Trust, unpublished data), while Spray and Milne (1988) found that 10.5 percent of 57 whooper swans found dead in Scotland during 1981–86 had died from gunshot wounds. They also found that 8.7 percent of mortality of 149 mute swans found dead during 1980–86 was attributable to this cause. Banko (1960) found that 15 percent of trumpeter swans carried embedded shot in North America, while necropsies of 72 individuals found dead in the western United States during 1976–87 revealed that 9.7 percent had died from gunshot wounds (Blus et al. 1989). Hunting mortality of trumpeter swans has decreased significantly in Minnesota following increased hunter education and public awareness (Degernes and Frank 1991). Autopsies of barnacle geese (*Branta leucopsis*) in the United Kingdom revealed embedded shot frequencies of 21 percent (n = 88) (Wildfowl & Wetlands Trust, unpublished data), although this species is also protected throughout its range.

Habitat Protection, Restoration, and Modification

The value-added incentive placed upon wetlands via recreational waterfowl hunting drives many habitat management, restoration, and protection programs. Nowhere is this more evident than in the United States, primarily because hunting-related activities currently fund about 75 percent of the cost of wildlife management at the state level (Sparrowe 1993) and a number of quarry species have been declining in this region for many years (USFWS 1994). It has been estimated that 40 million ha of wetland habitat in North America have been protected as a direct result of waterfowl hunting (Heitmeyer et al. 1993b). Such protection is obviously a substantial benefit to biodiversity. However, dense and diverse waterfowl assemblages are associated with eutrophic wetlands (e.g., Hoyer and Canfield 1994; Suter 1994), and in some areas such wetlands receive a disproportionate amount of protection, to the detriment of oligotrophic/mesotrophic wetlands (e.g., Manuel 1987).

Habitat restoration is a core program of some hunting organizations. In Canada, for example, the Prairie CARE Program of Ducks Unlimited (a nongovernmental hunting organization) has restored large areas of native grassland and wetland habitat (Heitmeyer et al. 1993b), while a similar program (Valley CARE) in California (USA) also involves the restoration of wetlands (DU 1994). Although there have been no critical analyses of these activities, they seem to offer substantial benefits to biodiversity.

One of the principal activities of the National Wildlife Refuge System (totaling more than 37 million ha) of the United States is waterfowl management (Wentz and Reid 1992; USFWS 1993). Such management is usually evaluated on the basis of the number of days of use a wetland receives by waterfowl, and a principal objective of many states is to hold waterfowl within their boundaries so as to improve the hunting potential for their licensed sportsmen (Linduska 1982). Smith et al. (1989) provide many examples of the type of habitat modification (or enhancement) practices employed, but in general the principal aim is to provide an abundance of appropriate food, open water for resting, and optimal nesting conditions. Although no studies have critically evaluated the effects of such habitat manipulation for waterfowl on biodiversity per se, the following account provides a review of the most widely applied management prescriptions, and attempts an evaluation based on current knowledge.

The hydrological process is a principal factor controlling the biodiversity of wetlands. Studies have shown that the early "hemimarsh" stage of a wetland (i.e., about an equal cover interspersion of emergent hydrophytes and open water) supports the greatest density and diversity of dabbling duck species (Weller and Fredrickson 1974; Weller 1978; Kaminski and Prince 1981; Britton 1982; Bethke and Nudds 1993). Since dabbling ducks comprise the majority of the duck harvest, various water control structures are often incorporated into wetlands in order to retain or create the hemimarsh stage. This involves draining the wetland for the summer period every one to five years (known as "drawdown"), depending on climate.

As might be expected, changes in vegetation are dramatic following drawdowns, generally producing a more complex structure of emergent species, but causing a reduction in submergent and floating-leaved species (Kadlec 1962; Meeks 1969; Weller 1978; Rundle and Fredrickson 1981). Changes in aquatic invertebrate communities can also be dramatic under frequent drawdown management, causing a decline in carnivores such as species of Tricladida, Hirundinea, Odonata, Hemiptera, and Coleoptera, and an increase in detritivores and herbivores, such as species of Diptera, Amphipoda, and Isopoda (biomass generally increases, but diversity decreases) (Kadlec 1962; Hubert and Krull 1973; Wegener et al. 1974; Voights 1976; Kaminski 1979; Britton 1982; Reid 1985). The virtual extirpation of fish populations is invariably associated with drawdowns (Weller 1978), and so clearly there are fundamental alterations to biodiversity when drawdown prescriptions are employed. Whether changes in biodiversity are positive or negative will depend on the naturalness of the site and the landscape in which it is located. Where such management disrupts the hydrological processes of natural wetlands, substantial degradation of biodiversity can be expected to occur. Since about the 1940s, this has commonly occurred in North America (see, e.g., Knighton 1985), but figures concerning the extent of this phenomenon are unavailable. Furthermore, the alteration of natural hydrological processes can favor the establishment of alien species (Macdonald et al. 1989).

The impact of drawdowns on seminatural and artificial wetlands is more abstruse, and is dependent on whether a more natural or unnatural hydrology is attained. Drawdowns can be an effective tool in restoring or emulating natural hydrological processes, although waterfowl managers employ them in an attempt to manipulate waterfowl productivity and give little consideration to natural water regimes. Consequently, water level management can become standardized within an area heavily managed for waterfowl hunting, thus causing a reduction in the diversity of hydrological dynamics and an associated loss of biodiversity. Though managers are beginning to purposefully restore or emulate natural hydrological processes (see, e.g., Heitmeyer et al. 1993b), the methods and results have not been adequately documented. Rundle and Fredrickson (1981) and Helmers (1992) provide guidelines for modifying drawdown prescriptions to benefit wading birds, although these merely address the timing and extent of mud exposure in accordance with the migratory passage of birds. It is notable that drawdowns can result in large amounts of dead organic material and shallow, warm water, and these factors are important contributors to outbreaks of avian botulism, perhaps the most significant cause of disease-related mortality in waterbirds (Locke and Friend 1987).

The farming of cereals, such as corn (*Zea mays*) and browntop millet (*Panicum ramosum*), to produce food for waterfowl is a common practice in North America and is a major endeavor of the national refuge system (Braun et al. 1978; Goodwin 1979; Linde 1985; Wentz and Reid 1992). This is often combined with intensive water-level management, whereby crops are grown during the drawdown phase. Such management has been very successful in concentrating birds and frequently too much so, as witnessed by problems with short-stopping birds before traditional wintering areas, crop damage problems adjacent to refuges, and outbreaks of density-dependent diseases (e.g., avian cholera and DVE). More general problems have been associated with the use of pesticides and herbicides (Braun et al. 1978).

Large beds of tall emergent vegetation (e.g., *Phragmites australis* and *Typha* spp.) support relatively low numbers of dabbling ducks, and so various management prescriptions are often employed in order to reduce their area and increase the area of open water. Herbicides (e.g., dalapon and glyphosate) are commonly used, often over large areas (Rollings and Warden 1964; Beule 1979; Thomas 1982; Linde 1985). Although long-term effects are poorly understood, short-term effects include dramatic changes to water chemistry and to faunal and floral communities (Newbold 1975; Beule 1979). Recolonization by the tall emergent vegetation is usually rapid, and so herbicides need to be applied regularly (Kadlec and Smith 1992), thus compounding problems.

Burning is also commonly used to reduce the area of tall emergent vegetation. This activity releases nutrients, opens the canopy and detrital layer, and allows for increased insolation and resultant earlier warming of soils (USFWS 1986b). The effect on waterfowl numbers is little understood, and studies have produced conflicting data (see USFWS 1986b; Kadlec and Smith 1992). Studies of the effects of burns on faunal and floral communities associated with *Phragmites australis* beds in the United Kingdom show that if burns are carefully timed and controlled there is no discernable effect on the invertebrate community, plant diversity increases, and the rate of litter decomposition remains unaltered, although bryophytes are severely affected and there is some litter loss (Cowie et al. 1992; Ditlhogo et al. 1992).

Alterations to the hydrology of salt marshes to produce less saline habitats more favored by target waterfowl species is a common practice along the coast of the United States (Palmisano 1972; Goodwin 1979). Although undocumented, substantial erosion of biodiversity, including the degradation of halophytic communities, may result.

Considerable controversy has surrounded the use of domestic animals in wetland management. Grazing continues to be widely used in an attempt to improve both wintering and breeding conditions for waterfowl (Braun et al. 1978; Kantrud 1985). Although herbivores can be critical to the maintenance of biodiversity in wetlands (Gordon et al. 1990), little quantitative research has investigated the use of domestic livestock for such purposes. Ironically, the vast majority of observations suggest that grazing is detrimental to breeding waterfowl (Braun et al. 1978), while overgrazing or grazing by inappropriate organisms or at inappropriate times of the year can be extremely detrimental to biodiversity (Reimold et al. 1975; Bassett 1980; Ovington 1984). Several studies have documented an increase in plant and bird species diversity following the employment of light stocking rates under a rotational system (see Kantrud 1985), but even a very sensitive grazing regime can cause a substantial increase in water turbidity, eutrophication, compaction of soil, and dramatic changes in community structure (Braun et al. 1978; Morris and Plant 1983; Rawes 1983; Kantrud 1985; Knopf et al. 1988; Gordon et al. 1990; Gibson et al. 1992a,b). Thus, although grazing is potentially a vital tool of biodiversity conservation, its use is little understood, and its popularity, no doubt due to the financial benefits, has probably caused the erosion of biodiversity in many areas. Mechanical mowing is often employed as a substitute for herbivores, but this produces a less diverse community. Also, if mowing or grazing is employed too early, substantial losses of nesting birds can occur (Braun et al. 1978; Green 1986, 1988).

In relatively high densities, certain species of fish (e.g., Esocids, Percids, Cyprinids, and Salmonids) can negatively affect some waterfowl species by competing for food, restricting food availability (e.g., by increasing water turbidity), or preying upon ducklings (Solman 1945; Eriksson 1979; McAllister-Eadie and Keast 1982; Hill et al. 1987; Giles 1991; McNicol and Wayland 1992). Consequently, fish are considered undesirable in wetlands managed for waterfowl, and where drawdowns are not possible, other measures are often taken to reduce fish biomass, such as the application of piscicides (e.g., rotenone) or by trawling (Kadlec and Smith 1992; Giles in press). The effect of such activities is site dependent, but in small, shallow, eutrophic wetlands, there is, in general, a reduction in turbidity, productivity, phytoplankton, and the rates of certain nutrient cycles (e.g., phosphorus), while the structure of invertebrate and macrophyte communities alters substantially, corresponding with an increase in their biomass and the duck species that consume them (Hayne and Ball 1956; Brooks and Dobson 1965; Andersson et al. 1978; Stenson et al. 1978; Morin 1984; Post and Cucin 1984; Giles 1994, in press; Hanson and Butler 1994). Conversely, in large, stratified wetlands there may often be little long-term effect, as fish populations can soon return to former densities (DeMelo et al. 1992; Matena et al. 1994). In hypereutrophic wetlands, the reduction of fish biomass can be a useful tool in restoring biodiversity when used in conjunction with attempts to reduce nutrient loads. However, when the control of fish populations is used purely as a tool to increase waterfowl productivity, the effects on biodiversity may often be negative, for example, by causing declines or local extinctions of those organisms associated with the fish fauna or by disrupting certain nutrient cycles.

The control of predators of waterfowl eggs and incubating females has long been practiced in an attempt to increase nesting success. Studies have shown that if predator densities are reduced substantially, then waterfowl production can be increased in certain areas (Balser et al. 1968; Duebbert and Kantrud 1974; Duebbert and Lokemoen 1980; Greenwood 1986). Conversely, there may often be little measurable effect. For example, Broyer et al. (in press) assessed the effect of a massive removal of carrion crow (*Corvus corone*) on the breeding success of waterfowl, and concluded that although predation rates by the crows decreased substantially, the benefit to duck production was low owing to the influence of other predators. The impact of predator control measures on biodiversity per se has never been investigated. Where such management involves the intensive control of introduced predators, the effect on native biodiversity should be positive. However, native species are more often the target of such programs, and when practiced with sufficient regularity and intensity, erosion of biodiversity can be expected to occur. Problems of predation may often be the result of management aimed at increasing waterfowl numbers, as this will inevitably attract predators.

The provision of nest boxes has been widely practiced in attempts to increase waterfowl populations for hunting, such as the wood duck (North America), mallard (North America and Europe), chestnut teal (Australia), and grey teal (*Anas gibberifrons*) (Australia and New Zealand). Increases in local numbers of some species can occur, but often no measurable effect is apparent (McLaughlin and Grice 1952; Strange et al. 1971; Norman and Riggert 1977; Haramis and Thompson 1985; Gauthier 1988; Savard 1988; Zicus 1990; McCoy et al. 1992; Newton 1994). No studies have investigated affects on nontarget species that utilize the boxes, such as other birds, mammals, and insects, although it seems safe to assume that, in areas with reduced natural cavities, biodiversity will benefit. Such management is costly, however, and money may often be better spent conserving habitat (Norman and Riggert 1977).

As noted earlier, dense and diverse waterfowl assemblages are associated with eutrophic wetlands. Recent research in Canada has aimed at increasing duck populations on natural oligotrophic wetlands through the application of fertilizers (Melanson and Payne 1988; Gabor et al. 1994; Murkin et al. 1994). Experiments so far have been rather unsuccessful, but there are proposals to conduct large-scale experiments incorporating long-term nutrient additions (Gabor et al. 1994). Such single-value actions will inevitably cause a substantial alteration or erosion of biodiversity, as clearly indicated by the initial experiments (e.g., effects have included reduced oxygen levels and substantial alterations to algal and invertebrate communities).

It is quite apparent that the manipulation of habitat within the discipline of waterfowl management is aimed exclusively at increasing the hunters' yield, regardless of biodiversity conservation. This ought to be rectified so that the fundamental goal is the restoration, protection, or emulation of pristine ecological conditions to whatever extent possible. Constitutional to such an approach is (1) the employment of a more scientific basis to manipulation techniques, so that effects can be evaluated and adapted; and (2) the adoption of the precautionary principle, so that damage to biodiversity can be minimized. Such management will often involve the recovery of lost or degraded ecosystem values and functions, and this will require managers to focus above the species level and adopt more of a holistic perspective (see, e.g., Eiseltová 1994). Encouragingly, such an approach is gaining popularity among resource managers (Pickett et al. 1992).

Over the last two decades, the quality of water resources in the United States has been assessed increasingly on the basis of its ability to support and maintain an integrated, adaptive community of organisms having a species composition and functional organization comparable to that of natural habitat of the region (Karr 1991). Such an approach could form the basis for assessing manipulation techniques in waterfowl management.

Selective Harvesting

Skewed harvesting is often considered a problem among game populations (e.g., Ryman et al. 1981; Ginsberg and Milner-Gulland 1994), and data suggest that waterfowl are vulnerable to being killed by hunters relative to age, sex, species, color phase, and condition (Cooch 1961; Olson 1965; Stott and Olson 1972; Aldrich 1973; Giroux and Bedard 1986; Greenwood et al. 1986; Hepp et al. 1986; Miller et al. 1988; Reinecke and Shaiffer 1988; Heitmeyer et al. 1993a). Although considerable concern has been directed toward this phenomenon of waterfowl hunting, available information suggests that selection does not occur at an intensity that causes significant alteration or erosion of genetic diversity within species.

Conclusions and Recommendations

The conservation of biodiversity is integral to the sustainability of recreational waterfowl hunting. On a local scale, this will require the formulation and implementation of detailed management plans for each hunting area. Encouragingly, management plans promoted by the British Association for Shooting and Conservation are aimed at integrating hunting and nonhunting interests on estuarine sites of high conservation importance (BASC 1991). There is a desperate need for such plans to be more widely incorporated into hunting activities. On a broader scale, recommendations considered instrumental to ecological sustainability are provided in the following sections.

Releases of Waterfowl

The release of waterfowl for hunting is commonly practiced in many countries, and this activity has been instrumental to the establishment of substantial feral populations of waterfowl over large areas. Particular problems and concerns related to such populations include hybridization with native waterfowl, the spread of various diseases, increased eutrophication, and alterations to the structure of wetland communities. The following actions are recommended:

1. Introductions of waterfowl should be stopped, and populations that have arisen from this activity should be controlled in problem areas.
2. Stocking of waterfowl should be discouraged, and populations that have arisen from this activity should be controlled in problem areas.
3. Where releases continue, they should be sensitively managed and mandatory inspection of stock for disease prior to release should be considered.

Lead Poisoning

Lead poisoning of waterbirds has so far been recorded in 27 countries, although the phenomenon is undoubtedly more widespread. It is known to cause the death of significant numbers of waterbirds (both quarry and nonquarry) and their associated predators. Some countries have taken action to combat this problem, while in others laissez-faire prevails. The following actions are recommended:

1. Governments, in collaboration with hunting organizations and industry, should implement well-defined schedules for transition to nontoxic shot in wetland areas where no such action has yet been taken.
2. Taxation of lead or subsidization of nontoxic substitutes should be considered to promote transition.

Shooting Disturbance

For some species of quarry and nonquarry birds (particularly herbivores), disturbance caused by shooting can have a powerful influence on their distribution, not only on a local but also on a regional, national, and even international scale. Shooting disturbance can greatly reduce species density and diversity, substantially alter migration behavior, and reduce survival and productivity. The redistribution of birds can cause significant alterations to community composition. Modifications to shooting activities can be effective in alleviating these problems, without compromising hunting opportunities. The following actions are recommended:

1. Development of refuge areas within hunting zones should be promoted, and research should focus on optimal configuration.
2. Integrated within (1), research should aim to identify the level of shooting disturbance a typical site can support before bird distribution is altered substantially.
3. Night shooting should be banned in those areas where it is still practiced.
4. The timing of hunting seasons should be revised so that they are ecologically compatible across political borders.

Hunting Mortality of Protected Species

Hunting mortality of protected species is a significant problem, and in some areas these species are exposed to a level of hunting pressure similar to that of quarry species. The following actions are recommended:

1. Proficiency tests should precede the issuance of hunting licenses.
2. Where protected populations are threatened by hunting mortality owing to confusion with similar game species, temporary bans on hunting should be imposed in problem areas.

Habitat Protection, Restoration, and Modification

Significant benefits to biodiversity can be assumed from habitat protection and restoration programs. However, the manipulation of habitat within these and other areas often conflicts with biodiversity conservation and is usually based on tradition. A more holistic approach to habitat manipulation, aimed at conserving pristine ecological conditions, seems to be the surest policy for ensuring a diverse and healthy waterfowl assemblage that will help meet future hunting needs. The following actions are recommended:

1. Habitat management should focus on the protection, restoration, or emulation of pristine ecological conditions to the extent possible.

2. Management should be based on comprehensive scientific data, employing a systems-oriented perspective.

Acknowledgments

Curtis H. Freese provided much help and advice during this study, while many other people have also provided valuable contributions, in particular: Bruce D. J. Batt, Martin J. Brown, Grant S. Dumbell, Diane R. Eggeman, Chris Elphick, Anthony D. Fox, Leigh H. Fredrickson, Mickey E. Heitmeyer, Steve Howell, Richard M. Kaminski, Richard T. Kingsford, Tony Laws, Jasper Madsen, James D. Nichols, F. Ian Norman, Brian Parkes, Kevin Peberdy, Hugo Phillipps, Simon Pickering, Graham W. Smith, Mark Underhill, Murray Williams, Pauline Wong, and Ute Zillich. In addition, three anonymous referees provided constructive criticism of an earlier draft.

References

Ákoshegyi, I. 1994. Lead poisoning of wild ducks caused by lead shots. *Magyar Állatorvosok Lapja* 49:345–348.

Aldrich, J. W. 1973. Disparate sex ratios in waterfowl. In *Breeding biology of birds,* ed. D. S. Farner, 482–489. Washington, D.C.: National Academy of Sciences.

Allan, J. R., J. S. Kirby, and C. J. Feare. 1996. The biology of the Canada goose *Branta canadensis* in relation to the management of feral populations. *Wildlife Biology* 1:129–143.

Amiard-Triquet, C., D. J. Pain, G. Mauvais, and L. Pinault. 1992. Lead poisoning in waterfowl: Field and experimental data. *Impact of Heavy Metals on the Environment* 2:219–245.

Anderson, D. R. 1975. *Population ecology of the mallard. V. Temporal and geographic estimates of survival, recovery and harvest rates.* Resource Publication 125. Washington, D.C.: United States Fish and Wildlife Service.

Anderson, D. R., and K. P. Burnham. 1976. *Population ecology of the mallard. VI. The effect of exploitation on survival.* Resource Publication 128. Washington, D.C.: United States Fish and Wildlife Service.

Anderson, D. R., and S. P. Havera. 1989. Lead poisoning in Illinois waterfowl (1977–1988) and the implementation of non-toxic shot regulations. *Illinois Natural History Survey Biological Notes* 133:1–35.

Anderson, W. L., and G. C. Sanderson. 1979. Effectiveness of steel shot in 3-inch 12 gauge shells for hunting Canada geese. Unpublished report. Illinois Department of Conservation, Springfield.

Andersson, G., H. Berggren, G. Cronberg, and C. Gelin. 1978. Effects of planktivorous and benthivorous fish on organisms and water chemistry in eutrophic lakes. *Hydrobiologia* 59:9–15.

Angermeier, P. L. 1994. Does biodiversity include artificial diversity? *Conservation Biology* 8:600–602.

Ankney, C. D., D. G. Dennis, and R. C. Bailey. 1987. Increasing mallards, decreasing American black ducks: Coincidence or cause and effect? *Journal of Wildlife Management* 51:523–529.

Ankney, C. D., D. G. Dennis, and R. C. Bailey. 1989. Increasing mallards, decreasing American black ducks: No evidence for cause and effect. A reply. *Journal of Wildlife Management* 53:1072–1075.

Artmann, J. W., and E. M. Martin. 1975. Incidence of lead shot in Sora Rails. *Journal of Wildlife Management* 39:514–519.

Balharry, E., B. W. Staines, M. Marquiss, and H. Kruuk. 1994. *Hybridization in British mammals.* JNCC Report 154. Peterborough, England: Joint Nature Conservation Committee.

Balser, D. S., H. H. Dill, and H. K. Nelson. 1968. Effect of predator reduction on waterfowl nesting success. *Journal of Wildlife Management* 32:669–682.

Banko, W. E. 1960. The trumpeter swan, its history, habits and population in the United States. *North American Fauna* 63. Washington, D.C.: United States Fish and Wildlife Service.

Banks, R. C. 1972. A systematist's view. In *Role of hand-reared ducks in waterfowl management: A symposium,* ed. A. T. Studholme, 117–120. Dundee, Illinois: Max McGraw Wildlife Foundation.

BASC (The British Association for Shooting and Conservation). 1991. *Management plans for shooting and conservation on estuarine sites of high conservation importance.* Project Handbook 6. Wrexham, Wales: BASC.

Bassett, P. A. 1980. Some effects of grazing on vegetation dynamics in the Camargue, France. *Vegetatio* 43:173–184.

Batt, B.D.J., and J. W. Nelson. 1990. The role of hand-reared mallards in breeding waterfowl conservation. *Transactions of the North American Wildlife and Natural Resources Conference* 55:558–568.

Bayle, P., F. Dhermain, and G. Keck. 1986. Trois cas de saturnisme chez le Flamant rose (*Phoenicopterus ruber*) dans la région de Marseille. *Bulletin de la Société Linneenne de Provence* 38:95–98.

Behan, M. J., and T. B. Kinraide. 1979. Lead accumulation in aquatic plants from metallic sources including shot. *Journal of Wildlife Management* 43:240–244.

Belanger, L., and J. Bedard. 1990. Energetic cost of man-induced disturbance to staging snow geese. *Journal of Wildlife Management* 54:36–41.

Bell, D. V., and M. Owen. 1990. Shooting disturbance: A review. In *Managing waterfowl populations,* ed. G.V.T. Matthews, 159–171. Special Publication 12. Slimbridge, England: International Waterfowl and Wetlands Research Bureau.

Bellrose, F. C. 1959. Lead poisoning as a mortality factor in waterfowl populations. *Illinois Natural History Survey Bulletin* 27:235–288.

Bethke, R. W., and T. D. Nudds. 1993. Variation in the diversity of ducks along a gradient of environmental variability. *Oecologia* 93:242–250.

Beule, J. D. 1979. *Control and management of cattails in southeastern Wisconsin wetlands.* Technical Bulletin 112. Madison, Wisconsin: Department of Natural Resources.

Billings, I. 1990. Lead poisoning in waterfowl: A review of the European situation. Unpublished report. Eurogroup for Animal Welfare, Brussels.

Birkhead, M. 1983. Factors affecting the breeding success of the mute swan (*Cygnus olor*). *Journal of Animal Ecology* 52:727–741.

Bishop, R. A., and R. G. Howing. 1972. Re-establishment of the giant Canada goose in Iowa. *Proceedings of the Iowa Academy of Science* 79:14–16.

Bishop, R. A., and W. C. Wagner II. 1992. The US Cooperative Lead Poisoning Control Information Program. In *Lead poisoning in waterfowl,* ed. D. J. Pain, 42–45. Special Publication 16. Slimbridge, England: International Waterfowl and Wetlands Research Bureau.

Blanco, J. C., S. Reig, and L. Cuesta. 1992. Distribution, status, and conservation problems of the wolf *Canis lupus* in Spain. *Biological Conservation* 60:73–80.

Blus, L. J., R. K. Stroud, B. Reiswig, and T. McEneaney. 1989. Lead poisoning and other mortality factors in trumpeter swans. *Environmental Toxicology and Chemistry* 8:263–271.

Borden, R., and H. A. Hochbaum. 1966. Gadwall seeding in New England. *Transactions of the North American Wildlife and Natural Resources Conference* 31:79–88.

Boyd, H. 1983. *Intensive regulation of duck hunting in North America: Its purpose and achievements.* Occasional Paper no. 50. Ottawa, Ontario, Canada: Canadian Wildlife Service.

Braithwaite, L. W., and B. Miller. 1975. The mallard, *Anas platyrhynchos,* and mallard-black duck, *Anas superciliosa rogersi,* hybridization. *Australian Wildlife Research* 2:47–61.

Brakhage, G. K. 1953. Migration and mortality of ducks hand-reared and wild-trapped in Delta, Manitoba. *Journal of Wildlife Management* 17: 465–477.

Brand, C. J. 1987. Duck plague. In *Field guide to wildlife diseases: General field procedures and diseases of migratory birds,* ed. M. Friend, 117–128. Resource Publication 167. Washington, D.C.: United States Fish and Wildlife Service.

Brand, C. J., and D. E. Docherty. 1988. Post-epizootic surveys of waterfowl for duck plague (duck virus enteritis). *Avian Diseases* 32:722–730.

Braun, C. E., K. W. Harmon, J. A. Jackson, and C. D. Littlefield. 1978. Management of national wildlife refuges in the United States: Its impacts on birds. *Wilson Bulletin* 90(2):309–321.

Brister, B. 1992. Steel shot: Ballistics and gunbarrel effects. In *Lead poisoning in waterfowl,* ed. D. J. Pain, 26–28. Special Publication 16. Slimbridge, England: International Waterfowl and Wetlands Research Bureau.

Britton, R. H. 1982. Managing the prey fauna. In *Managing wetlands and their birds,* ed. D. A. Scott, 92–97. Slimbridge, England:

Brodsky, L. M., and P. J. Weatherhead. 1984. Behavioral and ecological factors contributing to American black duck-mallard hybridization. *Journal of Wildlife Management* 48(3):846—852.

Brooks, J. L., and S. I. Dobson. 1965. Predation, body size, and composition of plankton. *Science* 150:28–35.

Broyer, J., J. Fournier, and P. Varagnat. In press. Carrion Crow (*Corvus corone*) predation rates of simulated nests in Dombes (France). *Gibier Faune Sauvage.*

Buchanan, I., and R. Barker. 1993. A survey of gamebird harvest and hunter effort in the fish and game region, 1993. Unpublished report. New Zealand Fish and Game Council, Wellington, New Zealand.

Buffington, J. D., and D. E. MacLauchlan. 1993. Sustaining the diversity of birds: Intercontinental experiences. *Biopolicy International* 15:32. Nairobi, Kenya: Acts Press, African Centre for Technological Studies.

Burgess, E. C., J. Ossa, and T. M. Yuill. 1979. Duck plague: A carrier state in wildfowl. *Avian Diseases* 23:940–949.

Burgess, E. C., and T. M. Yuill. 1983. Superinfection in ducks persistently infected with duck plague virus. *Avian Diseases* 26:40–46.

Burnham, K. P., and D. R. Anderson. 1984. Tests of compensatory vs. additive hypotheses of mortality in mallards. *Ecology* 65:105–112.

Burnham, K. P., G. C. White, and D. R. Anderson. 1984. Estimating the effect of hunting on annual survival rates of adult mallards. *Journal of Wildlife Management* 48:350–361.

Butler, D. 1990. The incidence of lead shot ingestion by waterfowl in Ireland. *Irish Naturalist Journal* 23:309–312.

Byers, S. M., and J. R. Cary. 1991. Discrimination of mallard strains on the basis of morphology. *Journal of Wildlife Management* 55:580–586.

Caithness, T., M. Williams, and J. D. Nichols. 1991. Survival and band recovery rates of sympatric grey ducks and mallards in New Zealand. *Journal of Wildlife Management* 55:111–118.

Callaghan, D. A., and A. J. Green. 1993. Wildfowl at risk, 1993. *Wildfowl* 44:149–169.

Calle, P. P., D. F. Kowalczyk, F. J. Dein, and F. E. Hartman. 1982. Effects of hunters' switch from lead to steel shot on potential for oral lead poisoning in ducks. *Journal of the American Veterinary Medical Association* 181:1299–1301.

Campredon, P. 1979. Quelques données concernant l'hivernage des limicoles sur le Bassin d'Arcachon (Gironde). *L'Oiseau* 49:113–131.

Caughley, G. 1985. Harvesting of wildlife: Past, present and future. In *Game harvest management,* ed. S. L. Beasom and S. F. Robertson, 3–14. Kingsville, Texas: Caesar Kleberg Wildlife Research Institute.

Cheng, K. M., R. N. Shoffner, R. E. Phillips, and F. B. Lee. 1978. Mate preference in wild and domesticated (game-farm) mallards (*Anas platyrhynchos*). I. Initial preference. *Animal Behaviour* 26:996–1003.

Cheng, K. M., R. N. Shoffner, R. E. Phillips, and F. B. Lee. 1979. Mate preference in wild and domesticated (game-farm) mallards (*Anas platyrhynchos*). II. Pairing success. *Animal Behaviour* 27:417–425.

Clark, T. W. 1993. Creating and using knowledge for species and ecosystem conservation: Science, organizations and policy. *Perspectives in Biology and Medicine* 36:497–525.

Clausen, B., and C. Wolstrup. 1979. Lead poisoning in game from Denmark. *Danish Review of Game Biology* 11(2):3–22.

Conroy, M. J., and D. G. Krementz. 1990. A review of the evidence for the effects of hunting on American black duck populations. *Transactions of the North American Wildlife and Natural Resources Conference* 55: 501–517.

Cooch, G. 1961. Ecological aspects of the blue snow goose complex. *Auk* 78:72–89.

Cowie, N. R., W. J. Sutherland, M.K.M. Ditlhogo, and R. James. 1992. The effects of conservation management of reed beds. II. The flora and litter disappearance. *Journal of Applied Ecology* 29:277–284.

Cowx, I. G. 1994. Stocking strategies. *Fisheries Management and Ecology* 1:15–30.

Cramp, S., and K.E.L. Simmons, eds. 1977. *Handbook of the birds of Europe, the Middle East, and North Africa. The birds of the Western Palearctic,* vol. 1. Oxford: Oxford University Press.

Davis, B. 1988. Recent recorded incidents of lead poisoning among wildfowl in Scotland. Unpublished report. Aberdeen, Scotland: Nature Conservation Council.

Degernes, L. A., and R. K. Frank. 1991. Causes of mortality in trumpeter swans *Cygnus buccinator* in Minnesota 1986–1989. *Wildfowl Supplement* 1:352–355.

Delany, S. N. 1993. Introduced and escaped geese in Britain in summer 1991. *British Birds* 86:591–599.

Del Bono, G., G. Braca, and S. Rindi. 1975. Lead poisoning in ducks: Effects on the fertility and fecundity. *Proceedings of the 20th World Veterinary Congress* 3:2355–2356.

DeMelo, R., R. France, and D. J. McQueen. 1992. Biomanipulation: Hit or myth? *Limnology and Oceanography* 37:192–207.

Dennis, D. G., K. L. Fischer, and G. B. McCullough. 1984. The change in status of mallards and black ducks in southwestern Ontario. In *Waterfowl studies in Ontario, 1973–1981,* ed. S. G. Curtis, D. G. Dennis, and H. Boyd, 27–30. Occasional Paper of the Canadian Wildlife Service. Ottawa, Ontario: Canadian Wildlife Service.

Ditlhogo, M.K.M., R. James, B. R. Laurence, and W. J. Sutherland. 1992. The effects of conservation management of reed beds. I. The invertebrates. *Journal of Applied Ecology* 29:265–276.

Dobrowolski, K., and K. Dmowski. In press. Hunting pressure on ducks and geese in Poland. *Gibier Faune Sauvage.*

Dobson, A. P., and R. M. May. 1986. Disease and conservation. In *Conservation biology: The science of scarcity and diversity,* ed. M. E. Soulé, 345–366. Sunderland, Massachusetts: Sinauer Associates.

Dobson, A. P., and R. M. May. 1991. Parasites, cuckoos and avian population dynamics. In *Bird population studies: Relevance to conservation and management,* ed. C. M. Perrins, J. D. Lebreton, and G.J.M. Hirons, 391–412. Oxford: Oxford University Press.

Doty, H. A., and A. D. Kruse. 1972. Techniques for establishing local breeding populations of wood ducks. *Journal of Wildlife Management* 36:428–435.

DU (Ducks Unlimited). 1994. *Valley CARE: Bringing conservation and agriculture together in California's Central Valley,* vol. 1, June. Sacramento, Califorinia: Ducks Unlimited, Western Regional Office.

Duebbert, H. F., and H. A. Kantrud. 1974. Upland duck nesting related to land use and predator reduction. *Journal of Wildlife Management* 38: 257–265.

Duebbert, H. F., and J. T. Lokemoen. 1980. High duck nesting success in a predator-reduced environment. *Journal of Wildlife Management* 44: 428–437.

Duggins, D. O. 1980. Kelp beds and sea otters: An experimental approach. *Ecology* 61:447–453.

Eiseltová, M. 1994. *Restoration of lake ecosystems: A holistic approach.* Special Publication 32. Slimbridge, England: International Waterfowl and Wetlands Research Bureau.

Elder, W. H. 1954. The effect of lead poisoning on the fertility and fecundity of domestic mallard drakes. *Journal of Wildlife Management* 18:315–323.

Eriksson, M.O.G. 1979. Competition between freshwater fish and Goldeneye *Bucephala clangula* (L.) for common prey. *Oecologia* 41:99–107.

Estes, J. A., and J. F. Palmisano. 1974. Sea otters: Their role in structuring nearshore communities. *Science* 185:1058–1060.

Evans, M. E., N. A. Wood, and J. Kear. 1973. Lead shot in Bewick's swans. *Wildfowl* 24:56–60.

Fabricius, E. 1983. *The Canada goose in Sweden.* Solna: Swedish Nature Conservancy.

Fabricius, E., A. Bylin, A. Ferno, and T. Radesater. 1974. Intra- and interspecific territorialism in mixed colonies of the Canada goose *Branta canadensis* and the greylag goose *Anser anser. Ornis Scandinavica* 5:25–35.

Faralli, L. In press. Hunting effects on Anatidae in a typical Italian wetland: An attempt to assess the impact of shooting. *Gibier Faune Sauvage.*

Fawcett, D. 1996. *Lead poisoning in waterfowl: International update report 1995.* Slimbridge, England: International Waterfowl and Wetlands Research Bureau.

Feierabend, J. S. 1983. *Steel shot and lead poisoning in waterfowl: An annotated bibliography of research 1976–1983.* National Wildlife Federation Scientific and Technical Series 8. Washington, D.C.: National Wildlife Federation.

Figley, W. K., and L. W. VanDruff. 1982. The ecology of urban mallards. *Wildlife Monograph* 81:1–39.

Fog, J. 1971. Survival and exploitation of mallards (*Anas platyrhynchos*) released for shooting. *Danish Review of Game Biology* 6(4):1–12.

Follestad, A. In press. Hunting and the change in autumn migration period of the greylag goose (*Anser anser*) in Norway. *Gibier Faune Sauvage.*

Frederick, R. B., W. H. Clark, and E. E. Kaas. 1987. Behaviour, energetics, and management of refuging waterfowl: A simulation model. *Wildlife Monograph* 96:1–35.

Gabor, T. S., H. R. Murkin, M. P. Stainton, J. A. Boughen, and R. D. Titman. 1994. Nutrient additions to wetlands in the Interlake region of Manitoba, Canada: Effects of a single pulse addition in spring. *Hydrobiologia* 279/280:497–510.

Gabuzov, O. S. 1990. Prospects of the introduction of *Branta canadensis* in the USSR. In *Managing waterfowl populations,* ed. G.V.T. Matthews, 66–69. Special Publication 12. Slimbridge, England: International Waterfowl and Wetlands Research Bureau.

Gauthier, G. 1988. Factors affecting nest-box use by buffleheads and other cavity-nesting birds. *Wildlife Society Bulletin* 16:132–141.

Gerdes, K., and H. Reepmayer. 1983. Zur räumlichen Verteilung überwinternder Saat- und Blessgänse (*Anser fabalis* und *A. albifrons*) in Abhängigkeit von naturschutzschädlichen und fordernden Einflüssen. *Die Vogelwelt* 104:54–67.

Gere, G., and S. Andrikovics. 1992. Effects of waterfowl on water quality. *Hydrobiologia* 243/244:445–448.

Gere, G., and S. Andrikovics. 1994. Feeding of ducks and their effects on water quality. *Hydrobiologia* 279/280:157–161.

Geroudet, P. 1967. L'évolution du stationnement des Anatides dans une réserve de chasse sur le Rhône en aval de Genève. *Nos Oiseaux* 24:141–153.

Gibson, C.W.D., V. K. Brown, L. Losito, and G. C. McGavin. 1992a. The response of invertebrate assemblages to grazing. *Ecography* 15:166–176.

Gibson, C.W.D., C. Hambler, and V. K. Brown. 1992b. Changes in spider (Araneae) assemblages in relation to succession and grazing management. *Journal of Applied Ecology* 29:132–142.

Giles, N. 1991. Fish removal from lakes increases food for waterfowl. In *Wetland management and restoration,* ed. C. M. Finlayson and T. Larsson, 80–86. Proceedings of a workshop, Solna, Sweden, September 12–15, 1990. Swedish Environmental Protection Agency Report. Solna, Sweden: Swedish Environmental Protection Agency.

Giles, N. 1994. Tufted duck (*Aythya fuligula*) habitat use and brood survival increases after fish removal from gravel pit lakes. *Hydrobiologia* 279/280:387–392.

Giles, N. In press. *Feeding relationships between waterfowl and fish.* Special Publication. Slimbridge, England: International Waterfowl and Wetlands Research Bureau.

Gillespie, G. D. 1985. Hybridization, introgression, and morphometric differentiation between mallard *Anas platyrhynchos* and grey duck *Anas superciliosa* in Otago, New Zealand. *Auk* 102:459–469.

Ginsberg, J. R., and E. J. Milner-Gulland. 1994. Sex-biased harvesting and population dynamics in ungulates: Implications for conservation and sustainable use. *Conservation Biology* 8:157–166.

Giroux, J. F. 1981. Ducks nesting in association with Canada geese. *Journal of Wildlife Management* 45:778–782.

Giroux, J. F., and J. Bedard. 1986. Sex-specific hunting mortality of greater snow geese along firing lines in Quebec. *Journal of Wildlife Management* 50:416–419.

Godin, A. J. 1967. *Test of grit types in alleviating lead poisoning in mallards.* Special Scientific Report 107. Patuxent, Maryland: United States Fish and Wildlife Service.

Golovatch, O. In press. The diet of ducks on Kiev Reservoir (Ukraine) and the incidence of lead shot. *Gibier Faune Sauvage.*

Gomes, R. 1981. The effect of wildfowling on the waterfowl at Gayton Sands, 1980–1981. Unpublished report. Royal Society for the Protection of Birds, Sandy, England.

Gomes, R. 1982. The effect of wildfowling on the waterfowl at Gayton Sands, 1981–1982. Unpublished report. Royal Society for the Protection of Birds, Sandy, England.

Goodwin, T. M. 1979. Waterfowl management practices employed in Florida and their effectiveness on native and migratory waterfowl populations. *Biological Sciences* 3:123–129.

Gordon, I. J., P. Duncan, P. Grillas, and T. Lecomte. 1990. The use of domestic herbivores in the conservation of the biological richness of European wetlands. *Bulletin of Ecology* 21(3):49–60.

Gough, R. E. 1984. Laboratory-confirmed outbreaks of duck virus enteritis (duck plague) in the United Kingdom from 1977 to 1982. *Veterinary Record* March 17:262–265.

Gough, R. E., and D. J. Alexander. 1990. Duck virus enteritis in Great Britain, 1980 to 1989. *Veterinary Record* June 16:595–597.

Grandy, J. W. 1983. The North American black duck (*Anas rubripes*): A case study of 28 years of failure in American wildlife management. *International Journal of the Study of Animal Problems Supplement* 4:1–35.

Green, A. J. 1992. Wildfowl at risk, 1992. *Wildfowl* 43:160–184.

Green, R. E. 1986. The management of lowland wet grassland for breeding waders. Unpublished report, Royal Society for the Protection of Birds, Sandy, England.

Green, R. E. 1988. Effects of environmental factors on the timing and success of breeding of common snipe *Gallinago gallinago* (Aves: Scolopacidae). *Journal of Applied Ecology* 25:79–93.

Greenwood, H., R. G. Clark, and P. J. Weatherhead. 1986. Condition bias of hunter-shot mallards (*Anas platyrhynchos*). *Canadian Journal of Zoology* 64:599–601.

Greenwood, R. J. 1986. Influence of striped skunk removal on upland duck nesting success in North Dakota. *Wildlife Society Bulletin* 14:6–11.

Greig, J. C. 1980. Duck hybridization: A threat to species integrity. *Bockmakierie* 32(3):88–89.

Grier, J. W. 1980. Modelling approaches to bald eagle population dynamics. *Wildlife Society Bulletin* 8:316–322.

Grinnell, G. B. 1894. Lead poisoning. *Forest & Stream* 42:117–118.

Grumbine, M. E. 1994. What is ecosystem management? *Conservation Biology* 8:27–38.

Haddon, M. 1984. A re-analysis of hybridization between mallards and grey ducks in New Zealand. *Auk* 101:190–191.

Hall, S. L., and F. M. Fisher, Jr. 1985. Lead concentrations in tissues of marsh birds: Relationships of feeding habits and grit preference to spent shot ingestion. *Bulletin of Environmental Contamination and Toxicology* 35:1–8.

Hanson, M. A., and M. G. Butler. 1994. Responses of food web manipulation in a shallow waterfowl lake. *Hydrobiologia* 279/280:457–466.

Haramis, G. M., and D. Q. Thompson. 1985. Density-reproduction characteristics of box-nesting wood ducks in a northern greentree impoundment. *Journal of Wildlife Management* 49:429–436.

Harper, M. J., and M. Hindmarsh. 1990. Lead poisoning in magpie geese *Anseranas semipalmata* from ingested lead pellet at Bool Lagoon Game Reserve (South Australia). *Australian Wildlife Research* 17:141–145.

Harradine, J. 1985. Duck shooting in the United Kingdom. *Wildfowl* 36:81–94.

Haux, C., and A. Larsson. 1982. Influence of inorganic lead on the biochemical blood composition in the rainbow trout *Salmo gairdneri*. *Ecotoxicology and Environmental Safety* 6:28–43.

Havera, S. P., C. S. Hine, and M. M. Georgi. 1994. Waterfowl hunter compliance with nontoxic shot regulations in Illinois. *Wildlife Society Bulletin* 22:454–460.

Hayne, D. W., and R. C. Ball. 1956. Benthic productivity as influenced by fish predation. *Limnology and Oceanography* 1:162–175.

Heggberget, T. M. 1987. Development of the Norwegian population of Canada geese *Branta canadensis* up to 1984. *Fauna* 40:1–9.

Heitmeyer, M. E., L. H. Fredrickson, and D. D. Humburg. 1993a. Further evidence of biases associated with hunter-killed mallards. *Journal of Wildlife Management* 57:733–740.

Heitmeyer, M. E., J. W. Nelson, B.D.J. Batt, and P. J. Caldwell. 1993b. *Waterfowl conservation and biodiversity.* Midwest Fish and Wildlife Conference 1993. Columbia, Missouri.

Helmers, D. L. 1992. *Shorebird management manual.* Manomet, Massachusetts: Western Hemisphere Shorebird Reserve Network.

Hepp, G. R., R. J. Blohm, R. E. Reynolds, J. E. Hines, and J. D. Nichols. 1986. Physiological condition of autumn-banded mallards and its relationship to hunting vulnerability. *Journal of Wildlife Management* 50: 177–183.

Heusmann, H. W. 1974. Mallard-black duck relationships in the northeast. *Wildlife Society Bulletin* 2:171–177.

Heusmann, H. W. 1981. Movements and survival rates of park mallards. *Journal of Field Ornithology* 52:214–221.

Heusmann, H. W. 1988. The role of parks in the range expansion of the mallard in the northeast. In *Waterfowl in winter,* ed. M. W. Weller, 405–412. Minneapolis: University of Minnesota Press.

Hill, D. A., R. Wright, and M. Street. 1987. Survival of mallard *Anas platyrhynchos* ducklings and competition with fish for invertebrates on a flooded gravel quarry in England. *Ibis* 129:159–167.

Hill, K. 1994. It's the law. *Florida Wildlife* 48:41.

Hirons, G., and G. Thomas. 1993. Disturbance on estuaries: RSPB nature reserve experience. *Wader Study Group Bulletin* 68:72–78.

Holland, M. M. 1988. Wise use of wetlands: Consideration of the biosphere reserve concept. In *Convention relative aux zones humides d'importance internationale particulièrement comme habitats des oiseaux d'eau,* 365–373. Procès-verbaux de la troisième session de la conférence des parties contractantes. Regina, Saskatchewan, Canada, May 27–June 5, 1987.

Holmes, J. 1994. *1992 duck season in Victoria.* ARI Technical Report 132. Heidelberg, Victoria, Australia: Arthur Rylah Institute for Environmental Research.

Holmes, J. C. 1982. Impact of infectious disease agents on the population growth and geographical distribution of animals. In *Population biology of infectious diseases,* ed. R. M. Anderson and R. M. May, 37–51. New York: Springer-Verlag.

Hoyer, M. V., and D. E. Canfield, Jr. 1994. Bird abundance and species richness on Florida lakes: Influence of trophic status, lake morphology, and aquatic macrophytes. *Hydrobiologia* 297/280:107–119.

Hubert, W. A., and J. N. Krull. 1973. Seasonal fluctuations of aquatic macroinvertebrates in Oakwood Bottoms Greentree Reservoir. *American Midland Naturalist* 90:351–364.

Humburg, D. D., and K. M. Babcock. 1982. *Lead poisoning and lead/steel shot.* St. Louis, Missouri: Missouri Department of Conservation Technical Report (Terrestrial Series) 10.

Imber, M. J. 1971. The identity of New Zealand's Canada geese. *Notornis* 18:253–261.

Jepsen, P. U. 1983. Vildtreservaterns som st i den samlede naturforvaltning. *Proceedings of the 3rd Nordic Ornithological Conference 1981,* 133–142. Meltaus, Sweden: Nordic Council for Wildlife Research.

Johansson-Sjobeck, M. L., and A. Larsson. 1979. Effect of inorganic lead on delta-aminolevulinic acid dehydratase activity and hematological variables in the rainbow trout *Salmo gairdneri. Archives Environmental Contamination and Toxicology* 8:419–431.

Johnsgard, P. A. 1967. Sympatry changes and hybridization incidence in mallards and black ducks. *American Midland Naturalist* 77:51–63.

Johnsgard, P. A., and R. Di Silvestro. 1976. Seventy-five years of changes in mallard-black duck ratios in eastern North America. *American Birds* 30:905–908.

Johnson, M. A., ed. 1983. *Transactions of the Canada goose symposium.* Bismarck, North Dakota: The Wildlife Society.

Jones, J. C. 1939. On the occurrence of lead shot in stomachs of North American gruiformes. *Journal of Wildlife Management* 3:353–357.

Jordan, J. S., and F. C. Bellrose. 1951. Lead poisoning in wild waterfowl. *Illinois Natural History Survey Biological Notes* 26:1–27.

Kadlec, J. A. 1962. Effects of a drawdown on a waterfowl impoundment. *Ecology* 43:267–281.

Kadlec, J. A., and L. M. Smith. 1992. Habitat management for breeding areas. In *Ecology and management of breeding waterfowl,* ed. B. D. Batt, A. D. Afton, M. G. Anderson, C. Davison-Ankney, D. H. Johnson, J. A. Kadlec, and G. L. Krapu, 590–610. Minneapolis and London: University of Minnesota Press.

Kaiser, G. W., K. Fry, and J. G. Ireland. 1980. Ingestion of lead shot by dunlin. *The Murrelet* 61:37.

Kaminski, R. M. 1979. Dabbling duck and aquatic invertebrate responses to manipulated wetland habitat. Doctoral dissertation, Michigan State University, Ann Arbor.

Kaminski, R. M., and H. H. Prince. 1981. Dabbling duck activity and foraging responses to aquatic macroinvertebrates. *Auk* 98:115–126.

Kantrud, H. A. 1985. *Effects of vegetation manipulations on breeding waterfowl in prairie wetlands: A literature review.* Technical Report 3. Washington, D.C.: United States Fish and Wildlife Service.

Karr, J. R. 1991. Biological integrity: A long neglected aspect of water resource management. *Ecological Applications* 1:66–84.

Kear, J. 1990. *Man and wildfowl.* London: T. and A. D. Poyser.

Kennedy, J. A., and T. Nadeau. 1993. *Lead shot contamination of waterfowl and their habitats in Canada.* Canadian Wildlife Service Technical Report Series 164. Ottawa: Canadian Wildlife Service.

Kiørboe, T. 1980. Distribution and production of submerged macrophytes in Tipper Grund (Ringkøbing Fjord, Denmark), and the impact of waterfowl grazing. *Journal of Applied Ecology* 17:675–687.

Knighton, M. D., ed. 1985. *Water impoundments for wildlife: A habitat management workshop.* St. Paul, Minnesota: United States Department of Agriculture, Forest Service, North Central Forest Experimental Station.

Knopf, F. L., J. A. Sedgwick, and R. W. Cannon. 1988. Guild structure of a riparian avifauna relative to seasonal cattle grazing. *Journal of Wildlife Management* 52:280–290.

Koh, T. S., and M. J. Harper. 1988. Lead poisoning in black swans, *Cygnus atratus,* exposed to spent lead shot at Bool Lagoon Game Reserve, South Australia. *Australian Wildlife Research* 15:395–403.

Koranda, J., K. Moore, M. Stuart, and C. Conrado. 1979. *Dietary effects of lead uptake and trace element distribution in mallard ducks dosed with lead shot.* Lawrence Livermore Laboratory Report. Livermore, California: Lawrence Livermore Laboratory.

Lampio, T. 1983. Waterfowl hunting in Europe, North America and some African and Asian countries in 1980–81. *Finnish Game Research* 40:5–35.

Landry, P. 1990. Hunting harvest of waterfowl in the Western Palearctic and Africa. In *Managing waterfowl populations,* ed. G.V.T. Matthews, 120–121. Special Publication 12. Slimbridge, England: International Waterfowl and Wetlands Research Bureau.

Larsson, T. 1994. Lead poisoning update: Sweden. *IWRB News* 12:11. Slimbridge, England.

Lecocq, Y. 1993. Wise use of waterfowl: A European perspective. In *Waterfowl and wetland conservation in the 1990s: A global perspective,* ed. M. Moser, R. C. Prentice, and J. van Vessem, 85–86. Special Publication 26. Slimbridge, England: International Waterfowl and Wetlands Research Bureau.

Lee, F. B., C. H. Schroeder, T. L. Kuck, L. J. Schoonover, M. A. Johnson, H. K. Nelson, and C. A. Beauduy. 1984. *Rearing and stocking giant Canada geese in the Dakotas.* Bismarck, North Dakota: North Dakota Game and Fish Department.

Lévesque, H., B. Collins, and A. M. Legris. 1993. *Migratory game birds harvested in Canada during the 1991 hunting season.* Progress Notes 204. Ottawa, Ontario, Canada: Canadian Wildlife Service.

Linde, A. F. 1985. Vegetation management in water impoundments: Alternatives and supplements to water-level control. In *Water impoundments for wildlife: A habitat management workshop,* ed. M. D. Knighton, 51–

60. General Technical Report NC-100. St. Paul, Minnesota: United States Department of Agriculture, Forest Service, North Central Experimental Station.

Linduska, J. P. 1982. Sanctuaries in waterfowl management in the USA. In *Managing wetlands and their birds,* ed. D. A. Scott, 319–324. Slimbridge, England: International Waterfowl and Wetlands Research Bureau.

Locke, L. N., and M. Friend. 1987. Avian botulism. In *Field guide to wildlife diseases: General field procedures and diseases of migratory birds,* ed. M. Friend, 83–94. Resource Publication 167. Madison, Wisconsin: National Wildlife Health Center.

Locke, L. N., and M. Friend. 1992. Lead poisoning of avian species other than waterfowl. In *Lead poisoning in waterfowl,* ed. D. J. Pain, 19–22. Special Publication 16. Slimbridge, England: International Waterfowl and Wetlands Research Bureau.

Lund, M., J. Davis, and F. Murray. 1991. The fate of lead from duck shooting and road run-off in three western Australian wetlands. *Australian Journal of Marine and Freshwater Resources* 42:139–149.

McAllister-Eadie, J., and A. Keast. 1982. Do goldeneye and perch compete for food? *Oecologia* 55:225–230.

McCoy, M. B., J. M. Rodríguez Ramírez, and J. L. Altuve Marenco. 1992. Reproductive success and population increase of black-bellied whistling ducks (*Dendrocygna autumnalis*) in newly placed artificial nests in a tropical freshwater marsh. In *Wildlife 2001: Populations,* ed. D. R. McCullough and R. H. Barrett, 653–664. Barking, England: Elsevier Science.

Macdonald, I.A.W., L. L. Loope, M. B. Usher, and O. Hamann. 1989. Wildlife conservation and the invasion of nature reserves by introduced species: A global perspective. In *Biological invasions: A global perspective,* ed. J. A. Drake, H. A. Mooney, F. di Castri, R. H. Groves, F. J. Kruger, M. Rejmánek, and M. Williamson. Chichester, New York, Toronto, and Singapore: John Wiley & Sons.

McLaughlin, C. L., and D. Grice. 1952. The effectiveness of large-scale erection of wood duck boxes as a management procedure. *Transactions of the North American Wildlife and Natural Resources Conference* 17: 242–259.

McNicol, D. K., and M. Wayland. 1992. Distribution of waterfowl broods in Sudbury area lakes in relation to fish, macroinvertebrates and water chemistry. *Canadian Journal of Fisheries and Aquatic Sciences* 49(suppl. 1):122–133.

Madsen, J. 1985. Impact of disturbance on field utilization of pink-footed geese in west Jutland, Denmark. *Biological Conservation* 33:53–63.

Madsen, J. 1988. Autumn feeding ecology of herbivorous wildfowl in the Danish Wadden Sea and impact of food supplies and shooting on movements. *Danish Review of Game Biology* 13:1–32.

Madsen, J. 1993. Managing hunting disturbance for wise use of Danish waterfowl. In *Waterfowl and wetland conservation in the 1990s: A global perspective,* ed. M. Moser, R. C. Prentice, and J. van Vessem, 93–96. Special Publication 26. Slimbridge, England: International Waterfowl and Wetlands Research Bureau.

Madsen, J. 1995. Impacts of disturbance on migratory waterfowl. *Ibis* 137:S67-S74.

Madsen, J., and A. Andersson. 1990. Status and management of *Branta canadensis* in Europe. In *Managing waterfowl populations,* ed. G.V.T. Matthews, 66–69. Special Publication 12. Slimbridge, England: International Waterfowl and Wetlands Research Bureau.

Madsen, J., and P. U. Jepsen. 1992. Passing the buck: Need for a flyway management plan for the Svalbard pink-footed goose. In *Waterfowl and agriculture: Review and future prospects of the crop damage conflict in Europe,* ed. M. van Roomen and J. Madsen, 109–110. Special Publication 21. Slimbridge, England: International Waterfowl and Wetlands Research Bureau.

Maher, M. 1982. Response by waterfowl to hunting pressure: A preliminary study. *Australian Wildlife Research* 9:527–531.

Manny, B. A., W. C. Johnson, and R. G. Wetzel. 1994. Nutrient additions by waterfowl to lakes and reservoirs: Predicting their effects on productivity and water quality. *Hydrobiologia* 279/280:121–132.

Manuel, P. 1987. The failure of wetland conservation programs to accommodate peatland environments: A case study of the maritime provinces. In *Proceedings from wetlands/peatlands symposium,* ed. C.D.A. Rubec and R. B. Overend, Edmonton Convention Centre, Alberta, Canada.

Marchant, S., and P. J. Higgins, eds. 1990. *Handbook of Australian, New Zealand and Antarctic birds,* vol. 1, *Ratites to ducks.* Oxford, Aukland, and New York: Oxford University Press.

Marion, L., P. Clergeau, L. Brient, and G. Bertru. 1994. The importance of avian-contributed nitrogen (N) and phosphorus (P) to Lake Grande-Lieu, France. *Hydrobiologia* 279/280:133–147.

Martin, E. M., and P. I. Padding. 1993. *Preliminary estimates of waterfowl harvest and hunter activity in the United States during the 1992 hunting season.* Administrative report. Patuxent, Maryland: United States Fish and Wildlife Service.

Martindale, J. 1984. *Counts of freckled duck Stictonetta naevosa in eastern Australia during January–February 1983.* Royal Australian Ornithologists' Union Report 13. Victoria, Australia: RAOU.

Master, T. L., and C. S. Oplinger. 1984. Nesting and brood rearing ecology of an urban waterfowl population (*Anas platyrhynchos* and *Branta canadensis*) in Allentown, Pennsylvania. *Proceedings of the Pennsylvania Academy of Science* 58:175–180.

Matena, J., V. Vyhnálek, and K. Šimek. 1994. Food web management. In *Restoration of lake ecosystems: A holistic approach*, ed. M. Eiseltová, 97–102. Special Publication 32. Slimbridge, England: International Waterfowl and Wetlands Research Bureau.

May, R. M. 1994. Disease and the abundance and distribution of bird populations: A summary. *Ibis* 137:S85-S86.

Mayhew, P. W. 1988. The daily energy intake of European wigeon in winter. *Ornis Scandinavica* 19:217–223.

Mazourek, J. C., and P. N. Gray. 1994. The Florida duck or the mallard? We can't have both. *Florida Wildlife* 48:29–31.

Meeks, R. L. 1969. The effects of drawdown date on wetland plant succession. *Journal of Wildlife Management* 33:817–821.

Meire, P., and E. Kuijken. 1991. Factors affecting the number and distribution of wintering geese and some implications for their conservation in Flanders, Belgium. *Ardea* 79:143–158.

Meister, V. B., and J. Kösters. 1981. Weitere Untersuchungen zur Bleivergiftung bei Greifvögeln. *Der Praktische Tierarzt* 10:870–879.

Melanson, R., and F. Payne. 1988. *Wetlands habitat management through nutrient enrichment.* Nova Scotia Department of Lands and Forests Occasional Paper 1. Halifax, Novia Scotia.

Meltofte, H. 1978. Skudeffektivitet ved intensiv kystfuglejagt i Danmark: En pilotundersøgelse. *Dansk Ornithologisk Forenings Tidsskrift* 72: 217–221.

Meltofte, H. 1982. Jagtlige forstyrrelser af svømme- og vadefugle. *Dansk Ornithologisk Forenings Tidsskrift* 76:21–35.

Meltofte, H. 1989. *Dansk Rastepladser for Vadefugle. Vadefuletaellinger i Danmark 1974–1978*. Vadefuglegruppen, Dansk Ornithologisk Forening Miljoministeriet. Meltaus, Finland: Nordic Council for Wildlife Research.

Meltofte, H. In press. A new Danish hunting and wildlife management law: The result of mutual understanding and agreement between hunters and non-hunters. *Gibier Faune Sauvage.*

Merendino, M. T., and C. D. Ankney. 1994. Habitat use by mallards and American black ducks breeding in central Ontario. *Condor* 96:411–421.

Merendino, M. T., C. D. Ankney, and D. G. Dennis. 1993. Increasing mallards, decreasing American black ducks: More evidence for cause and effect. *Journal of Wildlife Management* 57:199–208.

"Mexican Duck" not a Mexican Duck. 1978. *Oryx* 14:307.

Mikkola, H., and E. A. Lind. 1974. Hailuodon sorsista ja niiden kayttaytymisesta metastyskauden alkaessa. *Soumen Riista* 25:20–28.

Miller, M. R., J. Beam, and D. P. Connelly. 1988. Dabbling duck harvest dynamics in the Central Valley of California: Implications for recruitment. In *Waterfowl in winter,* ed. M. W. Weller, 553–569. Minneapolis: University of Minnesota Press.

Mitchell, C., R. L. Cromie, and M. J. Brown. 1993. *Mallard as potential disease vectors: A case study at Slimbridge.* JNCC Report 162. A report by the Wildfowl and Wetlands Trust to the Joint Nature Conservation Committee. Peterborough, England. Unpublished report.

Mooij, J. 1991. Numbers and distribution of grey geese (genus *Anser*) in the Federal Republic of Germany, with special reference to the Lower Rhine region. *Ardea* 79:125–134.

Morehouse, K. A. 1992a. Crippling loss and shot-type: The United States experience. In *Lead poisoning in waterfowl,* ed. D. J. Pain, 32–37. Special Publication 16. Slimbridge, England: International Waterfowl and Wetlands Research Bureau.

Morehouse, K. A. 1992b. Lead poisoning of migratory birds: The US Fish & Wildlife Service position. In *Lead poisoning in waterfowl,* ed. D. J. Pain, 51–55. Special Publication 16. Slimbridge, England: International Waterfowl and Wetlands Research Bureau.

Morin, P. J. 1984. The impact of fish exclusion on the abundance and species composition of larval odonates: Results of short-term experiments in a North Carolina farm pond. *Ecology* 65:53–60.

Morozov, V. In press. Status, distribution and trends of the lesser white-fronted goose (*Anser erythropus*) population in Russia. *Gibier Faune Sauvage.*

Morris, M. J., and R. Plant. 1983. Response of grassland invertebrates to management by cutting. V. Changes in Hemiptera following cessation of management. *Journal of Applied Ecology* 20:157–177.

Mudge, G. P. 1983. The incidence and significance of ingested lead pellet poisoning in British wildfowl. *Biological Conservation* 27:333–372.

Mudge, G. P. 1989. *Night shooting of wildfowl in Great Britain: An assessment of its prevalence, intensity and disturbance impact.* A report by the Wildfowl and Wetlands Trust to the Nature Conservancy Council, Peterborough, England.

Mudge, G. P. 1992. Options for alleviating lead poisoning: A review and assessment of alternatives to the use of non-toxic shot. In *Lead poisoning in waterfowl,* ed. D. J. Pain, 23–25. Special Publication 16. Slimbridge, England: International Waterfowl and Wetlands Research Bureau.

Murkin, H. R., J. B. Pollard, M. P. Stainton, J. A. Boughen, and R. D. Titman. 1994. Nutrient additions to wetlands in the Interlake region of Manitoba, Canada: Effects of periodic additions throughout the growing season. *Hydrobiologia* 279/280:483–495.

Newbold, C. 1975. Herbicides in aquatic systems. *Biological Conservation* 7:97–118.

Newton, I. 1994. Experiments on the limitation of bird breeding densities: A review. *Ibis* 136:397–411.

Nichols, J. D., and J. E. Hines. 1983. The relationship between harvest and survival rates of mallards: A straightforward approach with partitioned data sets. *Journal of Wildlife Management* 47:334–348.

Nichols, J. D., M. J. Conroy, D. R. Anderson, and K. P. Burnham. 1984. Compensatory mortality in waterfowl populations: A review of the evidence and implications for research and management. *Transactions of the North American Wildlife and Natural Resources Conference* 49: 535–554.

Nichols, J. D., M. Williams, and T. Caithness. 1990. Survival and band recovery rates of mallards in New Zealand. *Journal of Wildlife Management* 54:629–636.

Nilsson, L. 1980. Wintering diving duck populations and available food resources in the Baltic. *Wildfowl* 31:131–143.

Norman, F. I. 1971. Rearing, release, and recovery of hand-reared chestnut teal *Anas castanea* (Eyton) in Victoria. *Search* 2:138–140.

Norman, F. I. 1976. The incidence of lead shotgun pellets in waterfowl (Anatidae and Rallidae) examined in southeastern Australia between 1957 and 1973. *Australian Wildlife Research* 3:61–71.

Norman, F. I. 1987. *The ducks of Macquarie Island.* Australian National Antarctic Research Notes 42. Kingston, Tasmania, Australia: Australian National Anarctic Research Expeditions.

Norman, F. I. 1990. Macquarie Island ducks: Habitats and hybrids. *Notornis* 37:53–58.

Norman, F. I., and T. L. Riggert. 1977. Nest boxes as nest sites for Australian waterfowl. *Journal of Wildlife Management* 41:643–649.

Norriss, D. W., and H. J. Wilson. 1988. Disturbance and flock size changes in Greenland white-fronted geese wintering in Ireland. *Wildfowl* 39:63–70.

Olson, D. P. 1965. Differential vulnerability of male and female canvasbacks to hunting. *Transactions of the North American Wildlife and Natural Resources Conference* 30:121–134.

Osbourne, P. 1982. Some effects of Dutch elm disease on nesting farmland birds. *Bird Study* 29:2–16.

Ovington, J. D. 1984. Ecological processes and national park management. In *National parks, conservation and development: The role of protected areas in sustaining society,* ed. J. A. McNeely and K. R. Miller, 60–64. Washington, D.C.: International Union for the Conservation of Nature and Natural Resources/Smithsonian Institution Press.

Owen, M. 1993. The UK shooting disturbance project. *Wader Study Group Bulletin* 68:35–46.

Owen, M., and J. M. Black. 1989. Factors affecting the survival of barnacle geese on migration from the breeding grounds. *Journal of Animal Ecology* 58:603–618.

Owen, M., and J. M. Black. 1990. *Waterfowl ecology.* Glasgow and London: Blackie & Son.

Owen, M., J. M. Black, and H. Liber. 1988. Pair bond duration and timing of its formation in barnacle geese *Branta leucopsis.* In *Wildfowl in winter,* ed. M. W. Weller, 257–269. Minneapolis: University of Minnesota Press.

Owen, M., J. S. Kirby, and D. G. Salmon. In press. Canada geese in Great Britain: History, problems and prospects. *Journal of Wildlife Management.*

Owen, M., and D. G. Salmon. 1988a. Introductions of wildfowl in Europe: Problems and prospects. Unpublished report. The Wildfowl and Wetlands Trust, Slimbridge, England.

Owen, M., and D. G. Salmon. 1988b. The importance and management of intertidal areas for wintering and passage wildfowl. Unpublished report. The Wildfowl and Wetlands Trust, Slimbridge, England.

Owen, M., and G. J. Thomas. 1979. The feeding ecology and conservation of wigeon wintering at the Ouse Washes, England. *Journal of Applied Ecology* 16:795–809.

Owens, N. W. 1977. Responses of wintering Brent geese to human disturbance. *Wildfowl* 28:5–14.

Pain, D. J. 1990. Lead shot ingestion by waterbirds in the Camargue, France: An investigation of levels and interspecific differences. *Environmental Pollution* 66:273–285.

Pain, D. J. 1991a. L'intoxication saturnine de l'avifaune: Une synthèse des travaux français. *Gibier Faune Sauvage* 8:79–92.

Pain, D. J. 1991b. Why are lead-poisoned waterfowl rarely seen? The disappearance of waterfowl carcasses in the Camargue, France. *Wildfowl* 42:118–122.

Pain, D. J. 1992a. Lead poisoning of waterfowl: A review. In *Lead poisoning in waterfowl,* ed. D. J. Pain, 7–13. Special Publication 16. Slimbridge, England: International Waterfowl and Wetlands Research Bureau.

Pain, D. J. 1992b. National reports. In *Lead poisoning in waterfowl,* ed. D. J. Pain. Special Publication 16. Slimbridge, England: International Waterfowl and Wetlands Research Bureau.

Pain, D. J., and C. Amiard-Triquet. 1993. Lead poisoning of raptors in France and elsewhere. *Ecotoxicology and Environmental Safety* 25:183–192.

Pain, D. J., C. Amiard-Triquet, C. Bavoux, G. Burneleau, L. Eon, and P. Nicolau-Guillaumet. 1993. Lead poisoning in wild populations of marsh harriers *Circus aeruginosus* in the Camargue and Charente-Maritime, France. *Ibis* 135:379–386.

Pain, D. J., C. Amiard-Triquet, and C. Sylvestre. 1992. Tissue lead concentrations and shot ingestion in nine species of waterbirds from the Camargue (France). *Ecotoxicology and Environmental Safety* 24:217–233.

Pain, D. J., and G. I. Handrinos. 1990. The incidence of ingested lead shot in ducks of the Evros Delta, Greece. *Wildfowl* 41:167–170.

Paine, R. T. 1974. Intertidal community structure: Experimental studies on the relationship between a dominant competitor and its principal predator. *Oecologia* 15:93–120.

Paine, R. T. 1980. Food webs: Linkage interaction strength and community structure. *Journal of Animal Ecology* 49:667–685.

Palmer, R. S., ed. 1976. *Handbook of North American birds,* vol. 2, part 1. New Haven and London: Yale University Press.

Palmisano, A. W., Jr. 1972. The effects of salinity on the germination and growth of plants important in the Gulf Coast marshes. *Proceedings of the Annual Conference of the South Eastern Association of Game and Fish Commissioners* 25:215–223.

Paton, J. B., R. Storr, L. Delroy, and L. Best. 1992. Patterns to the distribution and abundance of mallards, Pacific black ducks and their hybrids in South Australia in 1987. *South Australian Ornithology* 31:103–110.

Pattee, O. H., and S. K. Hennes. 1983. Bald eagles and waterfowl: The lead shot connection. *Transactions of the North American Wildlife and Natural Resources Conference* 48:230–237.

Pfister, C., B. A. Harrington, and M. Levine. 1992. The impact of human disturbance on shorebirds at a migration staging area. *Biological Conservation* 60:115–126.

Pickett, S.T.A., V. T. Parker, and P. L. Fielder. 1992. The new paradigm in ecology: Implications for conservation above the species level. In *Conservation biology: The theory and practice of nature conservation, preservation and management,* ed. P. L. Fielder and S. K. Jain, 65–88. New York and London: Chapman & Hall.

Pierce, D. 1988. Wise use of wetlands: A position paper. In *Convention relative aux zones humides d'importance internationale particulièrement comme habitats des oiseaux d'eau,* 401–402. Procès-verbaux de la troisième session de la conférence des parties contractantes. Regina, Saskatchewan, Canada, May 27–June 5, 1987.

Pletscher, D. H., and R. L. Hutto. 1991. Wildlife management and the maintenance of biological diversity. *Western Wildlands* Fall:8–12.

Post, J. R., and D. Cucin. 1984. Changes in the benthic community of a small pre-Cambrian lake following the introduction of yellow perch (*Perca flavescens*). *Canadian Journal of Fisheries and Aquatic Sciences* 41:1496–1501.

Prevett, J. P., and C. D. MacInnes. 1980. Family and other social groups in snow geese. *Wildlife Monograph* 71:1–46.

Ramsar. 1990. Report of the working group on criteria and wise use. Unpublished report. Ramsar Bureau, Gland, Switzerland.

Ratcliffe, D. 1979. The end of the large blue butterfly. *New Scientist* 8:457–458.

Raveling, D. G., W. E. Crews, and W. P. Klimstra. 1972. Activity patterns of Canada geese during winter. *Wilson Bulletin* 84:278–295.

Rawes, M. 1983. Changes in two high altitude blanket bogs after cessation of sheep grazing. *Journal of Ecology* 71:219–235.

Reichel, W. L., S. K. Schmeling, E. Cromarthie, T. E. Kaiser, A. J. Krynitsky, T. G. Lanout, B. M. Mulhern, R. M. Prouty, C. J. Stafford, and D. M. Swineford. 1984. Pesticide, PCB, and lead residues and necropsy data for bald eagles from 32 states: 1978–1981. *Environmental Monitoring and Assessment* 4:395–403.

Reichholf, J. 1973. Begründung einer ökologischen Strategie der Jagd auf Enten (Anatinae). *Anzeiger der Ornithologischen Gesellschaft in Bayern* 12:237–247.

Reid, F. A. 1985. Wetland invertebrates in relation to hydrology and water chemistry. In *Water impoundments for wildlife: A habitat management workshop,* ed. M. D. Knighton. General Technical Report NC-100. St. Paul, Minnesota: United States Department of Agriculture, Forest Service, North Central Experimental Station.

Reimold, R. J., R. A. Linthurst, and P. L. Wolf. 1975. Effects of grazing on a salt marsh. *Biological Conservation* 8:105–125.

Reinecke, K. J., and C. W. Shaiffer. 1988. A field test for differences in condition among trapped and shot mallards. *Journal of Wildlife Management* 52:227–232.

Rogers, J. P., and J. H. Patterson. 1984. The black duck population and its management. *Transactions of the North American Wildlife and Natural Resources Conference* 49:527–534.

Rollings, C. T., and R. L. Warden. 1964. Weedkillers and waterfowl. In *Waterfowl tomorrow,* ed. J. P. Linduska, 593–598. Washington, D.C.: United States Fish and Wildlife Service.

Rudfeld, L. 1990. *25 års beskyttelse af Vadehavet.* Copenhagen: Miljoministeriet, Skov-og Naturstyrelsen.

Rundle, W. D., and L. H. Fredrickson. 1981. Managing seasonally flooded impoundments for migratory rails and shorebirds. *Wildlife Society Bulletin* 9:80–87.

Rusch, D. H., C. D. Ankney, H. Boyd, J. R. Longcore, F. Montalbano III, J. K. Ringelman, and V. D. Stotts. 1989. Population ecology and harvest of the American black duck: A review. *Wildlife Society Bulletin* 17:379–406.

Ryman, N., R. Baccus, C. Reuterwall, and M. H. Smith. 1981. Effective population size, generation interval, and potential loss of genetic variability in game species under different hunting regimes. *Oikos* 36:257–266.

Salwasser, H. 1990. Conserving biological diversity: A perspective on scope and approaches. *Forest Ecology and Management* 35:75–90.

Sanderson, G. C., and F. C. Bellrose. 1986. *A review of the problem of lead poisoning in waterfowl.* Illinois Natural History Survey Special Publication 4. Champaign, Illinois: Illinois Natural History Survey.

Sanderson, G. C., H. W. Norton, and S. S. Hurley. 1981. *Effects of ingested lead-iron shot on mallards.* Illinois Natural History Survey Biological Notes 116. Champaign, Illinois: Illinois Natural History Survey.

Savard, J. L. 1988. Use of nest boxes by Barrow's goldeneyes: Nesting success and effect on the breeding population. *Wildlife Society Bulletin* 16:125–132.

Schladweiler, J. L., and J. R. Tester. 1972. Survival and behaviour of hand-reared mallards released in the wild. *Journal of Wildlife Management* 36:1118–1127.

Schmitz, R. A., A. Alonso Aguirre, R. S. Cook, and G. A. Baldassare. 1990. Lead poisoning of Caribbean flamingos in Yucatan, Mexico. *Wildlife Society Bulletin* 18:399–404.

Schneider-Jacoby, M., P. Frenzel, H. Jacoby, G. Knötzsch, and K. H. Kolb. 1991. The impact of shooting disturbance on a protected species, the whooper swan *Cygnus cygnus* at Lake Constance. *Wildfowl Supplement* 1:378–382.

Schricke, V. In press. An investigation of lead poisoning in ducks in France. *Gibier Faune Sauvage.*

Sellers, R. A. 1973. Mallard releases in under-stocked prairie pothole habitat. *Journal of Wildlife Management* 37:10–22.

Seymour, N. R. 1990. Forced copulations in sympatric American black ducks and mallards in Nova Scotia. *Canadian Journal of Zoology* 68:1691–1696.

Shoffner, R. N. 1972. A summary of the genetic implications of hand-reared birds introduced into wild populations. In *Role of hand-reared ducks in waterfowl management: A symposium,* ed. A. T. Studholme, 113–116. Dundee, Illinois: Max McGraw Wildlife Foundation.

Simenstad, C. A., J. A. Estes, and K. Kenyon. 1978. Aleuts, sea otters, and alternate stable-state communities. *Science* 200:403–411.

Smit, C. J., and G.J.M. Visser. 1993. Effects of disturbance on shorebirds: A summary of existing knowledge from the Dutch Wadden Sea and Delta. *Wader Study Group Bulletin* 68:6–19.

Smith, L. M., R. L. Pederson, and R. M. Kaminski. 1989. *Habitat management for migrating and wintering waterfowl in North America.* Lubbock, Texas: Texas Technical University Press.

Solman, V.E.F. 1945. The ecological relations of pike, *Esox lucius* L., and waterfowl. *Ecology* 26:157–170.

Soutiere, E. C. 1989. Survival rates of hand-reared mallards released on two private farms. *Journal of Wildlife Management* 53:114–118.

Sparrowe, R. D. 1993. What is wise use of waterfowl populations? In *Waterfowl and wetland conservation in the 1990s: A global perspective,* ed. M. Moser, R. C. Prentice, and J. van Vessem, 85–86. Special Publication 26. Slimbridge, England: International Waterfowl and Wetlands Research Bureau.

Spray, C. J., and H. Milne. 1988. The incidence of lead poisoning among whooper and mute swans *Cygnus cygnus* and *C. olor* in Scotland. *Biological Conservation* 44:265–281.

Stendell, R. C., J. W. Artmann, and E. Martin. 1980. Lead residues in Sora rails from Maryland. *Journal of Wildlife Management* 44:525–527.

Stenson, J.A.E., T. Bohlin, L. Henrikson, B. I. Nilsson, H. G. Nymsan, H. G. Oscarson, and P. Larsson. 1978. Effects of fish removal from a small lake. *Verh. Int. Verein. Limno.* 20:794–801.

Stott, R. S., and D. P. Olson. 1972. Differential vulnerability patterns among three species of sea ducks. *Journal of Wildlife Management* 36:775–783.

Strange, T. H., E. R. Cunningham, and J. W. Goertz. 1971. Use of nest boxes by wood ducks in Mississippi. *Journal of Wildlife Management* 35:786–793.

Studholme, A. T., ed. 1972. *Role of hand-reared ducks in waterfowl management: A symposium.* Dundee, Illinois: Max McGraw Wildlife Foundation.

Suter, W. 1994. Overwintering waterfowl on Swiss lakes: How are abundance and species richness influenced by trophic status and lake morphology? *Hydrobiologia* 279/280:1–14.

Swift, J. A., and A. R. Laws. 1982. The release of waterfowl into the wild with particular reference to restocking programmes. In *Managing wetlands and their birds,* ed. D. A. Scott, 283–287. Slimbridge, England: International Waterfowl and Wetlands Research Bureau.

Tamisier, A. 1985. Hunting as a key environmental parameter for the Western Palearctic duck populations. *Wildfowl* 27:19–32.

Tamisier, A., and T. Saint-Gerand. 1981. Stationnements d'oiseaux d'eau et chasse de nuit dans les départements côtiers de France. *Alauda* 49:81–93.

Taris, J.-P., and Y. Bressac-Vaquer. 1987. *Oiseaux migrateurs transcontinenteaux: Cas particulier de la barge à queue noir, Limosa limosa* (L.), *en Camargue.* Rapport Office National de la Chasse. Paris: ONC.

Tewari, H., T. S. Gill, and J. Pant. 1987. Impact of chronic lead poisoning on the hematological and biochemical profiles of a fish, *Barbus conchonius* (Ham). *Bulletin of Environmental Contamination and Toxicology* 38:748–752.

Thomas, G. J. 1975. Ingested lead pellets in waterfowl at the Ouse Washes, England, 1968–1973. *Wildfowl* 26:43–48.

Thomas, G. J. 1978. Breeding and feeding ecology of waterfowl at the Ouse Washes, England. Doctoral dissertation, University of East Anglia, Norwich, England.

Thomas, G. J. 1982. Management of vegetation at wetlands. In *Managing wetlands and their birds,* ed. D. A. Scott, 21–37. Slimbridge, England: International Waterfowl and Wetlands Research Bureau.

Thomas, V. G. 1994. Lead shot in the environment: Its role in toxicosis and remediation of the problem. *OECD Workshop on Lead Products, 12–15 September 1994, Toronto, Canada.*

Trost, R., K. Dickson, and D. Zavaleta. 1993. Harvesting waterfowl on a sustained yield basis: The North American perspective. In *Waterfowl and wetland conservation in the 1990s: A global perspective,* ed. M. Moser, R. C. Prentice, and J. van Vessem, 106–112. Special Publication 26. Slimbridge, England: International Waterfowl and Wetlands Research Bureau.

Trouvilliez, J., R. Sparrowe, and F. Duverney. In press. A global review of the status of hunting of Anatidae. *Gibier Faune Sauvage.*

Tulgat, R., and G. B. Schaller. 1992. Status and distribution of wild Bactrian camels *Camelus bactrianus ferus. Biological Conservation* 62:11–19.

Ueno, O. 1989. The hunting regulation system in Japan. In *Symposium of Wildlife in Japan,* ed. I. S. Won, 10–15. Tokyo: Wildlife Society of Japan.

Urbánek, B. In press. The mallard (*Anas platyrhynchos*): An endangered species in central Europe. *Gibier Faune Sauvage.*

USFWS (United States Fish and Wildlife Service). 1985. *Steel: Final environmental statement. Proposed use of steel shot for hunting waterfowl in the United States.* Washington, D.C.: USFWS.

USFWS. 1986a. *Final environmental impact statement: Use of lead shot for hunting migratory birds in the United States.* Washington, D.C.: USFWS.

USFWS. 1986b. *Effects of vegetation manipulation on breeding waterfowl in prairie wetlands: A literature review.* Washington, D.C.: USFWS.

USFWS. 1988. *Final supplemental environmental impact statement: Issuance of annual regulations permitting the sport hunting of migratory birds.* Washington, D.C.: USFWS.

USFWS. 1993. *Draft environmental impact statement. Refuges 2003: A plan for the future of the National Wildlife Refuge System.* Washington, D.C.: USFWS.

USFWS. 1994. *Waterfowl: Population status, 1994.* Washhington, D.C.: USFWS.

Vikberg, P., and P. Moilanen. 1985. Introduction of the Canada goose in Finland. *Suomen Riista* 32:50–56.

Voights, D. K. 1976. Aquatic invertebrate abundance in relation to changing marsh conditions. *American Midland Naturalist* 95:313–322.

Wallace, B. M., R. J. Warren, and G. D. Gaines. 1983. Lead shot incidence in sandhill cranes collected from Alaska, Canada and Texas. *Prairie Naturalist* 15:155–156.

Wegener, W., V. Williams, and T. D. McCall. 1974. Aquatic macroinvertebrates responses to an extreme drawdown. *Proceedings of the Annual Conference of the Southeastern Association of Fish and Game Commissioners* 28:126–144.

Weller, M. W. 1969. Potential dangers of exotic waterfowl introductions. *Wildfowl* 20:55–58.

Weller, M. W. 1978. Management of freshwater marshes for wildlife. In *Freshwater wetlands,* ed. R. Good, D. Whigham, and R. Simpson, 267–284. New York: Academic Press.

Weller, M. W., and L. H. Fredrickson. 1974. Avian ecology of a managed glacial marsh. *Living Bird* 12:269–291.

Weller, M. W., and P. Ward. 1959. Migration and mortality of hand-reared redheads (*Aythya americana*). *Journal of Wildlife Management* 23:427–433.

Wentz, W. A., and F. A. Reid. 1992. Managing refuges for waterfowl purposes and biological diversity: Can both be achieved? *Transactions of the North American Wildlife and Natural Resources Conference* 57:581–585.

Whitehead, P. E., and J. E. Elliot. 1993. Incidence of lead contamination in waterfowl and of lead poisoning of bald eagles from British Columbia, 1988–1991. In *Lead shot contamination of waterfowl and their habitats in Canada,* ed. J. A. Kennedy and S. Nadeau, 29–46. Canadian Wildlife Service Technical Report Series 164. Ottawa, Canada: Canadian Wildlife Service.

Whitehead, P. J., and K. Tschirner. 1991. Lead shot ingestion and lead poisoning of magpie geese *Anseranas semipalmata* foraging in a northern Australian hunting reserve. *Biological Conservation* 58:99–118.

Williams, M. J. 1994. Progress in the conservation of New Zealand's threatened waterfowl. *Threatened Waterfowl Research Group Newsletter* 5:3–6. Slimbridge, England.

Williams, M. J., and C. Roderick. 1973. Breeding performance of the grey duck *Anas superciliosa,* mallard *Anas platyrhynchos* and their hybrids in captivity. *International Zoo Yearbook* 13:62–69.

Windingstad, R. M., S. M. Kerr, and L. N. Locke. 1984. Lead poisoning in sandhill cranes (*Grus canadensis*). *Prairie Naturalist* 16:21–24.

Woollhead, J. 1994. Birds in the trophic web of Lake Esrom, Denmark. *Hydrobiologia* 279/280:29–38.

Zicus, M. C. 1990. Nesting biology of the hooded merganser using nest boxes. *Journal of Wildlife Management* 54:637–643.

Ziegler, G., and W. Hanke. 1988. Entwicklung von Stockenten (*Anas platyrhynchos*): Bestanden in der Haverner Marsch unter dem Einfluss der Jagd. *Die Vogelwelt* 109:118–124.

FIFTEEN

Striped Bass Management and Conservation along the Atlantic Coast of the United States

Harold F. Upton

Striped bass (*Morone saxatilis*) fisheries of the New England and the Mid-Atlantic regions are among the oldest and most valuable in the United States. They are also among the most difficult to manage because of the number and diversity of resource users, the extensive migratory range of the species, the large number of jurisdictions that are involved with striped bass management, large annual variations in reproductive success resulting in periods of population decline and abundance, and inherent uncertainties due to our incomplete understanding of striped bass biology. Two main interrelated considerations involve (1) conservation and management of a highly mobile resource throughout its range, and (2) the distribution of benefits among different jurisdictions and resource users. The resulting dilemma involves the common need for all concerned public and private entities to work in concert while each attempts to maximize benefits.

Although striped bass catches have tended historically to fluctuate widely, landings declined precipitously in the late 1970s and early 1980s. Faced with the prospect that striped bass commercial and recreational fisheries might be entirely lost, user groups and state and federal governments came together in support of strict long-term management measures. Quotas, size limitations, and, in some areas, closure of fisheries were instituted in the early and mid-1980s.

Management jurisdiction over striped bass resides with the states for fisheries which occur within three miles* of the coastline. The Atlantic States Marine Fisheries Commission's (ASMFC) Interstate Management Plan for Striped Bass was created in 1981 to coordinate state management actions, and in 1984 the Congress passed the Atlantic Striped Bass Conservation Act to ensure compliance with the interstate plan. Under the act, the secretaries of the Department of the Interior and Department of Commerce may impose a moratorium on fishing for striped bass in states that are found to be out of compliance with the ASMFC management plan. The management plan has succeeded in arresting and reversing striped bass population declines and has been the key to the recovery of commercial and recreational fisheries. This study addresses the market, socioeconomic, institutional, and biological factors that affect the striped bass fisheries along the Atlantic coast of the United States from Maine to North Carolina.

The Use and the Market

Striped bass have been exploited by humans for both subsistence and commercial utilization, first by Native Americans and later by colonists beginning in the 1600s. Indians caught and smoked striped bass in large numbers (Fearing 1903, in Setzler et al. 1980). Colonists also landed large numbers of striped bass, and in addition to fresh consumption, they smoked, pickled, and salted striped bass. Together with salted cod, striped bass were traded in the West Indies (Raney 1952). Captain John Smith wrote that "the Basse is an excellent fish, both fresh & salte. . . . They are so large the head of one will give a good eater a dinner, & for daintinesse of diet they excell the Marybones of Beefe. There are such multitudes that I have seene stopped in the river close adjoining to my house with a sande at one tide as many as will loade a ship of 100 tonnes" (Jordan and Evermann 1902, p. 374).

The first conservation measures in colonial America were enacted for striped bass and codfish. "In 1639, the General Court of the Mas-

*English units of measure are used throughout this chapter since they are the standard units for striped bass regulations and markets.

sachusetts Bay Colony passed a law that neither fish could be sold as fertilizer" (Setzler et al. 1980, p. 51). By 1776, both Massachusetts and New York had passed laws that prohibited sales of striped bass and cod in the winter months. In 1670, an act of the Plymouth Colony required income from the capture of striped bass, herring, and mackerel be used for a free school which became the first public school in the New World.

Throughout the 1800s and into the 1900s, striped bass declines were noticed, but marked with periods of abundance from dominant year-classes. In Maryland during the 1930s, striped bass was the top fish species harvested by value and the third by weight (Vladykov and Wallace 1952). To the present day, striped bass have remained in high demand for both the dinner table and for the fight they give at the end of a fishing line.

Commercial Fisheries

Historically, in the United States striped bass have been harvested commercially in each of the 12 coastal states from Maine to North Carolina and in the Potomac River.* Striped bass commercial activities have been divided into four major regions, which are still somewhat appropriate today (Norton et al. 1984). The regions include New England (Maine, New Hampshire, Massachusetts, Rhode Island, and Connecticut), the Mid-Atlantic (New York, New Jersey, and Delaware), the Chesapeake (Maryland and Virginia), and the South Atlantic (North Carolina). Each region's contribution to total landings has varied over time, but generally one-half to two-thirds of the total has been landed in the Chesapeake region, with the remainder somewhat equally divided among the other three regions. Seasonal fishing patterns in New England and the Mid-Atlantic were largely dependent on seasonal migration of Hudson River and Chesapeake Bay striped bass populations. Striped bass commercial capture techniques depend on region, season, and regulations. Major gear types include rod and reel, pound net, trawl net, gill net, and haul seine.

Total commercial landings fluctuated around an annual average of 10.6 million pounds during the period 1966–75 (ASMFC 1990)

*The Potomac River is considered a separate jurisdiction for the purposes of fishery management. Management is administered by the Potomac River Fisheries Commission.

Table 15.1. 1992 Estimated Striped Bass Landings (Pounds) by Harvest Category along the Atlantic Coast of the United States[a]

State	Commercial landings	Percentage commercial	Recreational landings	Charter boat landings	Trophy season	Percentage recreational	Totals
Maine	N/A	0	63,913	N/A	N/A	1.26	63,913
New Hampshire	N/A	0	43,459	N/A	N/A	0.86	43,459
Massachusetts	239,100	14.13	1,286,131	N/A	N/A	25.44	1,525,231
Rhode Island	39,033	2.31	272,347	N/A	N/A	5.39	311,380
Connecticut	N/A	0	173,554	N/A	N/A	3.43	173,554
New York	226,611	13.39	874,362	N/A	N/A	17.30	1,100,973
New Jersey	N/A	0	760,478	N/A	2,168	15.09	762,646
Delaware	17,795	1.05	30,919	N/A	N/A	0.61	48,714
Maryland	808,980	47.82	844,973	229,259	22,834	21.07	1,906,046
District of Columbia	N/A	0	420	N/A	N/A	0.01	420
Potomac River Fisheries Commission	127,398	7.53	193,738	28,378	N/A	4.39	349,514
Virginia	205,192	12.13	214,455	10,364	N/A	4.45	430,011
North Carolina	27,702	1.64	3,626	N/A	N/A	0.07	31,328
Total	1,691,811		4,762,375	268,001	25,002		6,747,189
Coastwide losses due to incidental capture							252,384
Coastwide losses to hook and line, poaching, and scientific use							2,549,566
Total							9,549,139

Source: Dubovsky and Laney 1993.

[a]N/A, not available.

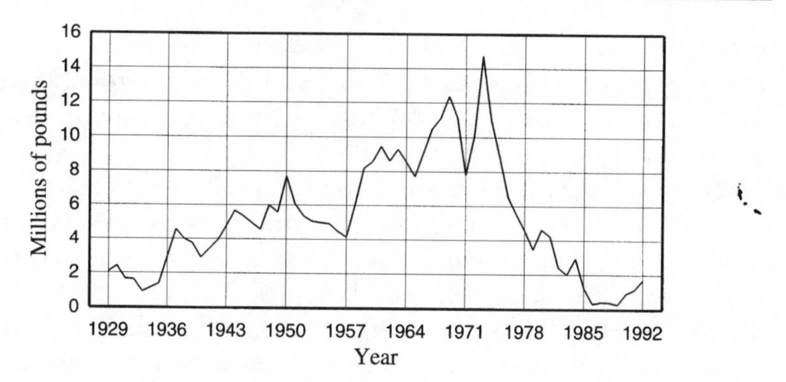

Figure 15.1. East Coast striped bass landings by year from 1929 to 1992. Before 1950, averages of adjacent years were used to fill in for missing years in 1934, 1936, 1941, 1943, and 1946. The North Carolina contribution to total landings was not available in 1933, 1935, 1942, 1944, 1947, 1948, and 1949. (*Source:* ASMFC 1996.)

(Fig. 15.1). Landings peaked in 1973 at 14.7 million pounds and then dropped to 335,000 pounds in 1986 (ASMFC 1990). Landings remained at low levels through 1989 because of low levels of abundance and strict regulation. Though commercial landings have fluctuated widely throughout the last half century, declines have never been as rapid or continuous as that which occurred in the late 1970s and 1980s (Fig. 15.1) (ASMFC 1990). Striped bass are still taken commercially in seven states and in the Potomac River. Commercial fishing is not allowed in the Hudson River due to PCB contamination. In 1992, total commercial landings from North Carolina to Maine were estimated at nearly 1.7 million pounds (Table 15.1).

Before 1981, most of the striped bass taken in the Chesapeake Bay consisted of premigratory fish of ages 2–4 years, most of which are less than 17 inches in length. In 1985, a moratorium was initiated by Maryland on all striped bass commercial fishing in the Maryland area of the Chesapeake Bay. It was reopened in 1990 with interim management measures that included a cap on landings, gear regulations, and a minimum size of 18 inches. Coastal areas are governed by larger minimum size limits of 28 to 36 inches during their commercial seasons.

Although most information collected on striped bass markets spanned the years before the fishery closure of the mid-1980s, much of it is still relevant as commercial fisheries have been reopened and markets reestablished. Striped bass markets include both local and central market outlets. The flow of product is affected by seasonality of landings, regulations, quantity, and size. Fish pass from fishermen to primary wholesalers to local outlets, central markets, and buyers outside the region (Norton et al. 1984). From 1972 to 1978, 65 percent of reported landings passed through central markets in New York and Baltimore (Norton et al. 1984). Central markets are important in determination of prices and as distribution centers. Northeast markets generally handled large fish and fillets, while southern markets were known for pan-size fish. Ex-vessel price was, in part, dependent on the quantity landed, which was usually contingent on availability of fish to fishermen, weather, and work habits. In 1982, ex-vessel or dockside prices were US$1–$2 per pound, wholesale prices were US$2–$3 per pound, and retail prices were US$3–$4 per pound. The wholesale marketing margin or markup in the Northeast and Mid-Atlantic was generally 10–20 percent, depending on market conditions (Norton et al. 1984). Little processing was done before fish were sent for retail sales. Usually, whole striped bass were shipped fresh on ice in boxes. Larger fish were usually filleted, and sometimes headed and gutted for restaurants. The point of final sale was primarily restaurants, fish markets, and supermarkets.

Several major changes have taken place in the reopened fishery under Amendment 4 to the management plan. First, especially in the Chesapeake, landings are now concentrated during the fall and winter commercial seasons of Virginia, Maryland, and the Potomac. The 18-inch minimum size in producer areas has increased average size of fish in the commercial catch and eliminated the market for small pan-size fish.

Second, New York requires extensive sampling for PCB testing of striped bass before they can be sold within the state. This has effectively closed New York to the sale of striped bass that are landed outside its borders. The Fulton Fish Market in New York City, one of the major fish marketing centers in the Northeast, no longer plays a leading role in commercially caught striped bass markets.

Third, culture of the hybrid obtained from crossing striped bass and white bass has made major strides during the last decade. U.S. farm production in 1991 was estimated at 3.5 million pounds, of which 2.3 million pounds were produced in the southeastern United States (Smith and Jenkins 1992). Production in the Southeast is projected to more than double during the 1990s. Market demand for striped bass is not fully satisfied by the aquaculture product. The availability of the cultured product has not driven market prices below the costs of production in the capture fisheries. Wild fish can also be harvested at a larger size, which is preferred for fillets served in restaurants. Cultured hybrid striped bass have the advantage of being available outside the wild commercial season and may actually increase demand for striped bass because it will allow for consistent supplies of product.

Recreational Fisheries

The bulk of striped bass landings, approximately 75 percent by weight, are taken by the recreational sector. Striped bass recreational anglers number more than half a million. Recreational fishing is generally defined as fishermen who are motivated by sport and consumption of fish (Zavolta et al. 1987). The commercial sale of recreationally caught fish is now illegal. However, before the mid-1980s it was common in New England and New York, where many hook-and-line sportsmen sold their catch. The striped bass recreational fishery is difficult to characterize due to its diversity, both with respect to its geography and the characteristics of its participants. Recreational activity includes charter and party boat operations that take paying customers on fishing trips, individuals fishing from noncharter boats, and shore and surf fishermen. Charter trips are marketed in the Chesapeake Bay, along most of the Mid-Atlantic, and in New England. In the Chesapeake Bay region, the striped bass season is a major contributor to charter boat income. In Maryland alone, more than four hundred boats are licensed to participate in the striped bass charter fishery. Due to the diversity and the scattered nature of landings, it is relatively difficult to obtain accurate statistics and economic values for these activities.

All 13 jurisdictions that manage striped bass allow recreational fisheries, with season and size limit varying by region. Recreational landings have equaled or surpassed commercial landings in most years. Landings likely peaked in 1970 at several times the commercial level (Deuel 1973). In the first year of the National Marine Fisheries Service recreational survey in 1979, an estimated 1.2 million striped bass, weighing approximately 6 million pounds, were landed (i.e., caught but not released) by recreational fishermen (Norton et al. 1984). Recreational

Table 15.2. Estimated Total Recreational Catch (1,000 Fish) of Striped Bass, Including Both Harvested and Released Fish, Maine to North Carolina, 1979–1991

State	1979	1980	1981	1982	1983	1984	1985	1986	1987	1988	1989	1990	1991
Maine	<30	<30	<30	<30	<30	177	<30	<30	<30	<30	<30	16	69
New Hampshire	0	<30	<30	0	<30	0	<30	0	<30	<30	<30	15	7
Massachusetts	66	<30	<30	129	68	132	123	655	113	302	236	453	582
Rhode Island	31	<30	<30	<30	<30	72	50	<30	98	31	47	82	46
Connecticut	81	42	<30	555	45	41	41	<30	80	30	111	159	371
New York	733	59	37	<30	36	101	95	149	219	146	376	286	800
New Jersey	<30	<30	40	151	210	84	<30	43	63	95	287	246	222
Delaware	<30	0	0	<30	<30	<30	<30	0	<30	<30	<30	18	39
Maryland	1,005	377	174	40	155	148	102	502	145	182	152	631	1,422
Virginia	0	0	0	<30	<30	<30	<30	<30	<30	36	98	60	344
Potomac River Fisheries Commission	[a]	[a]	[a]	[a]	[a]	[a]	[a]	[a]	[a]	[a]	[a]	64	55
North Carolina	57	<30	576	0	<30	<30	<30	<30	0	<30	<30	0	<1
Total	2,005	548	892	911	568	626	618	1,399	761	840	1,334	2,030	3,957

Sources: USDOI and USDOC 1993.

Note: Before 1990, precise numbers for catches of less than 30,000 were not reported for individual states.

[a]Prior to 1990 the Potomac River Fisheries Commission catch was included with Maryland and Virginia.

fishing effort targeted on striped bass was estimated at 2 million striped bass fishing trips (Norton et al. 1984). Landings fell to 1.7 million pounds in 1984, and total numbers of individuals caught (both landed and released) declined from 2 million in 1979 to 600,000 in 1985 (ASMFC 1990) (Table 15.2). The importance of striped bass relative to other species taken along the Atlantic coast and the contribution of striped bass to the total finfish catch also dropped significantly during this time period. Between 1960 and 1970, striped bass accounted for 2–3 percent of total recreational landings and fell to only 0.73 percent by 1979 and 0.25 percent in 1983 (ASMFC 1990). The relatively low contribution of striped bass continued into the late 1980s because of low abundance and strict regulations, including a moratorium in Maryland. Before strict regulations were implemented during the mid-1980s, most recreationally caught striped bass (61%) were taken in internal waters such as the Chesapeake Bay, but this accounted for only 41 percent of landings by weight because of lower average size in these areas (ASMFC 1990). Fish taken from coastal areas under state jurisdiction, 0–3 miles measured from the coastline, accounted for 16 percent of the harvest, but 30 percent by weight because of the larger average size of fish found in these areas (ASMFC 1990). The Exclusive Economic Zone (EEZ), 3–200 miles from shore, accounted for only 4 percent by number and 10 percent by weight (ASMFC 1990). Coastwide landings for the reopened recreational fishery have increased to more than 5 million pounds in 1992 (Dubovsky and Laney 1993).

Future Demand

Both recreational and commercial demand for striped bass should remain high in the foreseeable future. Although there are several possible recreational substitute species for striped bass, the geographic range and peak abundance of possible substitutes are often different. Striped bass also have a unique appeal to anglers because of the angling challenge they provide and the large size they may attain. The increasingly popular fly fishing fishery, as well as surf and shore fishing, put the fishery within reach of most anglers. Major sport species such as bluefish (*Pomatomus saltatrix*) and weakfish (*Cynoscion regalis*) have somewhat similar appeal to many recreational fishermen, but recent declines in abundance and impending regulations make them poor substitutes at the present time. Striped bass have become a more desirable recreational target species as many species in the Northeast have declined during the last two decades.

It is likely that disposable income in the United States will remain high and that consumer tastes and preferences for seafood products will continue to grow. Per capita consumption of seafood has risen by 25 percent since the late 1970s, and the growing numbers of health-conscious consumers should maintain this trend. High levels of disposable income should also support growth in the recreational sector for both charter and private fishing trips.

Commercial substitutes for consumption, such as tilefish (*Lopholatilus chamaeleonticeps*) in the Mid-Atlantic, are available, but most finfish species of similar quality are fully or overexploited. As the populations of potential substitute species decrease, these fisheries will probably be subject to more regulation and rising prices. Farm-raised striped bass production is increasing, but relatively high production costs and smaller size at harvest of the farm-raised product should allow the wild-caught product to retain its current markets. Historical demand and current market conditions indicate that demand exists for additional striped bass commercial landings and recreational opportunities.

Striped Bass Biology and Management

Striped bass, especially those of the Mid-Atlantic region, are anadromous. Adults return to spawn in freshwater from extensive migrations which may range from North Carolina to the southeastern provinces of Canada. Striped bass management is largely dependent on the following biological characteristics of the species: (1) large year-to-year fluctuations in year-class strength that may result from a combination of physical and biological factors; (2) the survival of early life stages, especially prejuveniles, which determines the relative strength of a given year-class; (3) migratory behavior and mixing of populations in coastal areas; (4) the relatively long-lived nature of striped bass; and (5) the importance of estuarine habitat for early life stages (such as spawning) and nursery areas.

Few if any fish species have been studied more intensively than striped bass. Even before the population declines of the late 1970s and 1980s, there was passionate interest in the population and life history characteristics of the species. Although researchers have made progress, varying degrees of uncertainty exist with regard to stock composition, migration, life history, population dynamics, ecology, and genetic diversity.

Striped Bass Biology

Distribution

Striped bass are native to the Atlantic and Gulf of Mexico coasts. In the Atlantic, they range from the St. Lawrence Estuary in the north to the St. Johns River in Florida in the south (ASMFC 1990). In the Gulf of Mexico, striped bass are present in tributaries from the western coast of Florida to Louisiana but rarely in coastal waters (Setzler et al. 1980). Striped bass may have spawned in all major Atlantic rivers before the construction of dams restricted access to many spawning areas and before pollution degraded them (Merriman 1941).

Striped bass have also been introduced into many inland lakes and reservoirs, and were introduced into the West Coast of the United States in 1879 and 1881. This introduction was quite successful, resulting in a commercial fishery with landings that peaked in the early 1900s at nearly 1.8 million pounds (Hassler 1988).

Striped bass populations south of Cape Hatteras, in Gulf of Mexico rivers, and in the St. Lawrence River are riverine and exhibit little coastal migration (Raney 1957). North of Cape Hatteras to New England, striped bass usually exhibit coastal migratory behavior (Merriman 1941). The major estuaries in the United States contributing to the coastal population of the Atlantic include the Hudson River, Delaware River, Chesapeake Bay, and the Roanoke River–Albemarle Sound system of North Carolina (Fig. 15.2).

Life History

The following description is a generalized account of striped bass life history in the Mid-Atlantic region, including parts of North Carolina and New England. As anadromous species, striped bass spawn and remain as juveniles in freshwater or brackish water, but spend most of their adult life in marine coastal areas. Spawning takes place once a year in April, May, or June in fresh or slightly brackish water with salinities of 0–5 parts per thousand (ppt) (Fay et al. 1983). The degree to which fish return to their natal streams or segregate by river system

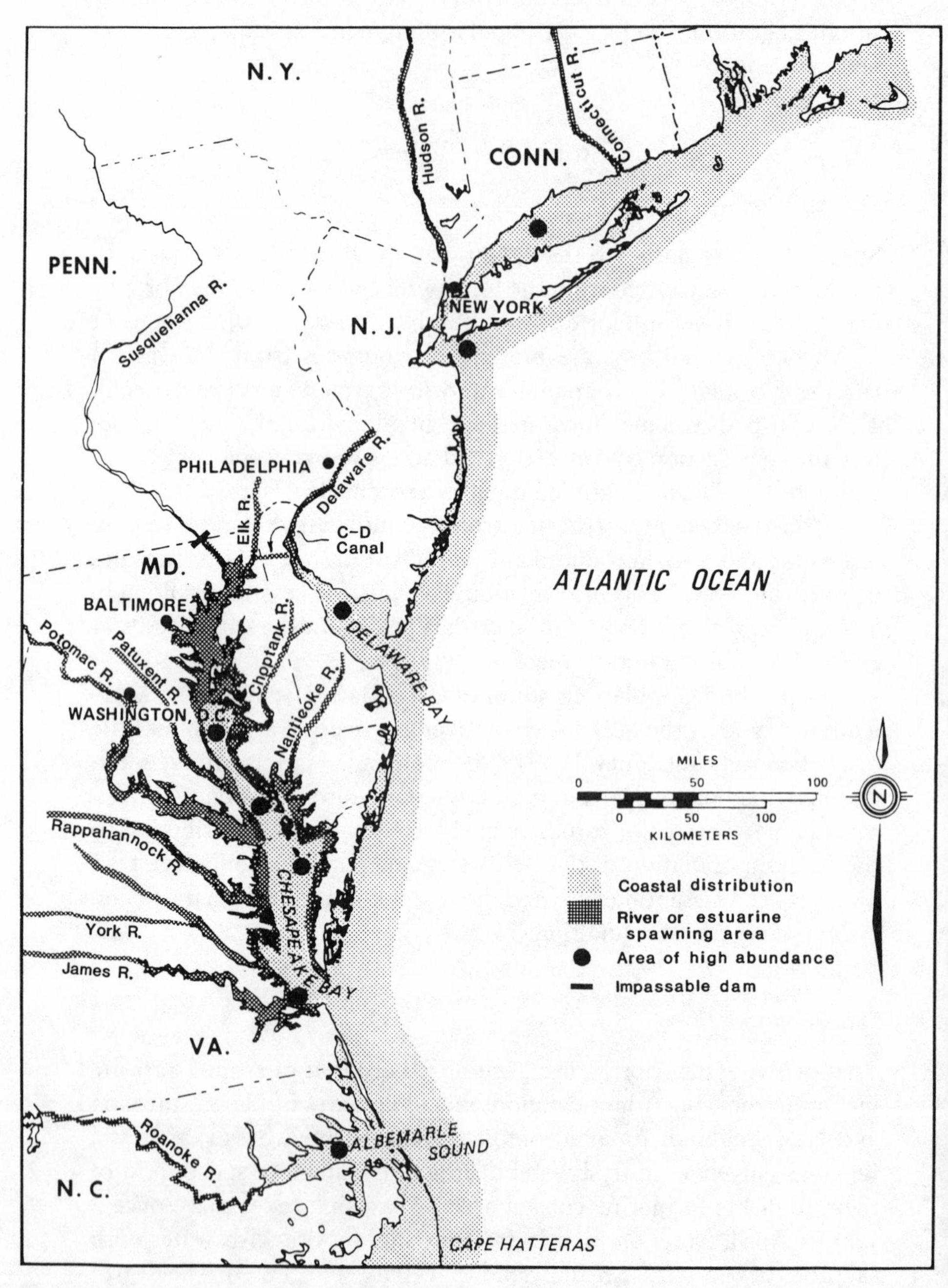

Figure 15.2. Mid-Atlantic distribution of striped bass. (*Source:* Fay et al. 1983, p. 2.)

is poorly understood. However, mitochondrial DNA studies suggest that homing fidelity is high for some river systems (Wirgin et al. 1993). The number of eggs produced by a given female varies widely and is primarily dependent on size. Estimates range from 15,000 eggs to more than 4 million eggs per female (Setzler et al. 1980). Depending on temperature, striped bass eggs hatch 29–80 hours after fertilization (Setzler et al. 1980).

Survival of striped bass during early life stages is critical to the success of a given year-class (Fay et al. 1983). Survival of eggs and larvae are affected by physical conditions such as temperature, salinity, current velocities, and oxygen concentration (Setzler et al. 1980). Spawning, egg, and larval stages all occur in nearly the same section of the river (Setzler-Hamilton et al. 1981). A critical biological factor for larval survival is sufficient concentrations of prey, especially during the first several days of feeding (Setzler et al. 1980). Colder than normal winters, greater than average spring water flows (Hassler 1988; Boynton et al. 1977), and high food availability (Heinle et al. 1976) are factors that have been suggested to coincide with dominant year-classes.

After the larval stages, at approximately 50–60 days old and about 1 inch in length, striped bass are considered to be juveniles. Juvenile striped bass usually remain in their natal stream for the first year with general downstream and shoreward movement. Juveniles also exhibit schooling behavior, numbering from a few fish to thousands of individuals (Westin and Rogers 1978).

In the Mid-Atlantic region, striped bass males may mature as early as age one, but generally mature at age 2–3 years with a minimum length at maturity of 7 inches total length (Clark 1968). Females mature later at age 4–6 years with a minimum length at maturity of 17 inches (Westin and Rogers 1978). Recent work indicates that females mature at older ages than previously thought. Full maturity of a given year-class may not occur until age 8 (Berlinsky et al. 1988).

Striped bass may live to at least 30 years of age and grow to large size. Several weighing more than 125 pounds were taken off Edenton, North Carolina, in 1891 (Bigelow and Schroeder 1953). Today, fish in the 50- to 70-pound range are not unusual on the spawning grounds of the Chesapeake Bay or Hudson River estuary. Females generally grow larger and live longer than males.

Migration

After spawning and moving to coastal areas, striped bass migrate in a northerly direction to the New England coastline in the summer, and south to the Virginia and North Carolina coast in the winter (ASMFC 1990). During the winter, striped bass are also found in deep portions of estuaries throughout the Northeast (ASMFC 1990). As is the case with juveniles, the migrations of adults vary according to river system. In addition, many fish do not adhere to the general model and may not migrate at all. Tagging studies have indicated that some fish remained in an area all year, while others were recaptured more than 600 miles away (Fay et al. 1983). Young adult fish are more likely to remain in estuaries throughout the year, while large females have been shown to undertake the longest migrations (Bigelow and Schroeder 1953). Historically, females have been shown to dominate the catch at 85–90 percent in waters off New England and Long Island (Bigelow and Schroeder 1953; Schaefer 1968). However, during the 1980s the male-to-female ratio off Long Island approached 1:1, perhaps because of decreasing mortality of males in the Chesapeake Bay (Young 1987). It appears that the Chesapeake and Hudson populations of striped bass have similar ranges and are the widest-ranging populations on the East Coast (Waldman et al. 1990). Tagged Chesapeake and Hudson River striped bass have been recovered as far north as New Brunswick, Canada, and as far south as the coast of North Carolina.

Stock Composition

The coastal population of striped bass is composed of several races of different estuaries of origin. Presently the two major sources for the coastal population are the Chesapeake Bay and the Hudson River. The Roanoke River, Delaware River, and several other rivers in New England and Canada may have had historical significance, but the current contribution of these systems is relatively small. The origin of fish at a given coastal area depends on location, sex of fish, season, and the relative year-class strength of different systems that contribute to the coastal population. For example, in 1975, the Hudson River contribution to coastal striped bass landings was 40–50 percent of the 1965 year-class but only 10 percent or less of the 1966, 1968, and 1969 year-classes (Van Winkle and Kumar 1982). Also in 1975, stock composition analysis for coastal areas from North Carolina to Maine concluded that more than 90 percent of striped bass taken in this region were of Chesapeake origin, but this was in part attributable to the dominant 1970 Chesapeake year-class (Berggren and Lieberman 1978). Later studies have shown that the proportion of Chesapeake striped bass is lower in southern New England and areas adjacent to the Hudson River (USDOI and USDOC 1991).

Ecological Relationships

The striped bass is a predatory species that feeds on a wide variety of prey items. Prey depends on striped bass life stage and prey availability. Larvae consume zooplankton and copepods. Juvenile striped bass consume insect larvae, polychaetes, larval fish, mysids, and amphipods (ASMFC 1990). As juveniles grow older, small fish become a more important part of their diet. By age 2 years, fish comprise most of the striped bass diet. The most important prey species include alosids, menhaden, and anchovies, but crustaceans and mollusks may also be taken in significant amounts (Fay et al. 1983).

Striped bass have different competitors depending on life stage. Larvae may be susceptible to competition and predation. If prey densities are reduced by feeding of other species of fish such as blueback herring (*Alosa aestivalis*), then reduced survival of striped bass larvae could result (ASMFC 1990). White perch (*Morone americana*), which also share a common nursery area with striped bass, usually occur in greater numbers and may compete for food resources (Mihursky et al. 1976).

The copepod (*Cyclops bicuspidatus*) preys on striped bass larvae (Smith and Kernehan 1981), and several species of fish prey on juveniles (McFadden et al. 1978). Potential adult competitors are predatory species such as bluefish and weakfish, which have similar food preferences and occupy the same area during certain times of the year.

Population Declines and Factors Influencing Abundance

The reasons for the population declines of the late 1970s and early 1980s belong to two major categories: (1) habitat or ecological factors, and (2) overexploitation. In the early 1980s the following main hypotheses were put forward to explain the decrease in Chesapeake and Roanoke striped bass populations: (1) toxic contaminants, (2) starvation of larval fish, (3) overexploitation, (4) predation on fry, (5) occurrence of unfavorable natural climatic events, (6) recent modifications of water use practices, (7) competition with other species for food and space, and (8) reduction of water quality due to agricultural and (9) sewage treatment practices (USDOI and USDOC 1982). In 1979, the Emergency Striped Bass Study (ESBS) was established. Under the ESBS, hypotheses were grouped into general research categories, and funding and coordination were provided by the National Marine Fisheries Service and the U.S. Fish and Wildlife Service.

Habitat and Ecological Factors

Striped bass survival is highly dependent on the natural variability of the estuarine ecosystem (Rago 1992). During the last two centuries, especially during the last 50 years, humans have introduced new variations into ecosystems that are usually detrimental to most life stages of striped bass. In the Chesapeake basin, nutrient loading from municipal and agricultural sources, increased contaminant loadings, and increased sedimentation have resulted in eutrophication, losses of submerged vegetation, lowered pH, and other environmental impacts. According to Rago (1992, p. 105), "most estuarine species can compensate for such changes but there are obvious limits to adaptation, beyond which the integrity of the system begins to break down." Yet detection of changes in survival due to anthropogenic factors is confounded by the high natural variability of striped bass reproduction, fishing mortality, and compensatory mechanisms. Long-term studies (in this case 10 years) would be needed to detect a 0.5 conditional change in mortality rate with a 70 percent probability of being correct (Rago 1992).

The sources of organic and inorganic contaminants are rainwater, point source pollution, and runoff. Concentrations of contaminants quickly decrease as one moves away from the point source in the James River or Baltimore Harbor (Rago 1992). The James, Potomac, and Susquehanna rivers were found to be the sources of cadmium, chromium, copper, iron, lead, and zinc. Toxic organic compounds were also present but not identified (USDOI and USDOC 1987). Contaminants are also found throughout the Chesapeake Bay, likely as the result of atmospheric deposition (Rago 1992). Contaminants and related episodic low pH conditions may have had local effects, but do not appear to have been a major cause of the striped bass population decline.

The reduction of water quality due to nutrient overloading and eutrophication is another concern, especially in the Chesapeake Bay and Albemarle Sound–Roanoke River systems. Increased levels of benthic respiration produce anoxic (zero oxygen concentration) and hypoxic (oxygen concentrations below optimum) conditions in regions of the Chesapeake Bay during the summer (Rago 1992). Several researchers have hypothesized that adult striped bass remaining in the Chesapeake Bay may be stressed by these conditions, which result in reduced egg quantity and quality (Coutant and Benson 1987). However, in the Chesapeake Bay a large proportion of mature females are not subject to these conditions because they migrate to the Atlantic Ocean during the summer season. In the Albemarle Sound–Roanoke system these conditions may have greater impact if a smaller proportion of the striped bass population leaves the system.

Eutrophication is also a factor in the decline of submerged aquatic vegetation (SAV), which in 1978 reached its lowest levels in the Chesapeake in 40 years (Rago 1992). Although SAV provides detrital matter that supports the Chesapeake Bay's trophic structure, no direct relationship with striped bass nursery areas or survival has been demonstrated (Hershner and Wetzel 1987).

Although food availability, predation, and competition are likely to affect striped bass abundance in a given year, the complexity of ecological relationships, lack of long-term data, and the difficulty in quantifying the strength of these relationships preclude definitive conclusions about their effects. It is unlikely, however, that these factors were directly responsible for the striped bass declines during the 1980s (USDOI and USDOC 1982, 1987).

In the upper Chesapeake Bay and in the Roanoke River, dam construction and water flow alterations during the 1950s, 1960s, and early 1970s may have affected striped bass spawning and larval stages (ASMFC 1990). Power plant entrainment and impingement effects have also been a major issue, especially on the Hudson River. However, the impacts of these changes in water use remain unknown.

It is likely that year-class success in striped bass spawning areas such as the Potomac River is linked to extrinsic, density-independent environmental factors (Ulanowicz and Polgar 1980). For example, in 1976, a sudden water temperature drop in the Hudson River caused high mortality among those larvae that were spawned before the temperature change (Boreman and O'Brien 1983). Although environmental conditions are important factors, no consistent trends related to climatic events can be directly linked to the declines of striped bass populations.

Fishing Mortality

The major cause of the striped bass decline has been identified as fishing mortality, though it has been difficult to distinguish from other factors (USDOI and USDOC 1991). The recovery of striped bass populations after the restriction of the fisheries is probably the most compelling evidence for this. During the 1970s, there were few restrictions on the commercial and recreational capture of striped bass other than minimum sizes. Although not well documented, recreational and commercial fishing effort grew because of increased numbers of participants and improved fishing technologies (Florence 1980). Average age of fish caught in the commercial fishery decreased during the period between 1960 and the early 1980s (Goodyear 1985). Modeling of Chesapeake Bay striped bass indicated that (1) low size limits and high fishing mortality of the mid-1970s resulted in growth overfishing (Goodyear 1984);* (2) the relative importance of increased fishing mortality was greater than decreased first-year survival (Cohen et al. 1983; Goodyear et al. 1985); and (3) regardless of underlying causes, reductions in fishing mortality would yield immediate benefits to the population's reproductive potential (USDOI and USDOC 1992). Further work showed that fishing mortality rates during the period of the decline were 0.8 (USDOI and USDOC 1992).‡ Population abundance is governed by both the quality of habitat and the rate at which fish are removed from the population, but evidence showed that the most direct short-term solution was to decrease fishing mortality.

*Growth overfishing occurs when high fishing mortality results in the removal of fish before yield can be maximized. A decrease in average size may indicate growth overfishing.

‡Fishing mortality (*F*) is expressed as an instantaneous rate at which fish are removed from the population due to fishing activities. For example, an *F* of 0.8 is equivalent to the removal of 56 percent of the population that has been recruited into the fishery.

Population Management

The striped bass population is typically made up of large numbers of fish from few year-classes or dominant year-classes (ASMFC 1990). For example, record commercial catches in 1973 were largely supported by the 1970 year-class, the second largest on record (Rago and Dorazio 1988). Researchers have been unable to quantify a relationship between parental stock size and subsequent recruitment (ASMFC 1990). However, below certain levels of spawning stock size, such as the very low levels of the early 1980s, the chances of good recruitment become quite small.

Density-independent factors such as environmental variation play a large role in the success of striped bass recruitment (USDOI and USDOC 1989). This led the Atlantic States Marine Fisheries Commission's (ASMFC) Scientific and Statistical Committee to take an adaptable approach to management when the fishery reopened in 1990. The overall management goals of the program are not targeted on a given harvest level, but on the restoration and maintenance of the population at a historical level of abundance (ASMFC 1990). Allowable fishing mortality rates are set accordingly. Dorazio and Rago (1988a) predicted that at a fishing mortality rate of 0.5 and a 18-inch minimum size in the Chesapeake Bay and the Hudson River, and a 28-inch minimum size along the coast, striped bass stocks can be expected to remain stable over the long term. "The maximum allowable base fishing mortality rate is weighted by the fraction of each age class's length distribution that exceeds the minimum size limit. Therefore, age specific fishing mortality rates do not all equal the maximum base *F*" (Dorazio and Rago 1988b, in ASMFC 1990, p. D-4). For example, if the length limit in coastal areas is kept at 26 inches, three of the four abundant age classes in coastal areas will receive substantial protection from fishing (Dorazio and Rago 1988b). Under Amendment 4 of the ASMFC management plan, the population was managed under an interim fishing mortality rate of 0.25 for the Chesapeake stock and 0.5 for the Hudson stock with minimum sizes of 18 inches for estuaries such as the Hudson and Chesapeake, and 34 inches in coastal waters. Tagging studies in 1993 and 1994 provided mortality estimates at or near the 0.25 target (ASMFC 1994). Regulations in North Carolina's internal waters are implemented separately. Under Amendment 5 to the plan, the fully recovered Chesapeake Bay population will be fished at a rate of 0.5. A model for estimation of Spawning Stock Biomass (SSB) has been developed to determine when stock recovery has occurred. The model will use SSB levels from the early 1970s as a benchmark to assess the relative health of the population.

Current Population Condition

Striped bass populations within the management unit have shown remarkable progress since the management program began in 1981. The Hudson River population was considered by the ASMFC Technical Committee to have recovered in the early 1990s. Surveys in 1985 and 1986 showed reproduction to be at low levels, but 1987, 1988, and 1989 surveys indicated very high reproduction, with 1987 and 1988 the highest on record (ASMFC 1990). During the 1980s the catch per unit effort increased rapidly with a broad range of ages in the spawning stock (USDOI and USDOC 1992).

During most of this century, the Delaware River population remained at extremely low levels, most likely due to environmental degradation of the estuary. Recent water quality improvement in the Delaware estuary coupled with coastwide regulation of striped bass fisheries has resulted in rapid population growth during the late 1980s. Abundance in the Delaware River juvenile striped bass survey illustrates an increasing trend since the origin of the survey in 1980. The last two years of the survey, 1992 and 1993, are the highest in the survey's history (Baum 1994).

The Chesapeake Bay population reached full recovery in 1995 as defined by spawning biomass levels of the early 1970s (ASMFC 1995a). The juvenile index has improved from successive poor years of the late 1970s and most of the 1980s (Fig. 15.3). Virginia's juvenile survey recorded the best recruitment indices on record from 1987 to 1990 (USDOI and USDOC 1992).

Maryland's juvenile survey, which began in 1954, is the longest continuous time series available. The juvenile index has been characterized by highly variable recruitment with strong year-classes produced every two to five years (USDOI and USDOC 1992). Two weak years followed the exceptionally strong 1989 year-class. The 1993 year-class was the strongest on record, and 1994 was well above average (MDDNR 1994a).

Maryland spawning surveys of the mid-1980s showed large numbers of relatively young males and few surviving females from the year-

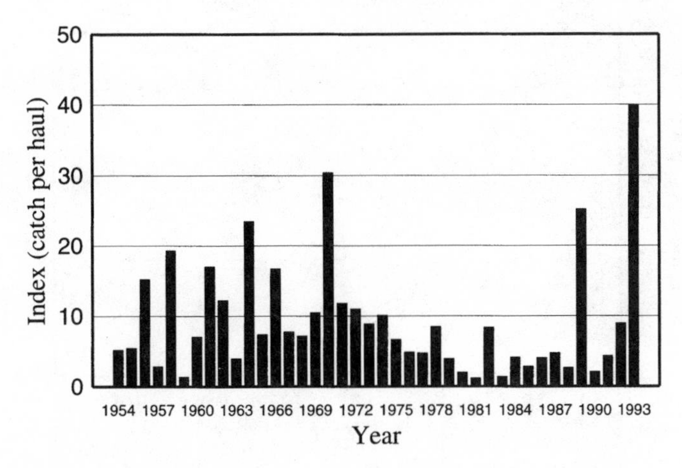

Figure 15.3. Maryland striped bass juvenile index by year, 1954–93. (*Source:* MDDNR 1994b.)

classes of the 1970s. As the 1982 year-class matured,* the ratio of females increased on the spawning grounds during the late 1980s (USDOI and USDOC 1992). Striped bass stock biomass has also risen steadily since extremely low levels of the mid-1980s (Fig. 15.4). Other surveys from coastal states are also showing evidence of a recovering population. The Catch Per Unit Effort (CPUE) of striped bass in the Marine Recreational Fishery Statistics Survey (MRFSS) has steadily increased during the 1990s. CPUE in the Massachusetts hook-and-line commercial fishery and the Connecticut MRFSS and Volunteer Survey show similar results during the 1990s.

The condition of North Carolina's Albemarle Sound–Roanoke River stock is questionable, and probably marginal as of 1995. During the 1980s, fishing pressure was very high (ASMFC 1990). This resulted in the decline of large mature females in the spawning population. Juvenile abundance was generally poor during the 1980s, although 1988 and 1989 increased to levels near the long-term average (ASMFC 1990) and 1993 set a record for recruitment (Laney et al. 1994). Current population increases are largely due to the 1988 and 1989 year-classes. According to the North Carolina management plan,

*The 1982 year-class was one of the best in the Chesapeake Bay since 1974. As it matured, it would become the first year-class to receive stringent protection under the striped bass management plan.

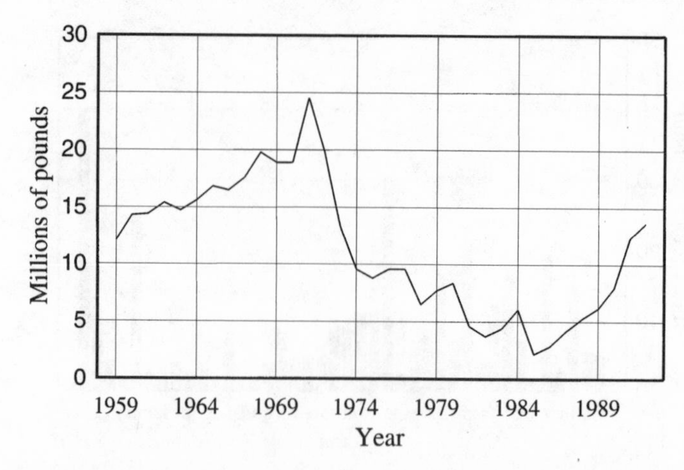

Figure 15.4. Striped bass stock biomass in Chesapeake Bay, 1959–92. (*Source:* ASMFC Technical Committee 1994.)

fishing mortality rates are currently too high to sustain population growth (Laney et al. 1994). Mortality associated with bycatch, poaching, and recreational landings during the late 1980s also cloud the picture regarding the status of the stock. It is also possible that changes in the flow rates due to the construction of dams on the Roanoke River affected recruitment in the early 1980s. A flow regime is currently being implemented that is intended to improve conditions for striped bass reproduction (ASMFC 1990).

Managers have considered the simultaneous exploitation of mixed populations during their coastal migration (ASMFC 1990). Coastal fishing mortality might be well within reason for a recovered population, but not for those populations that remain at low levels. All major systems have shown improvement under the striped bass management plan, largely because of stringent control of fishing mortality rates. It appears that large coastal size limits applied in both estuaries and adjacent marine areas of small recovering systems such as those in Maine will allow for continued population growth in these systems.

Nonharvest Losses and Enforcement

Nonharvest losses such as mortality caused by incidental capture, hook-release mortality, and poaching have potential impacts on the management program because of their contribution to total mortality. In

1991, total nonharvest losses may have exceeded the directed harvest of striped bass (USDOI and USDOC 1992).

In recreational fisheries, losses are caused by hook-release mortality and poaching which includes the retention of undersize fish, fish taken out of season, and fish retained in excess of creel limits. According to the National Marine Fisheries Service, Marine Recreational Fisheries Statistics Survey for 1991, 11 fish were released for each fish retained (USDOI and USDOC 1992). The probability of catching a legal fish is reduced by size limits, seasons, and creel limits, although the probability of catching a striped bass is not (USDOI and USDOC 1992). Therefore, the impacts on the striped bass population could be great, depending on the magnitude of hook-release mortality. Hook-release mortality has been estimated to range between 2.6 percent (Harrell 1987) and 17.5 percent (Diodati and Hoopes 1991). The Striped Bass Technical Committee uses an 8 percent mortality rate for hooked and released fish (USDOI and USDOC 1992). Coastwide losses due to hook-release mortality were estimated at 151,300 in 1990 and 280,600 in 1991, by using the 8 percent estimate (USDOI and USDOC 1992).

Illegal recreational fishing is difficult to quantify. Law enforcement reports have been used for several jurisdictions, yielding citation rates for law enforcement inspection ranging between 0.7 and 1.7 percent of fishermen checked (USDOI and USDOC 1992). Shepard (1992) estimated a rate of 1.3 percent that would result in losses of 46,200 fish in 1991. This is a relatively small proportion of nonharvest losses when compared to the probable hook-release mortality, although it should be emphasized that there is a great amount of uncertainty associated with recreational rates of noncompliance.

Nonharvest losses in commercial fisheries include mortality of undersize fish caught in directed striped bass fisheries, striped bass caught in fisheries targeting other species, and poaching of undersize or out-of-season striped bass. The mortality of commercial bycatch that is discarded depends on gear type and the manner in which it is fished. For example, gill nets cause greater mortality than pound or trap nets, which may cause little if any. Discard mortality rates have been estimated at 8 percent for drift gill nets and 47 percent for anchor gill nets (Seagraves and Miller 1989). However, gill nets can be very selective gear that can be targeted on specific sizes and species, resulting in fewer discards. Significant bycatch losses were sustained in the Massachusetts hook and line, in many trawl fisheries, Maryland white perch gill net, and Virginia shad gill net fisheries (USDOI and USDOC 1992).

Little information exists regarding compliance with regulations in commercial fisheries. Massachusetts assumes a poaching rate of 3.5 percent of the legal catch (Diodati and Hoopes 1991). In 1991, Maryland recorded 15 violations in 989 inspections (Rugolo et al. 1991). Coastwide, illegal commercial losses in 1991 were estimated at 285,242 fish weighing 629,060 pounds (USDOI and USDOC 1992).

Enhancement

Striped bass stocking was first attempted along the Atlantic coast in the 1880s. These attempts were considered failures, either due to poor survival of releases or because there was no method by which the survival and hatchery contributions could be evaluated. In the 1980s, stocking programs were initiated for several North Carolina river systems, systems in both the Virginia and Maryland portions of the Chesapeake Bay, Delaware Bay, the Navesink River in New Jersey, the Hudson River, and several rivers in Maine. More than 13 million striped bass were released between 1983 and 1995 (ASMFC 1996). Most programs have been discontinued because of the current striped bass recovery (ASMFC 1996). The goals of the stocking programs were to (1) enhance or reestablish spawning populations, and (2) fulfill agreements with power companies that impact populations through the use of river water in spawning areas. The Chesapeake program was considered a pilot project to investigate the feasibility of striped bass enhancement. Most hatchery fish are marked with Binary Coded Wire Tags (BCWTs) or other tagging methods to determine their contribution to commercial and recreational fisheries and to the striped bass population.

During the early 1990s, hatchery fish made up approximately 5 percent of the harvest in Chesapeake Bay fisheries and more than 10 percent of some year-classes (ASMFC 1995b). Hatchery fish have been encountered on all major spawning grounds in the Chesapeake Bay and include approximately 35 percent of fish found on the Patuxent River spawning grounds. Chesapeake Bay releases have been found as far north as Nova Scotia and as far south as the North Carolina coast. Although enhancement has contributed to the current recovery of several striped bass populations, including the Chesapeake, it appears that fishing mortality reductions are responsible for most of the current recovery. Enhancement efforts have had the greatest impact on small, severely depleted river systems, such as the Patuxent River in Maryland, where significant population increases can be attributed to stocking activities (ASMFC 1995b).

Stocking activities are guided by recommendations of the ASMFC Striped Bass Stocking Committee. These recommendations include concerns related to genetic integrity, tagging and evaluation, state coordination, disease, and stocking strategies.

Summary

After nearly 15 years of intensive studies by coastal states, the federal government, and academic institutions under the Emergency Striped Bass Study and other investigations, our understanding of striped bass biology appears adequate for management. Although uncertainties exist, it is likely that returns on research investments are declining relative to costs, especially with respect to biological information. As striped bass fisheries are reopened, fishery-related concerns such as monitoring of mortality rates and the potential impact of recreational and commercial nonharvest losses appear to have become priorities. Though habitat degradation of the last century has affected the productivity of the striped bass population, its effects are difficult to detect as well as reverse (Rago 1992). Recent reproductive success indicates that many estuaries contain productive striped bass nursery habitat. There is little doubt that the best short-term solution to the striped bass decline was a large reduction in fishing mortality. Strict fishing regulations have promoted recovery of the striped bass populations of the Hudson River and Chesapeake Bay to levels of the early 1970s. The long-term goal of conserving and restoring the quality of striped bass habitat may further reestablish striped bass populations and increase potential future yields.

The Impact of Fisheries on Biological Diversity

Impacts of striped bass fisheries and management on biological diversity are difficult to assess due to our fragmentary knowledge of striped bass ecology and genetics, and the complexity of these topics. Although definitive answers may not be forthcoming, it is possible to examine general impacts and their implications.

Genetic Diversity

Concerns with the possible impacts of harvest mortality on genetic diversity of striped bass may be valid, yet are probably insignificant when compared to the last three centuries of other anthropogenic impacts. Merriman (1941, p. 16) noted, for example, that "there can be little doubt that striped bass in early times entered and spawned in every river of any size, where the proper conditions existed, along the greater part of the Atlantic coast, and that as cities were built and dams and pollution spoiled one area after another, the number of rivers that were suitable for spawning became fewer and fewer." If Merriman is correct, relatively small, genetically distinct populations may have spawned in estuaries of New Jersey, Connecticut, Massachusetts, and Maine. Further, striped bass reproduction completely or nearly ceased in the Navesink, Delaware, Androsoggin, and Kennebec rivers until environmental conditions improved and, in some cases, stocking was initiated (ASMFC 1990).

It is generally accepted that most fish return to their natal river or system to spawn, including such major systems as the Chesapeake, Hudson River, and Albemarle Sound–Roanoke (ASMFC 1990). The degree to which striped bass stray determines whether there are discernable genetic differences between populations of different river systems (Schill and Dorazio 1990). Striped bass from the Hudson showed strong differences in blood serum albumin from striped bass taken from spawning grounds in the Chesapeake and Roanoke (Schill and Dorazio 1990). Morphometric and electrophoretic methods have also been used to differentiate these populations in studies used to determine origin of mixed stocks (Fabrizio 1987). Kriete et al. (1978) suggested that spawners segregate themselves by river systems in the Chesapeake, and work by Morgan et al. (1973) indicated that populations of specific systems in the upper Chesapeake Bay could be detected by electrophoresis. However, subsequent work could not duplicate these results. Chesapeake Bay river systems as far south as the Rappahannock River demonstrated apparent genetic homogeneity among upper Chesapeake Bay populations (Sidell et al. 1980). Recent mitochondrial DNA studies have shown differences between striped bass from different spawning areas such as the southern stocks of the Chesapeake and the Hudson River, and the Shubenacadie River in New Brunswick, Canada (Wirgin et al. 1993).

In recent years, genetic integrity has become a major concern of stocking programs. According to Schill and Dorazio (1990), "unfortunately, hatchery stocking practices have altered the genetic composition of all U.S. stocks." Yet it is unknown whether these changes have been significant or if they might impact population viability and behavior. In the Chesapeake Bay and other regions, only brood stock from the system to be stocked with progeny are used for hatchery spawning. It is currently unknown whether hatchery fish are more apt to stray than wild fish to other spawning areas within the Chesapeake Bay or to other systems along the coast. One must weigh concerns related to coastwide genetic diversity against increases in species and ecosystem diversity gained through stocking individual systems where the striped bass populations may have been entirely lost. Of greater concern was the stocking of the hybrid resulting from the crossing of striped bass and white bass hybrids in the Susquehanna and Patuxent River watersheds during the mid-1980s. This hybrid cross is considered to be sterile, but the capture of gravid hybrids and ripe striped bass males on the spawning grounds has raised concerns over possible backcrosses (Harrell et al. 1993). According to recent investigations, white bass genes are being introgressed into the striped bass population (Harrell et al. 1993). The Striped Bass Stocking Committee has recommended against the use of hybrids because of possible escapement. In both Pennsylvania and Maryland, the stocking of hybrids has been discontinued in areas where escapement to striped bass spawning areas is likely. Committee recommendations regarding the maintenance of biological diversity were endorsed and included in Amendments 4 and 5 of the Striped Bass Fishery Management Plan.

Effects on Other Species

In most striped bass fisheries, the incidental capture of other species (or bycatch) is usually a small part of the total catch. Pound nets and traps, and hook and line allow for the release of species that are captured incidentally. Drift gill nets are used in the striped bass Chesapeake Bay commercial fishery. Drift gill nets in the Maryland fishery are size selective and take minimal incidental catch. The current regulation of fishing effort by season and gear is also much more restrictive than historical fishing within the striped bass range. The types of fishing gear utilized by recreational and commercial fishermen are unlikely to cause undue harm to the ecosystem. One of the few concerns would be ghost fishing due to lost gill nets. Current Maryland regulations require the tending of gill nets; therefore the loss of gill nets or other gear is unlikely unless they are intentionally discarded.

Striped bass are one of the top predators of estuaries and coastal marine areas. A change in the abundance of striped bass could impact the abundance of prey or competing species. Preliminary data suggest that predation by the resurgent striped bass population has negatively impacted populations of the Chesapeake Bay blue crab (*Callinectes sapidus*) and particularly the Connecticut River shad (*Alosa sapidissima*). Currently, however, the strength of these relationships is unknown.

Summary

Except in reference to enhancement activities, biological diversity is not one of the stated concerns of the Striped Bass Management Plan. However, it appears that present and future fishing levels under the plan should not result in significant depletion of any river population. Stocking activities, by reestablishing striped bass populations, may have actually increased diversity at the species and ecosystem levels. It is likely that striped bass genetic diversity has decreased during the last four hundred years as populations were lost from specific river systems due to human impacts other than fishing. There is evidence of both genetic homogeneity, especially among rivers of the upper Chesapeake Bay, and genetic differences elsewhere, such as between the Shubenacadie River and other systems. Striped bass are generalists when one considers the variety of habitats in which they thrive and the relatively recent period during which they recolonized many East Coast rivers (15,000 years since the last glacial retreat). Finally, incidental catch in striped bass fisheries appears to have a negligible impact on other species. However, we remain largely ignorant of the effects that changes in striped bass populations have on other species and ecosystem structure/function.

Incentives and Controls

The striped bass is a scarce resource. Fewer striped bass are available than consumers such as recreational fishermen and the fish-eating public would like. This may seem to be a trivial point, but if the supply of striped bass were unlimited or if striped bass were a less desirable species, such as carp, elaborate management schemes, biological investigations, and countless meetings would become unnecessary.

The Management Program

Management Institutions

The Atlantic States Marine Fisheries Commission (ASMFC) was created in 1942 by an act of the United States Congress (Public Law 539 of the 77th Congress, 1942) and by the Atlantic coast states. Its purpose is to recommend management measures and coordinate interjurisdictional fisheries management along the Atlantic coast. There are 15 member states, each of which is represented by three commissioners: the executive officer of the state's marine fisheries agency, a member of the state's legislature, and a governor's appointee (ASMFC 1992).

The striped bass resource is managed under the Interstate Fisheries Management Program (ISFMP). A grant agreement between the National Marine Fisheries Service (NMFS) and the ASMFC was reached in 1981 to foster the cooperative management of marine, estuarine, and anadromous fisheries in Atlantic coast state waters.

The ASMFC policy board, consisting of the marine fisheries administrator from each member state and representatives of the NMFS and the U.S. Fish and Wildlife Service (FWS), provides oversight of the ISFMP. Technical input and recommendations are supplied to the policy board by the Striped Bass Technical Committee, which consists of state and federal fishery scientists who consider state management proposals and striped bass population assessments. Historically, ISFMP fishery management plans are implemented and enforced on a voluntary basis in each member state. However, in the case of striped bass, compliance is further encouraged by the Atlantic Striped Bass Conservation Act.

State jurisdiction includes internal waters such as the Chesapeake Bay, and the territorial sea 0–3 miles from shore. The National Marine Fisheries Service (NMFS) promulgates regulations in coordination with ASMFC member states in the Exclusive Economic Zone (EEZ), 3–200 miles from shore.

Management Objectives

The general goal of the ASMFC interstate striped bass management program in 1986 is "to perpetuate the striped bass resource throughout its range so as to generate optimal social and economic benefits to the nation from its commercial and recreational harvest and utilization over time" (ASMFC 1992). The goal of the Maryland fishery management plan was almost identical except it referred specifically to state waters and benefits to the state. The main concern of ASMFC management was recovery of striped bass populations and their maintenance. More specific goals concerning economic and allocation decisions are generally made at the state level, although under Amendment 4, commercial harvest was limited to 20 percent of the average annual level which occurred in that jurisdiction between 1972 and 1979. Biological diversity has not been a management priority, although genetic considerations have been of concern to some researchers.

The management unit has been defined as all striped bass stocks north of South Carolina within state internal and territorial waters 0–3 miles from shore. Management in all coastal areas and the Chesapeake Bay, Hudson River, and Delaware Bay has conformed to the ASMFC management plan. However, North Carolina estuarine areas such as the Albemarle Sound–Roanoke River system have not been brought fully under the plan.

Management Program History

The striped bass population(s) declined to extremely low levels in the late 1970s and early 1980s. The first legislative response to this decline was the Emergency Striped Bass Study (ESBS), passed by the U.S. Congress in 1979 as an amendment to the Anadromous Fish Conservation Act (P.L. 96-118, U.S.C. 757g). The purpose of the study was to research and identify causes of the striped bass population decline. Because there was no interjurisdictional coordination of striped bass management, states from North Carolina to Maine regulated their fisheries primarily through the use of size limits. Then, in 1981, the ASMFC Interstate Striped Bass Management Plan (FMP) was developed. Its original objectives were to "maintain a spawning stock and minimize recruitment failure, to promote harmonious use of the resource among various components of the fishery, to provide for continued collection of economic, social and biological data required for the effective monitoring and assessment of management efforts, to promote research that will improve understanding of the species and its fisheries, to promote adoption of environmental standards necessary for the maximum natural production of the species, and to establish a system for management effort coordination" (ASMFC 1981, in USDOI and USDOC 1992, p. 30). The ASMFC plan called for a minimum size of 14 inches total length in producing areas such as the Chesapeake Bay and 24 inches along the coast (USDOI and USDOC 1992). During the spawning season, major rivers and spawning areas were to be closed to fishing. Since the Striped Bass Fishery Management Plan was only a collection of recommendations, states only gradually adopted its regulations. During the early 1980s, the lack of consistency among states was still a major problem. If a state unilaterally increased its minimum size, other states could retain smaller minimum sizes and reap the benefits of the conservation measure.

In 1984, the Atlantic Striped Bass Conservation Act (P.L. 98-613) was passed by Congress to ensure compliance with the ASMFC Interstate Striped Bass Management Plan. The law imposes a federal moratorium on fishing in any state found out of compliance with the plan. State compliance with respect to required regulations and enforcement is determined by the ASMFC. Upon an ASMFC finding of state noncompliance, the secretaries of the interior and commerce are notified and a determination regarding compliance is made jointly. The secretaries of each department may then declare a federal moratorium on striped bass fishing in the concerned state.

The management plan was amended three times in 1984 and 1985. In 1984, Amendment 1 permitted greater flexibility by allowing states to substitute alternate management measures as long as the same reduction in mortality could be obtained. However, substitutions for size limits were not allowed. Recruitment remained poor throughout the early 1980s except for 1982. Amendment 2 required the reduction of fishing mortality by 55 percent on Chesapeake Bay stocks (USDOI and USDOC 1992).

Amendment 3 required protection of the 1982 and subsequent year-classes through successive increases in minimum size. The objective of the amendment was to allow 95 percent of the sizable 1982 year-class the opportunity to spawn (USDOI and USDOC 1992). Coastal states were allowed to maintain a limited fishery as long as the increase in size limits stayed ahead of the 1982 year-class. By 1990, minimum sizes in coastal areas had increased to 36 inches. The Chesapeake Bay is considered a producer area where most fish are of premigratory ages. Therefore, its fishery is dependent on younger ages and smaller-sized individuals than in coastal areas. As the 1982 year-class matured, the only way to guarantee protection was to stop fishing. In 1985, Maryland implemented a moratorium on commercial and recreational striped bass fishing to conform to the fishery management plan. The original plan and the first three amendments concentrated on "strategies to restore and maintain self-sustaining stocks of striped bass and minimize recruitment failure" (USDOI and USDOC 1992, p. 30).

Amendment 4 was implemented in 1990 to govern the transition period between a recovering and recovered striped bass population. States were required to monitor stock status including reproductive success, mortality rates, and the age composition of the spawning stock (USDOI and USDOC 1992). The main features of Amendment 4 are adaptive fishery management, which allows for flexibility in the face of changing stock status, and the priority of stock recovery over allocation and user benefits. The most important benchmark of Amendment 4 is a target fishing mortality rate (F) of 0.25. This level of fishing has allowed populations to increase.

Amendment 5 will be implemented when the Chesapeake population reaches full recovery. The target fishing mortality rate will be 0.5 for all mortality due to fishing activities. Initially, fishing mortality rates will be relaxed from 0.25 to a rate of 0.33. Depending on monitoring results, recreational and commercial fisheries may be further expanded to a rate of 0.4. A nonharvest mortality rate of 0.1 will be added to account for bycatch mortality, poaching, and hook-and-release mortality (ASMFC 1995a).

In 1988, two amendments were made to the Atlantic Striped Bass Conservation Act. Section 5, the North Carolina striped bass study, was promulgated because of continued striped bass declines in the Albemarle Sound–Roanoke River system. This section outlined a major study to determine the significance of different factors contributing to the decline. Authorization was for US$1 million. Section 6 directed the secretary of commerce to promulgate striped bass regulations in the Exclusive Economic Zone (EEZ), 3–200 miles from shore.

Research and Monitoring

The federal government has provided funding and a framework for striped bass research under Section 7 of the Anadromous Fish Conservation Act, formerly known as the Emergency Striped Bass Study (ESBS). The funding, administered by the NMFS and the FWS, was made available to states, universities, and private institutions. NMFS was given primary responsibility for overseeing monitoring programs, while the FWS was given responsibility for investigating suspected causes of striped bass declines. From 1980 to 1995, ESBS/Section 7 funding has totaled nearly US$9 million. The Emergency Striped Bass Study was discontinued in 1996, although funding for research on anadromous species, including the striped bass, is still availbale under the act.

Research and monitoring of the striped bass population and fisheries are required of participating states under the ASMFC Striped Bass Fishery Management Plan. Funding includes contributions from Section 7, the Sports Fish Restoration Act, and state governments. In many cases, projects are supported by a combination of federal grants and matching state funding. All seven states with commercial fisheries and the Potomac River Fisheries Commission (PRFC) are required to collect catch composition data and catch/effort data (Stephan 1992b). States with significant recreational fisheries are also required to collect catch composition data and catch/effort data.

Total public costs for monitoring, management, and research in states from North Carolina to Maine by both federal and state agencies were generally US$3–$4 million per year, with total costs since 1980 approaching US$60 million. Some states require recreational and commercial fishing permits, but fees are nominal, usually covering only the administrative costs of licensing programs. One can assume that funding will continue to decrease in the mid-1990s from its peak in the late 1980s as the striped bass picture brightens and other species become priorities.

Conflict and Compromise

Implementation of the striped bass program has required a remarkable amount of cooperation between concerned states and the federal government. The plan is dependent on a distribution of benefits (striped bass harvest) that is considered to be equitable to all jurisdictions within the management unit. Sources of contention include individual state prerogatives, distribution of harvest between coastal and producer states, and uncertainties related to nearly all aspects of striped bass biology.

Compliance has been good, because of both the stick provided by the Striped Bass Conservation Act and the understanding that it is in the interest of all parties to cooperate. All states have been reported to be in compliance by the ASMFC annually except for the District of Columbia in 1986 and New Jersey in 1986, 1988, 1989, and 1990 (USDOI and USDOC 1992). In all cases except for New Jersey in 1990, appropriate regulatory changes were made before federal action could be taken. In 1990, a moratorium was imposed in New Jersey, but appropriate regulations were instituted after five days and the moratorium was lifted. Without doubt, as serious conservation decisions were made during the 1980s and early 1990s, the Atlantic Striped Bass Conservation Act helped to maintain cohesion and coordination among concerned East Coast states.

There is also a perception that although flawed, the management plan and process include provisions that make it as fair as possible. An example is the designation of producer status to states with major striped bass spawning areas. Under the plan, the Chesapeake estuary was closed to exploitation during spawning, and size limits were increased from historical limits to protect premigratory striped bass found in these areas. Producer status has allowed states such as Maryland the opportunity to exploit premigratory fish at prescribed levels of fishing mortality, 0.25 under Amendment 4. Although the specifics concerning the producer/coastal division of the harvest are somewhat contentious, this system has provided the process through which dialogue can begin and compromise can be reached.

Another difficult area that affects the program is the uncertainty related to nearly all aspects of management such as assessments and striped bass biology in general. Although uncertainties often slow the management process, the depleted status of the resource created a sense of urgency and the conviction that uncertainty should not be an excuse for inaction.

Striped Bass Economics

Fishery management is still the province of biologists. Only one major economic study was undertaken under the striped bass study, and there has been no systematic study of economic trends since the fishery reopened in 1990. As fishery management plans have moved forward, it has become evident that economic, social, and political considerations may have greater impacts on the management process than biological considerations. Most fisheries management involves the management of human activities, not fish populations. Although government at federal and state levels promoted and implemented the ASMFC Striped Bass Management Plan, public recognition of the need for conservation gave scientists and managers the latitude to take actions. Constituent support also played a major role in federal and state legislation that enacted funding and management. Much of this support is rooted in the value of the resource, a net economic value of US$12 million in 1980.*

Commercial Fishery Economics

Although the commercial fishery has changed since the ESBS report of 1984, this summary is based on many of its principal findings which are still relevant today. During this century the striped bass commercial fishery has been characterized by three major cycles of declining catch and upswings, each of greater amplitude and shorter duration (Strand et al. 1980). The last decline has been the greatest, with landings dropping from 14.7 million pounds in 1973 to less than 290,000 pounds in 1989. Declines in commercial landings took place in states throughout the striped bass range (Fig. 15.1).

Initial effects of lower volume of striped bass did not immediately impact the harvesting sector. Revenues were maintained because price increased as product volume decreased (Table 15.3), and thus a strong incentive remained for fishermen to continue fishing as stocks declined. From 1944 to 1974, the real price of striped bass fluctuated, but remained relatively stable when compared to the large price increases

*According to Norton, the net economic value or benefits are "the value of consumption in excess of the opportunities foregone in production." In other words, it is the difference between the willingness to pay or the dollar value of a good, and the costs of production (Norton et al. 1983, p. 33).

Table 15.3. Changes in Commercial Catch and Revenues in Striped Bass Fisheries within the Four Regions of the Atlantic Coast of the United States, 1970–1978

Region	Percentage of total catch, average 1970–1978	Percentage change in catch, 1970–1978	Percentage change in revenue, 1970–1978
New England (Maine, Rhode Island)	17	–39	+39
Mid-Atlantic (New Jersey, New York)	17	–23	+88
Chesapeake (Maryland, Virginia)	50	–71	–23
South Atlantic (North Carolina)	16	–70	–26

Source: Strand et al. 1980.

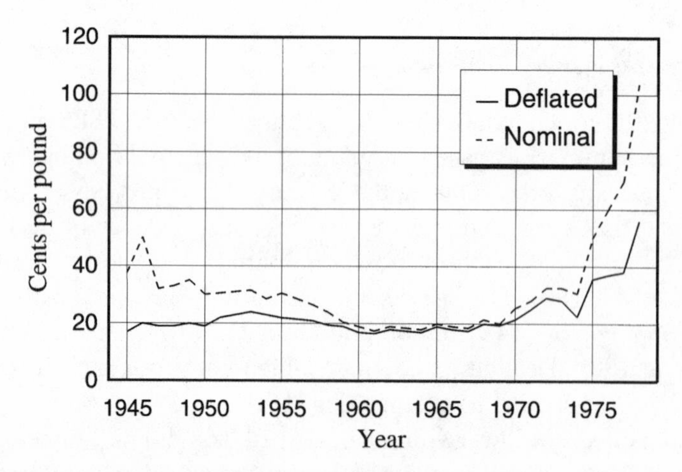

Figure 15.5. Atlantic Coast striped bass prices (nominal and deflated), 1944–78. Deflated 1967 = 100. (*Source:* Strand et al. 1980.)

during the late 1970s. Due to the record harvest of 1973, prices decreased. However, between 1974 and 1978, landings decreased by 69 percent and prices increased by 200 percent (Fig. 15.5) (Strand et al. 1980). The ex-vessel value, the value paid to fishermen at the dock, increased from US$3.3 million in 1974 to US$5.6 million in 1981 (ASMFC 1990).

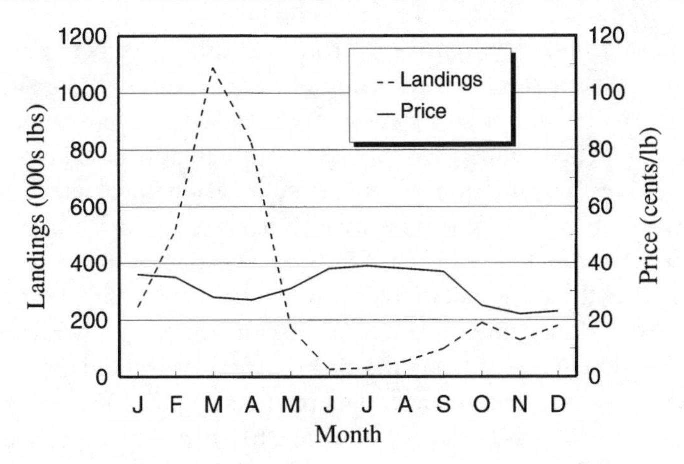

Figure 15.6. Monthly striped bass landings and prices in Maryland, average 1971–73. (*Source:* Strand et al. 1980.)

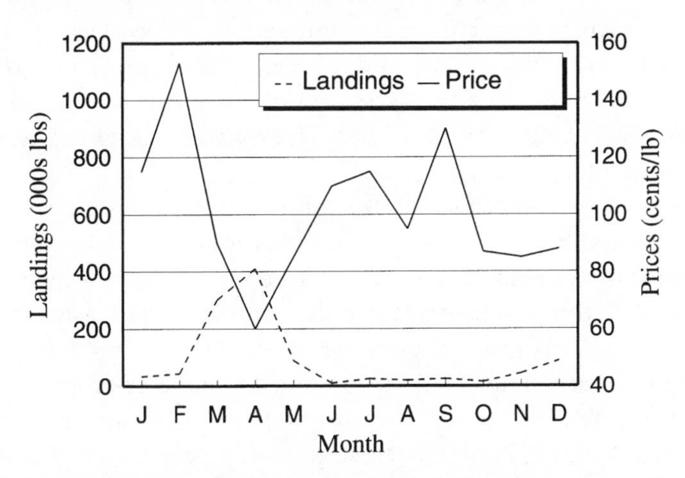

Figure 15.7. Monthly striped bass landings and prices in Maryland, average 1977–79. (*Source:* Strand et al. 1980.)

From 1971 to 1979, there was also an increase in price variation within each year, especially in the Chesapeake region (Figs. 15.6 and 15.7) (Strand et al. 1980). As landings declined, central markets in major cities were lost, and dependence on local markets grew. When landings peaked, the fixed capacity of local markets resulted in their saturation and dramatic drops in price (Strand et al. 1980). Fishermen's motives are usually dependent on revenues, not abundance. "The fisherman's economic incentive is profit, the difference between revenues and costs. If revenues are rising at a rate greater than costs, the watermen will continue to fish despite the declining abundance. Furthermore, most currently used management regulations will not effectively reverse this situation. This can be a serious problem because there is not an economic disincentive to harvest fish as abundance is reduced" (Strand et al. 1980, p. 55). This was one of the main problems as striped bass populations dropped during the late 1970s and early 1980s. Management regulations became increasingly restrictive until Amendment 3 resulted in the closing of the Maryland fishery and drastic decreases in coastwide commercial landings by 1985.

Another problem before and during the striped bass decline was the open-access nature of the fishery. No property rights were established in any of the major commercial states. Entry and exit from the fishery depended on price, alternatives such as fishing other species, other employment opportunities, abundance of striped bass, and tradition or lifestyle choice. Entry and exit are probably greater in striped bass fisheries because capital investment requirements are low relative to many fisheries. Sometimes striped bass gear and the vessel are also employed in other fisheries.

Although flexibility may be desirable in this situation because of the large swings in striped bass abundance, the entry of fishermen during periods of abundance continues until economic profits have been dissipated. At this point, returns on the capital and labor invested in the fishery no longer generate economic profits. This means that greater returns on fishery inputs could be obtained in another economic sector. At this point the fishery is overcapitalized; it has attracted more investment than is needed to harvest the available resource. When the abundance of the striped bass declines, fishermen may continue fishing and actually increase fishing effort due to the higher prices mentioned above. Fishermen that enter the fishery during the good times may find themselves trapped when the population declines, especially in cases where alternatives do not exist. The schooling nature of many fish species such as striped bass may also provide incentives to keep fishing. For those fishermen fortunate enough to encounter a school, the payoffs could be exceptional.

Another factor involves technological improvements, which have generally decreased costs relative to revenues during this century. The use of electronics, synthetic fibers for gill nets, and improved vessels have lowered costs of production. This allows fishermen to continue to fish at lower levels of stock abundance.

During the striped bass decline, commercial alternatives were becoming more limited in the Chesapeake Bay region. In the last two decades, there have been large decreases in the abundance of other major Chesapeake commercial species. White perch, American oyster, softshell clam, American shad, and Atlantic croaker landings all decreased during the period because of a combination of factors related to habitat degradation, overfishing, and natural climatic cycles (Jones et al. 1988). Notable exceptions were increased landings of blue crab, menhaden, and catfish. Fishing effort in the blue crab fishery has also increased during the last two decades. According to Zavolta et al. (1987), crab and clam operations are competitive with striped bass operations. It was found that crab and striped bass licenses were substitute products and that the current number of crab licenses influenced the demand for gill net licenses (Zavolta et al. 1987). Before the management plan, it appears that a combination of these factors provided incentives to increase rather than decrease fishing effort during the most recent striped bass population decline.

When Amendment 5 is implemented with a fishing mortality increase from 0.25 to 0.50, commercial quotas will increase approximately twofold. Commercial fisheries are now monitored much more closely and make up a much smaller contribution to total East Coast landings than recreational fisheries. Under Amendment 4, commercial fisheries are capped at 20 percent of average historical levels between 1972 and 1979. All commercially caught striped bass must have a unique numbered tag attached to them. In Maryland, all striped bass must also be checked in at approved Department of Natural Resources check stations. Monitoring through the use of tags is required in all states with commercial striped bass fisheries.

The Maryland gill net fishery reopened in 1991. It is now nearly three months long, running from December to February. Participation and fish landed per day have increased as the quota has increased. The 1991 quota was 302,000 pounds, of which only 130,000 pounds was harvested, but in 1994 the quota was increased to more than 600,000 pounds, all of which was harvested. The number of eligible fishermen in the fishery jumped from 453 in 1991 to more than 900 in 1994.

Although the official season is nearly three months long, in 1994 the number of days open to fishing during each month were decreased by nearly half because the quota was caught so quickly. Although it appears that the trend has moved toward shorter seasons and higher daily landings from 1991 to 1994, the entire quota was not filled during the 1995 and 1996 seasons, probably because of a combination of fish distribution and weather conditions. The state of Maryland has now limited the issuance of additional commercial striped bass permits.

Prices in the reopened Maryland commercial fishery have been unstable during the winter gill net season. After the first week of fishing in 1994, prices fell from US$2.50 per pound to US$1.25 per pound as local markets were likely glutted by Maryland gill net landings. The fishery closure of the 1980s and the seasonality of the current fishery probably make the development of more distant markets difficult. Another marketing constraint involves New York state PCB regulations and the Fulton Fish Market closure to out-of-state striped bass. The price swings observed during the 1994 gill net season may be greater than those experienced during the 1970s because of the loss of distant markets that could help stabilize price.

In many states, the allocation of a limited number of commercial tags to individuals at the beginning of the striped bass season has worked to control entry to the fishery, allocate quotas to individual fishermen, and track the season quota. This could allow fishing when market conditions are best as long as fishermen have tags allocated by the concerned state. Most commercial striped bass states such as Delaware are moving in this direction, but many issues remain. Should tags be transferable? Should all tags be allocated at the beginning of the season? Who should receive tags and how many? Most fisheries do not operate in isolation, but involve a complex set of interactions with other fisheries and sectors of the economy. Fishermen are both consciously and unconsciously making decisions related to their business which have implications for resource management and social welfare. Biological controls can conserve striped bass populations, but economic and social concerns may not be well served. A better understanding of user group motivations and goals, and the impact of these on the fisheries, is necessary for improving management, lowering costs of programs, and increasing social welfare.

Recreational Fishery Economics

The value of recreational fisheries cannot be determined directly from markets. In 1979, Norton et al. (1984) investigated the consumer surplus, the difference between what a consumer is willing to pay for a commodity and the amount that is actually paid in the market, in the four regions cited earlier for the commercial fishery. Results were different for each region because of different cultural or social backgrounds of fishermen, different fishing methods, and different sizes of fish caught. The average daily consumer surplus was US$169 per trip in the Mid-Atlantic, US$115 per trip in the South Atlantic region, US$86 per trip in New England, and US$36 per trip in the Chesapeake region. Total value of striped bass trips was determined by multiplying the average consumer surplus per trip by the number of trips in a given region. The coastwide recreational fishery totaled just under US$200 million in 1979 (Table 15.4).

Because many sportsmen seek more than one species of fish during a fishing trip and also gain satisfaction from esthetics associated with trips, the entire value of these trips cannot be attributed to striped bass. To account for this, the marginal value of striped bass was multiplied by the number of striped bass caught in each state. The marginal value is the amount a recreational fisherman is willing to pay for catching an additional striped bass. This generated an estimated net value of US$8 million from the recreational capture of striped bass in 1979 (Table 15.4) (Norton et al. 1984). Expenditures of recreational

Table 15.4. Value (Thousands of U.S. Dollars) of Striped Bass Recreational Fishing Trips, by Region[a]

Region	Value of recreational fishing trips	Estimated recreational net value	Ex-vessel value of commercial landings	Recreational expenditures
New England	44,456	2,337	842	25,547
Mid-Atlantic	103,089	2,157	580	35,863
Chesapeake	36,131	3,487	1,747	22,861
South Atlantic	12,536	51	316	5,051
Total	196,212	8,032	3,485	89,262

Source: Norton et al. 1984.

[a]Based on 1979 total trips and 1980 striped bass participation rates (value of commercial landings for 1979 is provided for comparison).

fishermen at US$89 million in 1980 illustrate the magnitude of related economic activities. However, since these expenditures represent spending that is related to recreational fishing activity, they should not be confused with the net value of the fishery, which is the difference between user benefits and costs of production.

User Group Motivations and Management

The decline in striped bass populations was costly for recreational and commercial fishermen and coastal communities. From 1974 to 1980, the declines in catch cost the region 7,500 jobs and US$220 million in lost economic activity (Norton et al. 1984). Yet in 1980, "commercial and recreational fisheries still supported 5,600 jobs, $89 million in spending, and $200 million in related economic activity" (Norton et al. 1984, p. xi). The net value of the resource was estimated at US$11.5 million annually (US$8 million recreational and US$3.5 million commercial) (Norton et al. 1984). At the low point of the striped bass fishery in the mid- to late 1980s, the value of the fishery was reduced to only a fraction of the figure cited for 1980. Since the fishery reopened in Maryland in 1990, landings and associated economic value have grown steadily. There are no definitive studies on economic value of the fishery in 1994, but it is certain that the fishery will surpass its 1980 value in constant dollars when it reopens under Amendment 5 as a recovered fishery.

As is the case with most valuable resources, there is intense and growing competition among user groups concerning striped bass allocation. Recreational and commercial groups have been seeking political power and larger allocations for decades. In New Jersey, where striped bass have gamefish status, there is no commercial allocation and striped bass can be taken only by recreational fishermen. The relatively low level of striped bass landings in Maine, New Hampshire, Connecticut, Pennsylvania, and the District of Columbia have supported only recreational fisheries during the last decade.

Several recreational groups have supported gamefish status for striped bass on both the state and federal levels. In 1990, a bill was introduced to make Atlantic striped bass a gamefish. In Maryland, similar efforts have been undertaken by the Maryland Saltwater Sportsman's Association (MSSA), which advocates a buyout of commercial striped bass fishermen. This resulted in the introduction of

S.B. 603, the Rockfish/Gamefish bill in 1989. Neither bill has gained the necessary support for passage.

The Maryland recreational/commercial allocation was determined by cutting the allowable catch down the middle with 42.5 percent recreational, 42.5 percent commercial, and 7.5 percent from each sector going to the charter boat fishery. Although somewhat arbitrary, this was one of the few, if any, "equitable" and politically feasible solutions.

Whether caught and retained by recreational or commercial fishermen, striped bass conservation hinges on exploitation rates that are sustainable. Once the biological benchmarks such as mortality rates and spawning biomass levels have been defined, then allocation decisions can proceed. Ultimately the political system should reflect society's values. Best use may reflect recreational, commercial, preservation, or biological diversity values, or most likely some combination of these. Probably the biggest problem revolves around whether political systems are capable of an unbiased evaluation of these values.

Within the last 20 years, a greater recognition of the need to conserve fishery resources has taken root. This is reflected in a growing conservation ethic among striped bass fishermen. Many commercial and recreational fishermen as well as recreational fishing organizations supported the Maryland moratorium. Many groups voted to delay reopening of the striped bass fishing in Maryland (Walters 1990). Recently, at a public hearing in Massachusetts, recreational fishermen spoke in opposition to lowering the 36-inch size limit to 34 inches. In both Maine and Massachusetts, catch-and-release fishing has become more popular.

In addition to the resource itself, striped bass habitat may also benefit from the striped bass harvesting program. Slowly, sport and commercial groups are starting to put differences aside in order to speak out on habitat concerns. Fishermen Involved in Saving Habitat (FISH) is a coalition formed for this purpose. As the link between environmental quality and robust populations is made, these efforts should become stronger. Unfortunately, to date, greater effort has been expended on questions of allocation than on habitat conservation.

The future challenge will be to keep harvest levels within safe limits while recognizing that motivations of user groups will not always work toward conservation of the resource. Public expenditure on research, monitoring, and management were estimated to be US$3.8 million in 1991 (Stephan 1992a), as compared to a net economic value of

US$10–$20 million per year. As public funding becomes more scarce, program priorities must be refocused, especially in the areas of monitoring, enforcement, and work with user groups. To date, ASMFC and state management programs have concentrated on biology and succeeded in arresting the striped bass decline and in rebuilding the severely depleted coastal stocks. However, there is a need to improve understanding of user motivations and to design management systems that work for conservation of the fisheries, resource users, and efficiency of management regulations.

Conclusion

The demand for striped bass by commercial and recreational fishermen is greater than the natural population can produce. High demand coupled with increasing numbers of fishermen and improved fishing gear and technologies resulted in overfishing by both recreational and commercial fishermen in the late 1970s and early 1980s. It is likely that high disposable incomes, the lack of substitutes, and health-conscious consumer preferences for seafood products will contribute to continued or increased demand for striped bass in both recreational and commercial fisheries. Degradation of estuarine and coastal habitat has also contributed to declines in the population size that can be supported by the environment, especially during the last century.

Although recruitment is highly variable, striped bass is a manageable species. A large spawning population does not guarantee good recruitment; it provides an opportunity for good—sometimes tremendous—year-classes given the right combination of physical and biological conditions. These year-classes present the opportunity for significant harvest levels. The main challenge is to limit harvest so that the population can be maintained during below-average years.

The current plan has been successful in the areas of monitoring, research, enforcement, and coordination of these elements. However, perhaps the greatest success has been the ability to separate conservation and allocation decisions. Allocation decisions involving producer and coastal catch regulations, commercial and recreational users, and proposed state plans have been and will continue to be contentious. However, consensus regarding biological determinations such as fishing mortality rates and minimum state regulatory standards has been maintained. This is in part due to recognition that all parties will benefit from continued cooperation, and to the threat of a fishing moratorium that could be imposed under the Striped Bass Conservation Act.

Finally, striped bass habitat conservation is an issue on which consensus among most fishermen is possible. Unfortunately, because it is a long-term concern, short-term questions such as allocation have received more attention from fishermen and fishing organizations. Development pressure continues to be great in areas adjacent to rivers, estuaries, and the coast. A productive and valuable resource such as striped bass is another strong argument for conservation of these areas.

References

ASMFC (Atlantic States Marine Fisheries Commission). 1981. *Interstate fisheries management plan for the striped bass.* Fisheries Management Report no. 1. Washington, D.C.

ASMFC. 1990. *Source document for the supplement to the striped bass FMP —Amendment 4.* Fisheries Management Report no. 16. Washington, D.C.

ASMFC. 1991. *Interstate fisheries of the Atlantic Coast.* Washington, D.C.

ASMFC. 1992. *Addendum II of the striped bass fishery management plan.* Washington, D.C.

ASMFC. 1994. Unpublished material of the ASMFC Striped Bass Technical Committee. Distributed at the Striped Bass Board Meeting, April 27, 1994, Norfolk, Virginia.

ASMFC. 1995a. *Amendment 5 to the fisheries management plan for Atlantic striped bass.* Washington, D.C.

ASMFC. 1995b. *Proceedings of a workshop on the striped bass binary coded wire tag recovery program.* Special Report no. 43. Washington, D.C.

ASMFC. 1996. *Recommendations concerning the stocking of striped bass in Atlantic Coastal waters.* Special Report no. 50. Report from the Striped Bass Stocking Subcommittee of ASMFC. Washington, D.C.

Baum, T. 1994. Delaware River striped bass young of the year recruitment survey. Presented at the 1994 Striped Bass Study Annual Workshop, Alexandria, Virginia.

Berggren, T. J., and J. T. Lieberman. 1978. Relative contribution of Hudson, Chesapeake, and Roanoke striped bass, *Morone saxatilis,* stocks to the Atlantic coast fishery. *United States Fish and Wildlife Service Fishery Bulletin* 76:335–345.

Berlinsky, D. L., J. F. O'Brien, and J. L. Specker. 1988. *Reproductive status of female striped bass.* Interim Report. Department of Zoology, University of Rhode Island.

Bigelow, H. B., and W. C. Schroeder. 1953. The striped bass. In *Fishes of the Gulf of Maine. United States Fish and Wildlife Service Fishery Bulletin* 53:389–404.

Boreman, J., and J. O'Brien. 1983. Stock composition of striped bass landed in the Rhode Island trap net fishery. Paper presented at the 1983 Northeast Fish and Wildlife Conference, May 15–18, 1983, Mt. Snow, Vermont.

Boynton, W. R., E. M. Setzler, K. V. Wood, H. H. Zion, M. Homer, and J. A. Mihursky. 1977. *Final Report of Potomac River fisheries study: ichthyoplankton and juvenile investigations.* Solomons, Maryland: University of Maryland, Chesapeake Biological Laboratory Ref. no. 77-169.

Clark, J. R. 1968. Seasonal movements of striped bass contingents of Long Island Sound and the New York Bight. *Transactions of the American Fisheries Society* 97:320–343.

Cohen, J. E., S. W. Christensen, and C. P. Goodyear. 1983. A stochastic age-structured population model of striped bass (*Morone saxatilis*) in the Potomac River. *Canadian Journal of Fisheries and Aquatic Sciences* 40:2170–2183.

Coutant, C. C., and D. L. Benson. 1987. *Linking estuarine water quality and impacts on living resources: Shrinking striped bass habitat in Chesapeake Bay and Albemarle Sound.* Environmental Sciences Division, Publication no. 2972. Oak Ridge, Tennessee: Oak Ridge National Laboratory.

Deuel, D. G. 1973. *1970 salt-water angling survey.* United States Department of Commerce, Current Fishery Statistics 6200. Washington, D.C.: United States Department of Commerce.

Diodati, P. J., and T. B. Hoopes. 1991. *Fisheries monitoring report for the Massachusetts 1990 striped bass fisheries.* Report to the Atlantic States Marine Fisheries Commission Technical Committee. May 1991.

Dorazio, R. M., and P. J. Rago. 1988a. *Maximum allowable fishing mortality rates for anadromous striped bass of Chesapeake Bay: Revised predictions and further comments.* Report to the Statistical and Scientific Committee of the Atlantic States Marine Fisheries Commission, Hartford, Connecticut, April 1988.

Dorazio, R. M., and P. J. Rago. 1988b. *Effects of noncatch fishing mortality and maximum size limits on prediction of threshold F values for anadromous striped bass of Chesapeake Bay.* Report to the Statistical and Scientific Committee of the Atlantic States Marine Fisheries Commission, Norfolk, Virginia, June 1988.

Dubovsky, C. H., and W. Laney. 1993. Supplement to the striped bass FMP Amendment 5 management information document (draft). Atlantic States Marine Fisheries Commission, Washington, D.C.

Fabrizio, M. C. 1987. Contribution of Chesapeake and Hudson River stocks of striped bass to Rhode Island coastal waters as estimated by isoelectric focusing of eye lens proteins. *Transactions of the American Fisheries Society* 116:588–593.

Fay, C. W., R. J. Neves, and G. B. Pardue. 1983. *Species profiles: Life histories and environmental requirements of coastal fishes and invertebrates (Mid-Atlantic)—striped bass.* United States Fish and Wildlife Service, Division of Biological Services, FWS/OBS-82/11.8. United States Army Corps of Engineers, TR EL-82-4.

Fearing, D. B. 1903. Some early notes on striped bass. *Transactions of the American Fisheries Society* 32:90–98.

Florence, B. M. 1980. Harvest of northeastern coastal striped bass stocks produced in the Chesapeake Bay. In *Proceedings of the Fifth Annual Marine Recreational Fisheries Symposium,* ed. H. Clepper, 29–44. Washington, D.C.: Sport Fishing Institute.

Goodyear, C. P. 1984. Analysis of potential yield per recruit for striped bass produced in the Chesapeake Bay. *North American Journal of Fisheries Management* 4:488–496.

Goodyear, C. P. 1985. Relationship between reported commercial landings and abundance of young striped bass in the Chesapeake Bay, Maryland. *Transactions of the American Fisheries Society* 114:92–96.

Harrell, R. M. 1987. Catch and release mortality of striped bass with artificial lures and baits. *Proceedings of the Annual Conference of the Southeastern Association of Fish and Wildlife Agencies* 41:70–75.

Harrell, R. M., X. L. Xu, and B. Bly. 1993. Evidence of introgressive hybridization in Chesapeake Bay *Morone. Molecular Marine Biology and Biotechnology* 2:291–299.

Hassler, T. J. 1988. *Species profiles: Life histories and environmental requirements of coastal fishes and invertebrates (Pacific Southwest)—striped bass.* United States Fish and Wildlife Service Biology Report 82(11.82). United States Army Corps of Engineers, TR EL-82-4.

Heinle, D. R., D. A. Flemer, and J. F. Ustach. 1976. Contribution of tidal marshlands to Mid-Atlantic estuarine food chains. In *Estuarine processes,* vol. 3, ed. M. L. Wiley, 309–320. New York: Academic Press.

Hershner, C., and R. L. Wetzel. 1987. Submerged and emergent aquatic vegetation of the Chesapeake Bay. In *Contaminant problems and management of living Chesapeake Bay resources,* ed. S. K. Majumdar, L. W. Hall Jr., and H. M. Austin, 116–133. Pennsylvania Academy of Science.

Jones, P. W., H. J. Speir, N. H. Butowski, R. O'Reilly, L. Gillingham, and E. Smoller. 1988. *Chesapeake Bay fisheries: Status, trends, priorities, and data needs.* Annapolis, Maryland: Maryland Department of Natural Resources Tidewater Administration; Norfolk, Virginia: Virginia Marine Resources Commission.

Jordan, D. S., and B. W. Evermann. 1902. *American food and game fishes.* New York: Doubleday, Page & Co.

Kriete, W. H., Jr., J. V. Merriner, and H. M. Austin. 1978. *Movement of the 1970 year-class striped bass between Virginia, New York, and New England.* Gloucester Point, Virginia: Virginia Institute of Marine Science.

Laney, W. R., J. C. Benton, L. T. Henry, H. Johnson, J. W. Kornegay, K. L. Nelson, S. D. Taylor, and S. E. Winslow. 1994. *North Carolina estuarine striped bass fishery management plan.* Prepared by the North Carolina Division of Marine Fisheries and North Carolina Wildlife Resources Commission, Division of Boating and Inland Fisheries, in cooperation with the United States Fish and Wildlife Service. Morehead City, North Carolina.

McFadden, J. T.; Texas Instruments, Incorporated; and Lawler, Matusky, and Skelly, Engineers. 1978. *Influence of the proposed Cornwall pumped storage project and stream electric generating plants on the Hudson River estuary with emphasis on striped bass and other fish populations,* revised. Prepared for Consolidated Edison Company of New York, Inc.

MDDNR (Maryland Department of Natural Resources). 1994a. *History of striped bass management in Maryland.* Annapolis, Maryland: Tidewater Administration.

MDDNR. 1994b. *Investigation of striped bass in Chesapeake Bay.* USFWS Federal Aid Project F-42-R-7 1993–1994. Annapolis, Maryland: Tidewater Administration.

Merriman, D. 1941. Studies on the striped bass (*Roccus saxatilis*) of the Atlantic Coast. *United States Fish and Wildlife Service Fishery Bulletin* 50:1–17.

Mihursky, J. A., W. R. Boynton, E. M. Setzler, K. V. Wood, H. H. Zion, E. W. Gordon, L. Tucker, P. Pulles, and J. Leo. 1976. *Final report on Potomac estuary fisheries study: Ichthyoplankton and juvenile investigations.* Solomons, Maryland: University of Maryland Chesapeake Biological Laboratory ref. no. 76-12-CBL.

Morgan, R. P., II, T.S.Y. Koo, and G. E. Krantz. 1973. Electrophoretic determination of populations of the striped bass, *Morone saxatilis,* in the Upper Chesapeake Bay. *Transactions of the American Fisheries Society* 102:221–232.

Norton, V., T. Smith, and I. Strand, eds. 1984. *$tripers: The economic value of the Atlantic coast commercial and recreational striped bass fisheries.* University of Maryland Sea Grant Publication no. UM-SG-TS-83–12.

Rago, P. J. 1992. Chesapeake Bay striped bass: The consequences of habitat degradation. In *Stemming the tide of coastal fish habitat loss: A symposium on conservation of coastal fish habitat,* ed. R. H. Stroud, 105–116. Savannah, Georgia: National Coalition for Marine Conservation.

Rago, P. J., and R. M. Dorazio. 1988. *Use of historical trends in recruitment and exploitation to forecast the consequences of management measures on striped bass abundance, catch, and young of the year.* Report to the Statistical and Scientific Committee of the Atlantic States Marine Fisheries Commission, Hartford, Connecticut, August 1988.

Raney, E. C. 1952. The life history of the striped bass. *Bulletin of the Bingham Oceanographic Collection, Yale University* 14:5–97.

Raney, E. C. 1957. Subpopulations of the striped bass in tributaries of Chesapeake Bay. *United States Fish and Wildlife Service. Special Scientific Report. Fisheries* 208:85–107.

Rugolo, L. J., P. W. Jones, C. M. Stagg, and H. Speir. 1991. *Maryland's 1990 striped bass harvest and background fishing losses.* Report to the Atlantic States Marine Fisheries Commission Striped Bass Technical Committee. May 1991.

Schaefer, R. 1968. Size, age composition, and migration of striped bass from the surf waters of Long Island. *New York Fish and Wildife Journal.* 5(1): 1–51.

Schill, W. B., and R. M. Dorazio. 1990. Immunological discrimination of Atlantic striped bass stocks. *Transactions of the American Fisheries Society* 119:77–85.

Seagraves, R. J., and R. W. Miller. 1989. *Striped bass bycatch in Delaware's commercial shad fishery.* Delaware Division of Fish and Wildlife Report. Dover, Delaware: Department of Natural Resources and Environmental Control.

Setzler, E. M., W. R. Boynton, K. V. Wood, H. H. Zion, L. Lubbers, N. K. Mountford, P. Frere, L. Tucker, and J. A. Mihursky. 1980. *Synopsis of biological data on striped bass, Morone saxatilis (Walbaum).* National Oceanic and Atmospheric Administration Technical Report FAO Synopsis no. 121.

Setzler-Hamilton, E. M., W. R. Boynton, J. A. Mihursky, T. T. Polgar, and K. V. Wood. 1981. Spatial and temporal distribution of striped bass eggs, larvae, and juveniles in the Potomac estuary. *Transactions of the American Fisheries Society* 110:121–136.

Shepard, G. 1992. *Revised estimates of poaching of striped bass from MFRSS Survey data, 1987–1991.* Report to the Striped Bass Technical Committee, Atlantic States Marine Fisheries Commission, Philadelphia, Pennsylvania, April 1992.

Sidell, B. D., R. G. Otto, D. A. Powers, M. Karweit, and J. Smith. 1980. Apparent genetic homogeneity of spawning striped bass in the Upper Chesapeake Bay. *Transactions of the American Fisheries Society* 109:99–107.

Smith, R. E., and R. J. Kernehan. 1981. Predation by the free living copepod *Cyclops bicuspidatus* on larvae of striped bass and white perch. *Estuaries* 21(4):32–48.

Smith, T. I., and W. E. Jenkins. 1992. Regional development of hybrid striped bass aquaculture in the southeastern United States. Paper presented at the World Fisheries Congress, Athens, Greece, May 1992.

Stephan, C. D. 1992a. *1992 review of the Atlantic States Marine Fisheries Commission Fisheries management plan for Atlantic striped bass (Morone saxatilis)*. Washington, D.C.: Atlantic States Marine Fisheries Commission.

Stephan, C. D. 1992b. *Striped bass monitoring and research coordination.* First year completion report. Washington, D.C.: Atlantic States Marine Fisheries Commission.

Strand, I. E., V. J. Norton, and J. G. Adriance. 1980. Economic aspects of commercial striped bass harvest. In *Fifth Annual Marine and Recreational Fisheries Symposium,* ed. H. Clepper, 51–62. Boston, Massachusetts: Sport Fishing Institute.

Trent, W. L., and W. W. Hassler. 1968. Gill net selection, migration, size and age composition, sex ratio, harvest efficiency, and management of striped bass in the Roanoke River, North Carolina. *Chesapeake Science* 9(4):217–232.

Ulanowicz, R. E., and T. T. Polgar. 1980. Influence of anadromous spawning behavior and optimal environmental conditions upon striped bass (*Morone saxatilis*) year-class success. *Canadian Journal of Fisheries and Aquatic Science* 37:143–154.

USDOC (United States Department of Commerce, National Oceanic and Atmospheric Administration). 1987. *Maryland striped bass research.* Project no. NA86EA-D000009. Grant no. AFC-15.

USDOI and USDOC (United States Department of the Interior and United States Department of Commerce). 1982. *Emergency striped bass research study. 1981 annual report.* Washington, D.C.

USDOI and USDOC. 1987. *Emergency striped bass research study. 1986 annual report.* Washington, D.C.

USDOI and USDOC. 1989. *Emergency striped bass research study. 1988 annual report.* Washington, D.C.

USDOI and USDOC. 1991. *Emergency striped bass research study. 1989 annual report.* Washington, D.C.

USDOI and USDOC. 1992. *Emergency striped bass research study. 1990 annual report.* Washington, D.C.

USDOI and USDOC. 1993. *Emergency striped bass research study. 1991 annual report.* Washington, D.C.

Valadykov, V. D., and D. H. Wallace. 1952. Studies of the striped bass, *Roccus saxatilis* (Walbaum), with special reference to the Chesapeake Bay region during 1936–1938. *Bulletin of the Bingham Oceanographic Collection, Yale University* 14:132–177.

Van Winkle, W., and K. D. Kumar. 1982. *Relative stock composition of the Atlantic coast striped bass population: Further analysis.* NUREG/CR-2563; URNL/TM-8217. Oak Ridge, Tennessee: Oak Ridge National Laboratory.

Waldman, J. R., D. J. Dunning, Q. E. Ross, and M. T. Mattson. 1990. Range dynamics of Hudson River striped bass along the Atlantic coast. *Transactions of the American Fisheries Society* 119:910–919.

Walters, K. 1990. *Chesapeake stripers.* Bozman, Maryland: Aerie House.

Westin, D. T., and B. A. Rogers. 1978. *Synopsis of the biological data on the striped bass.* University of Rhode Island Marine Technical Report no. 67.

Wirgin, I. I., T. Ong, L. Maceda, J. R. Waldman, D. Moore, and S. Courtenay. 1993. Mitochondrial DNA variation in striped bass (*Morone saxatilis*) from Canadian rivers. *Canadian Journal of Fisheries and Aquatic Sciences* 50:80–87.

Young, B. H. 1987. *A study of the striped bass in the marine district of New York IV.* New York State Department of Environmental Conservation, Completion Report, Project AFC-13–2, Stony Brook, New York.

Zavolta, S. M., I. E. Strand, and D. G. Swartz. 1987. Supply response in the harvest of striped bass. Paper presented at the Conference on Economics of Chesapeake Bay Management III, Annapolis, Maryland, May 28–29, 1987.

SIXTEEN

Sustainable Use of Salmon

Its Effects on Biodiversity and Ecosystem Function

Robert C. Francis

The thesis that human harvest of salmon alone is capable of inflicting lasting damage on salmon populations inhabiting large drainage basins is not consistent with history and geography of salmon populations. While it is clearly physically possible to reduce or eliminate anadromous species such as salmon from small drainages by excluding them from their freshwater spawning and rearing grounds, the history of fishing does not provide examples of fishing operations coinciding with extirpation of salmon populations from broad geographic areas where freshwater spawning and rearing habitats have been left intact. The present status of salmon in large watersheds capable of bearing salmon in North America such as the Sacramento, Klamath, Columbia, Fraser, Skeena, Copper, Kvichak, Yukon, Penobscot, St. John, and St. Lawrence appears to be more closely related to human population density and land-use practices than to the history and extent of exploitation. . . . Ecosystem management is not merely a stylish term, it is imperative for the long-term persistence of all species of salmon throughout their ranges. (From a prepublication draft of Mundy 1996)

This chapter is a personal journey through a very complex and, to a great degree, subjective field. Essentially, the overall objective of the contributions in this volume is to examine the effects of sustainable use on biodiversity and ecosystem function. My task has to do specifically with sustainable use of salmon. In essence, we are asked to better understand what kinds of trade-offs, if any, in biodiversity and ecosystem function may have to be accommodated and watched out for under conditions of sustainable use (of salmon) where there is a premium on maximizing the production of target (salmon) populations. The foci here are on sustainable uses, salmonid ecosystems, and biodiversity preservation.

To approach this task I focus on the Northeast Pacific salmon fisheries ranging from Alaska to California, with an emphasis on the marine environment and its effects on salmonid populations and ecosystems. I therefore provide considerable groundwork in the following areas:

1. Taxonomic and biogeographic scope.
2. What is a salmonid ecosystem?
3. What is biodiversity and how does it apply to salmonid ecosystems?
4. How do the concepts of health and carrying capacity apply to salmonid ecosystems?
5. What is sustainable use of salmon? What is its history in the Northeast Pacific?

Next I present some thoughts and references on two scientific areas which are of general interest in considering the problem being posed:

6. Sustainable salmonid fishery management;
7. Some biological aspects of sustainable salmon use.

Finally, I express my views on what we know about the effects of sustainable use on biodiversity and ecosystem function. In so doing, I discuss not only what we already seem to know but how we need to change the way we do science and management in order to better understand and deal with this important issue.

The real difficulty in this task is to isolate the effects of sustainable exploitation on salmonid ecosystems. In my initial interpretation, sustainable use of salmon has two components: harvest and enhancement. To tease out the effects of these components on ecosystem structure and function, one must analyze conditions where their effects can be isolated. Alaska is the place to look at the effects of sustainable harvest. And the best and most intensely studied case in Alaska is Bristol Bay, where an enormous sockeye salmon fishery has been operating on wild populations for almost a century in nearly pristine habitat conditions. The effects of enhancement are more difficult to isolate, since enhancement is usually undertaken in response to the negative impacts of various forms of habitat loss (e.g., dams being built or forestry). Two locations where these effects have been studied and stand some chance of being teased out are Prince William Sound, Alaska (where hatcheries were built in response to reductions in fishery production not due to habitat loss) and the Washington-Oregon-California coast where enhancement has been undertaken on a fairly grand scale.

Taxonomic and Biogeographic Scope

Pacific salmon (genus *Oncorhynchus*) are an important biological, economic, and cultural resource of the countries and cultures of the North Pacific rim. Their geographic distribution extends from the California coast northward along the Canadian and Alaskan coasts to rivers draining into the Arctic Ocean, and southward along the Asian coastal areas of Russia, Japan, and Korea (Groot and Margolis 1991).

Life History and Distributional Overviews

There are eight species of Pacific salmon. Six species—sockeye (*Oncorhynchus nerka*), pink (*O. gorbuscha*), chum (*O. keta*), chinook (*O. tshawytscha*), coho (*O. kisutch*), and steelhead (*O. mykiss*)—reproduce on both the Asian and North American continents. Two, masu (*O. masou*) and amago (*O. rhodurus*), occur only in Asia. The most comprehensive and up-to-date description of Pacific salmon (excluding steelhead) life histories is given by Groot and Margolis (1991). What follows is a brief overview of life history and distributional factors important to this chapter.

Pacific salmon spawn in the fall, usually in freshwater and normally in the place where they originated. They die soon after spawning. The eggs, which are buried in the gravel, develop during the winter and the fry emerge in the spring. The remainder of the salmon life cycle varies both among species and among populations within species. Pacific salmon generally migrate to sea after an early freshwater life (ranging from weeks for some pink salmon populations to several years for some sockeye salmon populations). Once in the sea, they become widely distributed over the North Pacific Ocean and Bering Sea, most performing extensive migrations while at sea. Some populations tend to inhabit coastal waters, while others migrate extensively on the high seas. Upon maturation, after one to four years in the sea, Pacific salmon usually return to their home rivers to spawn.

Since young salmon have no contact with their parents, their ability to complete their life cycle successfully must be entirely inherited. It is most likely that each generation inherits the abilities to survive, grow, and reproduce that were most advantageous for the environment of their parents. Thus changes in environment will likely cause changes in population dynamics and structure (Rogers 1987).

The following is a brief description of the "typical" life history of the five species of Pacific salmon that originate in North American watersheds and that are examined in this chapter (the steelhead is excluded from the analysis).

Sockeye salmon exhibit a greater variety of life history patterns than other members of the genus *Oncorhynchus* (Burgner 1991), and characteristically make more use of lake-rearing habitat (one to two years) in the juvenile stages. Anadromous sockeye spend from one to four years in the ocean before returning to freshwater to spawn. The adaptations of sockeye to lake-rearing environments appear to require more precise homing to spawning areas, both as to time and location, than is found in other species of Pacific salmon. Very large sockeye concentrations are found both in the large lake systems that drain into Bristol Bay (E. Bering Sea) and in the lakes of the Fraser River system (southern British Columbia).

Pink salmon is the most abundant of the seven species of salmon in the North Pacific. In general, they are associated with small to intermediate-sized coastal rivers, generally spend their freshwater lives within a few kilometers of the sea, and have the simplest and most specialized life cycle within the genus. Upon emergence, pink salmon fry migrate quickly to the sea and grow rapidly as they make extensive oceanic feeding migrations. After approximately 18 months in the ocean, maturing fish return to their river of origin to spawn and die (Heard 1991). Because of the fixed two-year life cycle, pink salmon spawning in a particular river system in odd and even years are reproductively isolated from each other and have developed into different genetic lines. Historically, pink salmon have been more abundant in Asia than in North America (Rogers 1987).

Chum salmon have the widest natural geographic distribution of all Pacific salmon species (Salo 1991). Chum salmon migrate to the sea in the spring or early summer in which they emerge from the gravel. They return to spawn after two to four years at sea. Common to virtually every region of chum salmon distribution is the occurrence of early and late returning stocks to the natal stream. In general, early run spawners spawn in main stems of streams and late spawners seek out spring water that tends to have more favorable temperatures through the winter. Since the mid-1970s, the largest concentrations of chum salmon have most likely originated in hatcheries in northern Japan.

Coho salmon typically spawn in tributary streams in river systems (Rogers 1987). After emergence, juveniles spend one to two years in the stream (predominantly one year in the south and two years in the north) before migrating to the sea. Hatcheries currently produce a significant proportion of coho migrants, which may average two or three times heavier than natural migrants (Rogers 1987). Coho salmon spend one year in the ocean but generally return later in the fall than other species (Sandercock 1991).

Chinook salmon are the largest, the oldest at maturity, and the least abundant of the Pacific salmon species. Chinook migrate to sea during their first or second year and commonly return to spawn after two to four years at sea. Individual spawning populations are relatively small. A large part of the variation in chinook life history apparently derives from the fact that the species occurs in two behavioral forms. "Stream-type" chinook spend one or more years as fry or parr in freshwater before migrating to sea, perform extensive offshore oceanic migrations, and tend to return to their natal river in the spring or summer, several months prior to spawning. Stream-type chinook are typical of Asian populations and of northern populations and headwater tributaries of southern populations in North America. "Ocean-type" chinook migrate to sea during their first year of life, normally within three months after emergence, spend most of their ocean life in coastal waters, and return to their natal river in the fall, a few days or weeks before spawning (Healey 1991). Ocean-type chinook are typical of populations on the North American coast south of latitude 56° N.

Based on estimates derived from Rogers (1994) and prorated (over Southeast Alaska, British Columbia, and Washington-Oregon-California) catch statistics described in Francis and Hare (1994), the origin and abundance of populations of these five species over the last four decades are as follows. Pink salmon is the most abundant species in the North Pacific (average total 1951–88 run of 168 million) with around 60 percent of the run originating in Asia. Chum salmon is next (average total 1951–88 run of 67 million) with almost 70 percent of the run originating in Asia (and much of the recent run from hatcheries of northern Japan). Sockeye salmon is next (average total 1951–88 run of 53 million) with almost 45 percent of the run originating in Western Alaska (Bristol Bay). Coho salmon is next (average total 1951–88 run of 18 million) with over 50 percent of the run originating in British Columbia and Washington-Oregon-California. Chinook salmon is the least abundant species in the North Pacific (average total 1951–88 run of 6 million) with almost 60 percent of the run originating in Washington-Oregon-California.

Interestingly, the fraction of North America-bound salmon has been low in the Japanese high seas fishery of the last 40 years. Most of the harvest in those fisheries was of Asia-bound pink, chum, and sockeye salmon.

Recoveries of tagged fish (Myers et al. 1990) indicate that, in general, pink, chum, and sockeye tend to be oceanic in the marine component of their life histories, with stocks of North American origin occupying the Central Subarctic Domain (Ware and McFarlane 1989), whereas coho and chinook tend to be more coastal, with stocks of North American origin occupying the coastal region from Southeast Alaska to California. More detailed information on ocean migration by species is given in Groot and Margolis (1991).

Salmonid Ecosystems

The concept of a salmonid ecosystem is one that many seem to understand intuitively but few have ever attempted to define explicitly. After all, a salmonid ecosystem is the total environment within which salmon live. Let's see if we can be a little more specific.

The definition of an ecosystem by Tansley (1935) is one that has stood the test of time:

> the whole system (in the sense of physics) including not only the organism-complex, but also the whole complex of physical factors forming what we call the environment of the biome—the habitat factors in the widest sense.

Though the organisms may claim our primary interest, when we are trying to think fundamentally we cannot separate them from their special environment, with which they form one physical system. It is the systems so formed which, from the point of view of the ecologist, are the basic units of nature on the face of the earth.

Pomeroy et al. (1988) elaborate somewhat on this theme, and introduce the concept of scale as being critical to the understanding of ecosystems:

One feature of the biosphere that clearly emerges from ecological studies is the range of scales of space and time over which events occur. Each kind of event may have a characteristic time or half-time, ranging from nanoseconds for intracellular processes . . . to millions of years for the evolution of continents and ocean basins and their contained ecosystems. We are really dealing with a continuum, of functional response over a range of time scales, and it is often necessary to think about a substantial part of that range in order to properly describe and understand ecological processes.

Allen and Hoekstra (1992) describe the concepts of ecosystems as complex systems with self-organizing properties, capable of developing rapid shifts or dislocations in organization when processes occurring at very different rates interact. In essence, this type of "complex" behavior emerges when there is a mismatch of scale between processes which converge to control or constrain particular ecosystem entities. In particular, ecosystems are made up of processes or fluxes occurring at different rates in space and time. What is meant here is ecology in four dimensions. An ecosystem cannot be defined without a clear reference to time.

The essence of this theory is contained in the work of Ilya Prigogine (see, e.g., Nicolis and Prigogine 1977; Prigogine 1980) on chemical systems as interpreted by Allen (1985). They consider ecosystems to be thermodynamically open dissipative systems. Natural systems are open, usually have strong inputs and outputs, and have strong internal coupling between elements. Although they may appear to be stable (dynamic stability), they are in fact accompanied by incessant fluxes of energy and materials and, as a result, are maintained far from equilibrium. Prigogine coined the term *dissipative* for structures that are formed in response to the fluxes themselves. They are formed within the system and obtain the energy for their formation from the dissipation of substances or energy flowing through the system. A classic example of a dissipative structure in fluid dynamics is a weather system. In fact, the flow of substances through the system can cause tension in the organization of the system, similar to the tension that occurs between tectonic plates prior to an earthquake. When this tension or disequilibrium reaches a certain intensity, "then many amazing and surprising things can happen" (Allen 1985). The system is in a state of self-organized criticality (Bak and Chen 1991; Waldrop 1992), where a mechanism that leads to minor events can also lead to major events. In this way, systems can flip from one state to another causing quite drastic alterations in system structure and yet may be triggered by changes in system fluxes that are themselves relatively small.

And so it can be the tension within a system caused by relative fluxes of energy and material that create what we call order. The greater the tension, the greater the capacity is for rapid reorganization. What this implies is that order is emergent and that many important ecosystem processes are irreversible. And thus history plays a vital role in determining the nature of order in an ecosystem.

The reason I am taking so much space in discussing the concept of an ecosystem is that it is at the center of the fundamental question being posed in this chapter: in essence, how does sustainable exploitation affect salmonid ecosystems? *Salmonid ecosystem* refers to a collection of biological and physical processes and fluxes in the freshwater, estuarine, and marine environment of salmon, occurring at different rates in space and time, which are directly connected to the focal salmon populations of concern. As we shall see later, these ecosystems are continuously in states of disequilibrium and, as a result, are capable of manifesting rapid and unpredictable shifts in organization which make study by traditional reductionist scientific methods very difficult. Therefore answers provided in this chapter will be more at the periphery of the question than a frontal attack.

To facilitate scientific attention, salmonid ecosystems can be partitioned into logical components, based on environments occupied in various stages of their life histories: freshwater, estuarine, and marine. Unfortunately, convention often leads us to treat these three items as frameworks for separate and disconnected ecosystems. As the work of Vannote et al. (1980), Holtby and Scrivener (1989), and Cooney (1993), among others, clearly points out, connectedness both in time and space is very important in ecosystems. Because they occupy a massive space over their life histories, salmon may themselves be an important "glue" that holds their ecosystem together in that they serve as a nutrient pump from the marine to the freshwater parts of the system. Without them, nutrient fluxes within these systems are vastly altered.

Biodiversity and Salmonid Ecosystems

Biodiversity refers to the variety of life and its processes (Hughes and Noss 1992). According to Wilson (1994), "biodiversity is the totality of hereditary variation in life forms, across all levels of biological organization, from genes and chromosomes within individual species to the array of species themselves and finally, at the highest level, the living communities of ecosystems such as forests and lakes." Cairns and Lackey (1992) emphasize that biodiversity can be recognized at four levels in a biological hierarchy: genetic, species, ecosystem (relates to the variety of ecological processes, communities, and habitats within a region), and landscape. They further indicate that the principal causes behind the recent increase in loss of aquatic biodiversity include habitat alteration, fragmentation, and simplification. Unfortunately, as Hughes and Noss (1992) point out, "clear and generally accepted measures of biological diversity do not exist."

Ryman et al. (1994) point out that since fish are the only major human food source harvested from natural populations, there is a particularly strong need for better understanding of biodiversity, in particular genetic diversity, among fishes below the species level. They indicate that the three major categories of threat to genetic diversity within species are extinction, hybridization, and loss of genetic variation within populations. Further, they feel that the most immediate need for conservation of intraspecific genetic variability in fishes is to protect local populations of freshwater and anadromous species.

Biodiversity can be defined with respect to any spatial or temporal scale. In this case, my focus is on the variety of life associated with salmonid ecosystems. In particular, I ask the questions: What is the relationship between salmonid ecosystem biodiversity and the organization and dynamics of those ecosystems? Further, how does sustainable use (exploitation, enhancement) of salmonids affect that relationship?

The Concepts of Carrying Capacity and Health Applied to Salmonid Ecosystems

Carrying capacity refers to the maximum abundance that a particular species or population can attain within a given environment. In this case, the issue of carrying capacity boils down to one of the effects of ecological factors (density dependence) on survival and growth of Pacific salmon during various stages of their life histories. Holtby and Scrivener (1989), Pearcy (1992), Cooney (1993), Jeff Hard (NMFS, Northwest Fisheries Science Center, Seattle, pers. comm. 1994), and Francis and Brodeur (n.d.), among others, document current evidence for the existence of limits to salmonid production in freshwater, estuarine, nearshore marine, and high seas marine habitats. Perhaps the most important point to arise from recent studies is that complex physical-biological interactions are capable of affecting rapid and sometimes irreversible changes in the limits to the production of salmonid populations. These changes can occur in the freshwater environment at very localized scales (e.g., rapid shifts in sockeye salmon production in the Bristol Bay region of Alaska, discussed in more detail later) and in the marine environment at basinwide scales (e.g., rapid decadal scale shifts in the abundance of Alaska salmon production [Francis and Hare 1994; Hare and Francis 1995] with the opposite shifts occurring in the abundance of Pacific Northwest salmon [Pearcy 1992; Francis and Sibley 1991]). As will be seen later, both sustainable fishing and enhancement can significantly affect the nature of these relationships.

Ecosystem health is a term that some consider to be oxymoronic and others consider to be meaningful. Costanza et al. (1992) fall into the latter camp and address the issue of ecosystem health from a broad basis. In essence, they conclude that ecosystem health is closely linked to the concept of sustainability, which in turn is seen to be a comprehensive multiscale, dynamic measure of system resilience, organization, and vigor. Norton (1992) further elaborates on this point by stating that ecosystems are "self-organizing systems that provide the context for all human activity." To say that an ecosystem is "healthy" is to say that the overall system maintains sufficient complexity and flexibility to protect its self-organizing qualities. Lichatowich et al. (1995) extend this concept to the salmonid populations and say that "the important ecological relationships are those between the habitat template and intrapopulation life histories . . . the spatial/temporal diversity of life histories expressed within a complex habitat structure is an important determinant of productivity and self-organizing capacity in Pacific salmon."

Therefore the concept of ecosystem health as put forward by Costanza et al. (1992) and Lichatowich et al. (1995) recognizes the dynamic and evolutionary nature of ecosystems and the inextricable link between human activity and ecosystem evolution and organization. In this context, "management must have as a central goal the protection of the system's creativity" (Norton 1992). The extent to which this is done relates directly to the overall health of the system.

Sustainable Use of Salmon

No term receives more attention or seems to be less well understood than *sustainability*. The following three definitions may serve as the basis of the term as used in this chapter.

> Sustainability is a relationship between dynamic human economic systems and larger, dynamic, but normally slower-changing ecological systems, such that human life can continue indefinitely, human individuals can flourish, and human cultures can develop—but also a relationship in which the effects of human activities remain within bounds so as not to destroy the health and integrity of self-organizing systems that provide the environmental context for these activities. (Norton 1992)

> Sustainability, by my definition, refers to maintenance of the *potential* for our land and water ecosystems to produce the same quantity and quality of goods and services in perpetuity. . . . The basis of sustainability, in my view, lies in maintaining the physical and biological elements of productivity. (Franklin 1993)

> It begins to be useful as a guide to science, investment, and action when the definition of sustainability focuses on the social and economic development of a region with the goals to invest in the maintenance and restoration of critical ecosystem functions, to synthesize and make accessible knowledge and understanding for economies, and to develop and communicate the understanding that provides a foundation of trust for citizens. (Holling 1993)

What do these definitions have in common? First, human cultures and institutions change much faster and, generally, at a smaller spatial scale than the ecological systems they are related to. As Norton (1992) points out, *resource management* generally operates at an annual cycle of production, whereas *environmental management,* governed by constraints necessary to protect the self-organizing and self-regulatory system that provides the context of annual resource management decisions, must operate at much longer time scales. And so sustainability refers to the maintenance and preservation of options that function at different time and space scales. Thus there seems to be a one-to-one mapping between true sustainability and ecosystem health.

The sustainable use of salmonids therefore refers to maintenance of the health and integrity of salmonid ecosystems in all of their valuable contexts to human society (commodity, recreational, cultural, esthetic). How does one determine when sustainable use is, in fact, in effect? I attempt to make that determination by looking at long-term harvest and enhancement statistics, in terms of both the ecological and human components, for a number of regions of the Northeast Pacific where salmonids are exploited.

The History of Use of Northeast Pacific Salmonid Ecosystems

In this section, I examine the "use" of salmonids in the major regions of the Northeast Pacific in an attempt to determine if, over a span of decades, it is sustainable. Attention here will be paid to the mapping between sustainable use and ecosystem health.

Production and *catch* are largely interchangeable as used here. Unless otherwise specified, the terms refer to the combined total of wild and hatchery fish taken.

Alaska

Alaska is perhaps the easiest region to study from the sustainability point of view. For analysis of salmon fisheries, Alaska can be divided into three regions: Western Alaska, Central Alaska, and Southeastern Alaska (Fig. 16.1).

Western Alaska is a pristine region (no hatcheries, relatively little habitat loss) that is dominated by the production of sockeye (mean of 88% of total salmon production during 1910–93). Central Alaska

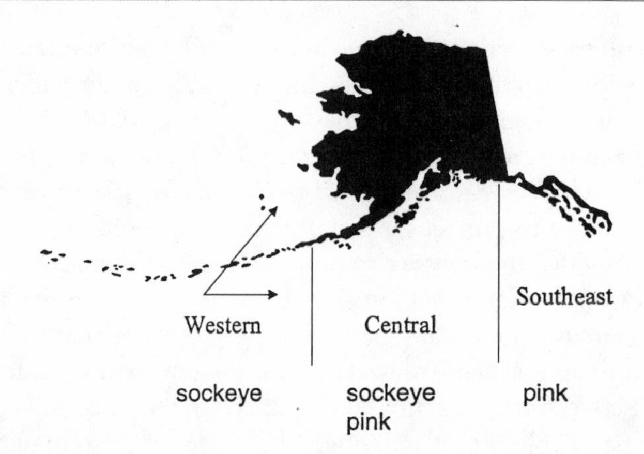

Figure 16.1. The three statistical subregions of Alaska and the predominant species harvested in those subareas.

salmon production is dominated by pink salmon (mean of 63% of total salmon production over 1910–93), followed by sockeye (mean of 22%). Until the early 1980s, there was very little significant hatchery production in the region. The rapid increase in hatchery production of Alaska pink salmon of the 1980s occurred mainly in the Prince William Sound region of Central Alaska. Southeastern Alaska salmon production is also dominated by pink salmon (78%) followed by chum salmon (12%). Significant hatchery releases of both pink and chum salmon occurred in Southeastern Alaska in the 1980s (J. Winton, School of Fisheries, University of Washington, pers. comm. 1995).

The production of salmon in Alaska has shown significant decadal scale fluctuations in response to large-scale climate forcing. Figure 16.2, taken from Francis and Hare (1994), shows fishery production for sockeye salmon in Western and Central Alaska and pink salmon in Central and Southeastern Alaska along with estimates of sudden climate-related changes in average production. As can be seen, a sudden increase in Alaska salmon production occurred in the late 1970s (e.g., an increase in fishery production of around 17 million for Western Alaska sockeye and around 25 million for Central Alaska pink) and a sudden decrease in the late 1940s (e.g., 5 million for Western Alaska sockeye and 7 million for Central Alaska pink). It is quite obvious that low-frequency climate shifts in the North Pacific affect major alterations in the productive capacities of salmon populations over a vast region and must be taken into account when considering the concept of sustainable use.

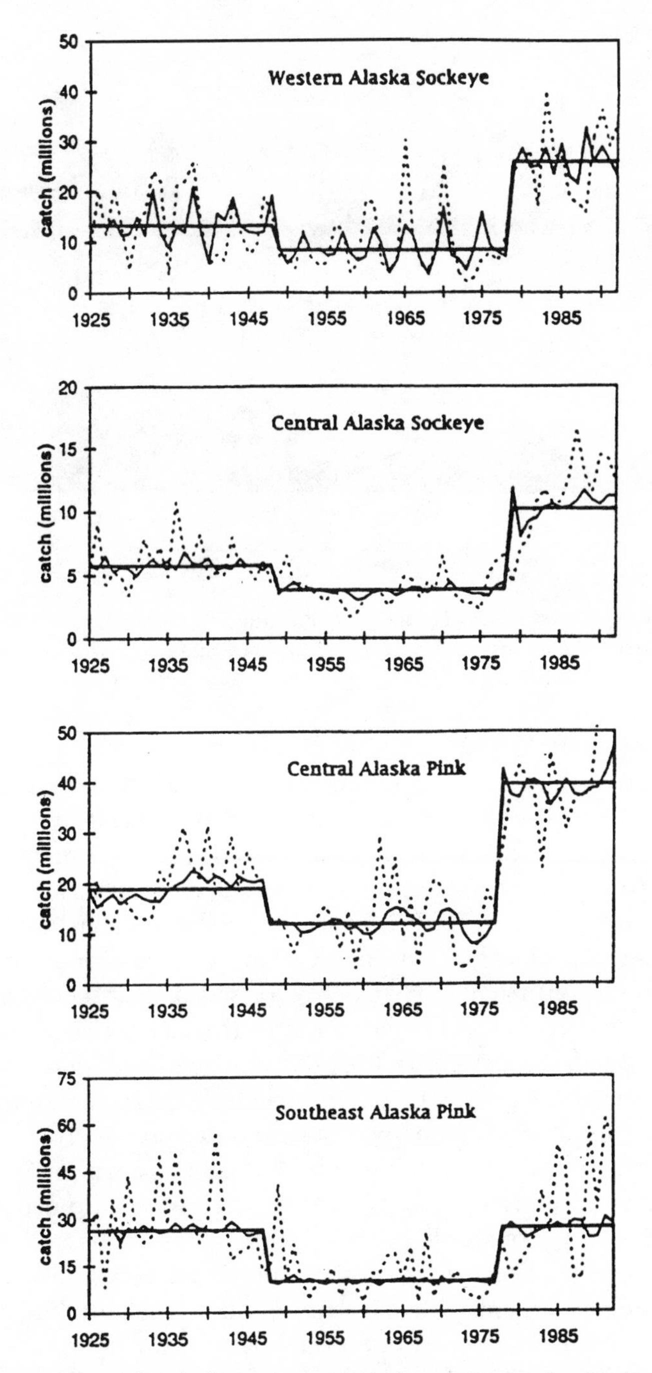

Figure 16.2. Decadal scale fluctuations in Alaska salmon production in response to climate forcing. Time history (dashed lines), intervention model fits (thin solid lines), estimated interventions (thick solid lines). (*Source:* Francis and Hare 1994.)

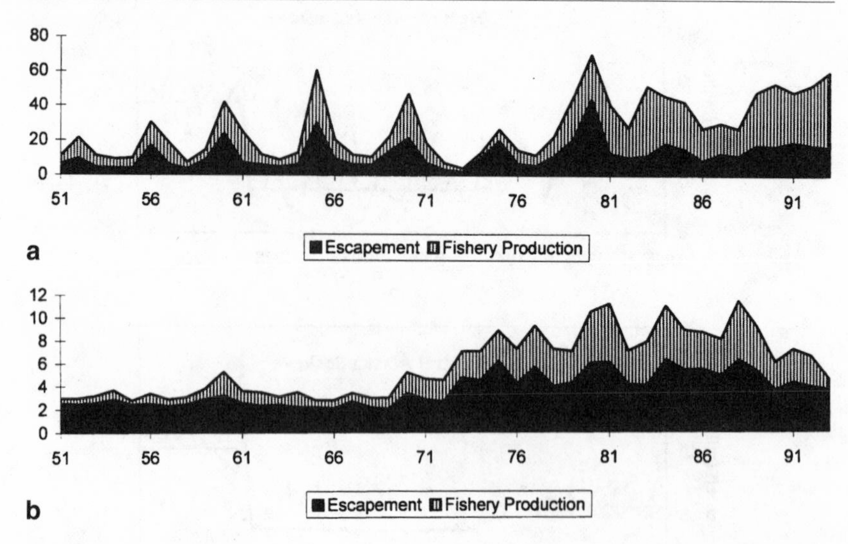

Figure 16.3. Western Alaska sockeye (a) and chum (b) escapement and total fishery production (millions), 1951–93. (*Sources:* Hare and Francis 1995; D. E. Rogers, pers. comm. 1995.)

Let us take the major regions of Alaska one at a time and see if we can determine whether salmon use has been sustainable over, at least, several decades. The assessment for Western Alaska is quite simple. The sockeye salmon fishery clearly rebounded in response to the climate shift of the late 1970s. Figure 16.3a shows estimates of total spawning escapement, fishery production, and run size (escapement plus production) from 1951 to 1993. It is clear that a fundamental change occurred in the use pattern after the increase in production was realized in the late 1970s. Both the average exploitation rate (fraction of the run harvested) and the spawning escapement increased (1969–78 mean exploitation was 0.45 and escapement 10.3 million, and 1979–88 mean exploitation was 0.61 and escapement 16.5 million). In essence, fishery managers have recently tried to depress the extremely strong four- to five-year cycle runs (unique to certain stocks of sockeye) and settle on as constant an annual escapement as possible. The bottom line is that the runs look to be as healthy as they have ever been. As a result, one would have to say that the Western Alaska sockeye salmon fishery is currently a classic example of sustainable use. The same can be said for the Western Alaska chum salmon fishery (Fig. 16.3b), which shows similar trends, although the major

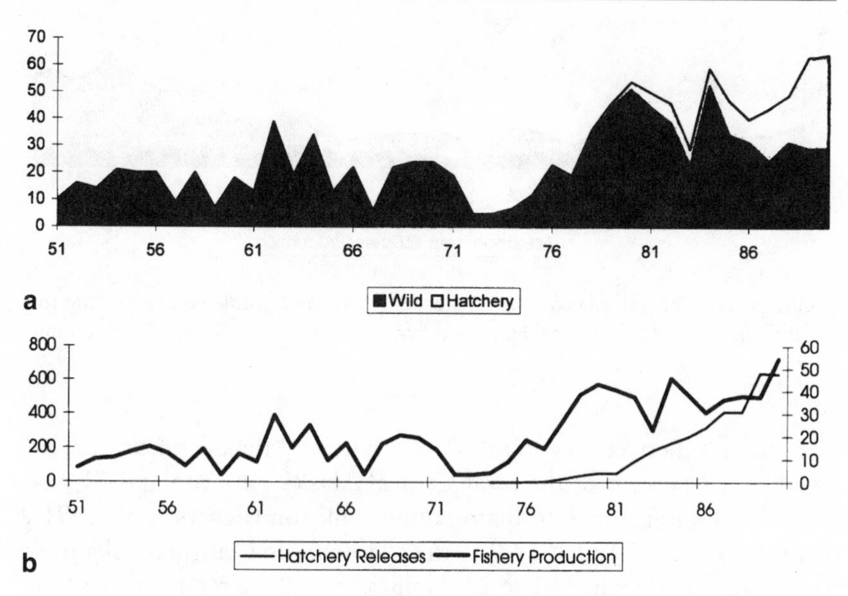

Figure 16.4. (a) Central Alaska pink salmon escapement and total fishery production (millions), 1951–90. (b) Central Alaska pink salmon hatchery releases and fishery production (millions), 1951–90. (*Sources:* Hare and Francis 1995; D. E. Rogers, pers. comm. 1995.)

increases in total fishery production and escapement occurred in the early 1970s. As is the case, managers must be careful to update policy continually in order to respond to possible changes in overall production.

The assessment for Central Alaska is a little more complex. The major species of concern is pink salmon, which makes up greater than 60 percent of the fishery production. The major problem area in Central Alaska is Prince William Sound where a significant enhancement effort was initiated for pink salmon in the 1980s. Figure 16.4a shows estimates of total run size (escapement plus fishery production) from 1951 to 1990. Figure 16.4b shows total pink salmon fishery production (year i) and hatchery releases (year $i - 1$) over the same time period. Several things are clear. First, as was the case all over Alaska, pink salmon production increased significantly in response to large-scale climate forcing in the late 1970s. Second, wild pink salmon escapement has likely increased since the climate shift. And third, in the 1980s there has been a significant reduction in the wild pink salmon run and associated fishery production corresponding to an increase in hatchery run and fishery production which is focused in Prince Wil-

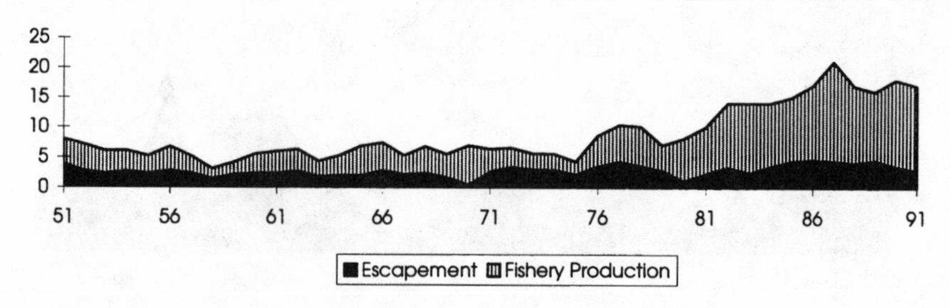

Figure 16.5. Central Alaska sockeye escapement and total fishery production (millions), 1951–91. (*Sources:* Hare and Francis 1995; D. E. Rogers, pers. comm. 1995.)

liam Sound. Between 1986 and 1990, hatchery pinks constituted, on average, 42 percent of the total Central Alaska pink run and 71 percent of the total Prince William Sound pink run (Eggers et al. 1991). Between 1981–85 and 1986–90, the average wild Central Alaska pink run was reduced from 34.7 to 26.3 million fish (24% reduction) and the wild Prince William Sound pink run from 19.3 to 7.7 million (60% reduction). Therefore, to the extent that sustainable use of salmon has two components—harvest and enhancement—the Central Alaska pink salmon hatchery program, focused in Prince William Sound, does not appear to be sustainable for the wild pink salmon populations of that region (more about this later). This hatchery program seems to have had a major impact on the maintenance of the health and integrity of this important component of the Prince William Sound salmonid ecosystem. One should point out that these systems are quite resilient and, as a result, would likely rebound once the perturbation was removed.

Figure 16.5 shows estimates of escapement and total fishery production for Central Alaska sockeye salmon, almost entirely wild, which made up 22 percent of the Central Alaska regional salmon fishery production from 1910 to 1993 (Hare and Francis 1995). As was the case with Western Alaska sockeye, the use of Central Alaska sockeye appears to be sustainable although fishery production is clearly variable.

Southeast Alaska is more difficult to assess in terms of sustainability because neither hatchery release nor total run statistics can presently be separated from estimates for other regions. Pink salmon is, by far, the most important species in Southeastern Alaska, accounting for 78 percent of the regional salmon fishery production from 1910 to

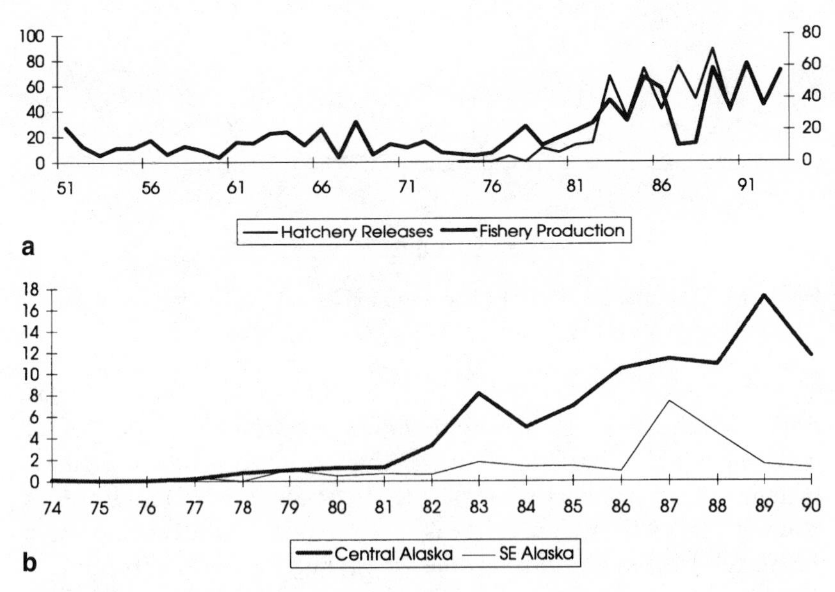

Figure 16.6. (a) Southeastern Alaska pink salmon hatchery releases and fishery production (millions), 1951–94. (*Sources:* Hare and Francis 1995; D. E. Rogers, pers. comm. 1995.) (b) Central and Southeastern Alaska ratios of pink salmon hatchery releases (year $i - 1$) to fishery production (year i), 1974–90.

1993 (Hare and Francis 1995). Figure 16.6a shows the Southeastern Alaska pink salmon fishery production and hatchery releases one year earlier for 1951–93. In order to assess the potential impact of hatchery releases on wild production, Figure 16.6b shows the ratio of pink salmon hatchery releases (year $i - 1$) to fishery production (year i) for both Central and Southeastern Alaska. It is clear that, with the exception of the 1987 and 1988 Southeastern returns which were very low, the potential for hatchery releases of pink salmon to impact wild production, at least on a regional basis, appears to be much lower in Southeastern Alaska than in Central Alaska. However, this may not be the case for chum salmon (Fig. 16.7). The tremendous increase in chum salmon fishery production of recent years is clearly a response to massive hatchery releases. The impact on wild populations of the region is difficult to assess with available data. It is interesting to compare this program of chum salmon enhancement in Southeastern Alaska with that in Japan. Pearcy (1992) reports that since 1979, around 1.5 billion chum juveniles have been released each year from northern Japan with survival averaging around 2 percent. In 1994, the fishery produc-

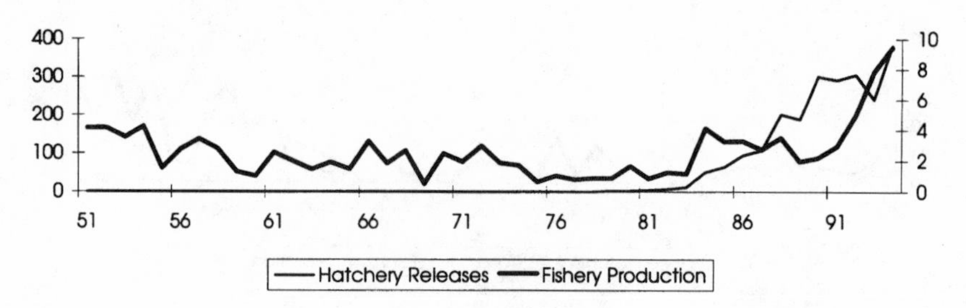

Figure 16.7. Southeastern Alaska chum salmon fishery production and hatchery releases three years earlier (millions), 1951–94.

tion of Southeastern Alaska chum salmon was around 2 percent of the hatchery releases three years earlier which reached almost 0.5 billion juveniles. One wonders what the effects of these massive releases of chum juveniles from the Northeast Pacific (one-third of the northern Japan releases) might have on the whole question of ocean carrying capacity for other Northeast Pacific salmon stocks as well as wild chum salmon production in Southeastern Alaska. As was the case with Central Alaska, the question of sustainable use of salmon is difficult to answer. The balance between pink salmon harvest, escapement, and enhancement may be, overall, sustainable under the present climatic forcing regime, which is very favorable to Alaska salmon production. The overall program for Southeastern Alaska chum salmon may have some serious problems because of the recent massive releases of hatchery juveniles.

British Columbia

Pink (44%) and sockeye (28%) dominate the harvest of British Columbia salmon, together accounting for greater than 70 percent of the 1910–93 average fishery production (Hare and Francis 1995). Fishery production has remained incredibly constant over the entire 1910–93 time period in terms of both species composition and spatial distribution. The only significant shifts in fishery production have been increases in pink production in the north and sockeye production in the south (Fraser River) in recent years. Recent major enhancement efforts have involved predominantly chum and chinook salmon. A good review of this is provided by Hilborn and Winton (1993).

Figure 16.8 shows the 1963–92 British Columbia chum and chinook fishery production and hatchery releases three years earlier. It appears

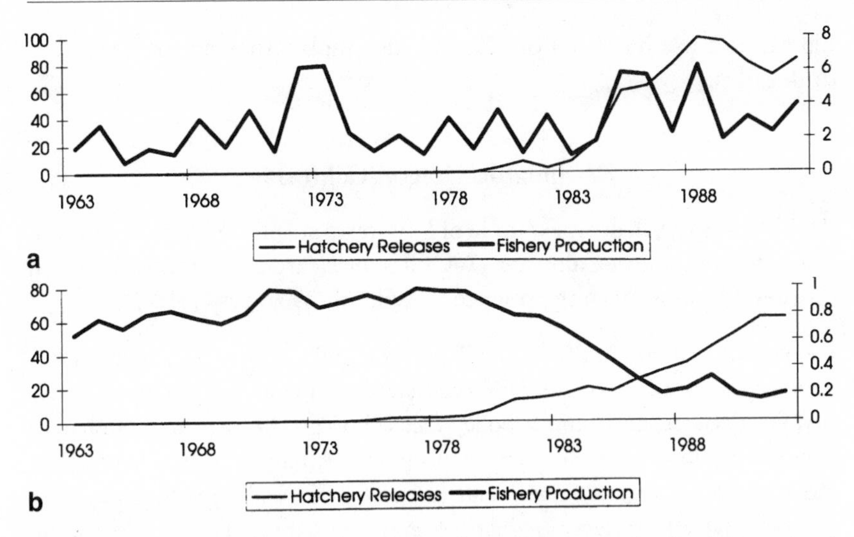

Figure 16.8. (a) British Columbia chum salmon fishery production in year *i* and hatchery releases in year *i* – 3 (millions), 1963–92. (b) British Columbia chinook salmon fishery production in year *i* and hatchery releases in year *i* – 3 (millions), 1963–92.

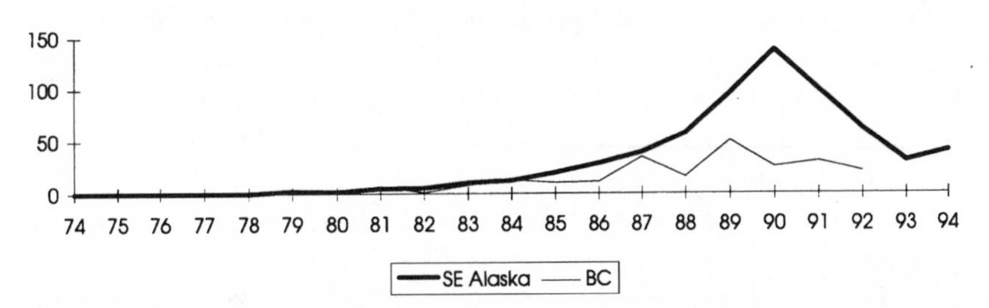

Figure 16.9. Southeastern Alaska and British Columbia ratios of chum salmon hatchery releases in year *i* – 3 to fishery production in year *i*, 1974–94.

that there may have been a slight increase in chum fishery production in response to significant hatchery releases in the late 1980s and 1990s. An alarming trend in chinook fishery production relates directly and inversely to hatchery releases. Figure 16.9 shows the ratio of chum salmon hatchery releases to fishery production three years later for both Southeastern Alaska and British Columbia. British Columbia wild chum salmon appear to have a much lower probability of being adversely affected by the magnitude of recent hatchery releases than Southeastern Alaska. With the exception of chinook salmon, British Columbia salmon use has been on a fairly sustainable footing for a number of decades.

Washington-Oregon-California

As Alaska is the clearest example of long-term sustainable use of salmon, Washington-Oregon-California (WOC) is the clearest example of a salmon disaster. Quoting from the preface to Wilderness Society (1993),

> the plight of salmon in the Pacific Northwest and California exemplifies this nation's failure to protect natural ecosystems and conserve biological diversity. For decades, government policy focused on producing salmon in hatcheries while allowing or even actively promoting extensive alteration and degradation of the region's rivers and streams. The results of this misguided approach have been severe reductions in the amount and quality of salmon habitat, significant weakening of the species' genetic foundation, and, inevitably, widespread extinctions and declining populations.

Taking a long-term view, Wilderness Society (1993) estimates that total WOC salmon production biomass may be less than half as great as it was a century or more ago (declining from a peak annual commercial salmon harvest of 103 million kg from 1864 to 1922 to 47 million kg from 1961 to 1979), and that most of the current population is produced in hatcheries.

Figure 16.10 shows estimates of WOC coho and chinook fishery production and hatchery releases (1977–82 missing) one and three years before, respectively (hatchery data from J. Winton, pers. comm. 1995). WOC coho fishery production shows a linear response to increases in hatchery production through the 1960s and 1970s until the late 1970s when there is an abrupt (> 30% between 1967–76 and 1977–86) decrease in fishery production. A number of factors have been implicated in the decline including changes in the marine environment, compensatory effects of increases in hatchery production, habitat loss due to numerous causes (e.g., dams, timber harvest, urbanization, agriculture), and overfishing. What is more perplexing is that, with the exception of a slight surge in the mid-1980s, WOC coho fishery production has shown a monotonic decline from the late 1970s through the present. WOC wild coho production has reached such a low level that the entire region's populations have been petitioned to be listed as endangered under the Endangered Species Act.

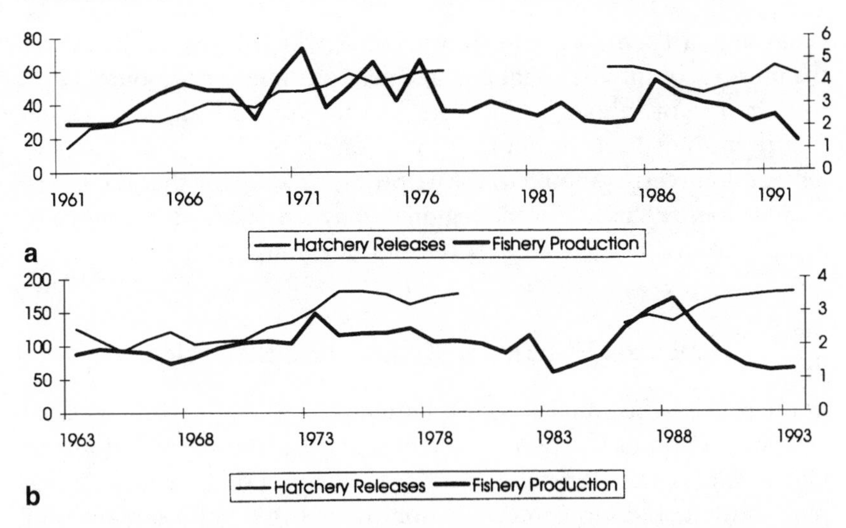

Figure 16.10. (a) Washington-Oregon-California coho salmon fishery production in year *i* and hatchery releases in year *i* – 1 (millions), 1961–94. (b) Washington-Oregon-California chinook salmon fishery production in year *i* and hatchery releases in year *i* – 3 (millions), 1963–94.

WOC chinook is not much better. Figure 16.10b shows that chinook fishery production increased with increases in hatchery production in the 1960s and early 1970s and then has declined since the mid-1970s with the exception of a mid-1980s surge similar to coho except a little later and a little more spread out. This surge appears to have been a response to favorable ocean conditions for year-classes of chinook and coho which entered the marine environment either in late 1985 or early 1986 (Francis et al. 1989).

What role might fishing itself have played in all of this? Estimates of fishery interceptions (catches in Alaska and British Columbia of fish bound for watersheds in WOC) and in-area catches for WOC coho and chinook from 1960–93 show that while the absolute levels of interceptions have remained relatively constant, as the total fishery production for both WOC coho and chinook declined in the late 1970s, 1980s, and 1990s, the fraction of that total which was harvested outside the region increased dramatically (Hare and Francis 1995). To the extent that total fishery production reflects total run size to the region, clearly the exploitation rate outside the region increased much more rapidly than that inside the region. It is interesting that the most rapid increases in interception rates have occurred since the inception of the

U.S./Canada Pacific Salmon Commission, which was established for the purpose of international management of these very resources.

The use of salmon in the Washington-Oregon-California region clearly has not been sustainable for a long time. Hatcheries, habitat loss, and harvest combine to create problems so great that the major salmon species "used" in the region, coho salmon, is on the verge of being listed as endangered over the entire region.

Three Case Histories: A Synthesis on Sustainable Use

In the previous regional sections, I attempted to take a truly regional view of salmon use, avoiding specific subareas. Now I would like to look at three cases—Bristol Bay sockeye salmon, Prince William Sound pink salmon, and Oregon Production Index (OPI) coho salmon—for which there is much more detailed information than that presented above and which tend to display characteristics that are particularly important to the issues concerning sustainable use of salmonids.

The Bristol Bay (Alaska) sockeye salmon fishery is the largest sockeye fishery in the world. It appears to be extremely well managed, with the major goal of management to keep escapement to the various watersheds at "optimum levels." Egegik is one watershed in Bristol Bay. From the 1950s through the 1980s, it accounted for less than one-third of the total sockeye production (total run) of the region. Then in the early 1990s, production suddenly exploded to where in 1992 and 1993 it accounted for well over 50 percent of the total sockeye production (Fig. 16.11). This all occurred under relatively constant long-term escapement levels (Rogers 1987). This seems to be a classic example of a rapid and unpredictable shift in the organization of an ecosystem (Bak and Chen 1991; Waldrop 1992), in this case the entire Bristol Bay salmon production system. It appears that it is also an example of density-dependent constraints operating from the top down (Apollonio 1994) and at the ecosystem level rather than the population level (Wilson et al. 1994). If one examines Figure 16.11 carefully, it appears that all systems in Bristol Bay responded to the climate-driven increase in production felt throughout Alaska starting in the late 1970s (Fig. 16.2), that most Bristol Bay production systems except Egegik declined in production in the early 1980s, and that Egegik really got the upper hand as production once again increased in the late 1980s, setting the stage for the explosion in Egegik production in

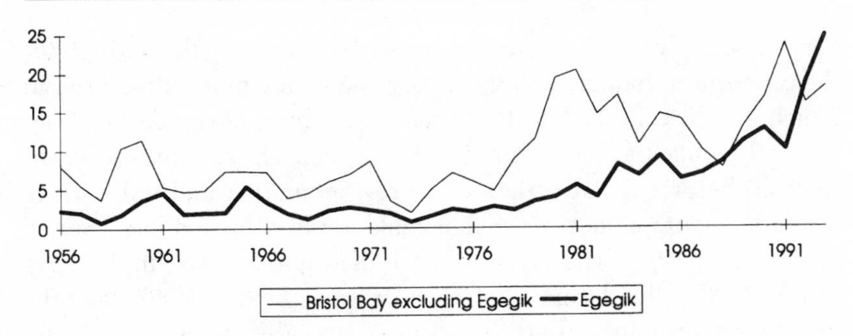

Figure 16.11. Bristol Bay sockeye salmon estimated total run sizes (millions), 1956–93. (*Source:* D. E. Rogers, pers. comm. 1995.)

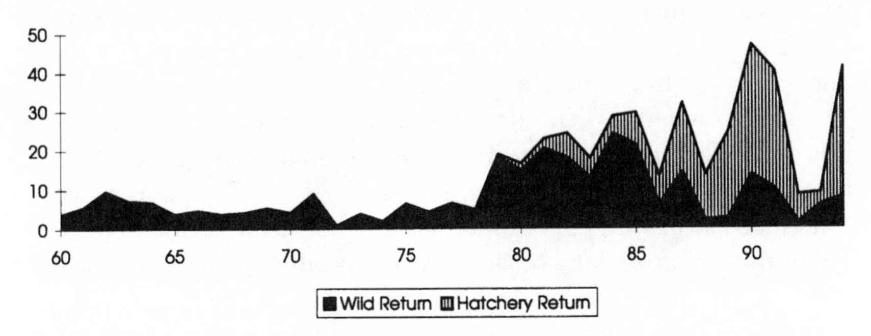

Figure 16.12. Prince William Sound, Alaska, pink salmon estimated run sizes (millions), 1960–94. (*Source:* D. E. Rogers, pers. comm. 1995.)

the early 1990s. The point here is that, in complex systems, history is important. A change in the constraint structure could have occurred in the early 1980s, which set the stage for the sudden outbreak at Egegik in the early 1990s. This is the essence of self-organized criticality (Bak and Chen 1991).

My guess is that a similar thing happened in Prince William Sound (Alaska) with very different results. Figure 16.12 shows the total Prince William Sound pink salmon wild and hatchery runs (catch plus escapement) from 1960 to 1994. To be brief, in the mid-1970s when catches and runs of pink salmon in Prince William Sound (and throughout the Gulf of Alaska) had been at an all-time low for a number of years, a group of fishermen and processors formed a nonprofit hatchery corporation with the idea of "enhancing" depressed levels of wild production. The hope was to create runs that would provide bountiful harvests even in years when wild runs were weak. By the mid-1980s, the consortium had created the largest manmade pink salmon run in North America. In the late 1980s, however, the wild run declined significantly, while the hatchery run stayed strong. The run (mostly hatchery fish) became so large that it outgrew its market in 1990, forcing the dumping of millions of fish that could not be sold, and then crashed in 1992 and 1993. The causes of both the rapid increase in the early to mid-1980s, the decline of the wild run in the late 1980s, and the precipitous fluctuations of the total run in the early 1990s are hotly debated issues. Candidates to take credit for the increase are the new hatchery system and the marine environment, and for the subsequent declines and fluctuations are the 1989 Exxon Valdez oil spill and its effects on both freshwater and marine habitats, overfishing, overproduction by the hatcheries, and changes in the marine environment.

As mentioned earlier, salmon production in Alaska increased significantly starting in the late 1970s due to a climate-induced shift in the marine environment (Francis and Hare 1994). There was a significant increase in wild pink salmon production in the region in the late 1970s that occurred long before hatchery production had become significant. The rapid decline in wild production in the late 1980s occurred at a time when hatchery production had become significant. This decline does not appear to be related to a reduction in overall wild pink escapement, which would result from "overharvest" of wild populations. As a matter of fact, the exploitation rate for wild pink salmon in Prince William Sound stayed well below that for hatchery pinks during the late 1980s and early 1990s. The likely cause of declines in wild pink production is due to competition between wild and hatchery juveniles when they first enter the marine environment (Cooney 1993). Hatchery smolts are generally released before wild smolts migrate from their natal streams into the nearshore marine environment. As a result, not only do they swamp the environment due to the recent quantities of releases, but they get a competitive jump on their wild counterparts in the timing of entry. This is speculation, but it appears to be the most likely scenario. The abrupt fluctuations in both wild and hatchery production which occurred in the early 1990s could reflect a reorganization of the system similar to what happened at approximately the same time in Bristol Bay. Prince William Sound could have been in a state of severe tension, far removed from equilibrium at that time and manifested a significant response in the early 1990s.

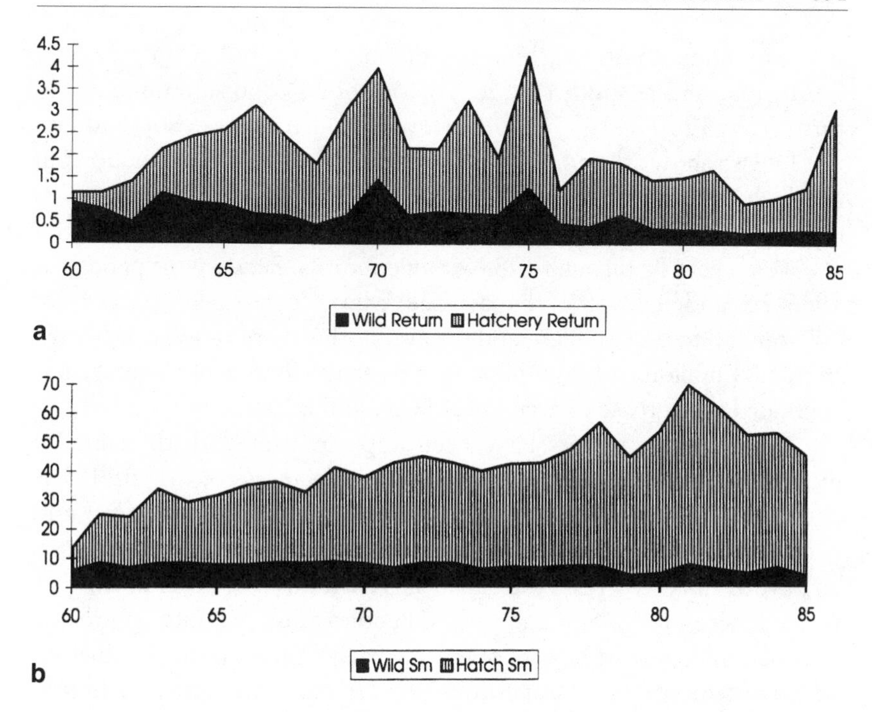

Figure 16.13. (a) OPI coho salmon estimated run size (millions), 1960–85. (b) OPI coho salmon estimated total smolt production (millions), 1960–85. (*Source:* Francis and Brodeur, n.d.)

The balance could have been tipped by an event as significant as the *Exxon Valdez* oil spill of 1989 and its ramifications through the system or as insignificant as a slight shift in marine climate or fishery dynamics. As Allen (1985) says, when this tension or disequilibrium reaches a certain intensity, "then many amazing and surprising things can happen." The difference between Prince William Sound and Bristol Bay, however, comes down to one (Bristol Bay) being considered to be within the realm of normal variability around sustainable use and the other (Prince William Sound) being considered a biological, social, and economic disaster. In neither case is cause directly attributable. However, in one (Bristol Bay), man's activities are clearly expected to ebb and flow with the vagaries of nature. In the other (Prince William Sound), blame gets quickly passed around from one human institution to another.

A third case involves the Oregon Production Index (OPI) coho salmon. (OPI refers to that portion of the Pacific coastal water bounded by Leadbetter Point, Washington, on the north and Monterey Bay, California, on the south.) Figure 16.13a shows coho salmon adult returns from 1960 to 1985 (Brodeur 1990; Emlen et al. 1990), and Figure 16.13b shows the corresponding increases in smolt production, all of which are due to increased hatchery output. However, there is a point in the mid-1970s when there was an abrupt decrease in OPI coho run size which persists through 1985 for wild returns. Between the periods of 1966–75 and 1976–85, the average wild run size decreased from 0.77 million to 0.36 million, and the average hatchery run size decreased from 2.04 million to 1.23 million (Emlen et al. 1990). Coho run size has continued to decrease to even lower levels in the 1990s.

A number of factors have been implicated in both the run size increases of the 1960s and early 1970s and decreases of the late 1970s and 1980s: changes in the marine environment, compensatory and depensatory effects of increases in hatchery production, habitat loss, and overfishing. For years the debate has been carrying on as to whether the declines in both wild and total OPI coho adult production are due to density-dependent factors operating early in the marine life history which are imposed by competition between smolts (overproduction of hatchery smolts) or shifts in constraints induced by the effects of large-scale climate change on the marine environment and its productive capacity (Nickelson 1986; Emlen et al. 1990; Pearcy 1992; Francis and Brodeur, n.d.). It is difficult to sort these factors out. As coho smolt production increased, due to increased hatchery output, ocean survival of both wild and hatchery populations decreased (Francis and Brodeur, n.d.). Also, there was a significant and abrupt decrease in ocean survival of coho in the late 1970s, which, with the exception of the 1985 year-class, has persisted into the 1990s. As noted, the late 1970s was when a climate-induced increase in Alaska salmon production occurred (Francis and Hare 1994). Although rigorous testing has yet to be done, many (e.g., Francis and Sibley 1991; Hollowed and Wooster 1992) believe that climate-driven effects on biological production tend to operate in opposite directions in the large California Current and Alaska Current oceanic ecosystems. As a result, one would expect to see a decrease in Washington-Oregon-California salmon production corresponding to the late 1970s increase in Alaska salmon production. The bottom line is that wild coho salmon populations in the OPI region are at very low levels and have been petitioned to be declared endangered under the Endangered Species Act.

How do these three specific cases relate to the issue of sustainable use of salmonids? I revert to Mundy (1996) for a few more words of wisdom: "What the currently prosperous salmon populations have, that the failed populations do not, is adequate habitat. In all cases where harvest of salmon apparently has been sustained indefinitely, there are two common elements: maintenance of adequate habitat, and the eventual implementation of rational limits on harvest." One can argue that all three cases have relatively rational limits on harvest, although a debate might arise in the case of OPI coho. What really distinguishes the three cases is habitat. Bristol Bay is essentially pristine. Prince William Sound is relatively pristine from the physical point of view. However, the existence of massive populations of hatchery juveniles and adults likely have had a serious impact on the productive capacities of the wild pink salmon populations of the region. In this case, artificially produced populations of pink salmon become part of the "habitat" for wild populations. The OPI coho have experienced a "double whammy." The region has had a significant reduction in coho spawning habitat as well as a significant increase in hatchery production. The habitats for wild coho have been systematically reduced in both the freshwater and nearshore marine environment. In addition, as both the wild and hatchery populations and fisheries have declined, the fraction of the catch taken out of the region (interceptions), the hardest component of a salmon fishery to control, has significantly increased. The end product has been disaster not only for the populations themselves but for the fisheries that depend on them.

Sustainable Salmonid Fishery Management

In this section, I discuss a number of important underlying concepts of salmonid fishery management and try to tease out principles that tend to lead toward sustainable use in conditions where habitat is available for populations to traverse their essential biological pathways. Thus, given a healthy ecosystem from a physical/biological point of view, how should management be undertaken to promote sustainability?

Management Models: The Concept of Stock and Recruit

The fundamental goal of salmon fishery management is to harvest the surplus from each run that is above escapement requirements for natural spawning and artificial production programs (Wright 1981). The analytic basis for decision making is a series of mathematical relationships that express the relationship between spawning biomass and subsequent returns of adults (recruits) to the vicinity of the river system where they were spawned. In the case of Pacific salmon, the most famous and frequently used is that of Ricker (1975). The rationale behind the use of this relationship for the management of Bristol Bay sockeye salmon fisheries is given by Eggers (1992). The Ricker curve shows maximum recruitment at intermediate spawning stock sizes and declining recruitment at low and high spawning stock sizes. This being the case, one can show that the harvestable surplus above replacement is maximized at a level of spawning biomass lower than that which would be expected under unfished conditions. The rationale for management based upon this particular relationship between spawning biomass and subsequent recruitment depends heavily on the assumption of limited carrying capacity of the spawning grounds and the resulting compensatory escapement-return relationship.

In essence, there are three ways to manage salmon fisheries using spawner-recruit theory. The first involves estimating "optimal escapement," which is the escapement that, theoretically, allows the maximum "surplus production" to be harvested from the stock. Attaining that escapement then becomes the goal of management. The second involves estimating "optimal harvest rate," which is the harvest rate that allows maximum "surplus production" to be harvested from the stock. Maintaining that harvest rate then becomes the goal of management. And the third involves estimating "optimal catch," which is the catch that, theoretically, allows maximum "surplus production" to be harvested from the stock. Although these three policies are identical in a universe that behaves according to Ricker's theoretical curve, in practice they have very different manifestations. Because of a number of factors, mostly unknown, the empirical relationship between stock and recruitment tends to be poorly determined. As a result, there is considerable uncertainty and variability in the empirical relationships between spawning biomass and subsequent recruitment. As is pointed out by Hilborn and Walters (1992), both constant catch and constant escapement policies tend to result in either highly variable harvests or spawning escapements, whereas constant harvest rate policies tend to be self-correcting in terms of the stock and have much lower overall variability in year-to-year harvests.

The bottom line with respect to stock-recruit theory and its application to salmonid fisheries management is that no matter what specific approach is taken (escapement, harvest rate, or catch management), as long as it is applied carefully and somewhat conservatively, as long as habitat remains intact, and as long as the negative impacts of artificial enhancement are minimized (Mundy 1996), wild salmonids should be able to flourish and be subjected to harvest on a sustainable basis.

Mixed-Stock Fisheries

One aspect of salmon fisheries which creates tremendous problems for stock assessment scientists and managers occurs when multiple stocks with varying rates of production or target harvest mortalities are indiscriminately harvested in a single fishery. Ricker (1973) questions why the runs of major river systems, when brought under the best available management, fail to produce at levels close to what is expected of them based on their past history. Walters (1980) points out what happens when natural and enhanced stocks are mixed in a fishery. In general, enhanced stocks can sustain much higher harvest rates than natural stocks. If those rates are applied over an extended period of time, significant declines will occur in the wild stocks. This clearly seems to be the case in Puget Sound where both hatchery and wild stocks run a gauntlet of multistock fisheries on their return to their streams of origin. Not only are target harvest rates different for wild and hatchery stocks, but the harvest of Puget Sound-origin salmon is high and difficult for Puget Sound salmon managers to control in Canadian fisheries.

The Management of Artificial Enhancement

Most management of artificial enhancement of salmonids occurs in the form of harvest management. As was mentioned earlier, it is clear that in a number of instances the release of hatchery salmonids has impacts on the productive capacities of wild stocks through biological interactions, most of which occur either in the initial freshwater stages or in the early marine life history. Francis and Brodeur (n.d.) developed a simulation model of coho salmon fishery production in the OPI area which incorporates empirical relationships in smolt to adult survival (1) between wild and hatchery fish, (2) under a range of ocean environmental conditions believed to influence early ocean survival, and (3) as a function of the total number of smolts entering the marine environment. They concluded that if smolt to adult survival is inversely related to the total number entering the marine environment, then (1) recent levels of OPI coho hatchery releases have been significantly in excess of those necessary to maximize total catch or to realize increases in wild adult production, and (2) the more one increases OPI coho wild spawning potential, the more one must reduce hatchery production in order to realize that potential in terms of wild adults. The bottom line of this analysis is that effective management of mixed hatchery and wild salmon fisheries involves control of both harvest and the number of smolts released from hatcheries.

Changes in the Size/Age Distribution of a Population or Stock

Concern has recently been expressed about a seemingly persistent decline in the average size at age of Pacific salmon. Bigler and Helle (1994) report a coastwide and multispecies trend of declining average weights (not corrected for age) from 1975 to 1993. They attribute this trend to concomitant net increases in the total biomass and numbers of salmon occupying the North Pacific and a density-dependent response in individual size due to a limitation to the salmon-sustaining resources of the ocean (the ocean carrying capacity hypothesis). Ishida et al. (1993) studied age composition and size of adult chum salmon from rivers in Japan, Russia, and Canada based on body weight and scale measurement data collected from 1953 to 1988. They found evidence for increases in average age at return and reductions in juvenile ocean growth for Asian chum salmon after 1970 when the number of Japanese hatchery chum salmon began to increase exponentially. No such relationships were found for Canadian chum salmon. Ricker (1973) reported that between 1950 and 1975, the size of all five species caught in Canadian fisheries decreased and that this was, for the most part, due to size decreases in the source populations. However, he makes the point that this is most likely a genetic effect caused by commercial fisheries selecting for fish of larger than average size.

The concept of ocean carrying capacity needs to be more regionally and age-specifically focused. Clearly ocean survival of most salmonid species is most significantly affected in their early marine lives, probably in their first several weeks or months in the ocean (Pearcy 1992). The average weight of returning adults, on the other hand, seems influenced much later in the ocean life history, if environmentally influenced, or affected by fundamental genetic changes in the source population. If increased releases of hatchery salmonids were having a density-dependent effect on overall salmonid production due to some carrying capacity mechanism, then the effect would be localized to either the coastal environment where the releases took place if survival was being affected, or to the high seas areas where various stocks overlap if size at return was being affected. These are very different situations and need to be distinguished if ocean carrying capacity is to be incorporated into sustainable fisheries management policy.

Biological Aspects of Sustainable Salmon Use

In this section, I examine several important biological impacts of sustainable salmon use. The descriptions and discussions are relatively short, quite limited, and certainly not exhaustive of the subject. The point I wish to make with these examples is that, in fact, we can speculate endlessly about the effects of sutainable use on such things as biodiversity and ecosystem function. But, in fact, there is little that we really know from a reductionist science point of view.

Changes in Habitat and Ecosystem Structure/Function

The concept of salmonid habitat is at the core of the message I hope to deliver in this chapter. This section attempts to fill in this important concept based on a number of key papers that have not yet been discussed.

Holtby and Scrivener (1989) used simulation modeling to compare the relative effects of climate (on both freshwater and marine life stages), logging, and fishing on interannual variability in adult chum and coho salmon returns to Carnation Creek, British Columbia (west coast of Vancouver Island). They conclude that, for both species, "most of the observed variation (1971–86) in adult numbers resulted from climatic variability in the stream and the ocean, and in roughly equal measure." The main problem with this analysis is the inability to explore the confounded effects of climate, logging, and fishing on salmon production. In fact, they found it difficult to link habitat change caused by logging to population level response.

Bisson et al. (1992) point out that trends in the abundance of individual populations are often of limited use in identifying the cumulative effects of forest management within a river system. They emphasize that changes in fish communities may provide more comprehensive evidence of the extent of environmental alteration. They see stream channel complexity and biodiversity as being a key target of preservation efforts, and that this can only be done by implementing measures to preserve physical and biological linkages between streams, riparian zones, and upland areas. This will insure delivery of woody debris, coarse sediment, and organic matter, the fundamentals of healthy salmonid habitat. Lichatowich et al. (1995) further emphasize this needed focus on a broader ecosystem context in the area of salmonid habitat restoration. They echo Norton (1992) in stating that ecosystem health is the maintenance of complexity and self-organizing capacity. Complexity is the distribution and abundance of habitat types and their connectivity throughout the salmon's range, and self-organization refers to the function of the exchange of genetic information between generations and the capacity to express that information through the life history diversity within a complex habitat. In essence, they say that restoration of salmon populations requires the reconstruction of life history-habitat relations.

Changes in Spawning Population Size

The term *overescapement* is often used by salmon fishery managers to refer to a condition where too many adult salmon return to the spawning ground and recruitment per spawner is reduced below what it would be at some "intermediate level of spawning biomass." The implications are that (1) fishing is "good" for the stock since it takes the "surplus production" and (2) salmon populations that evolved under no human harvest somehow did so under "suboptimal" conditions. Recent papers by Kline et al. (1990, 1993) and Bilby et al. (1996) indicate that salmonids serve as a source of marine nutrients, which they input in significant quantities to their natal watersheds. These nutrients have a broad influence on both riparian and lacustrine communities. All of these studies show enhanced nutrification of watersheds where anadromous salmonids are present and identify the importance of large escapements for maintaining inputs of marine-derived nutrients to the overall salmonid ecosystem nutrient pool. While it has generally been assumed that nutrients from spawning salmon are introduced into the trophic systems of streams through uptake by primary producers, Bilby et al. (1996) suggest that pathways other than autotrophic uptake may also be important. They suggest that, in the streams they studied, the incorporation of marine-derived nutrients into the stream biota occurs while the nitrogen and carbon are in organic form, prior to mineralization. They conclude that spawning salmon may be important in maintaining production regardless of the availability of nutrients from other sources. Within the salmon component of the ecosystem, they propose a feedback mechanism whereby increased spawning biomass means increased nutrients, increased growth, and overwinter survival of juvenile salmon, larger smolt size, increased marine survival, and so on.

The bottom line is that lower levels of returning (spawning) salmon, and the nutrients they carry, could have major consequences for the overall productivity of stream ecosystems. In the words of Bilby et al. (1996), "large-scale declines of coho populations in many coastal watersheds may represent significant losses of trophic productivity and nutrient capital in headwater streams, and reduced export of nutrients and organic matter from these streams may impact production in aquatic ecosystems." One should note that this effect would occur with any significant reduction in wild spawning populations, and that this is one of the major consequences of the massive hatchery programs that have been instituted in the Pacific Northwest and Alaska. In fact, these hatchery programs create a double-edged sword in that they not only replace wild spawners whose nutrients would be redistributed in their native aquatic ecosystems, but, because of their encouragement of grossly elevated harvest rates, they also tend to reduce further the wild populations that are left. Of course, as I have tried to point out in earlier sections of the report, the data that we do have indicate that these kinds of practices are not sustainable.

Finally, a recent paper by Montgomery et al. (1996) shows how salmon actively modify their habitat in a manner beneficial to their progeny. They show how streambed loosening during chum salmon spawning likely reduces the probability of streambed scour and excavation of buried salmon embryos during peak flows. This, in turn, should influence salmon survival-to-emergence because salmon embryo incubation within stream gravels coincides with high seasonal flows. This previously unrecognized positive feedback between mass spawning and bed mobility indicates that it may become increasingly difficult to maintain or rehabilitate declining populations of mass-spawning salmonids. In addition, although it has not been measured, this effect could influence scores of other stream bottom-dwelling organisms.

Salmonids as Keystone Species

Willson and Halupka (1995) discuss the concept of anadromous fish as keystone food resources for vertebrate predators and scavengers, forging a significant link between aquatic and terrestrial ecosystems. They make the point that the anadromous fish system is an extreme case in which prey is temporarily very abundant, spatially constrained, relatively easy to capture, and more or less predictable in time. They catalog mammal, bird, and fish predator/scavengers of salmon in or near freshwaters of Southeast Alaska. The real problem is that none of these relationships and interactions have been quantified. While salmon provide important trophic connections to numerous vertebrate predators and scavengers, the importance of these relationships is unknown. If one extended this to the importance of salmon to the maintenance of biodiversity in the freshwater ecosystems that they occupy, one would guess that they are important. However, the nature and magnitude of these relationships have not been studied and are unable to be quantified.

Changes in Genetic Diversity as a Result of Harvest and Enhancement

Clearly both harvest and artificial enhancement have major impacts on the genetic structures and variability of salmon populations. There is a vast literature on this subject, and I refer to only a few specific instances which I feel are most pertinent to the concept of sustainable use.

Nelson and Soule (1987) state that genetics is of central concern in problems of biological conservation. They note that the "loss of genotypic variance is essentially as irreversible as the loss of alleles from a population. One can no more sort out the genes into groups of genotypes resembling the original subpopulations than one can obtain the works of Shakespeare from a collection of monkeys and typewriters, and for similar reasons." They further state that the preservation of existing gene pools is a primary obligation of fisheries management.

Ryman et al. (1994) indicate that intraspecific genetic variability of fishes may be threatened in three distinct ways: extinctions that occur primarily through the losses or alterations of habitats, hybridization that occurs in salmonids due to releases associated with intentional stocking, ocean ranching, and escapes from net-pen rearings, and loss of genetic variability due to selection, genetic drift, and inbreeding. Much of the latter can be attributed to reductions in effective breeding population sizes.

Salmon aquaculture and its "management" have been identified as a major source of negative genetic impact on the productive capacities and genetic variability of wild populations. Hindar et al. (1991) call for strong restrictions on gene flow from cultured to wild populations. Waples and Teel (1990) identify the potential for rapid genetic drift in hatchery populations due, primarily, to the small effective breeding populations used by hatchery managers. Reisenbichler and McIntyre (1977) show that significant interbreedings between genetically distanced wild and hatchery populations can significantly lower the productive capacities (from a spawner-recruit point of view) of the wild stocks.

As Ricker (1981) and others have pointed out, genetic selective pressure exerted by salmonid fisheries can affect such population traits as individual fish growth and the timing of spawning. As Nelson and Soule (1987) point out, artificial selective pressures for an earlier or later breeding season can upset the delicate adjustment of subsequent early life history stages to other natural production cycles.

Finally, Scudder (1989) stresses the importance of marginal populations in the conservation of genetic resources of fish species. Referring specifically to salmonids, he states that marginal populations, due to their exposure to extreme environmental conditions, will evolve genotypes that allow them to deal with these extreme conditions. And it is these genotypes that need to be preserved for the survival of both metapopulations and the species itself. It is in marginal populations that unique adaptive traits are to be found, traits that need to be preserved. Unfortunately, it is often these very marginal populations that are the most vulnerable to fishing due to their (relative) inabilities to withstand high exploitation rates.

Discussion and Summary

What can we say about the effects of sustainable salmon use on biodiversity and ecosystem function? First we need to determine where it appears that salmon use is sustainable. It appears that most Alaskan salmon fisheries operate at sustainable levels with the exception of problems created in Prince William Sound by the massive pink salmon hatchery program and the potential problems in Southeastern Alaska by a significant chum salmon hatchery program. For the most part, wild salmon habitat is in almost pristine shape in Alaska with the exception of the areas in Southeastern Alaska that have been logged to a significant degree. In addition, with the exception of chinook salmon, British Columbia salmon use appears to be operating at a sustainable level. Finally, it is clear that salmon use in the Washington-Oregon-California region has not been sustainable for a long time.

Given that we can identify regions where salmon fisheries operate on a sustainable basis, what have been the effects of that use on biodiversity and ecosystem function? As I have pointed out earlier, I am not aware that anyone has studied that relationship from a reductionist scientific point of view. No one has measured "biodiversity" or "ecosystem function" in sustainably used salmonid ecosystems and compared those measures with similar systems that are not exploited. This could, perhaps, be done but would involve a massive effort for what, I believe, would be an insignificant payoff. For example, if one takes the sockeye salmon ecosystems of the Bristol Bay region of Alaska, and compares the amount of variability imposed on that system by climate regime scale changes in ocean survival, then one obtains a rough idea of one component of natural variability that the system is exposed to. Based on data in Figure 16.3a, the mean annual run size between 1951–76 and 1977–93, the periods of two distinct climate regimes (Francis and Hare 1994), increased from 19.0 million to 41.3 million. The net change in mean total run size between these two climate regimes is 22.3 million fish per year, which is approximately equal to the amount of catch removed annually from the system under the most favorable ocean environmental conditions. In addition, if one looks at the ability of this system to absorb short-term variability in run size, the mean difference between the largest run size and the smallest run size of a (four- or five-year) cycle is around 33 million fish (Fig. 16.3a). Thus nature appears to deal this system at least as much (natural) variability, in both the short and the long term, as the (sustainable) fishery has been able to remove at its peak. In essence, the system has evolved the capability of absorbing this level of variability over various time scales.

I believe that where things begin to break down is when the relationship between wild salmon and their complex habitat structure is severed in a significant way. Then ecosystem health is threatened and sustainable use is impeded. This seems to occur most often when freshwater habitat is destroyed or the salmonid suppliers of that habitat are replaced by the products of artificial production facilities.

Finally, one of the real dilemmas in the fisheries arena in dealing with what we term ecosystem issues has to do with the difficulties we always seem to encounter in attributing cause. Is it fishing? Is it habitat? Is it the ocean? Is it El Niño? I am becoming convinced that the real question is, is it any *one* thing at all? How do you separate the effects of man (something we can control) from those of the environment? Can you really pick things apart in a reductionist way and attribute cause? One is led to pose the question: *Is there a silver bullet?*

Certainly the machine metaphor (in the sense of Botkin 1990) encourages us to search for the silver bullet. If we can only find the one thing that controls what is "wrong" with the system, then we can either fix it or reengineer the system to enable us to overcome the problem. And so with regard to Prince William Sound salmon, we focus on whether it is the marine environment, the oil spill, overfishing, or hatcheries which might provide us with the silver bullet.

According to Holling (1993), this analytic worldview has led us to do our science in a particular way, and then assume that this is the only way for science to approach resource issues. Holling calls this traditional or first stream science and characterizes it as "disciplinary, reductionist, and detached from people, policies, and politics." The machine metaphor for nature pervades here. For example, the methodology of sequentially testing independent hypotheses which are either accepted or rejected reflects this orthogonal view of nature. Management is oriented to smoothly changing and reversible conditions, and operates under the view that one needs to know before taking action. If only we study the problem a little harder and obtain better information, we will be able to develop the operational knowledge to solve the problem effectively. Much of this arises from the way that first stream science has sold itself to the public. Certainty, predictability, reducibility are words that are associated with this approach to resource science.

Reality, however, leads us to answer our earlier question: *There is no silver bullet!* To put it simply, nature is not orthogonal. System properties do not necessarily result from the "lawful" interactions of independent components. In addition, historical contingency is an important aspect of natural processes. And as Gould (1989) so clearly points out, a historical explanation of a natural process does not rest solely on direct deductions from the laws of nature; it also takes into account an unpredictable sequence of antecedent states, where any major change in any step of the sequence would have altered the final result.

Holling (1993) argues that there is another stream of science more appropriate for approaching some of these complex ecosystem issues. Its premise is that knowledge will always be incomplete and, in fact, certain aspects are inherently unknowable and unpredictable. In terms of being a science for management, uncertainty and surprise must become an integral part of a sequence of actions, one dependent on the results of how the system responded to those that have come before. In addition, learning becomes one of the goals of management. Holling (1993) states the case beautifully when he says: "Not only is the science incomplete, the system itself is a moving target, evolving because of the impacts of management and the progressive expansion of the scale of human influences on the planet. . . . The essential point is that evolving systems require policies and actions that not only satisfy social objectives but, at the same time, also achieve continually modified understanding of the evolving conditions and provide flexibility for adaptation to surprises. Science, policy, and management then become inextricably linked."

In this context, uncertainty and surprises are expected and, as best possible, planned for. In particular, system responses are fundamentally nonlinear—they demonstrate those properties of nonlinear systems that have come to be termed chaotic—multiple quasi-stable states and abrupt discontinuities. As such, policies that rely on slow and smooth change and reversible conditions simply will not fly in this arena. Holling (1993) sums it up quite well when he says that we need to continue developing new science and management which openly acknowledges indeterminacy, unpredictability, and the historical nature of resource issues: "The problems are therefore not amenable to solutions based on knowledge of small parts of the whole, nor on assumptions of constancy or stability of fundamental relationships, ecological, economic, or social. . . . Therefore, the focus best suited for the natural science components is evolutionary, for economics and organizational theory is learning and innovation, and for policies is actively adaptive designs that yield understanding as much as they do product."

References

Allen, P. M. 1985. Ecology, thermodynamics, and self-organization: Towards a new understanding of complexity. In *Ecosystem theory for biological oceanography,* ed. R. E. Ulanowicz and T. Platt, 3–26. Canadian Bulletin of Fisheries and Aquatic Sciences 213. Ottawa: Department of Fisheries and Oceans.

Allen, T.F.H., and T. W. Hoekstra. 1992. *Towards a unified ecology.* New York: Columbia University Press.

Apollonio, S. 1994. The use of ecosystem characteristics in fisheries management. *Reviews in Fisheries Science* 2(2):157–180.

Bak, P., and K. Chen. 1991. Self-organized criticality. *Scientific American* January:46–53.

Bigler, B. S., and J. H. Helle. 1994. Decreasing size of North Pacific salmon (*Oncorhynchus* sp.): Possible causes and consequences. Document submitted to the annual meeting of the North Pacific Anadromous Fish Commission, Vladivostok, Russia, October 1994.

Bilby, R. E., B. R. Fransen, and P. A. Bisson. 1996. Incorporation of nitrogen and carbon from spawning coho salmon into the trophic system of small streams: Evidence from stable isotopes. *Canadian Journal of Fisheries and Aquatic Sciences* 53:164–176.

Bisson, P. A., T. P. Quinn, G. H. Reeves, and S. V. Gregory. 1992. Best management practices, cumulative effects, and long-term trends in fish abundance in Pacific Northwest river systems. In *Watershed management,* ed. R. Naiman, 189–232. New York: Springer-Verlag.

Botkin, D. B. 1990. *Discordant harmonies.* New York: Oxford University Press.

Brodeur, R. D. 1990. Feeding ecology of and food consumption by juvenile salmon in coastal waters, with implications for early ocean survival. Doctoral dissertation, School of Fisheries, University of Washington, Seattle.

Burgner, R. L. 1991. Life history of sockeye salmon (*O. nerka*). In *Pacific salmon life histories,* ed. C. Groot and L. Margolis, 3–117. Vancouver, British Columbia, Canada: UBC Press.

Cairns, M. A., and R. T. Lackey. 1992. Biodiversity and management of natural resources: The issues. *Fisheries* 17(3):6–10.

Cooney, R. T. 1993. A theoretical evaluation of the carrying capacity of Prince William Sound, Alaska, for juvenile Pacific salmon. *Fisheries Research* 18:77–87.

Costanza, R., B. G. Norton, and B. D. Haskell, eds. 1992. *Ecosystem health: New goals for environmental management.* Washington, D.C.: Island Press.

Eggers, D. M. 1992. The benefits and costs of the management program for natural sockeye salmon stocks in Bristol Bay, Alaska. *Fisheries Research* 14:159–177.

Eggers, D. M., L. R. Peltz, B. G. Blue, and T. M. Willette. 1991. Trends in abundance of hatchery and wild stocks of pink salmon in Cook Inlet, Prince William Sound and Kodiak, Alaska. Unpublished manuscript. Alaska Department of Fish and Game, Juneau, Alaska.

Emlen, J. M., R. R. Reisenbichler, A. M. McGie, and T. E. Nickelson. 1990. Density dependence at sea for coho salmon (O. *kisutch). Canadian Journal of Fisheries and Aquatic Sciences* 47:1765–1772.

Francis, R. C., and R. D. Brodeur. n.d. Production and management of coho salmon: A simulation model incorporating environmental variability. Unpublished manuscript.

Francis, R. C., and S. R. Hare. 1994. Decadal scale regime shifts in the large marine ecosystems of the northeast Pacific: A case for historical science. *Fisheries Oceanography* 3(4):279–291.

Francis, R. C., and T. H. Sibley. 1991. Climate change and fisheries: What are the real issues? *Northwest Environmental Journal* 7:295–307.

Francis, R. C., W. G. Pearcy, R. Brodeur, J. P. Fisher, and L. Stephens. 1989. Effects of the ocean environment on the survival of Columbia River juvenile salmonids. Report of Contract no. DE-AI79-88BP92866, Bonneville Power Administration, Portland, Oregon.

Franklin, J. F. 1993. The fundamentals of ecosystem management with applications in the Pacific Northwest. In *Defining sustainable forestry,* ed. G. Aplet, N. Johnson, J. Olson, and V. Sample, 127–144. Washington, D.C.: Island Press.

Gould, S. R. 1989. *Wonderful life: The Burgess shale and the nature of history.* New York: W. W. Norton.

Groot, C., and L. Margolis, eds. 1991. *Pacific salmon life histories.* Vancouver, British Columbia, Canada: UBC Press.

Hard, J. 1994. Production of hatchery salmon and ecological carrying capacity. Unpublished report.

Hare, S. R., and R. C. Francis. 1995. Climate change and salmon production in the Northeast Pacific Ocean. In *Climate change and northern fish populations,* ed. R. J. Beamish, 357–372. Canadian Special Publication of Fisheries Aquatic Sciences 121. Ottawa: Department of Fisheries and Oceans.

Healey, M. C. 1991. Life history of chinook salmon (*O. tshawytscha*). In *Pacific salmon life histories,* ed. C. Groot and L. Margolis, 313–393. Vancouver, British Columbia, Canada: UBC Press.

Heard, W. R. 1991. Life history of pink salmon (*O. gorbuscha*). In *Pacific salmon life histories,* ed. C. Groot and L. Margolis, 121–230. Vancouver, British Columbia, Canada: UBC Press.

Hilborn, R., and C. J. Walters. 1992. *Quantitative fisheries stock assessment.* New York: Chapman and Hall.

Hilborn, R., and J. Winton. 1993. Learning to enhance salmon production: Lessons from the salmonid enhancement program. *Canadian Journal of Fisheries and Aquatic Sciences* 50(9):2043–2056.

Hindar, K., N. Ryman, and F. Utter. 1991. Genetic effects of cultured fish on natural fish populations. *Canadian Journal of Fisheries and Aquatic Sciences* 48:945–957.

Holling, C. S. 1993. Investing in research for sustainability. *Ecological Applications* 3:552–555.

Hollowed, A. B., and W. S. Wooster. 1992. Variability of winter ocean conditions and strong year classes of Northeast Pacific groundfish. *ICES Marine Sciences Symposium* 195:433–444.

Holtby, L. B., and J. C. Scrivener. 1989. Observed and simulated effects of climatic variability, clear cut logging, and fishing on the numbers of chum salmon (*O. keta*) and coho salmon (*O. kisutch*) returning to Carnation Creek, British Columbia. In *Proceedings of the national workshop on effects of habitat alteration on salmonid stocks,* ed. C. D. Levings, L. B. Holtby, and M. A. Henderson, 62–81. Canadian Special Publication of Fisheries and Aquatic Sciences 105. Ottawa: Department of Fisheries and Oceans.

Hughes, R. M., and R. F. Noss. 1992. Biological diversity and biological integrity: Current concerns for lakes and streams. *Fisheries* 17(3):11–19.

Ishida, Y., S. Ito, M. Kaeriyama, S. McKinnell, and K. Nagasawa. 1993. Recent changes in age and size of chum salmon (*O. keta*) in the North Pacific Ocean and possible causes. *Canadian Journal of Fisheries and Aquatic Sciences* 50:290–295.

Kline, T. C., J. J. Goering, O. A. Mathisen, and P. H. Poe. 1993. Recycling of elements transported upstream by runs of Pacific salmon. $\delta^{15}N$ and $\delta^{13}C$

evidence in the Kvichak River watershed, Bristol Bay, Southwestern Alaska. *Canadian Journal of Fisheries and Aquatic Sciences* 50:1–16.

Kline, T. C., J. J. Goering, O. A. Mathisen, P. H. Poe, and P. L. Parker. 1990. Recycling of elements transported upstream by runs of Pacific salmon: $\delta^{15}N$ and $\delta^{13}C$ evidence from Sashin Creek, Southeastern Alaska. *Canadian Journal of Fisheries and Aquatic Sciences* 47:136–144.

Lichatowich, J., L. Mobrand, L. Lestelle, and T. Vogel. 1995. An approach to the diagnosis and treatment of depleted Pacific salmon populations in Pacific Northwest watersheds. *Fisheries* 20(1):10–18.

Montgomery, D. R., J. M. Buffington, N. P. Peterson, D. Schuett-Hames, and T. P. Quinn. 1996. Salmonid modifications of gravel streambeds: Implications for bed surface mobility and embryo survival. *Canadian Journal of Fisheries and Aquatic Sciences* 53:1061–1070.

Mundy, P. R. 1996. The role of harvest management in the future of Pacific salmon populations: Shaping human behavior to enable the persistence of salmon. In *Pacific salmon and their ecosystems: Status and future options,* ed. D. Stouder, P. A. Bisson, and R. J. Naiman, 315–330. New York: Chapman and Hall.

Myers, K. W., R. V. Walker, S. Fowler, and M. L. Dahlberg. 1990. *Known ocean ranges of stocks of Pacific salmon and steelhead as shown by tagging experiments, 1956–1989.* (INPFC Doc.) FRI-UW-9009. Fisheries Research Institute, University of Washington, Seattle.

Nelson, K., and M. Soule. 1987. Genetical conservation of exploited fishes. In *Population genetics and fishery management,* ed. N. Ryman and F. Utter, 345–368. Seattle, Washington: Washington Sea Grant Publication.

Nickelson, T. E. 1986. Influences of upwelling, ocean temperature, and smolt abundance on marine survival of coho salmon (*O. kisutch*) in the Oregon Production Area. *Canadian Journal of Fisheries and Aquatic Sciences* 43:527–535.

Nicolis, G., and I. Prigogine. 1977. *Self-organization in non-equilibrium systems.* New York: John Wiley & Sons.

Norton, B. G. 1992. A new paradigm for environmental management. In *Ecosystem health: New goals for environmental management,* ed. R. Costanza, B. G. Norton, and B. D. Haskell, 23–41. Washington, D.C.: Island Press.

Pearcy, W. G. 1992. *Ocean ecology of North Pacific salmonids.* Seattle, Washington: Washington Sea Grant Program.

Pomeroy, R., E. C. Hargrove, and J. J. Alberts. 1988. The ecosystem perspective. In *Concepts of ecosystem ecology,* ed. L. R. Pomeroy and J. J. Alberts, 1–18. New York: Springer-Verlag.

Prigogine, I. 1980. *From being to becoming: Time and complexity in the physical sciences.* New York: W. H. Freeman.

Reisenbichler, R. R., and J. D. McIntyre. 1977. Genetic differences in growth and survival of juvenile hatchery and wild steelhead trout, *Salmo gairdneri. Journal of the Fisheries Research Board of Canada* 34:123–128.

Ricker, W. E. 1973. Two mechanisms that make it impossible to maintain peak-periods yield from stocks of Pacific salmon and other species. *Journal of the Fisheries Research Board of Canada* 30:1275–1286.

Ricker, W. E. 1975. *Computation and interpretation of biological statistics of fish populations.* Bulletin, Fisheries Research Board of Canada 191. Ottawa: Fisheries and Marine Service, Department of the Environment.

Ricker, W. E. 1981. Changes in the average size and average age of Pacific salmon. *Canadian Journal of Fisheries and Aquatic Sciences* 38:1636–1656.

Rogers, D. E. 1987. Pacific salmon. In *The Gulf of Alaska: Physical environment and biological resources,* ed. D. W. Hood and S. T. Zimmerman, 461–476. Washington, D.C.: National Oceanic and Atmospheric Administration, Department of Commerce, and Mineral Management Service, Department of the Interior.

Rogers, D. E. 1994. Estimates of annual salmon runs from the North Pacific, 1951–1993. Unpublished report.

Ryman, N., F. Utter, and L. Laikre. 1994. Protection of aquatic biodiversity. In *The state of the world's fisheries resources,* ed. C. W. Voigtlander, 92–115. New Delhi, India: Oxford and IBH Publishing.

Salo, E. O. 1991. Life history of chum salmon (*O. keta*). In *Pacific salmon life histories,* ed. C. Groot and L. Margolis, 233–309. Vancouver, British Columbia, Canada: UBC Press.

Sandercock, F. K. 1991. Life history of coho salmon (*O. kisutch*). In *Pacific salmon life histories,* ed. C. Groot and L. Margolis, 397–445. Vancouver, British Columbia, Canada: UBC Press.

Scudder, G.G.E. 1989. The adaptive significance of marginal populations: A general perspective. In *Proceedings of the national workshop on effects of habitat alteration on salmonid stocks,* ed. C. D. Levings, L. B. Holtby, and M. A. Henderson, 180–185. Canadian Special Publication of Fisheries and Aquatic Sciences 105. Ottawa: Department of Fisheries and Oceans.

Tansley, A. G. 1935. The use and abuse of vegetational concepts and terms. *Ecology* 16:284–307.

Vannote, R. L., G. W. Minshall, K. W. Cummins, J. R. Sedell, and C. E. Cushing. 1980. The river continuum concept. *Canadian Journal of Fisheries and Aquatic Sciences* 37:130–137.

Waldrop, M. M. 1992. *Complexity: The emerging science at the edge of order and chaos.* New York: Touchstone.

Walters, C. J. 1980. Mixed-stock fisheries and the sustainability of enhancement production for chinook and coho salmon. In *Salmonid ecosystems of the North Pacific,* ed. W. J. McNeil and D. C. Himsworth, 109–115. Corvallis, Oregon: Oregon State University Press.

Waples, R. S., and D. J. Teel. 1990. Conservation genetics of Pacific salmon, I. Temporal changes in allele frequency. *Conservation Biology* 4(2):144–156.

Ware, D. A., and G. A. McFarlane. 1989. Fisheries production domains of the northeast Pacific Ocean. In *Effects of ocean variability on recruitment and evaluation of parameters used in stock assessment models,* ed. R. J. Beamish and G. A. McFarlane, 359–379. Can. Spec. Publ. Fish. Aquat. Sci. 108.

Wilderness Society. 1993. *Pacific salmon and federal lands: A regional analysis.* Washington, D.C.: Wilderness Society.

Willson, M. F., and K. C. Halupka. 1995. Anadromous fish as keystone species in vertebrate communities. *Conservation Biology* 9(3):489–497.

Wilson, E. O. 1994. *Naturalist.* Washington, D.C.: Island Press.

Wright, S. 1981. Contemporary Pacific salmon fisheries management. *North American Journal of Fisheries Management* 1:29–40.

Contributors

Anil Bakshi World Wide Fund for Nature–India, 172-B Lodi Estate, New Delhi 110 003, India. E-mail: wwfindel@unv.ernet.in

Richard E. Bodmer Tropical Conservation and Development Program, Latin American Studies, 319 Grinter Hall, University of Florida, Gainesville, Florida 32611-2037, USA. E-mail: bodmer@tcd.ufl.edu

Des A. Callaghan The Wildfowl & Wetlands Trust, Slimbridge, Gloucestershire GL2 7BT, United Kingdom. E-mail: des.callaghan@wwt.org.uk

Jason W. Clay World Wildlife Fund–U.S., 1250 24th Street NW, Washington, D.C. 20037, USA. E-mail: jason.clay@wwfus.org

Timothy M. Crowe FitzPatrick Institute, University of Cape Town, Rondebosch 7700, South Africa. E-mail: tmcrowe@botzoo.uct.ac.za

Avenash Datta World Wide Fund for Nature–India, 172-B Lodi Estate, New Delhi 110 003, India. E-mail: wwfindel@unv.ernet.in

Ruth M. Elsey Louisiana Department of Wildlife and Fisheries, Rockefeller Wildlife Refuge, 5476 Grand Chenier Highway, Grand Chenier, Louisiana 70643, USA

Tula G. Fang Facultad de Ciencias Biológicas, Universidad Nacional de la Amazonía Peruana, Plaza Serafín Filomeno, Iquitos, Peru

Robert C. Francis School of Fisheries WH-10, University of Washington, Seattle, Washington 98195, USA. E-mail: rfrancis@fish.washington.edu

Curtis H. Freese 8891 Bridger Canyon Road, Bozeman, Montana 59715, USA. E-mail: ubicf@msu.oscs.montana.edu

P. Gabriel Syndel Asia Sdn. Bhd., Subang New Village, 40150 Shah Alam, Selangor 40150, Malaysia. E-mail: gopin@pc.jaring.my

N. Gopinath Syndel Asia Sdn. Bhd., Subang New Village, 40150 Shah Alam, Selangor 40150, Malaysia. E-mail: gopin@pc.jaring.my

Andrew J. Hansen Biology Department, Montana State University, Bozeman, Montana 59717, USA. E-mail: ubiah@msu.oscs.montana.edu

S. Hugh High Department of Economics, University of Cape Town, Rondebosch 7700, South Africa. E-mail: hughhi@socsci.uct.ac.za

Baz Hughes The Wildfowl & Wetlands Trust, Slimbridge, Gloucestershire GL2 7BT, United Kingdom. E-mail: baz.hughes@wwt.org.uk

Ted Joanen 1455 Big Picture Road, Lake Charles, Louisiana 70605-9617, USA

Kurt A. Johnson 15250 SE 43rd Court, Apartment H202, Bellevue, Washington 98006, USA. E-mail: kurtbeth@delphi.com

Navin Kapoor Society for Promotion of Wastelands Development, Shri Ram Bharatiya Kala Kendra Building, 1 Copernicus Marg, New Delhi 110 001, India. E-mail: spwd@sdalt.ernet.in

Arvind Khare Society for Promotion of Wastelands Development, Shri Ram Bharatiya Kala Kendra Building, 1 Copernicus Marg, New Delhi 110 001, India. E-mail: spwd@sdalt.ernet.in

Michael J. Kiernan P.O. Box 33972, Washington, D.C. 20033, USA. E-mail: mjkiernan@aol.com

Jeff S. Kirby The Wildfowl & Wetlands Trust, Slimbridge, Gloucestershire GL2 7BT, United Kingdom. E-mail: jeff.kirby@wwt.org.uk

Robin M. Little Gamebird Research Unit, FitzPatrick Institute, University of Cape Town, Rondebosch 7700, South Africa. E-mail: rlittle@botzoo.uct.ac.za

Larry McNease Louisiana Department of Wildlife and Fisheries, Rockefeller Wildlife Refuge, 5476 Grand Chenier Highway, Grand Chenier, Louisiana 70643, USA

Thomas O. McShane World Wide Fund for Nature International, Avenue du Mont Blanc, CH-1196 Gland, Switzerland. E-mail: tmcshane@wwfnet.org

Erica McShane-Caluzi Chemin Pré du Marguiller, CH-1273 Arzier, Switzerland

Luis Moya I. PEDICP, Instituto Nacional de Desarrollo, Casilla Brasil 355-359, Iquitos, Peru

James W. Penn Tropical Conservation and Development Program, Latin American Studies, 319 Grinter Hall, University of Florida, Gainesville, Florida 32611-2037, USA. E-mail: jpenndc@nerdc.ufl.edu

Pablo Puertas Instituto Veterinario de Investigaciones Tropicales y de Altura, Apartado Postal 621, Iquitos, Peru

Sushil Saigal Society for Promotion of Wastelands Development, Shri Ram Bharatiya Kala Kendra Building, 1 Copernicus Marg, New Delhi 110 001, India. E-mail: spwd@sdalt.ernet.in

Samar Singh World Wide Fund for Nature–India, 172-B Lodi Estate, New Delhi 110 003, India. E-mail: wwfindel@unv.ernet.in

Bradley S. Smith Resource Economics Unit, FitzPatrick Institute, University of Cape Town, Rondebosch 7700, South Africa. E-mail: bsmith@botzoo.uct.ac.za

Mark A. Staton Mark Staton Co., P.O. Box 30985, Lafayette, Louisiana 70593, USA

James G. Teer Welder Wildlife Foundation, P.O. Box 1400, Sinton, Texas 78387, USA. E-mail: welderwf@aol.com

Harold F. Upton 2388 Post Road, R.R. 5, Wakefield, Rhode Island 02879, USA

Index

Library of Congress Cataloging-in-Publication Data

Harvesting wild species: implications for biodiversity conservation / edited by
Curtis H. Freese.

p. cm.

Includes bibliographical references and index.

ISBN 0-8018-5573-X. — ISBN 0-8018-5574-8 (pbk.)

1. Biological diversity conservation. 2. Sustainable development. I. Freese, Curtis H.

QH75.H38 1997

333.95'16—dc21 96-50972

CIP